PETROLEUM REFINING

Technology and Economics

Fifth Edition

James H. Gary
Glenn E. Handwerk
Mark J. Kaiser

CRC Press
Taylor & Francis Group
Boca Raton London New York

CRC Press is an imprint of the
Taylor & Francis Group, an informa business

CRC Press
Taylor & Francis Group
6000 Broken Sound Parkway NW, Suite 300
Boca Raton, FL 33487-2742

Printed in the United States of America on acid-free paper
10 9 8 7

International Standard Book Number-10: 0-8493-7038-8 (Hardcover)
International Standard Book Number-13: 978-0-8493-7038-0 (Hardcover)

Library of Congress Cataloging-in-Publication Data

Gary, James H., 1921-
Petroleum refining : technology and economics. -- 5th ed. / James H. Gary and Mark J. Kaiser.
p. cm.
Includes bibliographical references and index.
ISBN-13: 978-0-8493-7038-0 (acid-free paper)
ISBN-10: 0-8493-7038-8 (acid-free paper)
1. Petroleum--Refining. I. Kaiser, Mark J. II. Title.

TP690.G33 2007
665.5'38--dc22 2006027841

Visit the Taylor & Francis Web site at
http://www.taylorandfrancis.com

and the CRC Press Web site at
http://www.crcpress.com

Dedication

Glenn E. Handwerk

September 4, 1924 to February 29, 2004

Table of Contents

Preface

Since the fourth edition was published, the death of Glenn Handwerk has made it very difficult to prepare this edition, because his knowledge and experience in the industry were instrumental in maintaining the quality of the construction and operating costs for the processes included, as well as because of his practical experience in designing and supervising the construction and operation of refinery process units. He was a well-liked and highly regarded consulting engineer for many companies, and his work kept him up to date in all facets of the industry. He will be greatly missed.

Mark Kaiser is well experienced in preparing cost and utility data for the chemical and petroleum industry, and he uses his expertise to replace that lost by Glenn Handwerk's death. His help and the willingness of David Geddes to prepare Chapter 14, Refinery Economics and Planning, made it possible to complete this fifth edition. Their help is greatly appreciated.

Today, refiners are facing investments of billions of dollars in equipment to meet environmental requirements frequently set by political stipulation with little regard to true economic and environmental impacts. Guidelines set up by laws and regulations are changed frequently. Because the designing and building of new processing units entail several years of lead time, refiners are reluctant to commit millions or billions of dollars to constructing equipment that may no longer meet requirements when the units come on stream. For the period, much effort is being devoted to the development of reformulated fuels that have a minimal impact on degradation of the environment, and much work is being done to produce renewable fuels so the United States will be self-sustaining in its energy requirements.

It is essential that alternate renewable fuels be developed, but it is going to be a long-term project. Congress has mandated that by 2012, 7.5 billion gallons of our transportation fuel requirements be made from renewable fuels. This sounds like a large quantity, but at today's consumption in the United States, it will be enough for only 3 weeks of operation.

It is also desirable that transportation fuels not only be renewable but that they should be also nonpolluting. At the present time, the only nonpolluting fuels specified are solar and electric energy and hydrogen. This allows only a short time for the petroleum industry to recover the large investment required to meet the present legal requirements. It is apparent that the survivors of this period will be those companies utilizing the experience and skill of their engineers and scientists to the highest possible level of efficiency.

In writing this edition, we have taken the new environmental aspects of the industry into account, as well as the use of heavier crude oils and crude oils with higher sulfur and metal content. All these criteria affect the processing options and the processing equipment required in a modern refinery.

The basic aspects of current petroleum refining technology and economics are presented in a systematic manner suitable for ready reference by technical managers, practicing engineers, university faculty members, and graduate or senior students in chemical engineering. In addition, the environmental aspects of refinery fuels and the place of reformulated fuels in refinery product distribution are covered.

The physical and chemical properties of petroleum and petroleum products are described, along with major refining processes. Data for determination of typical product yields, investment, and operating costs for all major refining processes and for supporting processes are also given.

The investment, operating cost, and utility data given herein are typical average recent data. As such, this information is suitable for approximating the economics of various refining configurations. The information is not sufficiently accurate for definitive comparisons of competing processes.

The yield data for reaction processes have been extended to allow complete material balances to be made from physical properties. Insofar as possible, data for catalytic reactions represent average yields for competing proprietary catalysts and processes.

The material is organized to utilize the case-study method of learning. An example case-study problem begins in Chapter 4 (Crude Distillation) and concludes in Chapter 19 (Economic Evaluation). The appendices contain basic engineering data and a glossary of refining terms. Valuable literature references are noted throughout the book.

We have held responsible positions in refinery operation, design, and evaluation, and have taught practical approaches to many refinery problems. This publication relies heavily on our direct knowledge of refining in addition to the expertise shared with us by our numerous associates and peers.

Acknowledgments

Appreciation is expressed to the many people who contributed data and suggestions incorporated into this book.

Corporations that have been very helpful include the following:

Exxon Research and Engineering
Fluor Daniel
Stratco, Inc.
The M. W. Kellogg Co.
UOP LLC

Individual engineers who have contributed significant technical information to various editions of this book are listed below:

Robert W. Bucklin
Steve Chafin
D.A. Cheshire
Jack S. Corlew
Gary L. Ewy
P.M. Geren
Andy Goolsbee
Jeff G. Handwerk
Viron D. Kilewer
Jay M. Killen
David R. Lohr
James R. McConaghy
Jill Meister
James R. Murphy
Marvin A. Prosche
Ed J. Smet
Delbert F. Tolen
Donald B. Trust
William T. War
Diane York

Special credit is due James K. Arbuckle for his excellent drafting of graphs, to Pat Madison, Golden Software Co., for providing the Grapher 6 software to make the cost-curve figures, to Andrew Persichetti for preparing graphs and figures, and to Jane Z. Gary, who helped greatly in improving the clarity of presentation.

James H. Gary

Authors

Dr. James H. Gary is professor emeritus of chemical engineering and petroleum refining at the Colorado School of Mines in Golden. Previously, he was head of the department of chemical engineering and petroleum refining, dean of faculty, and vice president for academic affairs. He taught courses in petroleum refinery processing and petrochemicals and was principal investigator for research projects on nitrogen and sulfur removal from liquid hydrocarbons and processing of heavy oils. He was also director of the Colorado School of Mines Annual Oil Shale Symposium. Gary has written more than 40 publications in technical journals. He worked for the Standard Oil Co. (Ohio), now BP Oil Co., holds several patents in fuels and fuels processing, and has consulted for a number of petroleum companies. Gary received BS and MS degrees in chemical engineering from Virginia Polytechnic Institute and his PhD degree from the University of Florida. He is a fellow of the American Institute of Chemical Engineers and the American Association for the Advancement of Science, and is a member of Sigma Xi and the American Chemical Society. He is also a registered professional engineer in Colorado and Ohio.

Mark J. Kaiser is research professor and director, research and development, at the Center for Energy Studies, and adjunct professor in the Department of Petroleum Engineering and the Department of Environmental Studies at Louisiana State University, Baton Rouge. His research interests cover the oil, gas, and refining industry, cost estimation, fiscal analysis, infrastructure modeling, and energy policy. Prior to joining LSU in 2001, he held appointments at Auburn University, the American University of Armenia, and Wichita State University. Dr. Kaiser has served as a consultant and technical expert to government agencies and private firms, and is a member of the Society of Petroleum Engineers, the United States Association for Energy Economics, and the International Association for Energy Economics. Dr. Kaiser received the B.S. degree in agricultural engineering, and the M.S. and Ph.D. degrees in industrial engineering, all from Purdue University.

1 Introduction

Petroleum is a complex mixture of hydrocarbons that occur in the Earth in liquid, gaseous, or solid forms. The word *petroleum* is often restricted to the liquid form, but it also includes natural gas and the solid bitumen. The word *petroleum*, or "rock oil," derives from the Latin *petra* (rock or stone) and *oleum* (oil) and was first used in 1556 in a treatise published by Georg Bauer [1].

Small surface occurrences of natural gas and oil seeps have been known since ancient time. The ancient Sumerians, Assyrians, and Babylonians used crude oil and asphalt more than 5000 years ago. The Egyptians used oil as a weapon of war; early in the Christian era, the Arabs and Persians distilled crude oil to obtain flammable products; and probably as a result of the Arab invasion of Spain, distillation became available in Western Europe by the 12th century [1]. There was no widespread commercial use of oil before the 19th century, however, because no markets had developed for its application, and a steady supply could not be guaranteed. It was only with the discovery of kerosine that oil became a commercial commodity.

In 1846, Abraham Gesner discovered a process for making a liquid fuel out of coal, which he called "kerosine" (or coal oil) and marketed for street lighting in Halifax, Nova Scotia [2,3]. At the time, whale oil was a primary fuel for lamps, burning cleaner and having less of an odor than other animal fats. Whale oil was expensive, however, and by the mid-19th century whales were becoming harder to find, and mass killings were driving the population close to extinction. In 1854, George Bissel and a group of investors paid Benjamin Silliman Jr. to analyze the crude oil from seeps in Pennsylvania as a possible substitute for whale oil. Crude oil was known to burn, but it was not considered a good light source because it produced soot and smoke. Silliman found that by boiling and distilling oil at different temperatures, one of the fractions (which came to be known as kerosine) was a high-quality illuminating oil. Edwin Drake was charged to find a reliable supply of oil by the same group of investors, and in August 1859, he drilled the first commercial oil well in the United States in Titusville, PA.

The first refineries to process crude used existing coal oil refineries or were built where oil was found. Early refineries were simple devices that used large horizontal tanks to heat oil to separate the volatile components. The world's first oil refinery opened at Ploiesti, Romania, in 1856. In the United States, the first refinery opened in 1861. Over subsequent decades, the development of electricity and the advent of the internal combustion engine significantly impacted the demand for refined products, with ever-increasing amounts of gasoline and diesel fuels required in place of kerosine. Air transportation and World War II created the need for high-octane aviation gasoline and, later, jet fuel. Today, petroleum refining is a mature industry with a well-established infrastructure and technology base, employing a complex array of chemical and physical processing facilities to transform crude oil into products that consumers value.

1.1 REFINERY PROCESSES AND OPERATIONS

Crude oil in its natural state has no value to consumers and must be transformed into products that can be used in the marketplace. Various physical and chemical methods are used in refining processes. Heat, pressure, catalysts, and chemicals are applied under widely varying process designs, operating conditions, and chemical reactions to convert crude oil and other hydrocarbons into petroleum products.

Refining begins with distillation by boiling crude into separate fractions or cuts. All crude oils undergo separation processes through distillation, and so it is common to express the capacity of a refinery in terms of its distillation capacity. Two measures are commonly used: barrels per stream day (BPSD) and barrels per calendar day (BPCD). A barrel per stream day is the maximum number of barrels of input that a distillation facility can process when running at full capacity under optimal crude and product slate conditions with no allowance for downtime. A barrel per calendar day is the amount of input that a distillation facility can process under usual operating conditions, making allowances for the types and grades of products to be manufactured, environmental constraints, and unscheduled and scheduled downtime due to maintenance, repairs, and shutdown. Capacity expressed in BPSD is a few percentage points higher than BPCD capacity.

After crude oil is separated into its fractions, each stream is further converted by changing the size and structure of the molecules through cracking, reforming, and other conversion processes. The converted products are then subjected to various treatment and separation processes to remove undesirable constituents and improve product quality (see Figure 1.1).

Petroleum refining processes and operations are classified into five basic types [4]:

1. Distillation is the separation of crude oil in atmospheric and vacuum distillation columns into groups of hydrocarbon compounds based on molecular size and boiling-point ranges.
2. Conversion processes change the size or structure of hydrocarbon molecules by
   Decomposition: Breaking down large molecules into smaller molecules with lower boiling points through cracking and related processes.
   Unification: Building small molecules into larger molecules through alkylation, polymerization, and related processes.
   Reforming: Rearranging molecules into different geometric structures in isomerization, catalytic reforming, and related processes.
3. Treatment processes prepare hydrocarbon streams for additional processing and to prepare finished products using chemical or physical separation. Processes include desalting, hydrodesulfurization, solvent refining, sweetening, solvent extraction, and dewaxing.
4. Blending is the process of mixing and combining hydrocarbon fractions, additives, and other components to produce finished products with specific performance properties.

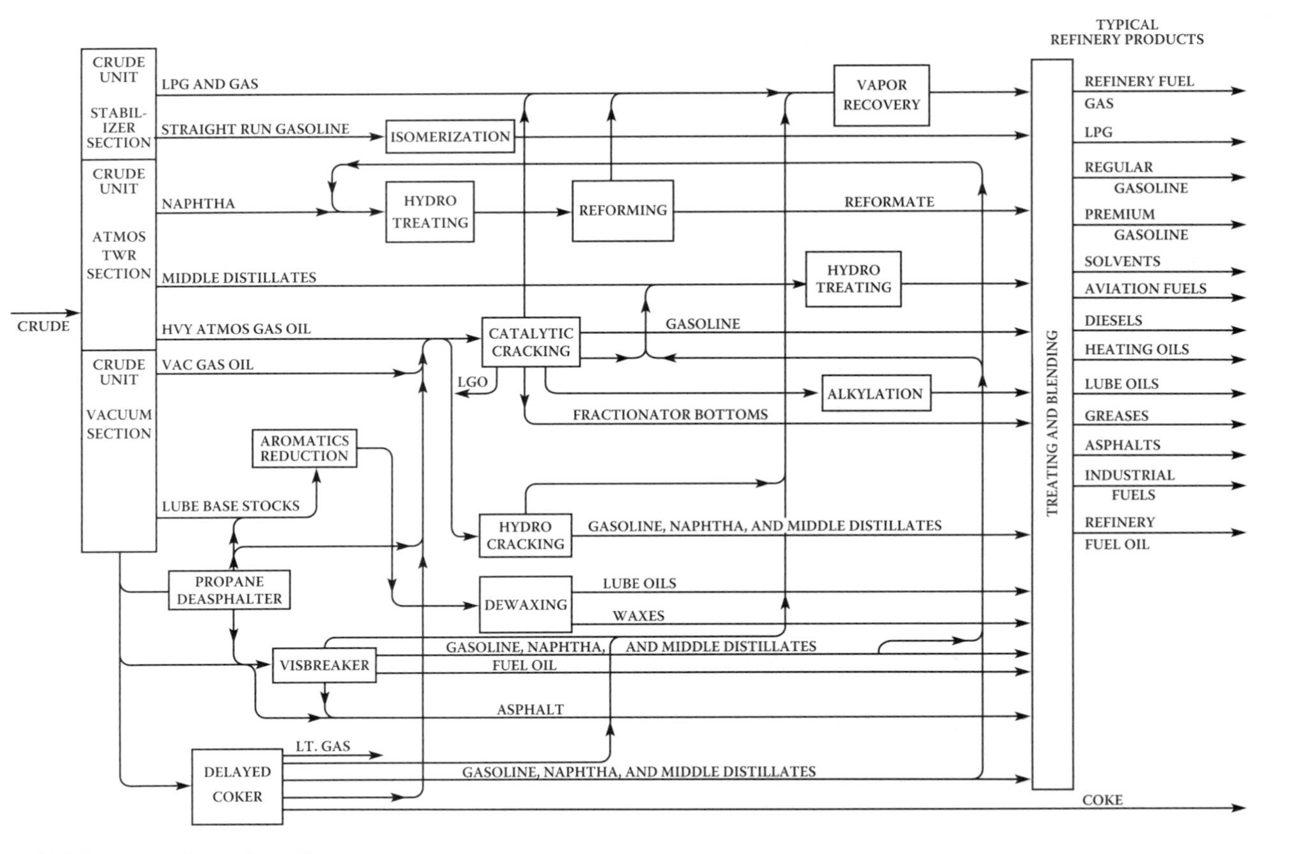

FIGURE 1.1 Refinery flow diagram.

5. Other refining operations include light-ends recovery, sour-water stripping, solid waste and wastewater treatment, process-water treatment and cooling, storage and handling, product movement, hydrogen production, acid and tail gas treatment, and sulfur recovery.

In general terms, refineries operate under the physical laws and engineering specifications of the system, the economic principles that guide investment and operating requirements, and the human-made rules governing production activities and product specifications. Complex interrelationships exist among the physical laws by which a system operates, the product demands required by the market, and the commercial rules and regulations established for the system.

1.2 HISTORIC OVERVIEW

The first refineries used horizontal, elevated tanks to heat the crude oil and vaporize its volatile components. The hot vapor would rise, cool, and condense in a batch operation. The process was repeated at different temperatures to separate the various fractions. The technology was simple, refineries were easy to set up, and before long, refining capacity exceeded crude supply. The need for improved product separation led to the use of fractionating columns, which allowed the different boiling-point cuts to be separated out in a continuous process. John D. Rockefeller sought to consolidate the U.S. refining business, and in 1870 he established the Standard Oil Co. with his partners [5]. By 1879, Standard Oil controlled 90% of the U.S. refining capacity [3].

The number of automobiles and the demand for gasoline greatly increased following the advent of mass production. Distillation processes, however, could only produce a certain amount of gasoline depending on the composition of the crude oil, and so scientists and engineers searched for new ways to increase the yield and performance of gasoline. Thermal cracking processes were first discovered in 1913, when heat and pressure was used to break down, rearrange, and combine hydrocarbon molecules (Table 1.1).

The introduction of catalytic cracking and polymerization processes in the mid- to late 1930s met the demand for higher octane gasoline. Eugene Houdry developed the first commercial process of cracking in the presence of clay mineral catalysts, which resulted in the large hydrocarbon molecules breaking apart; the products were converted to branched paraffins, naphthenes, and aromatics with desirable properties [6]. Visbreaking, another form of thermal cracking, was developed in the late 1930s to produce a more desirable and valuable product slate. In the 1940s, alkylation processes were developed to create high-quality aviation fuels for the war effort, which later were used extensively to produce gasoline blending stocks [7].

Isomerization was developed to convert straight-chain hydrocarbon molecules to branched-chained hydrocarbons to improve the octane of gasoline. In catalytic reforming, the presence of certain catalysts, such as finely divided platinum placed into a support of aluminum oxide, was discovered to reform and dehydrogenate straight-chain hydrocarbons with low octane into aromatics and their branched isomers with

TABLE 1.1
Brief History of Refining

Year	Process Name	Purpose	By-products, etc.
1862	Atmospheric distillation	Produce kerosine	Naphtha, tar, etc.
1870	Vacuum distillation	Lubricants (original) Cracking feedstocks (1930s)	Asphalt, residual coker feedstocks
1913	Thermal cracking	Increase gasoline	Residual, bunker fuel
1916	Sweetening	Reduce sulfur and odor	None
1930	Thermal reforming	Improve octane number	Residual
1932	Hydrogenation	Remove sulfur	Sulfur
1932	Coking	Produce gasoline base stocks	Coke
1933	Solvent extraction	Improve lubricant viscosity index	Aromatics
1935	Solvent dewaxing	Improve pour point	Waxes
1935	Catalytic polymerization	Improve gasoline yield and octane number	Petrochemical feedstocks
1937	Catalytic cracking	Higher octane gasoline	Petrochemical feedstocks
1939	Visbreaking	Reduce viscosity	Increased distillate, tar
1940	Alkylation	Increase gasoline yield and octane	High-octane aviation gasoline
1940	Isomerization	Produce alkylation feedstock	Naphtha
1942	Fluid catalytic cracking	Increase gasoline yield and octane	Petrochemical feedstocks
1950	Deasphalting	Increase cracking feedstock	Asphalt
1952	Catalytic reforming	Convert low-quality naphtha	Aromatics
1954	Hydrodesulfurization	Remove sulfur	Sulfur
1956	Inhibitor sweetening	Remove mercaptans	Disulfides
1957	Catalytic isomerization	Convert to molecules with high-octane number	Alkylation feedstocks
1960	Hydrocracking	Improve quality and reduce sulfur	Alkylation feedstocks
1974	Catalytic dewaxing	Improve pour point	Wax
1975	Residual hydrocracking	Increase gasoline yield from residual	Heavy residuals

Source: Adapted from *OSHA Technical Manual*, Section IV: Chapter 2, *Petroleum Refining Processes*, U.S. Department of Labor, Occupational Safety and Health Administration, Washington, D.C., 2005.

higher octane numbers. In the mid-1950s, a process called hydrotreating was developed to remove contaminants that would damage the catalyst used in catalytic reforming. Improved catalysts and processes such as hydrocracking were developed in the 1960s to further increase gasoline yields and improve antiknock characteristics.

1.3 PRODUCT DEMAND AND SUPPLY

Over the past 2 decades, the demand for petroleum products in the United States has risen steadily, due in part to a growing population, low fuel prices, Americans' preference for heavier and more powerful vehicles, and an increase in passenger and goods travels. In 2005, demand for refined products in the United States was 21 million barrels per day (BPD) [8], equivalent to a daily consumption rate of about

TABLE 1.2
Percentage Yields of Refined Petroleum Products from Crude Oil in the U.S., 1964–2003 (%)

	1964	1974	1984	1994	2003
Gasoline	44.1	45.9	46.7	45.7	46.9
Distillate fuel oil	22.8	21.8	21.5	22.3	23.7
Resid fuel oil	8.2	8.7	7.1	5.7	4.2
Jet fuel	5.6	6.8	9.1	10.1	9.5
Coke	2.6	2.8	3.5	4.3	5.1
Asphalt	3.4	3.7	3.1	3.1	3.2
Liquefied gases	3.3	2.6	1.9	4.2	4.2
Total	90.1	92.3	92.9	95.4	96.8

Source: U.S. Department of Energy, Energy Information Administration, *Petroleum Supply Monthly*, Washington, D.C., March 2005.

20 lb of petroleum per person. The transportation sector consumed about 14 million barrels per day, whereas the industrial, residential and commercial, and electric power sector consumed 4 million, 2 million, and 1 million BPD, respectively.

U.S. demand for crude oil and petroleum products exceeds domestic supply, and the United States imports a variety of intermediate and final petroleum products in addition to crude oil. About 60% of the U.S. petroleum requirements are currently imported, and although the United States is still one of the world's largest producers of crude oil, its long-term prospects are not encouraging, having only 3% of the world's proven reserves [9].

Raw crude and intermediate materials are processed at refineries into gasoline, distillate fuel oil (diesel fuel, home heating oil, industrial fuel), jet fuels (kerosine and naphtha types), residual fuel oil (bunker fuel, boiler fuel), liquefied petroleum gases (propane, ethane, butane), coke, and kerosine. About 90% of crude oil in the U.S. is converted to fuel products. Nonfuel products such as asphalt, road oil, lubricants, solvents, waxes and nonfuel coke, and petrochemicals and petrochemical feedstocks such as naphtha, ethane, propane, butane, ethylene, propylene, butylene, benzene, toluene, and xylene comprise the remaining crude conversions. Refinery production in the United States is dominated by gasoline, followed by distillate kerosine, jet fuel, diesel fuel (and home heating oils), and residual fuels (Table 1.2). Gasoline yield is nearly 50%, with distillate and residual fuels yielding 35% of refinery production. About half of gasoline production is conventional, with 25% satisfying various state and local requirements, and 25% produced according to federal reformulated gasoline (RFG) specifications. Approximately 60% of diesel fuel produced is ultra low-sulfur fuel (15 ppm maximum) for the on-road (highway) market, and 40% is higher sulfur fuel for the off-road market (e.g., construction, farming, locomotives, shipping, and mining). EPA specifications require that all diesel fuels except railroad and marine fuels meet the 15 ppm sulfur specification by 2011.

1.4 U.S. STATISTICS

1.4.1 CAPACITY

At the end of 2005, there were 149 operable refineries with 17.1 million BPCD crude charge capacity. Operable refineries are defined as those refineries that are in one of three states at the beginning of a given year: in operation; not in operation and not under active repair, but capable of being placed into operation within 30 days; or not in operation but under active repair that could be completed within 90 days [8]. Changes in refining capacity are due to plants being idled or shut down, expansion through bottleneck removal and more intensive use ("capacity creep"), and new grassroots construction.

In the late 1970s, the growth in product demand and subsidies under the U.S. Crude Oil Entitlement Program brought forth an expansion of refinery capacity, and by 1981, 324 refineries were operating at 19.4 million BPCD capacity. The Crude Oil Entitlement Program favored production from small, unsophisticated refineries, and the construction of simple hydro-skimming plants with less than 30,000 BPCD capacity accelerated [10]. By 1981, the U.S. government removed price and allocation controls on the oil industry [7], and many small refineries and inefficient plants could not compete and began to shut down. From 1981 to 2004, 171 refineries shut down or closed, most being small, inland facilities, with less than 100,000 BPCD and without access to water transportation. Economic rationalization, the lack of scale economies, domestic supply constraints, environmental regulations, and technological developments have all contributed to plant shutdowns [11,12].

1.4.2 CAPACITY CREEP

As small refineries close and existing facilities add to their capacity, the average capacity in the United States continues to increase. Over the past decade, the capacity of the refining industry has increased incrementally, at an average rate of about 1.7% per year (130,000 BPD/yr). About 0.2% per year refining capacity has been lost to refinery shutdowns. Expansion is often a more economic option than constructing a grassroots facility because the fixed costs associated with a new facility can be avoided and scale economies realized with existing units. In terms of regulatory requirements, expansion is also likely to be less costly, less time-consuming, and less subject to challenge by local groups [12].

1.4.3 UTILIZATION

The U.S. refining industry's ability to meet short-term increases in demand can be measured by the rate at which operable capacity is utilized. Utilization is defined as the gross inputs to crude oil distillation divided by operable capacity. Utilization fluctuates as refinery operations adjust to changes in demand. The amount of gross inputs to refineries has risen steadily throughout the past 2 decades in response to rising demand for petroleum products. The trend in recent years toward high utilization has been due to a decrease in the number of operating refineries and operations changes, including reducing refinery downtime and the number of unplanned outages, executing maintenance and revamps more efficiently, and extending run times through improved catalyst performance [13].

1.4.4 Industry Structure

The structure of the refining industry has undergone significant change over the past decade. Once led by a half-dozen vertically integrated majors, the industry is now characterized by a handful of super-majors and an array of midsize and small independents focused on refining and marketing within specific regions and product lines. Integrated oil companies are involved in all segments of the energy supply chain, from exploration and production ("upstream") to transportation ("midstream") to refining and marketing ("downstream"). Companies that operate in one or more but not all segments are called partially integrated or independents. Independent refiners and marketers are typically only involved in downstream activities. The traditional industry model of refining, based on ownership by vertically integrated oil companies and profitability viewed within the context of a linked supply chain, has been replaced by refineries operated in a stand-alone profit center mode.

Before 1980, nearly all U.S. refineries were held by integrated oil companies, whereas today, ownership structure is more diverse and concentrated (Table 1.3).

TABLE 1.3
Top 20 U.S. Refiners, 1985 vs. 2005

Company	1985 Capacity (MBPCD)	Company	2005 Capacity (MBPCD)
Chevron	1879	ConocoPhillips	2198
Exxon	1200	ExxonMobil	1847
Shell	1120	BP	1505
Amoco (Standard of Ind.)	982	Valero Energy	1450
Texaco	873	Chevron Texaco	1007
Mobil	755	Marathon Oil	948
Standard Oil (Ohio)	665	Sunoco	900
Atlantic Richfield	655	Premcor	768
U.S. Steel	493	Koch	763
Unocal	475	Motiva Enterprises	747
Sun	443	PDV America	719
DuPont	400	Royal Dutch Shell	597
Ashland	347	Tesoro	563
Phillips	300	Deer Park REFG Ltd.	334
Southland	297	Lyondell Chemical	270
Koch	280	Total SA	234
Coastal	261	Chalmette Refining	187
Union Pacific	220	Sinclair Oil	161
Kerr-McGee	165	Rosemore	155
Total	152	Murphy Oil	153

Source: API *Basic Petroleum Data Book* XXV(1), Washington, D.C., February 2005. U.S. Department of Energy, Energy Information Administration, *Petroleum Supply Monthly*, Washington, D.C., March 2005.

TABLE 1.4
Number of Refiners' Atmospheric Distillation Capacity (2005)

Number	Capacity (BPCD)	Total Capacity (BPCD)
26	> 100,000	16,288,755
12	30,001–100,000	669,970
9	10,001–30,000	180,458
8	< 10,000	45,687
55		17,124,870

Source: U.S. Department of Energy, Energy Information Administration, *Petroleum Supply Annual, 2004, Vol. 1*, Washington, D.C., 2004.

In 2005, the top 3 U.S. refiners processed 36% of total crude oil; the top 10 refiners processed 77%; and the top 20 refiners processed 92%. Independents currently own about 64% of U.S. refining capacity, versus 51% in 1990. Foreign ownership has risen from 19% of total capacity in 1990 to about 25% in 2005. Royal Dutch Shell, BP, Total, Saudi Aramco, and Petroleos de Venezuela SA are major foreign owners of U.S. refining capacity.

The majority of distillation capacity is currently concentrated in large, integrated companies with multiple refining facilities. Fifty-five firms, ranging in size from 880 BPCD to a combined refinery capacity of 1.8 million BPCD, comprise the industry (Table 1.4). About two thirds of firms are small operations producing less than 100,000 BPCD and representing about 5% of the total output of petroleum products. Large refiners often manage both large and small refineries (Table 1.5), whereas small operators mainly specialize in asphalt, lubricants, and other niche products.

Integrated firms such as ConocoPhillips, ExxonMobil, BP, and Chevron maintain a global portfolio of petroleum assets. Independent companies like Valero and Sunoco focus primarily on domestic refining, although they may also be involved in marketing and other operations. Independents grew to prominence in the mid-1990s as the majors restructured their downstream petroleum operations and sold their refining assets to new entrants [13]. Joint ventures and partnerships such as Motiva Enterprises also exist. Koch is a privately owned independent.

1.4.5 Regional Specialization

Refining capacity in the United States is concentrated in the Gulf Coast and West Coast regions, and along the corridor from Illinois to New Jersey. Table 1.6 depicts the top 10 refining states and total distillation capacity. Historically, refineries were located near the sources of oil, and over time, they developed in regions with easy access to water and shipping routes to minimize transport and logistical costs [14].

The Gulf Coast Petroleum Administration for Defense District (PADD) III is the nation's leading supplier of refined products, accounting for nearly half of the total

TABLE 1.5
Top 10 U.S. Refining Companies (2005)

Company	Capacity (1000 BPCD)	Number of Refineries	Average Size (BPCD)	Largest Plant (BPCD)	Smallest Plant (BPCD)
ConocoPhillips	2198	14	157,000	306,000	14,000
ExxonMobil	1847	6	308,000	557,000	60,000
BP	1505	6	251,000	437,000	13,000
Valero Energy	1450	13	112,000	210,000	6,000
Chevron	1007	6	168,800	325,000	45,000
Marathon	948	7	135,000	245,000	70,000
Sunoco	900	5	176,000	330,000	85,000
Premcor	768	4	192,000	255,000	158,000
Koch	763	3	254,000	288,000	210,000
Motiva Enterprises	747	3	249,800	285,000	227,000

Source: U.S. Department of Energy, Energy Information Administration, *Petroleum Supply Monthly*, Washington, D.C., March 2005.

TABLE 1.6
Top 10 Petroleum Refining States (2005)

State	Number of Operating Refineries	Distillation Capacity (MBPCD)
Texas	26	4,628
Louisiana	17	2,773
California	21	2,027
Illinois	4	896
Pennsylvania	5	770
New Jersey	5	666
Washington	5	616
Ohio	4	551
Oklahoma	5	485
Indiana	2	433
Subtotal 10	93	13,845
U.S. Total	144	17,125

Source: U.S. Department of Energy, Energy Information Administration, *Petroleum Supply Annual 2004, Vol. 1*, Washington, D.C., 2004.

capacity in the country (Table 1.7). Refineries in PADD III are large, integrated, and complex facilities. The Gulf Coast imports significant quantities of crude oil as refinery feedstocks and ships refined products to both the East Coast and to the Midwest (Table 1.8). The East Coast is the largest consuming area in the United States and imports over half of all the refined products that enter the United States because of its limited refining capability. The Pacific Northwest and Rocky Mountains

TABLE 1.7
Number and Capacity of Operable Petroleum Refineries by PAD District (2005)

PADD†	Number	Distillation Capacity (BPCD)	Refining Capacity (%)	Average Capacity (BPCD)
I	15	1,717,000	10.0	114,467
II	26	3,569,061	20.8	137,272
III	55	8,085,614	47.2	147,011
IV	16	587,550	3.4	36,722
V	36	3,165,645	18.5	87,935
U.S. Total	148	17,124,870	100.0	115,709

Note: †Petroleum Administration for Defense Districts (PADDs) were delineated during World War II to facilitate oil allocation. The PADD states are as follows:

I: CT, ME, MA, NH, RI, VT, DE, DC, MD, NJ, NY, PA, FL, GA, NC, SC, VA, WV
II: IL, IN, IA, KS, KY, MI, MN, MO, NE, ND, SD, OH, OK, TN, WI
III: AL, AR, LA, MS, NM, TX
IV: CO, ID, MT, UT, WY
V: AK, AZ, CA, HI, NV, OR, WA

Source: U.S. Department of Energy, Energy Information Administration, *Petroleum Supply Annual 2004, Vol. 1*, Washington, D.C., 2004.

TABLE 1.8
Movements of Petroleum Products by Pipeline, Tanker, and Barge Between PAD District, 2004 (Million Barrels)

PADD† (From):	(To): I	II	III	IV	V
I	—	123.2	1.6	0	0
II	27.1	—	80.8	20.1	0
III	1185	410.0	—	16.9	14.1
IV	0	23.0	51.6	—	12.1
V	0.1	0	0.1	0	—

Note: †Petroleum Administration for Defense Districts (PADDs) were delineated during World War II to facilitate oil allocation. The PADD states are as follows:

I: CT, ME, MA, NH, RI, VT, DE, DC, MD, NJ, NY, PA, FL, GA, NC, SC, VA, WV
II: IL, IN, IA, KS, KY, MI, MN, MO, NE, ND, SD, OH, OK, TN, WI
III: AL, AR, LA, MS, NM, TX
IV: CO, ID, MT, UT, WY
V: AK, AZ, CA, HI, NV, OR, WA

Source: U.S. Department of Energy, Energy Information Administration, *Petroleum Supply Annual 2004, Vol. 1*, Washington, D.C., 2004.

regions typically have small- to midsize facilities and only limited pipeline access from Gulf Coast and California refineries. Southern California, Nevada, and parts of Arizona have significant distribution infrastructure constraints and stringent local environmental regulations.

The Midwest and East Coast account for about 90% of the interregional flow of petroleum products, with the Gulf Coast by far the largest supplier, accounting for more than 80% of the flow. The Rockies and West Coast regions in contrast are isolated, due to the physical terrain and great distances involved, but also partially due to the stringent quality restrictions on gasoline and diesel requirements.

1.5 WORLD STATISTICS

1.5.1 Product Demand

Demand for gasoline, middle distillates, and heating oil varies with each region of the world. Asian demands for gasoline in the last decade have increased over 50%, with U.S. demands growing by about 15%. European gasoline demand has actually dropped due to high taxation rates and conversion of their transportation fleet to diesel fuel. For distillates, the demand pattern is consistent worldwide, with continued and sustained growth [15]. The primary use of diesel fuels is for transportation requirements.

1.5.2 Capacity

There are currently 662 refineries worldwide with a total capacity of 85.1 million BPCD (Table 1.9). The world's largest refining region is Asia (22.2 million BPCD) followed by North America (20.6 million). The top refining countries in the world are the United States, followed by China, Russia, and Japan. The ratio of refinery capacity to demand declined from 113% in 1990 to 103% in 2004, despite continued growth in global capacity, because demand accelerated at a faster rate.

The world's largest refining companies include ExxonMobil, Royal Dutch Shell, and BP (Table 1.10). There are also several national oil companies, including Sinopec, Petroleos de Venezuela SA (PDVSA), China National Petroleum Corp. (CNPC), Saudi Aramco, Petroleos Brasieiro (Petrobras), Petroleos Mexicanos (PEMEX), National Iranian Oil Co. (NIOC), and OAO Yukos. Petroleos de Venezuela's Paraguana refinery is currently the largest installation in the world with an installed capacity of 940,000 BPCD, followed by South Korea's SK Corp. with 817,000 BPCD capacity.

1.5.3 Configuration, Complexity, and Yield

Refining configuration varies widely on a regional basis, depending upon the demand for gasoline, diesel, and fuel oil (Table 1.11). Catalytic hydrotreating is the most widely applied process technology followed by catalytic cracking and catalytic reforming (Table 1.12). Western Europe has more diesel and jet fuel demand and lower gasoline and heating gas oil requirements due to the prevalence of diesel engines and use of natural gas for domestic heating. Western European refiners,

TABLE 1.9
Refining Operations on a Regional Basis — Thousand Barrels per Calendar Day (2005)

Region	No. of Refineries	Crude Distillation	Vacuum Distillation	Catalytic Cracking	Catalytic Reforming	Catalytic Hydrocracking	Catalytic Hydrotreating	Coke
Africa	45	3,230	514	211	476	62	906	1,841
Asia	155	22,206	4,102	2,646	1,927	739	8,556	20,200
Eastern Europe	93	10,245	3,742	929	1,494	320	4,264	12,570
Middle East	42	7,034	1,976	360	652	589	2,055	3,300
N. America	158	20,627	9,106	6,574	4,173	1,738	15,146	143,616
S. America	66	6,611	2,795	1,300	427	139	1,912	24,702
Western Europe	103	14,971	5,613	2,252	2,144	1,056	9,988	11,316
Total	662	85,127	27,848	14,272	11,292	4,647	42,827	217,545

Note: All capacities expressed in 1000 BPCD except coke, which is expressed in tonnes/day.

Source: O&GJ World Refinery Capacity Report, December 19, 2005.

TABLE 1.10
World's Largest Refiners (2005)

Rank (1/1/06)	Rank (1/1/05)	Company	Crude Capacity (1000 BPCD)
1	1	ExxonMobil	5690
2	2	Royal Dutch Shell	5172
3	3	BP	3871
4	4	Sinopec	3611
5	11	Valero Energy	2830
6	5	Petroleos de Venezuela	2792
7	6	Total	2738
8	7	ConocoPhillips	2659
9	13	China National Petroleum	2440
10	9	Saudi Aramco	2417
11	8	Chevron	2066
12	10	Petroleo Brasileiro	1953
13	12	Petroleos Mexicanos	1851
14	14	National Iranian Oil	1451
15	19	OAO Yukos	1182

Source: O&GJ World Refinery Capacity Report, December 19, 2005

TABLE 1.11
World Refinery Output by Area for Major Petroleum Products as Percentage of Total, 2003 (%)

Region	Gasoline	Distillate	Residual	Other
United States	46.4	21.4	4.2	28.0
Canada	37.0	28.8	6.5	27.5
Mexico	26.7	21.1	32.7	19.3
Central/South America	22.4	26.7	20.7	30.1
Western Europe	22.3	34.6	14.1	28.8
Middle East	13.7	28.4	25.1	32.6
Africa	18.2	26.7	26.1	28.8
Asia	18.4	30.7	14.3	36.5
FSU	19.3	30.0	25.0	25.5
World	26.1	28.4	14.8	30.5

Source: API *Basic Petroleum Data Book* XXV(1), Washington, D.C., February 2005. U.S. Department of Energy, Energy Information Administration, *Petroleum Supply Monthly*, Washington, D.C., March 2005.

therefore, have more hydrotreating and hydrocracking capacity and less catalytic cracking capacity. The conversion capacity of refineries as a percentage of distillation capacity is often used as a measure of refining complexity [16]. North America has the most complex refinery configurations, followed by Asia. Catalytic cracking capacity may reach 20% of crude capacity where there is a high demand for gasoline

TABLE 1.12
World Refining Configuration as a Percentage of Crude Distillation, 2005 (%)

Region	Catalytic Cracking	Catalytic Reforming	Catalytic Hydrocracking	Catalytic Hydrotreating
Africa	6.5	14.7	1.9	28.0
Asia	11.9	8.7	3.3	38.5
Eastern Europe	9.0	14.5	3.1	41.6
Middle East	5.1	9.2	8.3	29.2
N. America	20.2	31.9	8.4	73.4
S. America	19.6	6.4	2.1	28.9
Western Europe	15.0	14.3	7.0	66.7

TABLE 1.13
World Refining Investment Patterns, 2000–2005

	Million BPCD	Growth Rate	
Process	2000	2005	(%)
Atmospheric distillation	81.3	85.1	4.7
Coking	3.7	4.4	18.9
Coking, tons per day	176.7	217.5	23.1
Cracking	13.7	14.3	4.4
Hydrocracking	4.3	4.6	7.0
Hydrotreating	36.6	42.8	16.9
Sulfur, tons per day	56.1	71.1	26.7

Source: O&GJ World Refinery Capacity Report, December 19, 2005. O'Connor, T., Petroleum refineries: Will record profits spur investment in new capacity? Testimony before the House Government Reform Committee, Subcommittee on Energy and Resources, U.S. House of Representatives, October 19, 2005.

(North America, South America). Hydrocracking is required in regions with high diesel and jet fuel demand and is used to convert heavy fuel oils to high-quality transportation fuels. Coking is employed where there is low demand and value for heavy fuel oil.

1.5.4 Investment Patterns

Changes in global refinery capacity over the past 5 years are shown in Table 1.13. Investment in crude processing capacity has been limited. The primary growth in process expansions has been to reduce sulfur levels, which is shown by increases in hydrotreating and hydrocracking capacity, as well as in sulfur production capacity [15]. Sulfur reductions were mandated by government regulation, and refineries had to decide whether to invest to manufacture salable product or shut down operations. Capacity to process heavier and cheaper crude oils through coking has also been a major focus point. Investment in process heavy crude oils allows refiners to potentially improve their products by reducing the cost of their raw materials.

1.6 INDUSTRY CHARACTERISTICS

1.6.1 Each Refinery Is Unique

There does not exist a standard refinery configuration. The first refineries were simple, small, batch distillation units that processed hundreds of barrels of crude oil per day from one or a small collection of fields. Today, refineries are complex, highly integrated facilities that may contain as many as 15 to 20 process units capable of handling a dozen or more different crudes with capacity up to almost a million barrels a day. Refineries vary in size, sophistication, and cost depending on their location, the type of crude they refine, and the products they manufacture.

1.6.2 No Two Crude Oils Are the Same

Crude oil is not a homogeneous raw material. Each crude oil produced in the world has a unique chemical composition, which varies according to the manner of its formation. Currently, more than 150 crude grades are traded [17], and many of these are streams blended from two or more fields.

Crude oil is a complex mixture of hydrocarbon compounds and small quantities of materials such as oxygen, nitrogen, sulfur, salt, and water [18]. Carbon and hydrogen make up around 98% of the content of a typical crude oil. Crude oils are further classified as paraffinic, naphthenic, aromatic, or asphaltic based on the predominant hydrocarbon series molecules. The hydrocarbon series molecules dictate crude oil's physical and chemical properties.

The gravity, sulfur content, and Total Acid Number (TAN) are the most important properties of crude oil. Gravity is a measure of the density of the crude oil and is described in terms of degrees API. The higher the API number, the lighter the crude. Crude oils with low carbon, high hydrogen, and high API number are usually rich in paraffins and tend to yield greater proportions of gasoline and light petroleum products.

Sulfur is an undesirable impurity because of pollution and corrosion concerns [19]. Sulfur content is measured in terms of weight percentage. Crude oils that contain appreciable quantities of hydrogen sulfide or other reactive sulfur compounds are called "sour," whereas crudes low in sulfur are labeled "sweet." The term derives from the foul (or sour) odor of one of the sulfur species, the mercaptans. The split between sweet and sour crude was historically about 0.5%; today, the crossover is generally in the 0.5 to 1.5% range.

The Total Acid Number of crude oils is a measure of potential corrosivity. The TAN quantifies the number of milligrams of potassium hydroxide (KOH) needed to neutralize 1 g of crude oil. A crude with a TAN > 1 mg KOH/g is usually considered corrosive and labeled a High-TAN crude, but corrosion problems can occur with a TAN as low as 0.3 [20].

The gravity, sulfur, and TAN of crude oil have important economic and technical impacts on refining operations. Heavy, sour crude requires more sophisticated processes to produce lighter, more valuable products and is thus more expensive to manufacture. To attract buyers, heavy crude producers offer heavy oil at a significant discount to light crude. The size of the discount varies according to demand and supply conditions and other factors. Light, sweet crude is generally more valuable because it yields more of the lighter, higher-priced products than heavy crude and is cheaper to process. Acidity has also become increasingly important as the production of high-acid crudes, particularly from West Africa, has steadily increased. Very few refineries are able to refine High-TAN crude oil, and so these are sold at a substantial discount to marker crudes [20]. Refineries typically blend High-TAN crudes with other streams before refining.

1.6.3 Refinery Configuration Evolves over Time

Refineries are built at a given time and location using specific technologies; evolve with changes in market demand, feedstock, product specification, and environmental regulation; and will typically possess both very old and modern process units. There are many ways to classify and group refineries, but most schemes refer to the complexity of the process technology, which describes its capacity to convert heavy products, such as fuel oil and resid, to lighter products such as gasoline, diesel, and jet kerosine. Upgrading capacity comes in a number of different types of units, from the common catalytic cracking facilities to the more sophisticated and expensive hydrocracking and coking units.

1.6.4 Not All Refineries Are Created Equal

The simplest refinery, frequently referred to as a topping or hydroskimming refinery, consists of atmospheric distillation and typically one or more pretreatment facilities, catalytic reforming, and hydrodesulfurization. "Simple" refineries have limited capacity to change the composition of the crude oil input. A "complex" (cracking) refinery or "very complex" (coking) refinery is characterized by significant upgrading capacity and a high level of integration. Simple refineries have the lowest margins and are often operated by small, niche players. Complex refineries are able to change the composition of the crude oil input, taking low-value, heavy oils and converting them to high-value, light products, supplying the marginal barrel and realizing their profits from the base margins.

Cracking refineries are typically comprised of vacuum distillation; naphtha, distillate, kerosine, and vacuum gas oil hydrodesulfurization; catalytic reforming; fluid catalytic cracking; alkylation; and light naphtha isomerization units. Deep conversion refineries use a coking process to eliminate the vacuum residue, converting virtually the entire barrel of crude to valuable light products. Coking refineries enjoy the highest upgrading margins and are highly flexible. Unlike cracking refineries, the economics of coking refineries are driven largely by the light–heavy differential, the difference in price between light, sweet crude oils and heavy, sour crude oils. Complex refineries tend to make large margins when processing heavy, sour crudes and smaller margins when processing light crudes [21,22].

1.6.5 Refineries Are Capital-Intensive, Long-Lived, Highly Specific Assets

Refining projects are capital intensive, long-lived, and highly specific assets that require large initial investments before production can begin. The decision to invest in capacity is complex and depends primarily on expectations of return on investment (ROI). The capital-intensive nature means that refining investments are more exposed to financial risks, such as changes in interest rates and investment cycles, and will need to compete with other investment opportunities.

Refineries are designed and engineered by a combination of internal company resources and third-party firms. Feasibility studies may take several months, followed by construction times that last anywhere from 6 months to 2 years or more. Grass-roots construction projects in the United States may take 5 years or more to secure all the necessary permits.

Large-scale capital-intensive projects typically require that owners and contractors form an alliance. Engineering, procurement, installation, and construction contracts are typically handled on a lump-sum, turnkey basis. The contractor enlists subcontractors for work involving special units, pipelines, export facilities, and construction [23].

1.6.6 Refined Products Are Commodities

The main products of refining — gasoline (aviation and motor gasoline and light distillates), middle distillates (diesel fuel, jet fuel, and home heating oil), fuel oil, and other products (fuel gas, lubricants, wax, solvents, refinery fuels) — are commodities. Commodities are products that are undifferentiated from those of a competitor and sold on the basis of price, defined in competitive markets through the intersection of supply and demand curves [24]. The product slate is determined by the demand, inputs and process units available, and the result of intermediaries entering into the production of other products.

1.6.7 Refined Products Are Sold in Segment Markets

Refined products markets differ from crude markets in several respects, owing to the scale of operations, quality considerations, price differentials, and market size [24,25]. Refined products markets are more regional than crude oil markets, and no refiner or marketer is able to operate a completely balanced system. The function of the physical market is to redistribute the individual surpluses and deficits that arise at each location. There are regular flows of products from one region to another, and price levels are set accordingly.

1.6.8 Product Prices Are Volatile

In competitive markets, refined product prices are determined by supply, demand, and inventory conditions at a given location and time. The most fundamental economic relationship governing commodities is that quantity demanded is a function of price. Demand for commodities is generally inelastic in the short term, and inventories, when they exist in sufficient volumes, allow supply and demand to achieve an equilibrium of sorts, which acts to smooth out spot prices and reduce volatility [24,25]. Supply is defined by production and inventory. Balancing supply and demand occurs at both the regional and world levels, with prices in efficient markets reflecting the combined influence of all market information, including expectations of future supply and demand, seasonal factors, and inventory levels.

1.6.9 Product Prices Are Correlated to Crude Oil Prices

Crude oil prices are more volatile than other commodities, reflecting the influence of political and economic events, supply and demand conditions, perceptions about resource availability, and many other factors [26]. The price of oil is determined on a world market, and its availability can change dramatically within a very short period of time. The price of oil also strongly influences the prices of refined products and primary fuels (natural gas, liquefied natural gas, coal), depending on market conditions.

Crude oil prices are a fundamental determinant of gasoline prices. Crude oil and refinery petroleum products, especially gasoline and the middle distillates, have generally followed a similar path over the past 2 decades. Refined product prices have historically reacted to changes in the acquisition cost of crude oil in both directions, with falling crude prices leading to declines in refined product prices, and vice versa [27]. The price differential between gasoline and fuel oil and other products represents the supply–demand requirements for the fuel and the premium for production [28].

1.6.10 Refineries Are Price Takers

Most refineries use a mix of crude oils from various domestic and foreign sources, which changes based on the relative cost and availability. Refiners competitively determine the relative prices paid for various crude oil to reflect quality and location differentials, but do not establish absolute prices. Large crude oil producers that supply crude to U.S. refineries (Saudi Arabia, Mexico, Canada, Norway, Venezuela) employ models for U.S. refinery centers and develop sophisticated pricing strategies to minimize refiners' pricing power and attempts to game the system.

1.6.11 Gravity Down, Sulfur Up

Worldwide, the average gravity for crude oil is about 31°API and ranges from a high of 45° for light crude to a low of 8 to 15° for heavy crude and tar sands. The average API number of global production has changed noticeably over the past decade.

World average sulfur content of crude oil is about 1.25%, with commercial production ranging from 0.1 to 0.6%. Crude oil production on average is becoming heavier and sulfur content is higher, which is especially pronounced in the U.S. refinery crude input quality.

Heavy, high-sulfur crude is more difficult and expensive to process and yields more heavy products, which require additional processing. Capacity in the United

States that has been added over the past decade has primarily been in the form of downstream processing, particularly "bottom of the barrel" processing, to clean up the heavy crude and satisfy environmental regulations. The average U.S. Gulf Coast (USGC) refinery configuration, for example, has shifted toward coking refineries that process heavy, sour crude oil. The number of cracking refineries on the USGC has fallen by more than half over the past 20 years, whereas the number of coking refineries has increased by more than one third. U.S. Gulf Coast coking refineries currently comprise about 90% of total USGC crude capacity [29].

1.6.12 Refining Optimization Involves Multiple Trade-Offs

In most refineries, there are usually several ways to accomplish the same task. Trade-offs, experience, and site-specific factors thus govern decision making. A variety of operating variables — temperature, pressure, residence time, feed quality, cut points, recycle-gas ratio, space velocity, and catalyst — are used to balance feedstock, product, and quality. The limits of these variables are specific to the plant design and nature of the input and output requirements. For a specific feedstock and catalyst package, the degree of processing and conversion increases with the severity of the operation.

The subsystems that comprise a refinery are continually optimized, expanded, or idled to enhance operations. Because refining is an integrated, continuous operation, optimization requires a sophisticated, multidimensional, and time-varying formulation that links operating variables, product specifications, input and output prices, and environmental and economic constraints. Total site integration is used to optimize the interaction of the refinery units with the existing utility infrastructure. Optimizing a refinery is a matter of balance, because every benefit has a cost, every incremental gain has an offsetting loss, and every attempt to remove one unwanted product creates a new waste stream.

1.6.13 Petroleum Refining Is Energy Intensive

Petroleum refining is the most energy-intensive manufacturing industry in the United States. In 2002, the U.S. refining industry consumed 6.391 quads (quadrillion Btu, or 10^{15} Btu) of energy, accounting for about 28% energy consumption in U.S. manufacturing [30]. Process units consume or produce various utilities, such as steam, electric power, boiler feed, fuel gas, gas oil, cooling water, nitrogen, water, and plant air [31,32]. Electric power may be produced or purchased, but fuel requirements can usually be met internally. Each unit has its own specific utility requirements, depending on the design configuration and the need to integrate with existing facilities.

Processes that have the greatest throughput dominate energy consumption. Atmospheric and vacuum distillation account for 35 to 40% of total process energy consumed in the refinery (29), not because they are necessarily the most energy-intensive, but because every barrel of crude is subjected to an initial separation by distillation [32]. Similarly, because many refinery streams must be hydrotreated prior to entering downstream refining units and to achieve desired product quality, another 20% of energy consumption in a refinery is spent in hydrotreating.

1.6.14 Refining Operations and Products Impact the Environment

The oil and gas industry is one of the most heavily regulated industries in the United States, with regulations on the manner in which petroleum is produced, imported, stored, transported, and consumed [33]. Refining operations are affected by both environmental and political legislation. The manufacturing processes used to refine petroleum generate a variety of air emissions and other residuals, some of which are hazardous or toxic chemicals [34–36]. The environmental impacts of refining and the use of refined products have resulted in a number of environmental laws and regulations. Some of the most significant statutes focus on altering the product formulation to reduce air emissions generated by their use.

1.7 REFINERY ECONOMICS

1.7.1 Refinery Economics Is Complicated

The economic performance of a refinery is a function of many variables — the plant configuration, acquisition price of crude oil, product prices, strategic decisions, operating cost, and environmental requirements — and so simple economic models rarely provide sufficient guidance as an aid in decision making. Both crude oil and product prices change relative to each other, subject to the supply and demand forces in the market. The refinery has some control over the cost of production, but price structures exhibit a more complex relationship. Refiners attempt to maximize earnings from fluctuations in the market price of the crude oil and product slate. The potential for economic gain arises when there is a large price spread between light and heavy crude and light and heavy products. A refiner will buy heavy crude to minimize input costs and sell a light product mix to maximum product revenues.

1.7.2 Data Source

One of the best sources of public data on the financial and operating developments in the U.S. petroleum sector is the Energy Information Administration's (EIA's) *Performance Profiles of Major Energy Producers.* The EIA collects financial and operating data through a Financial Reporting System (FRS) Form, EIA-28, from a sample of U.S. major energy companies that are considered representative of the U.S. energy industry. A "major energy producing" company must control at least: (1) 1% of U.S. crude oil or natural gas liquids reserves or production, (2) 1% of U.S. natural gas reserves or production, or (3) 1% of U.S. crude oil distillation capacity or product sales, to be included within the annual survey.

The composition of the FRS group of companies changes over time, but the changes are usually incremental, and so year-to-year comparisons are considered meaningful. Information is collected for the corporate entity as well as by the lines of business within the company: petroleum, downstream natural gas, electric power, nonenergy, and other energy. The petroleum line of business is further segmented into exploration and production, refining and marketing, crude and petroleum product pipelines (for domestic petroleum), and international marine transport (for foreign petroleum).

In 2004, 29 major energy companies reported data through the FRS, representing operating revenues of $1.13 trillion, equal to about 15% of the $7.4 trillion in revenues of the Fortune 500 corporations [37]. About 94% of the FRS companies' operating revenues were derived from petroleum operations, and they accounted for 46% of total U.S. crude oil and natural gas liquid production, 43% of U.S. natural gas production, and 84% of U.S. refining capacity. Of the 29 major companies in the survey, 21 own refineries in the United States.

1.7.3 Boom and Bust Industry

The refining industry has historically been a high-volume, low-margin industry, characterized by low return on investments and volatile profits. Profitability is measured by return on investment, defined as the net income contributed by refining/marketing as a percentage of net fixed assets (net property, plant, and equipment plus investments and advances) in U.S. refining and marketing. The profitability of refining peaked in 1988 at 15%, averaged 2% from 1992–1995, peaked again in 2001 at 15%, and then plunged to –1.7% in 2002, the worst in the history of the FRS survey. In 2004, both domestic and foreign ROI exceeded 18%. High utilization capacity, fewer competitors, and increasing product demand usually make for a high-profit margin scenario, but it has only been in the last few years where this combination of conditions has resulted in high profits.

1.7.4 Refining Margins

Industry performance is also frequently assessed through the gross and net refining margins. The gross margin of a refinery is computed as the total revenues from product sales minus the cost of crude oil. Net margin is defined as the gross margin minus petroleum product marketing costs, internal energy costs, and other operating costs. The domestic refining/marketing gross margin increased to $8.68 per barrel in 2004, the highest since 1992. Operating costs have generally declined over time, resulting in a net refined product margin of $3.00 per barrel in 2004. Typical gross and net margins for USGC refineries are depicted in Table 1.14 in terms of refinery complexity.

1.7.5 Investment Decision Making

Capital expenditures in refining are used to enhance, upgrade, modify, and debottleneck existing configurations; to keep the refinery site and the products it produces in compliance with environmental standards; and for merger and acquisition activity.

The decision to invest is complex and depends on expectations of return on investment and several other factors. For the refining sector to attract funding, it must offer terms and rates of return that compare favorably with those offered in other sectors, taking into account the different risk profiles. In 2004, the FRS companies spent $86.5 billion on capital expenditures, with refining/marketing contributing 16% of the total [37]. Environmental investments incur expenses but do not produce revenues directly and do not result in lower cost or increased output, and so are expected to reduce ROI. All segments of the petroleum industry experienced mergers during the 1990s, with over 2600 transactions reported by John S. Herold, Inc. from

TABLE 1.14
Refinery Complexity and Typical Refining Margins

Refinery Complexity	U.S. Capacity (%)	Average Size (BPSD)	Gross Margin† ($/bbl)	Net Margin ($/bbl)
Simple	5.6	35,233	0.50–1.50	(0.50)–1.00
Cracking	28.7	92,552	3.00–4.50	0.00–2.50
Coking	65.7	196,936	5.00–7.00+	0.50–4.00+

†$1.00 per barrel equals approximately $35 million per year for a typical 100,000 BPSD refinery. Margins are for U.S. Gulf Coast refineries.

Source: O&GJ World Refinery Capacity Report, December 19, 2005.

O'Brien, J.B. and Jensen, S., A new proxy for coking margins — forget the crack spread, National Petrochemical and Refiners Associated, AM-09-55, San Francisco, March 13–15, 2005.

1991 to 2000 [39]. The upstream segment of the business accounted for the majority (85%) of these mergers, with 13% occurring in the refining and marketing segment. About half of downstream activity was asset mergers.

1.8 COST ESTIMATION

1.8.1 Data Sources

A wide variety of data sources are available to estimate the construction cost of a refinery unit, including government organizations, private and public companies, commercial databases, trade and academic publications, and press releases for licensors and companies.

Planned capacity expansions for refineries are reported annually in the October issue of the *Oil and Gas Journal* and *Hydrocarbon Processing*. The data coverage is similar in both surveys and provides information on the capacity that is expected to be added at each location by project type (increment of capacity added; total capacity after construction; revamping, modernization, or debottlenecking; or expansion); licensor, engineering company, and constructor; estimated completion date; estimated cost; and project status (abandoned, engineering, feed, completed, maintenance, planning, or under construction). The data are subject to availability, and because project descriptions are not provided, the surveys are of limited use for cost estimation. *Hydrocarbon Processing* also publishes a *Refining Handbook* that includes basic economic data for each process [40]. The quality of the information is somewhat better than the survey data, but remains of limited use for cost estimation because detailed process descriptions are not provided.

The best public sources of information for cost data are technical articles found in *Oil and Gas Journal*, *Hydrocarbon Processing*, and *Petroleum Technology Quarterly*; material presented at professional conferences such as the National Petrochemicals and Refining Association; and industry studies. Robert Meyers' *Handbook of Petroleum Refining Processes* [41] is an excellent source of process and economic data for a wide compendium of technologies. Maples [42], Raseev [43], Sadeghbeigi [44], and earlier editions of this text provide cost estimates for mid-1990 and earlier configurations and technologies.

Commercial databases from Baker & O'Brien (www.bakerobrien.com), Purvin and Gertz (www.purvingertz.com), Solomon Associates (www.SolomonOnline.com), Turner and Mason (www.turnermason.com), and other consultancies are widely used throughout industry for benchmarking and policy studies. The cost information incorporated in commercial sources is considered the best available, having been collected from projects and other assignments over an extended period of time. Each agency considers its database and modeling software proprietary, however, and so the quality, reliability, and consistency of commercial sources remain difficult to assess and compare.

1.8.2 Process Technologies

The process technologies employed in refining range from atmospheric and vacuum distillation to coking and thermal processes, catalytic cracking, catalytic reforming, catalytic hydrocracking, catalytic hydrotreating, alkylation, polymerization, aromatics extraction, isomerization, oxygenates, and hydrogen production (Table 1.15). For each process, one or more subcategories are defined, based on technology attributes, operating conditions, or feedstock.

TABLE 1.15
Refining Production Process Technologies

Process Operation	Technology
Coking	Fluid coking, delayed coking, other
Thermal process	Thermal cracking, visbreaking
Catalytic cracking	Fluid, other
Catalytic reforming	Semiregenerative, cycle, continuous regeneration, other
Catalytic hydrocracking	Distillate, upgrading residual, upgrading lube oil, other
Catalytic hydrotreating	Pretreatment of cat reformer feeds, other naphtha desulfurization, naphtha aromatics saturation, kerosine/jet desulfurization, diesel desulfurization, distillate aromatics saturation, other distillates, pretreatment of cat cracker feeds, other heavy gas oil hydrotreating, resid hydrotreating, lube oil polishing, post hydrotreating of FCC naphtha, other
Alkylation	Sulfuric acid, hydrofluoric acid
Polymerization/dimerization	Polymerization, dimerization
Aromatics	BTX, hydrodealkylation, cyclohexane, cumene
Isomerization	C_4 feed, C_5 feed, C_5 and C_6 feed
Oxygenates	MTBE, ETBE, TAME
Hydrogen production	Steam methane reforming, steam naphtha reforming, partial oxidation
Hydrogen recovery	Pressure swing adsorption, cryogenic, membrane, other

Source: O&GJ World Refinery Capacity Report, December 19, 2005.

1.8.3 Function Specification

The capital costs for refining units are frequently specified as a function of capacity and scaled using the power-law relation:

$$\frac{C(U_i, Q_1)}{C(U_i, Q_2)} = \left(\frac{Q_1}{Q_2}\right)^x,$$

where $C(U_i, Q_i)$ denotes the cost of process unit U_i of capacity Q_i. The value of x varies with each unit and is frequently assumed to vary between 0.5 and 0.7 [45,46]. In the text, the value of x is determined empirically through regression analysis based on data collected for units of comparable design and technology. For some units, it was not possible to develop useful curves from the feed capacity of the unit, whereas in other cases, if a large number of factors influence the design parameters, unrealistic uncertainty bounds may result. To manage these issues, we frequently cross-correlated cost data based on two or more process descriptors, and in cases where data were extremely sparse, (unit) cost data on a per barrel basis are presented. The assumptions used in each cost estimate are specified in a corresponding table.

1.8.4 Utility Requirements

The utility requirements for each process are presented in the text on a per barrel unit feed or product basis. Utility requirements correspond to "average" characteristics associated with the midpoint of the construction cost curve. Wide variability in utility values can be expected, depending upon the capacity of the unit and other process-specific factors. Typical utility cost data are provided in Table 1.16.

TABLE 1.16
Typical Utility Cost Data (2005)

Utility	English Units	SI Units
Steam, 450 psig	\$5.50/1000 lb	\$12.10/1000 kg
Steam, 150 psig	\$4.40/1000 lb	\$8.80/1000 kg
Steam, 50 psig	\$2.50/1000 lb	\$5.50/1000 kg
Electricity	\$0.04/kW-hr	\$0.04/kW-hr
Cooling water	\$0.05/1000 gal	\$0.013/m^3
Process water	\$0.05/1000 gal	\$0.13/m^3
Boiler feed water	\$1.50/1000 gal	\$0.40/m^3
Chilled water, 40°F	\$1.00/ton-day	\$3.30/GJ
Natural gas	\$5.40/1000 scf	\$0.226/scm
Fuel oil	\$0.75/gal	\$200/m^3

Source: Seider, W.D., Seader, J.D., and Lewin, D.R., *Product & Process Design Principles*, John Wiley & Sons, New York, 2004.

1.8.4.1 Normalization

The cost data of units of roughly comparable design and technology are normalized with respect to construction requirements, process specifications, location, and time of installation.

1.8.4.1.1 Dependent Variable

The normal basis in computing construction cost is the liquid-volume fraction of the crude that is fed to each process, but in several units (e.g., alkylation, polymerization, aromatics manufacture) a better descriptor is the barrels of product rather than the feed. In other units (e.g., isomerization, hydrotreating, catalytic reforming, hydrogen production), it is necessary to cross-correlate the cost with other factors, whereas in gas processing and sulfur manufacture, the liquid-volume basis needs to be replaced with the measures cubic feet (of gas) and long tons (of sulfur).

1.8.4.1.2 Project Type

Projects are classified according to new capacity, expansion of existing capacity, or a revamp or modernization of existing facilities. The cost estimates in the text pertain exclusively to grassroots construction, limited to equipment inside the battery limits (ISBL) of each process, and including materials and labor; design, engineering, and contractor's fees; overheads; and expense allowance.

1.8.4.1.3 Off-Site Expenses

Off-site expenses include the cost and site preparation of land, power generation, electrical substations, off-site tankage, or marine terminals. Off-site costs vary widely with the location and existing infrastructure at the site and depend on the process unit. ISBL costs do not include off-site expenses.

1.8.4.1.4 Location

It is common practice to state cost estimates relative to the USGC because this location has very favorable construction costs relative to other domestic and international markets. Design requirements, climate, regulations, codes, taxes, and availability and productivity of labor all influence and impact the cost of construction and, to some extent, the operating cost of a facility.

1.8.4.1.5 Time

The purchase cost of processing equipment in refining is generally obtained from charts, equations, or quotes from vendors at a particular date, usually month and year. Factors such as regulatory requirements, heavy feedstock, and the cost of materials may increase costs and inflate refinery investment over time, whereas other factors may act to lower cost, such as improvements due to a technological and process nature like improved catalyst, control and instrumentation, and materials technology. Time captures both long-term dynamics such as improvements in technology and operational efficiency, as well as local effects such as the cost of steel, permit requirements, and pollution control.

TABLE 1.17
Nelson-Farrar Refinery Construction Cost Index (1970–2005)

X	197X	198X	199X	200X
0	365	823	1226	1543
1	406	904	1253	1580
2	439	977	1277	1642
3	468	1026	1311	1710
4	523	1061	1350	1834
5	576	1074	1392	1918
6	616	1090	1419	
7	653	1122	1449	
8	701	1165	1478	
9	757	1196	1497	

Source: O&GJ, 1946 = 100. The Nelson-Farrar Refinery Construction Cost Index is published in the first issue of each month of *O&GJ*.

1.8.5 Nelson-Farrar Cost Indices

An estimate of the purchase cost at time t_2, $C(t_2)$, is obtained by multiplying the original (quoted) cost at time t_1, $C(t_1)$, by a ratio of cost indices:

$$C(t_2) = C(t_1)\left(\frac{I(t_2)}{I(t_1)}\right).$$

The Nelson-Farrar (NF) construction cost index normalizes cost during the time required to construct a process unit. The NF cost index is not suitable for determining the cost for refineries or process units that are more than 3 to 5 years old. The NF cost index also does not account for productivity attained in design, construction, or management skills. The NF construction cost index is published in the first issue of each month of *Oil and Gas Journal* (Table 1.17).

The NF operating cost indices are used to compare operating costs over time (Table 1.18). Unlike the construction index, the operating cost indices are normalized for the productivity of labor, changes in the amounts and kinds of fuel used, productivity in the design and construction of refineries, and the amounts and kinds of chemicals and catalysts employed [47,48]. Comparisons of operating indices can be made for any two periods of time.

1.8.6 Limitations of Analysis

Cost data are reported usually in terms of engineering estimates as opposed to actual (finished) cost. It would be preferable to use only actual cost in analysis, but the

TABLE 1.18
Nelson-Farrar Refinery Operating Cost Indices (2000–2005)

	2000	2001	2002	2003	2004	2005
Fuel and power cost	780	704	667	935	972	1360
Labor cost	249	221	211	201	192	202
Wage rate	1094	1007	968	972	984	1007
Productivity of labor	441	456	459	485	513	501
Investment related	589	594	620	643	687	716
Chemicals and catalyst	224	222	221	238	268	311
Refineries	444	429	433	465	487	542
Processes	554	521	514	613	638	787

Source: O&GJ, 1956 = 100. The Nelson-Farrar Refinery Operating Cost Index is published in the first issue of each month of *O&GJ*.

sample sets in most cases would be too sparse, and thus it is necessary when building cost curves to base the assessment on both engineering estimates and actual cost records. Because cost data exhibit normal scatter due to differing qualities of equipment fabrication, design differences, market conditions, vendor profit, and other considerations, a level of uncertainty is intrinsic to every cost estimate.

The cost curves in the text represent typical, or average, values and are presented as point estimates rather than in terms of intervals or ranges. Cost functions are meant to represent an "average" refinery, which of course does not exist as an actual plant, but is useful in developing conceptual cost estimates in the early stage of assessment and design. The cost curves can be assumed to have an accuracy limited to ±25%. Working capital, inventories, start-up expense, the cost of land, site preparation, taxes, licenses, permits, and duties are not considered in the estimation.

The level of uncertainty in cost estimation can only be reduced through a detailed front-end engineering design based on site-specific information. For definitive economic comparisons and estimation, other factors such as feedstock, production specifications, operating conditions, design options, and technology options must be considered.

1.9 REFINERY COMPLEXITY

1.9.1 Data Source

The *Oil and Gas Journal* publishes a refining survey in December of each year, which lists the capacity ratings of all refineries in the world by company and location in terms of charge and production capacities. The EIA also publishes data on U.S. refineries that are similar in format to the *Oil and Gas Journal* [49].

Crude distillation, vacuum distillation, coking, thermal processes, catalytic cracking, catalytic reforming, and catalytic hydrocracking are described in terms of charge capacity, which describes the input (feed) capacity of the facilities. Production capacity represents the maximum amount of product that can be produced; data are presented for alkylation, polymerization/dimerization, aromatics, isomerization, lubricants, oxygenates, hydrogen, coke, sulfur, and asphalt facilities.

1.9.2 Unit Complexity

Wilbur Nelson introduced the concept of complexity factor to quantify the relative cost of components that make up a refinery [50–52]. Nelson assigned a complexity factor of 1 to the atmospheric distillation unit and expressed the cost of all other units in terms of their cost relative to distillation. For example, if a crude distillation unit of 100,000 BPD capacity cost $10 million to build, then the unit cost per daily barrel of throughput would be $100 BPD. If a 20,000 BPD catalytic reforming unit cost $10 million to construct, then the unit cost is $500 BPD of throughput and the "complexity" of the catalytic reforming unit would be 500 / 100 = 5.

The complexity factor of process unit U_i of capacity Q_i and construction cost $C(U_i,Q_i)$ is defined as:

$$\gamma(U_i) = \frac{C(U_i, Q_i)/Q_i}{C(U_0, Q_0)/Q_0},$$

where $\gamma(U_i)$ denotes the complexity index of the unit and U_0 represents the atmospheric distillation unit.

Various methodological issues limit the use of complexity factors in cost estimation. Complexity factors do not account for the impact of capacity on cost, because the complexity factor is capacity invariant, and trends in complexity factors change slowly (or not at all) over time (Table 1.19), making their application suspect.

1.9.3 Refinery Complexity

Refinery complexity indicates how complex a refinery is in relation to a refinery that performs only crude distillation. The complexity index of refinery R is determined by the complexity of each individual unit weighted by its percentage of distillation capacity:

$$\gamma(R) = \sum_i \frac{Q_i}{Q_0} \gamma(U_i),$$

where $\gamma(R)$ represents the complexity index of refinery R, $\gamma(U_i)$ is the complexity index of unit U_i, and Q_i denotes the capacity of unit U_i. A simple refinery is typically defined by $\gamma(R) < 5$; a complex refinery by $5 < \gamma(R) < 15$; and a very complex refinery by $\gamma(R) > 15$. Refinery complexity is an often-cited industry statistic and is a useful tool in comparative analysis, being frequently used as a correlative or descriptive variable in marketing and valuation studies [53].

1.9.4 Example

ExxonMobil's charge and production capacities of its Louisiana facilities are shown in Table 1.20 and Table 1.21. The complexity index of the Baton Rouge refinery is computed as 13.4 (Table 1.22).

TABLE 1.19
Trends in Complexity Factors

Process Operation	1946	1961–1972	1976	1989
Atmospheric distillation	1	1	1	1
Vacuum distillation	2	2	2	2
Vacuum flash	1	1	1	1
Thermal process				
Thermal cracking	4.5	3	3	3
Visbreaking	2	2	2	2
Coking	5	5	5.5	5.5
Calcining		63	108	108
Catalytic cracking				
60% Conversion	4	5	5	5
80% Conversion		6	6	6
Catalytic reforming	5	4	5	5
Catalytic hydrocracking		6	6	6
Catalytic hydrorefining		4	3	3
Catalytic hydrotreating		2	2	1.7
Alkylation	9	9	11	11
Aromatics, BTX		40–70	20	20
Isomerization		3	3	3
Polymerization	9	9	9	9
Lubes				60
Asphalt	2		3.5	1.5
Hydrogen (MCFPD)				
Manufacturing		1.2	1.2	1
Recovery		0.7	0.7	1
Oxygenates				10
Sulfur (long ton per day)				
Manufacturing		85	85	85
Manufacturing from dilute gas		250–300	220–600	220–600

Source: Nelson, W.L., *Guide to Refinery Operating Cost (Process Costimating)*, 3rd ed., Petroleum Publishing, Tulsa, OK, 1976; Farrar, G.L., How Nelson cost indices are compiled, *O&GJ*, December 30, p. 145, 1985.

1.9.5 Generalized Complexity

The complexity index can be generalized across any level of aggregation, such as a company, state, country, or region, as follows:

$$\gamma(\Gamma) = \frac{1}{Q(\Gamma)} \sum_i Q(R_i)\gamma(R_i),$$

where $\gamma(\Gamma)$ represents the complexity measure of the category Γ under consideration, $\gamma(R_i)$ denotes the complexity of refinery R_i, $Q(R_i)$ denotes the distillation capacity of refinery R_i, and $Q(\Gamma) = \sum_i Q(R_i)$.

TABLE 1.20
Charge Capacity of ExxonMobil's Louisiana Facilities — Barrels per Calendar Day (2005)

Location	Crude	Vacuum Distillation	Delayed Coking	Thermal Operations	Catalytic Cracking	Catalytic Reforming	Catalytic Hydrocracking	Catalytic Hydrotreating
Baton Rouge	501,000	227,000	112,500	—	229,000	75,500	24,000	333,500
Chalmette	188,000	112,000	33,000	—	68,000	47,000	18,500	172,500
Total	689,000	335,000	145,500	—	297,000	122,500	42,500	506,000

Source: O&GJ World Refinery Capacity Report, December 19, 2005.

TABLE 1.21
Production Capacity of ExxonMobil's Louisiana Facilities — Barrels per Calendar Day (2005)

Location	Alkylation	Poly/Dim	Aromatics	Isomerization	Lubes	Oxygenates	Hydrogen (MMcfd)	Coke (tonnes per day)	Sulfur (tonnes per day)	Asphalt (tonnes per day)
Baton Rouge	140,000	9500	—	—	16,000	7000	12	5262	690	—
Chalmette	12,500	—	10,000	10,000	—	—	—	2050	920	—
Total	152,500	9500	10,000	10,000	16,000	7000	12	7312	1619	—

Source: O&GJ World Refinery Capacity Report, December 19, 2005.

TABLE 1.22
ExxonMobil's Baton Rouge Complexity Index (2005)

Process Operation	Capacity (BPSD)	Percent of Dist. Capacity	Complexity Factor	Complexity Index
Atmospheric distillation	501,000	100	1	1
Vacuum distillation	227,000	45.3	2	0.91
Coking	112,500	22.5	5.5	1.20
Catalytic cracking	229,000	45.7	6	2.74
Catalytic reforming	75,500	15.1	5	0.76
Catalytic hydrocracking	24,000	4.8	6	0.29
Catalytic hydrotreating	333,500	66.6	1.7	1.13
Alkylation	140,000	27.9	11	3.07
Polymerization	9,500	1.9	9	0.17
Lubes	16,000	3.2	60	1.92
Oxygenates	7,000	1.4	10	0.14
Hydrogen (MCFD)	12,000	2.4	1	0.02
Complexity Index				13.4

1.10 REFINERY FLOW SCHEME AND CHAPTER OUTLINE

1.10.1 Refinery Products and Feedstocks (Chapters 2, 3)

Crude oils exist in liquid phase in underground reservoirs and remain liquid at atmospheric pressure after passing through surface separating facilities. Crude oil also includes small amounts of hydrocarbons that exist in gaseous phase in reservoirs but become liquid after being recovered from oil well (casinghead) gas in lease separators, as well as small amounts of nonhydrocarbons such as sulfur and various metals, drip gases, and liquid hydrocarbons produced from tar or oil sands and oil shale. Crude oil has no value in its natural state and must be refined to produce products consumers demand. Chapter 2 reviews the properties of refinery products and gasoline specifications, whereas Chapter 3 defines the important properties of crude oil and their composition.

1.10.2 Crude Distillation (Chapter 4)

The first step in the refining process is the separation of crude oil into fractions by distillation. The crude oil is heated in a furnace to temperatures of about 600 to 700°F (315 to 370°C) and charged to a distillation tower, where it is separated into butanes and lighter wet gas, unstabilized light naphtha, heavy naphtha, kerosine, atmospheric gas, and topped (reduced) crude. The topped crude is sent to the vacuum distillation tower under reduced pressure and separated into vacuum gas oil and vacuum-reduced crude bottoms.

1.10.3 Coking and Thermal Processes (Chapter 5)

In thermal cracking, heat and pressure are used to break down, rearrange, and combine hydrocarbon molecules. Delayed coking, visbreaking, fluid coking, and flexicoking are thermal cracking processes.

The reduced crude bottoms from the vacuum tower are thermally cracked in a delayed coker to produce wet gas, coker naphtha, coker gas oil, and coke. Delayed coking decomposes heavy crude oil fractions to produce a mixture of lighter oils and petroleum coke. The light oils are processed in other refinery units to meet product specifications, whereas the coke can be used either as a fuel or in other applications such as the manufacturing of steel or aluminum. Without a coker, heavy resid would be sold for heavy fuel oil or (if the crude oil is suitable) asphalt. Visbreaking is a mild thermal cracking process in which heavy atmospheric or vacuum still bottoms are cracked at moderate temperatures to increase production of distillate products and reduce viscosities of the distillation residues. Fluid coking utilizes a fluidized solids technique to remove coke for continuous conversion of heavy, low-grade oils into lighter products. Flexicoking converts heavy hydrocarbons such as tar sands bitumen and distillation residues into light hydrocarbons and a fuel gas.

1.10.4 Catalytic Cracking (Chapter 6)

Catalytic cracking is a process that breaks down the larger, heavier, and more complex hydrocarbon molecules into simpler and lighter molecules by the action of heat and aided by the presence of a catalyst but without the addition of hydrogen. The atmospheric and vacuum crude unit gas oils and coker gas oil are used as feedstocks for the catalytic cracking or hydrocracking units. Heavy oils are converted into lighter products such as liquefied petroleum gas (LPG), gasoline, and middle distillate components. Catalytic cracking processes fresh feeds and recycle feeds. The unsaturated catalytic cracking products are saturated and improved in quality by hydrotreating or reforming. Fluid catalytic cracking is the most widely used secondary conversion process technology.

1.10.5 Catalytic Hydrocracking (Chapter 7)

Catalytic hydrocracking is a refining process that uses hydrogen and catalysts at relatively low temperature and high pressures for converting middle boiling points to naphtha, reformer charge stock, diesel fuel, jet fuel, or high-grade fuel oil. The process uses one or more catalysts, depending upon product output, and can handle high-sulfur feedstocks. Hydrocracking is used for feedstocks that are difficult to process by either catalytic cracking or reforming, because the feedstocks are usually characterized by a high polycyclic aromatic content or high concentrations of olefins, sulfur, and nitrogen compounds.

1.10.6 Hydroprocessing and Resid Processing (Chapter 8)

"Bottom of the barrel" is a term used to refer to the atmospheric tower bottoms boiling above 650°F (343°C) or vacuum tower bottoms boiling above 1050°F (566°C). Traditionally, this material has been blended into heavy or industrial fuel oil. There are several options for bottom-of-the-barrel processing, including delayed coking, visbreaking, and resid desulfurization. Hydroprocessing is used to reduce the boiling range of the feedstock as well as to remove impurities such as metals, sulfur, nitrogen, and high carbon forming compounds.

1.10.7 Hydrotreating (Chapter 9)

Distillation does not change the molecular structure of hydrocarbons, and so any impurities in the crude oils remain unchanged. Catalytic hydrotreating treats petroleum fractions in the presence of catalysts and substantial quantities of hydrogen. If contaminants are not removed from the petroleum fractions as they travel through the processing units, they can have detrimental effects on the equipment, the catalysts, and the quality of the finished products. Typically, hydrotreating is done prior to catalytic reforming, so that the catalyst is not contaminated by untreated feedstock. Hydrotreating is also used prior to catalytic cracking to reduce sulfur and improve product yields, and to upgrade middle-distillate petroleum fractions into finished kerosine, diesel fuel, and heating fuel oils. Hydrotreating results in desulfurization (removal of sulfur), denitrogenation (removal of nitrogen), and conversion of olefins to paraffins.

1.10.8 Catalytic Reforming and Isomerization (Chapter 10)

The heavy naphtha streams from the crude tower, coker, and cracking units are fed to the catalytic reformer to improve their octane numbers. Catalytic reforming converts low-octane naphthas into high-octane gasoline blending components called reformates. Reforming represents the total effect of cracking, dehydrogenation, and isomerization taking place simultaneously. Depending on the properties of the naphtha feedstock and catalysts, reformates can be produced with very high concentrations of benzene, toluene, xylene, and other aromatics useful in gasoline blending and petrochemical processing. Hydrogen is separated from the reformate for recycling and use in other processes.

Isomerization alters the arrangement of atoms without adding or removing anything from the original material. Isomerization is used to convert normal butane into isobutane, an alkylation process feedstock, and normal pentane and normal hexane into isopentane and isohexane, high-octane gasoline components. The light naphtha stream from the crude tower, coker, and cracking units is sent to an isomerization unit to convert straight-chain paraffins into isomers that have higher octane numbers.

1.10.9 Alkylation and Polymerization (Chapter 11)

Alkylation is a process for chemically combining isobutane with olefin hydrocarbons through the control of temperature and pressure in the presence of an acid catalyst, usually sulfuric acid or hydrofluoric acid. The product alkylate, an isoparaffin, has high-octane value and is blended into motor and aviation gasoline to improve the antiknock value of the fuel. The olefin feedstock is primarily a mixture of propylene and butylene.

Polymerization converts light olefin gases, including propylene and butylene, into hydrocarbons of higher molecular weight and higher octane number that can be used as gasoline blending stocks. The process is accomplished in the presence of a catalyst at low temperatures.

1.10.10 Product Blending (Chapter 12)

Blending is the physical mixture of a number of different liquid hydrocarbons to produce a finished product with certain desired characteristics. Products can be blended in-line through a manifold system or batch blended in tanks and vessels. In-line blending of gasoline, distillates, jet fuel, and kerosine is accomplished in the main stream where turbulence promotes thorough mixing. Additives including octane enhancers, metal deactivators, antioxidants, antiknock agents, gum and rust inhibitors, and detergents are added during or after blending to provide specific properties not inherent in hydrocarbons.

1.10.11 Supporting Processes (Chapter 13)

The wet gas streams from the crude tower, reformer, hydrocracker, and hydrotreaters are separated in the vapor recovery sections (saturated gas plants) into fuel gas, LPG, normal butane, and isobutane. The fuel gas is burned as a fuel in refinery furnaces, the normal butane is blended into gasoline or LPG, and the isobutane is used as a feedstock to the alkylation unit.

Unsaturated gas plants recover light olefinic hydrocarbons from wet gas streams and then compress and treat with amine to remove hydrogen sulfide either before or after they are sent to a fractionating absorber. The unsaturated hydrocarbons and isobutane are sent to the alkylation unit for processing.

Amine plants remove acid contaminants from sour gas and hydrocarbon streams. Gas and liquid hydrocarbon streams containing carbon dioxide or hydrogen sulfide are charged to a gas absorption tower or liquid contactor, where the acid contaminants are absorbed by counterflowing amine solutions. Sulfur recovery converts hydrogen sulfide in sour gases and hydrocarbon streams to elemental sulfur. The most widely used recovery system is the Claus process, which uses both thermal and catalytic conversion reactions.

Hydrogen recovery from catalytic reforming is often not enough to meet total refinery requirements, necessitating the manufacture of additional hydrogen or obtaining a supply from external sources. High-purity hydrogen is required for hydrosulfurization, hydrocracking, and petrochemical processes. The steam reformation of natural gas accounts for the majority of the hydrogen produced, but catalytic reforming and partial oxidation processes are also applied. Steam reformation and catalytic processes are generally suitable for reforming light hydrocarbons and naphtha, whereas partial oxidation processes are applied to a broader range of feedstocks, including heavy residual oils and low-value products.

1.10.12 Refinery Economics and Planning (Chapter 14)

Modern refinery operations are very complex because of complex feedstock, multiple sources of feedstock, sophisticated processing technology, and increasingly stringent product specifications. A sitewide model of the refinery is usually required in order to properly determine refinery economics. Typically, this economic decision analysis is assigned to refinery planning departments. A linear programming (LP) model illustration is used to show how this decision is reached.

1.10.13 Lubricating Oil Blending Stocks (Chapter 15)

In some refineries, the heavy vacuum gas oil and reduced crude from paraffinic or naphthenic base crude oils are processed into lubricating oils. The vacuum gas oil and deasphalting stocks are first solvent-extracted to remove the aromatic compounds and then dewaxed to improve the pour point. They are then treated with special clays or high-severity hydrotreating to improve their color and stability before being blended into lubricating oils. The purpose of solvent extraction is to prevent corrosion, protect the catalyst in subsequent processes, and improve finished products by removing unsaturated aromatic hydrocarbons from lubricant and grease stocks. Solvent dewaxing is used to remove wax from either distillate or residual basestocks at any stage in the refining process.

1.10.14 Petrochemical Feedstocks (Chapter 16)

Hydrocarbon feedstock aromatics, unsaturates, and saturates are utilized for petrochemical manufacture according to use and method of preparation. Aromatics are produced using the catalytic reforming units used to upgrade the octanes of heavy straight-run naphtha gasoline blending stocks. Most of the unsaturates are produced by steam cracking or low molecular polymerization. Saturates are recovered from petroleum fractions by vapor-phase adsorption or molecular sieves (normal paraffins), or by the hydrogenation of the corresponding aromatic compound (cycloparaffins).

1.10.15 Additives Production from Refinery Feedstocks (Chapter 17)

During the 1970s, various high-octane additives began to be blended into motor gasoline to maintain octane levels as the use of tetraethyl lead (TEL) was reduced. Methanol and ethanol were first used because of their availability, but these alcohols were gradually replaced with various ethers, such as methyl tertiary butyl ether (MTBE) and ethyl tertiary butyl ether (ETBE). Various commercial processes are available for ether production. As ethers are phasing out, the ether units are being converted to isooctene and isooctane production.

1.10.16 Cost Estimation (Chapter 18)

Cost estimation is an important aspect of process evaluation. Rule-of-thumb, cost curves, equipment factor, and definitive estimates are reviewed. An example problem illustrating the methods to estimate capital and operating costs is presented.

1.10.17 Economic Evaluation (Chapter 19)

Economic evaluations are carried out to determine if a proposed investment meets the profitability criteria of the company or to evaluate alternatives. A case study problem is presented that illustrates the basic calculations and procedures.

NOTES

1. Riva, J.P. Jr., Petroleum, in *Encyclopaedia Britannica*, 2006.
2. Knowles, R.S., *The First Pictorial History of the American Oil and Gas Industry 1859–1983,* Ohio University Press, Athens, OH, 1983.
3. Yergin, D., *The Prize: The Epic Quest for Oil, Money and Power,* Touchstone, New York, 1993.
4. OSHA technical manual, in *Petroleum Refining Processes*, U.S. Department of Labor, Occupational Safety and Health Administration, Washington, D.C., 2005, sec. IV, chap. 2.
5. Chernow, R., *Titan – The Life of John D. Rockefeller, Sr.*, Random House, New York, 1998.
6. Schobert, H.J., *Energy and Society*, Taylor & Francis, New York, 2002.
7. U.S. Department of Energy, Energy Information Administration, *Petroleum: An Energy Profile*, Washington, D.C., July 1999.
8. U.S. Department of Energy, Energy Information Administration, *Petroleum Supply Monthly*, Washington, D.C., March 2005.
9. BP, *Statistical Review of World Energy 2005*, 54th ed. Available at www.bp.com.
10. U.S. Department of Energy, Energy Information Administration, *The U.S. Petroleum Refining and Gasoline Market Industry*, Washington, D.C., June 1999.
11. Kumins, L., Parker, L., and Yacobucci, B., *Refining Capacity — Challenges and Opportunities Facing the U.S. Industry,* PIRINC, New York, 2004.
12. Pirog, R.L., Refining: Economic performance and challenges for the future, *CRS Report for Congress*, RL 32248, Congressional Research Service, Library of Congress, Washington, D.C., May 2005.
13. Peterson, D.J. and Mahnovski, S., *New Forces at Work in Refining, Industry Views on Critical Business and Operations Trends,* RAND Science and Technology, 2003.
14. Pratt, J.A., *The Growth of a Refining Region*, JAI Press, Greenwich, CT, 1980.
15. O'Connor, T., Petroleum refineries: Will record profits spur investment in new capacity? Testimony before the House Government Reform Committee, Subcommittee on Energy and Resources, U.S. House of Representatives, October 19, 2005.
16. ENI, *World Oil and Gas Review 2005*, 5th ed. Available at www.eni.it.
17. Energy Intelligence Research, *The International Crude Oil Market Handbook*, Energy Intelligence, New York, 2006.
18. Speight, J.G., *The Chemistry and Technology of Petroleum*, 3rd ed., Marcel Dekker, New York, 1998.
19. Ruschau, G.R. and Al-Anezi, M.A., Petroleum refining: Corrosion control and prevention, in *Corrosion Costs and Preventive Strategies in the United States, Appendix U*, U.S. Department of Transportation, 2001.
20. Bacon, R. and Tordo, S., Crude oil price differentials and differences in oil qualities: A statistical analysis, ESMAP Technical Paper 081, Energy Sector Management Assistance Program, Washington, D.C., October 2005.
21. Leffler, W., *Petroleum Refining in Nontechnical Language*, 3rd ed., PennWell, Tulsa, OK, 2000.
22. National Petroleum Council, *Observations on Petroleum Product Supply*, Washington, D.C., 2004.
23. Parkash, S., *Refining Process Handbook*, Elsevier, Amsterdam, 2003.
24. Geman, H., *Commodities and Commodity Derivatives*, John Wiley & Sons, West Sussex, England, 2005.

25. National Petroleum Council, *U.S. Petroleum Products Supply: Inventory Dynamics*, Washington, D.C., 1998.
26. Pirog, R.L., World oil demand and its effect on oil prices, CRS Report for Congress, RL 32530, Congressional Research Service, Library of Congress, Washington, D.C., June 2005.
27. Lichtblau, J., Goldstein, L., and Gold, R., Commonalities, uniqueness of oil as commodity explain crude oil, gasoline price behavior, *Oil Gas J.*, June 28, p. 18, 2004.
28. National Petroleum Council, *U.S. Petroleum Refining: Assuring the Adequacy and Affordability of Cleaner Fuels,* Washington, D.C., 2000.
29. O'Brien, J.B. and Jensen, S., A new proxy for coking margins — forget the crack spread, National Petrochemical and Refiners Associated, AM-09-55, San Francisco, March 13–15, 2005.
30. U.S. Department of Energy, Office of Industrial Technologies, *Manufacturing Energy Consumption Survey*, Washington, D.C., August 2004.
31. Ocic, O., *Oil Refineries in the 21st Century*, Wiley-VCH Verlag, Weinheim, 2005.
32. North, D., Refinery utilities, in Lucas, A.G. (Ed.), *Modern Petroleum Technology, Vol. 2 Downstream*, 6th ed., John Wiley & Sons, West Sussex, England, 2001.
33. U.S. Department of Energy, Office of Industrial Technologies, *Energy and Environmental Profile of the U.S. Petroleum Refining Industry*, Washington, D.C., December 1998.
34. U.S. Department of Energy, Energy Information Administration, *The Impact of Environmental Compliance Costs on U.S. Refining Profitability 1995–2001*, Washington, D.C., May 2003.
35. American Petroleum Institute, *Technology Roadmap for the Petroleum Industry*, Washington, D.C., February 2000.
36. Speight, J.G., *Environmental Analysis and Technology for the Refining Industry,* John Wiley & Sons, New York, 2005.
37. U.S. Department of Energy, Energy Information Administration, *Performance Profiles of U.S. Major Producers*, Washington, D.C., March 2005.
38. American Petroleum Institute, *U.S. Oil and Natural Gas Industry's Environmental Expenditures 1995–2004,* Washington, D.C., February 2006.
39. U.S. General Accounting Office, Energy markets effects of mergers and market concentration in the U.S. petroleum industry, in *GAO-04-96*, Washington, D.C., May 2004.
40. *Hydrocarbon Processing Refining Handbook*, Gulf Publishing, Houston, 2004.
41. Meyers, R.A., *Handbook of Petroleum Refining Processes*, 3rd ed., McGraw-Hill Book Co., New York, 2004.
42. Maples, R.E., *Petroleum Refining Process Economics*, 2nd ed., PennWell, Tulsa, OK, 2000.
43. Raseev, S., Thermal and Catalytic Processes in Petroleum Refining, Marcel Dekker, New York, 2003.
44. Sadeghbeigi, R., *Fluid Catalytic Cracking Handbook*, 2nd ed., Gulf Publishing, Houston, 2000.
45. Peters, M.S., Timmerhaus, K.D., and West, R.E., *Plant Design and Economics for Chemical Engineers*, 5th ed., McGraw-Hill Book Co., Boston, 2003.
46. Seider, W.D., Seader, J.D., and Lewin, D.R., *Product & Process Design Principles*, John Wiley & Sons, New York, 2004.
47. Farrar, G.L., How Nelson cost indices are compiled, *Oil Gas J.*, December 30, p. 145, 1985.
48. Farrar, G.L., Interest reviving in complexity factors, *Oil Gas J.*, October 2, p. 90, 1989.

49. U.S. Department of Energy, Energy Information Administration, *Annual Energy Review 2004*, DOE/EIA-0384 (2004), Washington, D.C., August 2005.
50. Nelson, W.L., *Guide to Refinery Operating Cost* (*Process Costimating*), 3rd ed., Petroleum Publishing, Tulsa, OK, 1976.
51. Nelson, W.L., How the Nelson refinery construction-cost indexes evolved, *Oil Gas J.*, November 29, p. 68, 1976.
52. Nelson, W.L., Here's how operating cost indexes are computed, *Oil Gas J.*, January 10, p. 86, 1977.
53. Neumuller, R., Method estimates U.S. refinery fixed costs, *Oil Gas J.*, September 19, p. 43, 2005.

2 Refinery Products

Although the average consumer tends to think of petroleum products as consisting of a few items such as motor gasoline, jet fuel, home heating oils, and kerosine, a survey by the American Petroleum Institute (API) conducted at petroleum refineries and petrochemical plants revealed over 2000 products made to individual specifications [1]. Table 2.1 shows the number of individual products in 17 classes.

In general, the products that dictate refinery design are relatively few in number, and the basic refinery processes are based on the large-quantity products such as gasoline, diesel, jet fuel, and home heating oils. Storage and waste disposal are expensive, and it is necessary to sell or use all of the items produced from crude oil even if some of the materials, such as high-sulfur heavy fuel oil and fuel-grade coke, must be sold at prices less than the cost of fuel oil. Economic balances are required to determine whether certain crude oil fractions should be sold as is (i.e., straight-run) or further processed to produce products having greater value. Usually, the lowest value of a hydrocarbon product is its heating value or fuel oil equivalent (FOE). This value is always established by location, demand, availability, combustion characteristics, sulfur content, and prices of competing fuels.

Knowledge of the physical and chemical properties of the petroleum products is necessary for an understanding of the need for the various refinery processes. To provide an orderly portrayal of the refinery products, they are described in the following paragraphs in order of increasing specific gravity and decreasing volatility and API gravity.

The petroleum industry uses a shorthand method of listing lower-boiling hydrocarbon compounds, which characterizes the materials by the number of carbon atoms and unsaturated bonds in the molecule. For example, propane is shown as C_3 and propylene as $C_3^=$. The corresponding hydrogen atoms are assumed to be present unless otherwise indicated. The notation will be used throughout this book.

2.1 LOW-BOILING PRODUCTS

The classification *low-boiling products* encompasses the compounds that are in the gas phase at ambient temperatures and pressures: methane, ethane, propane, butane, and the corresponding olefins.

Methane (C_1) is usually used as a refinery fuel, but can be used as a feedstock for hydrogen production by pyrolytic cracking and reaction with steam. Its quantity is generally expressed in terms of pounds or kilograms, standard cubic feet (scf) at 60°F and 14.7 psig, normal cubic meters (Nm^3) at 15.6°C and 1 bar (100 kPa), or barrels fuel oil equivalent based on a lower heating value (LHV) of 6.05×10^6 Btu (6.38×10^6 kJ). The physical properties of methane are given in Table 2.2.

Ethane (C_2) can be used as refinery fuel or as a feedstock to produce hydrogen or ethylene, which are used in petrochemical processes. Ethylene and hydrogen are sometimes recovered in the refinery and sold to petrochemical plants.

TABLE 2.1
Products Made by the U.S. Petroleum Industry

Class	Number
Fuel gas	1
Liquefied gases	13
Gasolines	40
Motor	19
Aviation	9
Other (tractor, marine, etc.)	12
Gas turbine (jet) fuels	5
Kerosines	10
Middle distillates (diesel and light fuel oils)	27
Residual fuel oil	16
Lubricating oils	1156
White oils	100
Rust preventatives	65
Transformer and cable oils	12
Greases	271
Waxes	113
Asphalts	209
Cokes	4
Carbon blacks	5
Chemicals, solvents, miscellaneous	300
Total	2347

Source: Note 1. Reproduced courtesy of the American Petroleum Institute.

Propane (C_3) is frequently used as a refinery fuel but is also sold as a liquefied petroleum gas (LPG), whose properties are specified by the Gas Processors Association (GPA) [2]. Typical specifications include a maximum vapor pressure of 210 psig (1448 kPa) at 100°F (37.8°C) and 95% boiling point of –37°F (–38.3°C) or lower at 760 mmHg (1 bar, 100 kPa) atm pressure. In some locations, propylene is separated for sale to polypropylene manufacturers.

The butanes present in crude oils and produced by refinery processes are used as components of gasoline and in refinery processing as well as in LPG. Normal butane (nC_4) has a lower vapor pressure than isobutane (iC_4) and is preferred for blending into gasoline to regulate its vapor pressure and promote better starting in cold weather. Normal butane has a Reid vapor pressure (RVP) of 52 psi (358 kPa) as compared with the 71 psi (490 kPa) RVP of isobutane, and more nC_4 can be added to gasoline without exceeding the RVP of the gasoline product. On a liquid volume basis, gasoline has a higher sales value than LPG; thus, it is desirable from an economic viewpoint to blend as much normal butane as possible into gasoline. Normal butane is also used as a feedstock to isomerization units to form isobutane.

Regulations promulgated by the U.S. Environmental Protection Agency (EPA) to reduce hydrocarbon emissions during refueling operations and evaporation from

TABLE 2.2
Physical Properties of Paraffins

	C_n	Boiling point (°F)	Melting point (°F)	Specific gravity (60/60°F)	API gravity (°API)
Methane	1	–258.7	–296.5	0.30	340
Ethane	2	–128.5	–297.9	0.356	265.5
Propane	3	–43.7	–305.8	0.508	147.2
Butane					
Normal	4	31.1	–217.1	0.584	110.6
Iso	4	10.9	–225.3	0.563	119.8
Octane					
Normal	8	258.2	–70.2	0.707	68.7
2,2,4	8	210.6	–161.3	0.696	71.8
2,2,3,3	8	223.7	219.0	0.720	65.0
Decane, normal	10	345.5	–21.4	0.734	61.2
Cetane, normal	16	555.0	64.0	0.775	51.0
Eicosane, normal	20	650.0	98.0	0.782	49.4
Triacontane					
Normal	30	850.0	147.0	0.783	49.2
2,6,10,14,18,22	30	815.0	–31.0	0.823	40.4

Note: Generations:

1. Boiling point rises with increase in molecular weight.
2. Boiling point of a branched chain is lower than for a straight-chain hydrocarbon of the same molecular weight.
3. Melting point increases with molecular weight.
4. Melting point of a branched chain is lower than for a straight-chain hydrocarbon of the same weight unless branching leads to symmetry.
5. Gravity increases with increase of molecular weight.
6. For more complete properties of paraffins, see Table B.2 in Appendix B.

hot engines after ignition turn-off have greatly reduced the allowable Reid vapor pressure of gasolines during summer months. This results in two major impacts on the industry. The first is the increased availability of n-butane during the summer months, and the second is the necessity to provide another method of providing the pool octane lost by the removal of the excessive n-butane. The pool octane is the average octane of the total gasoline production of the refinery if the regular, mid-premium, and super-premium gasolines are blended together. N-butane has a blending octane in the 90s and is a low-cost octane improver of gasoline.

Isobutane has its greatest value when used as a feedstock to alkylation units, where it is reacted with unsaturated materials (propenes, butenes, and pentenes) to form high-octane isoparaffin compounds in the gasoline boiling range. Although isobutane is present in crude oils, its principal sources of supply are from fluid catalytic cracking (FCC) and hydrocracking (HC) units in the refinery and from natural gas processing plants. Isobutane not used for alkylation unit feed can be sold

TABLE 2.3
Properties of Commercial Propane and Butane

Property	Commercial propane	Commercial butane
Vapor pressure, psig		
70°F (21.1°C)	124	31
100°F (38°C)	192	59
130°F (54°C)	286	97
Specific gravity of liquid, 60/60°F	0.509	0.582
Initial boiling point at 1 bar, °F (°C)	–51 (–47.4)	15
Dew point at 1 bar, °F (°C)	–46 (–44.6)	24
Sp. ht. liquid at 60°F, 15.6°C		
Btu/(lb) (°F)	0.588	0.549
kJ/(kg) (°C)	2.462	2.299
Limits of flammability, vol% gas in air		
Lower limit	2.4	1.9
Upper limit	9.6	8.6
Latent heat of vaporization at b.p.		
Btu/lb	185	165
kJ/kg	430.3	383.8
Gross heating values		
Btu/lb of liquid	21,550	21,170
Btu/ft^3 of gas	2,560	3,350
kJ/kg of liquid	50,125	49,241
kJ/m^3 of gas	9,538	12,482

Source: Note 2.

as LPG or used as a feedstock for propylene (propene) manufacture. Some isobutane is converted to isobutylene, which is reacted with ethanol to produce ethyl tertiary butyl ether (ETBE).

When butanes are sold as LPG, they conform to the NGPA specifications for commercial butane. These include a vapor pressure of 70 psig (483 kPa) or less at 100°F (37.7°C) and a 95% boiling point of 36°F (2.2°C) or lower at 760 mmHg atm pressure (101.3 kPa). N-butane as LPG has the disadvantage of a fairly high boiling point [32°F (0°C) at 760 mmHg] and, during the winter, is not satisfactory for heating when stored outdoors in areas that frequently have temperatures below freezing. Isobutane has a boiling point of 11°F (–12°C) and is also unsatisfactory for use in LPG for heating in cold climates.

Butane–propane mixtures are also sold as LPG, and their properties and standard test procedures are also specified by the GPA.

Average properties of commercial propane and butane are given in Table 2.3.

2.2 GASOLINE

Although an API survey [1] reports that 40 types of gasolines are made by refineries, about 90% of the total gasoline produced in the United States is used as fuel in automobiles. Most refiners produce gasoline in three grades (unleaded regular, premium, and super-premium) and, in addition, supply a leaded regular gasoline to meet the needs of farm equipment and pre-1972 automobiles. The principal difference between the regular and premium fuels is the antiknock performance. In 2005, the posted method octane number (PON) of unleaded regular gasolines (see Section 2.3) was about 87 and that of premium gasolines ranged from 89 to 93. For all gasolines, octane numbers average about two numbers lower for the higher elevations of the Rocky Mountain states. Posted octane numbers (which are required by the EPA to be posted on dispensing pumps) are arithmetic averages of the motor octane number (MON) and research octane number (RON) and average four to six numbers below the RON.

Gasolines are complex mixtures of hydrocarbons, having typical boiling ranges from 100 to 400°F (38 to 205°C), as determined by the American Society for Testing and Materials (ASTM) method. Components are blended to promote high antiknock quality, ease of starting, quick warm-up, low tendency to vapor-lock, and low engine deposits. Gruse and Stevens [3] give a very comprehensive account of properties of gasolines and the manner in which they are affected by the blending components. For the purposes of preliminary plant design, however, the components used in blending motor gasoline can be limited to light straight-run (LSR) gasoline or isomerate, catalytic reformate, catalytically cracked gasoline, hydrocracked gasoline, polymer gasoline, alkylate, n-butane, and such additives as ETBE, TAME (tertiary amyl methyl ether), and ethanol. Other additives, for example, antioxidants, metal deactivators, and antistall agents, are not considered individually at this time, but are included with the cost of the antiknock chemicals added. The quantity of antiknock agents added, and their costs, must be determined by making octane blending calculations.

Light straight-run gasoline consists of the C_5-190°F (C_5-88°C) fraction of the naphtha cuts from the atmospheric crude still (C_5-190°F fraction means that pentanes are included in the cut but that C_4 and lower-boiling compounds are excluded and the TBP end point is approximately 190°F). Some refiners cut at 180 or 200°F (83 or 93°C) instead of 190°F, but, in any case, this is the fraction that consists of pentanes and hexanes and cannot be significantly upgraded in octane by catalytic reforming without producing too large a quantity of benzene. As a result, it is processed separately from the heavier straight-run gasoline fractions and requires only caustic washing, light hydrotreating, or, if higher octanes are needed, isomerization to produce a gasoline blending stock. For maximum octane with no lead addition, some refiners have installed isomerization units to process the LSR fraction and achieve blending octane number (BON) improvements of 13 to 20 octane numbers over that of the LSR.

Catalytic reformate is the C_5^+ gasoline product of the catalytic reformer. Heavy straight-run (HSR) and coker naphthas are used as feed to the catalytic reformer, and when the octane needs require, FCC and hydrocracked gasolines of the same boiling range may also be processed by this unit to increase octane levels. The processing conditions of the catalytic reformer are controlled to give the desired product antiknock properties in the range of 90 to 104 RON (85 to 98 PON) clear (lead free).

The FCC and HC gasolines are generally used directly as gasoline blending stocks, but in some cases they are separated into light and heavy fractions, with the heavy fractions upgraded by catalytic reforming before being blended into motor gasoline. This has been true since motor gasoline is unleaded and the clear gasoline pool octane is now several octane numbers higher than when lead was permitted. It is usual for the heavy hydrocrackate to be sent to the reformer for octane improvement.

The reformer increases the octane by converting low-octane paraffins to high-octane aromatics. Some aromatics have high rates of reaction with ozone to form visual pollutants in the air, and the EPA claims that some are potentially carcinogenic. Restrictions on aromatic contents of motor fuels will have increasing impacts on refinery processing as more severe restrictions are applied. This will restrict the severity of catalytic reforming and will require refiners to use other ways to increase octane numbers of the gasoline pool by producing more alkylate blending stock.

Polymer gasoline is manufactured by polymerizing olefinic hydrocarbons to produce higher molecular-weight olefins in the gasoline boiling range. Refinery technology favors alkylation processes rather than polymerization for several reasons: larger quantities of higher octane product can be made from the light olefins available, the alkylation product is paraffinic rather than olefinic, and olefins are highly photoreactive and contribute to visual air pollution and ozone production, and have a high sensitivity to driving conditions (about 16 numbers).

Alkylate gasoline is the product of the reaction of isobutene with propylene, butylene, or pentylene to produce branched-chain hydrocarbons in the gasoline boiling range. Alkylation of a given quantity of olefins produces twice the volume of high-octane motor fuel as can be produced by polymerization. In addition, the blending octane (BON) of alkylate is higher and the sensitivity (RON – MON) is significantly lower than that of polymer gasoline.

Normal butane is blended into gasoline to give the desired vapor pressure. The vapor pressure (expressed as the Reid vapor pressure) of gasoline is a compromise between a high RVP to improve economics and engine starting characteristics and a low RVP to prevent vapor-lock and reduce evaporation losses. As such, it changes with the season of the year and varies between 7.2 psi (49.6 kPa) in the summer and 13.5 psi (93.1 kPa) in the winter. Butane has a high blending octane number and is a very desirable component of gasoline; refiners put as much in their gasolines as vapor pressure limitations permit. Isobutane can be used for this purpose, but it is not as desirable because its higher vapor pressure permits a lower amount to be incorporated into gasoline than n-butane.

Concern over the effects of hydrocarbon fuels usage on the environment has caused changes in environmental regulations that impact gasoline and diesel fuel compositions. The main restrictions on diesel fuels limit sulfur and total aromatics contents, and gasoline restrictions include not only sulfur and total aromatics contents but also specific compound limits (e.g., benzene), limits on certain types of compounds (e.g., olefins), maximum Reid vapor pressures, and also minimum oxygen contents for areas with carbon monoxide problems. This has led to the concept of "reformulated gasolines" (RFGs). A reformulated gasoline specification is designed to produce a fuel for spark-ignition engines that is at least as clean burning as high methanol-content fuels. As more is learned about the relationship between fuels and the environment, fuel specifications are undergoing change. Here, the main items of

TABLE 2.4
Sources of Sulfur in Gasoline Pool

	Composition wt%	Contribution to pool, %
LSR naphtha	0.014	1.7
C_5-270°F (132°C) FCC gasoline	0.07	11.2
Heavy FCC gasoline	0.83	86.1
Light coker gasoline	0.12	1.0

Source: Note 4.

concern are discussed along with relative impacts on the environment. For current specifications of fuels, see ASTM specifications for the specific fuel desired.

The U.S. EPA requires gasoline to have less than 30 wppm sulfur. As shown in Table 2.4, the fluid catalytic cracker naphtha is the main source of sulfur in the refinery gasoline pool. For a given refinery crude oil charge, to meet the < 300 ppm sulfur specification, with no octane penalty, it is necessary to hydrotreat the FCC feedstock to reduce the sulfur level sufficiently to produce FCC naphthas with acceptable sulfur contents. The alternative is to hydrotreat the FCC naphtha, but this saturates the olefins in the naphtha and results in a blending octane reduction of two or three numbers. The sulfur in the FCC naphtha is concentrated in narrow boiling ranges. If these are separated and hydrotreated, the reduction in octane can be reduced to less than one number.

Some aromatics and most olefins react with components of the atmosphere to produce visual pollutants. The activities of these gasoline components are expressed in terms of reactivity with (OH) radicals in the atmosphere. The sources and reactivities of some of these gasoline components are shown in Table 2.5 and Table 2.6. Specifically, xylenes and olefins are the most reactive, and it may be necessary to place limits on these materials.

Producing motor fuels to reduce environmental impact will require refinery equipment additions as well as changes in catalysts and processing techniques.

In the United States during the late 1990s and early 2000s, states and the U.S. Congress became very concerned about the appearance of MTBE in groundwaters from which cities and individuals were obtaining their drinking water. Even though the MTBE concentrations were lower than those permitted by the EPA, there was enough MTBE in the water to give it an unpleasant taste and odor. A number of states outlawed the use of MTBE in gasolines, and court cases were filed against oil companies for damages. The fear of lawsuits and possible damage payments caused the general disuse of ethers to supply octane and oxygen requirements for gasoline blending. The only ether acceptable for use in gasolines is ethanol.

Even though the EPA no longer requires the addition of oxygenates to RFGs, because of tax incentives, Congressional requirements, and the requirements mandated by some states to include ethanol in their gasolines, the use of ethanol in gasoline will increase in the future.

TABLE 2.5
Aromatics and Olefins in Gasoline

Blend stock	Percentage of pool	Percentage of aromatics	Percentage of olefins
Reformate	27.2	63	1
LSR naphtha	3.1	10	2
Isomerate	3.7	1	0
FCC naphtha	38.0	30	29
Light coker naphtha	0.7	5	35
Light HC naphtha	2.4	3	0
Alkylate	12.3	0.4	0.5
Polymer	0.4	0.5	96
n-Butane	3.1	0	2.6

Source: Note 5.

TABLE 2.6
Reactivity and RVP of Gasoline

		RVP	
	Reactivity[a]	psi	kPa
n-Butane	2.7	60	414
i-Pentane	3.6	21	145
n-Pentane	5.0	16	110
i-Hexane	5.0	7	48
n-Hexane	5.6	5	34
Benzene	1.3	3	21
Toluene	6.4	0.5	3
m-Xylene	23.0	0.3	2
Butene-1	30.0	65	448
Butene-2	65.0	50	200
Pentene-1	30.0	16	110
2-Methyl, 2-butene	85.0	15	103
2-Methyl, 1-butene	70.0	19	131

[a] Reactivity with OH free radical in atmosphere.

Source: Note 6.

The U.S. Policy Act of 2005, through its renewable fuels standard (RFS), requires that by 2012, 78.8MMM* barrels of renewable fuels be used. At the present time, the only acceptable renewable fuel is ethanol. Ethanol has advantages but it also has some disadvantages when used in motor fuels. The disadvantages are (1) it has a lower heat of combustion than hydrocarbon fuels and gets only about 70% of the mileage, and (2) ethanol has a higher vapor pressure than hydrocarbon blending stocks

* *Note*: Throughout this text the symbol M is used to represent 1000 in accordance with standard U.S. engineering practice. For example, MBPSD means 1000 BPSD and MMBtu means 1,000,000 Btu.

and causes about a 1 psi increase in the Reid vapor pressure of the blend. In most cases, the EPA gives ethanol blends a 1.0 RVP waiver for conventional gasoline blends but does not give this privilege for RFG blends. However, the Chicago–Milwaukee market is given an allowance of about 0.3 psi for ethanol blends [7].

With today's federal tax subsidy ($0.51 per gallon of ethanol blended into gasoline or used to make TAME for gasoline blending) to encourage its use in motor fuels and the federal mandate to produce more renewable fuels, there is a probability that more ethanol than mandated will be used in the future.

Since the 1940s, motor gasoline has been the principal product of U.S. refineries, and in 2005, gasoline was the largest of any of the basic industries in the United States. The more than 1 million tons of gasoline produced per day exceeded the output of steel, lumber, and other high-volume products [8]. Of this production, over 90% was used in trucks and automobiles.

The aviation gasoline market is relatively small and accounts for less than 3% of the gasoline market. For this reason, it is usually not considered in the preliminary refinery design.

2.3 GASOLINE SPECIFICATIONS

Although there are several important properties of gasoline, the three that have the greatest effects on engine performance are the Reid vapor pressure, boiling range, and antiknock characteristics.

The Reid vapor pressure and boiling range of gasoline govern ease of starting, engine warm-up, rate of acceleration, loss by crankcase dilution, mileage economy, and tendency toward vapor-lock. Engine warm-up time is affected by the percentage distilled at 158°F (70°C) and the 90% ASTM distillation temperature. Warm-up is expressed in terms of the distance operated to develop full power without excessive use of the choke. A 2- to 4-mile (3- to 7-km) warm-up is considered satisfactory, and the relationship between outside temperature and percentage distilled to give acceptable warm-up properties is

	% dist. at 158°F (70°C)					
	3	**11**	**10**	**29**	**38**	**53**
Min. ambient temp.						
°F	80	60	40	20	0	20
°C	26.7	15.6	4.4	–6.7	–18.0	–29.0

Crankcase dilution is controlled by the 90% ASTM distillation temperature and is also a function of outside temperature. To keep crankcase dilution within acceptable limits, the volatility should be

Min. ambient temp.						
°F	80	60	40	20	0	20
°C	26.7	15.6	4.4	6.7	18.0	29.0
Min. ambient temp.						
°F	370	350	340	325	310	300
°C	188	177	171	163	154	149

The tendency to vapor-lock is directly related to the RVP of the gasoline. In order to control vapor-lock, the vapor pressure of the gasoline should not exceed the following limits:

Ambient temp.		Max. allowable RVP	
°F	°C	psia	kPa
60	15.6	12.7	87.6
70	21.1	11.0	75.8
80	26.7	9.4	64.8
90	32.2	8.0	55.2

The Reid vapor pressure is approximately the vapor pressure of the gasoline at 100°F (38°C) in absolute units (ASTM designation D-323).

Altitude affects several properties of gasoline, the most important of which are losses by evaporation and octane requirement. Octane number requirement is greatly affected by altitude and, for a constant spark advance, is about three units lower for each 1000 ft (305 m) of elevation. In practice, however, the spark is advanced at higher elevations to improve engine performance, and the net effect is to reduce the PON of the gasoline marketed by about two numbers for a 5000-ft (1524-m) increase in elevation. Octane requirements for the same model of engine will vary by 7 to 12 RON because of differences in tuneup, engine deposits, and clearances. Table 2.7 lists some typical effects of variables on engine octane requirements.

There are several types of octane numbers for spark-ignition engines, with the two determined by laboratory tests considered most common: those determined by the "motor method" (MON) and those determined by the "research method" (RON). Both methods use the same basic type of test engine but operate under different conditions. The RON (ASTM D-908) represents the performance during city driving when acceleration is relatively frequent, and the MON (ASTM D-357) is a guide to engine performance on the highway or under heavy load conditions. The difference between the research and motor octane numbers is an indicator of the sensitivity of the performance of the fuel to the two types of driving conditions and is known as the "sensitivity" of the fuel. Obviously, the driver would like for the fuel to perform

TABLE 2.7
Effects of Variables on Octane Requirements

Variable	Effect on octane requirements
Altitude	3 RON per 1000 ft (305 m) increase in altitude
Humidity	0.5 RON per 10% increase in relative humidity at 70°F (21.1°C)
Engine speed	1 RON per 300 rpm increase
Air temperature	1 RON per 20°F (11.1°C) rise
Spark advance	1.5 RON per 1° advance
Coolant temperature	1 RON per 10°F (5.6°C) increase
Combustion chamber deposits	1 to 2 RON per 1000 miles (1609 km) up to 6000 miles (9650 km)

equally well both in the city and on the highway; therefore, low-sensitivity fuels are better. Since the posting of octane numbers on the service station pump has been required in the United States, the posted octane number is the one most well known by the typical driver. This is the arithmetic average of the research and motor octane numbers [(RON + MON) / 2].

Ethanol is blended with reformulated gasoline to give a product called E85, which is a blend of 70 to 85% ethanol in gasoline. Its purpose is to reduce the dependency of the United States on imported crude oils and gasolines. Automobile manufacturers make spark-ignition engines that will operate on either reformulated gasolines or E85. However, because the heat of combustion of ethanol is less than that of gasoline, E85 gives only about 70% of the mileage of reformulated gasolines. The exhaust contains less carbon monoxide and more aldehydes and ketones than when using reformulated gasoline and, because of its higher vapor pressure, gives more volatile hydrocarbon emissions.

EPA specifications for the sulfur content of gasolines is < 30 wppm.

2.4 DISTILLATE FUELS

Distillate fuels can be divided into three types: jet or turbine fuels, automotive diesel fuels, railroad diesel fuels, and heating oils. These products are blended from a variety of refinery streams to meet the desired specifications.

The consumption of heating oils has ranked high in refinery production goals, but as a percentage of refinery products, it has been decreasing because of increases in gasoline, diesel, and jet fuels in recent years. Increasingly severe environmental restrictions on fuel emissions have caused some users of heating oils to convert to natural gas and LPG. Expansion of air and truck travel has increased diesel and jet fuel demands.

2.4.1 Jet and Turbine Fuels

Jet fuel is blended for use by both commercial aviation and military aircraft. It is also known as turbine fuel, and there are several commercial and military jet fuel specifications. For most refineries, the primary source of jet fuel blending stocks is the straight-run kerosine fraction from the atmospheric crude unit, because stringent total aromatic and naphthalene content and smoke-point specifications limit the amount of cracked stocks that can be included. For refineries with a hydrocracker, kerosine boiling range hydrocarbons from this unit can also meet jet fuel specifications and are a major contributor to jet fuel production. Usually, jet fuels sell at higher prices than diesel fuels and No. 1 and No. 2 heating oils, and it is more profitable for the refiner to blend the kerosine fractions from the atmospheric crude unit and the hydrocracker into jet fuel rather than other products.

Commercial jet fuel is a material in the kerosine boiling range and must be clean burning. The ASTM specifications for jet and turbine fuels are given in Table 2.8. Two of the critical specifications relate to its clean burning requirements and limit the total aromatics as well as the content of double-ring aromatic compounds. These

TABLE 2.8
Characteristics of Aircraft Turbine Fuels (ASTM D-1655 and DERD 2494)

Property	Jet A	JP-5	DERD 2494	JP-8
Aromatics, vol%, max.	20	25	22.0	22.0
Combustion prop.				
Smoke point, mm, min, or	25	19	—	25
Smoke point, mm, min, and	18	—	19	20
Naphthalenes, vol%, max.	3.0	—	3.0	3.0
Distillation, D-86, °F (°C)				
10% recovered °F (°C), max.	400 (205)	400 (205)	401 (205)	401 (205)
50% recovered °F (°C), max.	Report	Report	Report	Report
FBP, °F (°C), max.	572 (300)	554 (290)	572 (300)	572 (300)
Flash point, °F (°C), min.	100 (38)	140 (60)	100 (38)	100 (38)
Freeze point, °F (°C), max.	–40 (–40)	–51 (–46)	–52.6 (–47)	–52.6 (–47)
Sulfur, wt%, max.	0.30	0.4	0.3	0.3

are the smoke point, expressed in mm of flame height at which smoking is detected, and the volume percentage of total aromatics and naphthalenes. Specifications limit total aromatic concentration to 20% and the naphthalene content to 3.0%. Hydrocracking saturates many of the double-ring aromatics in cracked products and raises the smoke point. The freeze point specification is very low [–40 to –58°F max. (–40 to –50°C max.)], and hydrocracking is also used to isomerize paraffins and lower the freeze point. Hydrocracking normally products a very low (14 to 16 mm) smoke point jet fuel when the cracking is done in the presence of a small amount of hydrogen sulfide or ammonia.

Jet fuel is blended from low-sulfur or desulfurized kerosine, hydrotreated light coker gas oil, and hydrocracked blending stocks. The smoke point and percentage aromatics specifications limit the amount of cracked stocks that can be blended into jet fuels.

The two basic types of jet fuels are naphtha and kerosine. Naphtha jet fuel was produced primarily for the military and was a wide boiling-range stock that extended through the gasoline and kerosine boiling ranges. The naphtha-type jet fuel is more volatile and has more safety problems in handling, but in case of a national emergency, there would be a tremendous demand for jet fuels; to meet the requirements, both naphtha and kerosine production would be needed. The military is studying alternatives, and JP-8 jet fuel has been phased in. The jet fuels are blended from the various components to arrive at the lowest-cost blend that meets specifications.

Safety considerations limit commercial jet fuels to the narrower boiling-range product [350 to 550°F (177 to 288°C)], which is sold as Jet A, Jet A-1, JP-5, JP-8, or JP-50. The principal differences among these are freezing points, which range from –40 to –58°F (–40 to –50°C) maximum. In addition to freezing point, the limiting specifications are flash point [110 to 150°F (43 to 66°C)], distillation, smoke point, and aromatics content.

2.4.2 Automotive Diesel Fuels

Volatility, ignition quality (expressed as cetane number or cetane index, CI), viscosity, sulfur content, percentage of aromatics, and cloud point are the important properties of automotive diesel fuels. No. 1 diesel fuel (sometimes called super-diesel) is generally made from virgin or hydrocracked stocks having cetane numbers above 45. It has a boiling range from 360 to 600°F (182 to 316°C), a sulfur content of 15 wppm (max.), distillation range, cetane number or cetane index (40 min.), percentage aromatics, and cloud point.

No. 2 diesel fuel is very similar to No. 2 fuel oil and has a wider boiling range than No. 1. It usually contains cracked stocks and may be blended from naphtha, kerosine, and light cracked oils from the coker and the fluid catalytic cracking unit. Limiting specifications are flash point [125°F (52°C)], sulfur content (15 wppm max.), distillation range, cetane index (40 min.), percentage of aromatics, and cloud point.

The ignition properties of diesel fuels are expressed in terms of cetane number or cetane index. These are very similar to the octane number (except the opposite), and the cetane number expresses the volume percentage of cetane ($C_{16}H_{34}$, high-ignition quality) in a mixture with alpha-methyl-naphthalene ($C_{11}H_{10}$, low-ignition quality). The fuel is used to operate a standard diesel test engine according to ASTM test method D-613. Because many refineries do not have cetane test engines, a mathematical expression developed to estimate the cetane number is used. The number derived is called the cetane index and is calculated from the midboiling point and gravity of the sample. This equation uses the same parameters as the Watson or UOP correlation factor (K_W) and U.S. Bureau of Mines Correlation Index (CI) and is actually an expression of the hydrogen/carbon (H/C) ratio of the hydrocarbon components in the sample; the higher the H/C ratio, the better the burning characteristics (i.e., the higher the smoke point and the higher the cetane index).

To improve air quality, more severe restrictions are placed on the sulfur and aromatic contents of diesel fuels. As the cetane index is an indicator of the H/C ratio, it is also an indirect indicator of the aromatic content of the diesel fuel. Therefore, frequently a minimum cetane index specification is used as an alternative to maximum aromatics content. Lowering sulfur and aromatics contents specifications also lowers the particulate emissions from diesel engines.

2.4.3 Railroad Diesel Fuels

Railroad diesel engine fuel [6] is one of the significant markets for diesel fuels. Railroad diesel fuels are similar to the heavier automotive diesel fuels but have higher boiling ranges [up to 750°F (400°C) end point] and lower cetane numbers (30 min.). No. 4 diesel and No. 4 fuel oil have very similar specifications.

2.4.4 Heating Oils

Although the consumption of petroleum products for space heating ranks very high, the consumption varies widely according to locality and climate. In recent years,

TABLE 2.9
Heating Oil Specifications (ASTM D-396)

	No. 1	No. 2	No. 4	No. 6
Flash point, °F (°C), min.	100 (38)	100 (38)	130 (55)	140 (60)
Pour point, °F (°C), max.	9 (–18)	28 (–6)	28 (–6)	—
Distillation temp, °F (°C)				
10% recovered, max.	419 (215)	—	—	—
90% recovered, min.	—	540 (282)	—	—
max.	550 (288)	640 (338)	—	—
Viscosity, mm²/s				
at 104°F (40°C)				
Min.	1.3	1.9	> 5.5	—
Max.	2.1	3.4	24.0	—
at 212°F (100°C)				
Min.	—	—	—	15.0
Max.	—	—	—	50.0
Density, kg/m³ 60°F (15°C)				
Max. (°API min.)	850 (35)	876 (30)	—	—
Ramsbottom carbon residue on 10% btms, wt%, max.	0.15	0.35	—	—
Ash, wt%, max.			0.10	—
Sulfur, wt%, max.	0.50	0.50	—	—
Water and sediment, vol%, max.	0.05	0.05	0.50	2.00

the proportional demand for heating oils has decreased as LPG usage has increased. The ASTM specifications for heating oils are given in Table 2.9. The principal distillate fuel oils consist of No. 1 and No. 2 fuel oils. No. 1 fuel oil is very similar to kerosine, but generally has a higher pour point and end point. Limiting specifications are distillation, pour point, flash point, and sulfur content.

No. 1 fuel is used in furnaces that use a pot-type burner, which vaporizes the fuel oil when it comes in contact with the hot metal of the pot. It requires high volatility for it to vaporize as quickly as it is fed to the burner.

No. 2 fuel oil is very similar to No. 2 diesel fuel, contains cracked stock, and is blended from naphtha, kerosine, diesel (atmospheric gas oil), and cracked gas oils. Limiting specifications are sulfur content, pour point, distillation, and flash point. It is used for furnaces with atomizing-type burners, which spray the fuel into the combustion area, and the small droplets produced burn completely.

2.5 RESIDUAL FUEL OILS

No. 4 fuel oil is usually a light residual oil and is used for furnaces with burners that can atomize oils with higher viscosities than No. 2. It may require preheating in order to vaporize it in a spray-nozzle burner when stored at low temperatures. No. 5 fuel oil is a residual oil with higher viscosities than No. 4 fuel oil and may require preheating for atomizing and handling. No. 6 fuel oil is a high-viscosity residual oil that requires preheating for storage, handling, and atomizing. Bunker C oil is a heavy residual oil usually used for ships, and its specifications are determined by contract between the seller and user.

Most of the heavy residual fuel oil used in the United States is imported. It is composed of the heaviest parts of the crude and is generally the fractionating tower bottoms from vacuum distillation. It sells for a very low price (historically about 70% of the price of crude from which it is produced) and is considered a by-product. Critical specifications are viscosity and sulfur content. Sulfur content specifications are generally set by the locality in which it is burned. Currently, only low-sulfur fuel oils can be burned in some areas, and this trend will continue to expand. Heavy fuel oils with very low sulfur contents are much in demand and sell at prices near those of the crude oils from which they are derived.

NOTES

1. *Amer. Petrol. Inst. Inform. Bull.* No. 11, Philadelphia, 1958.
2. Gas Processors Assoc. Publication 2140, Liquefied Petroleum Gas Specifications and Test Methods, Tulsa, OK.
3. Gruse, W.A. and Stevens, D.R., *Chemical Technology of Petroleum*, 3rd ed., McGraw-Hill Book Co., New York, 1960, pp. 424–472.
4. NPRA Survey, 1989.
5. Gas Processors Assoc. Publication 2140, Liquefied Petroleum Gas Specifications and Test Methods, Tulsa, OK.
6. Reno, M.E., Bozzano, U.S., and Tieman, W.C., UOP 1990 Technology Conference, UOP, Des Plaines, IL, 1990.
7. Unzelman, G.H., *Oil Gas J. 88*(15), pp. 43–48, 1990.
8. Jansen, S.D. and Tamm, D.C., RFS will require more blendstock production, *Oil Gas J.*, May 8, 2006.
9. Newsletter, *Oil Gas J. 97*(51), 3, 1999.
10. Bland, W.F. and Davidson, R.L. (Eds.), *Petroleum Processing Handbook*, McGraw-Hill Book Co., New York, 1967, pp. 11–39.

3 Refinery Feedstocks

The basic raw material for refineries is petroleum or crude oil, even though in some areas synthetic crude oils from other sources (Gilsonite, oil sands, etc.) and natural gas liquids are included in the refinery feedstocks. The chemical compositions of crude oils are surprisingly uniform even though their physical characteristics vary widely. Since 1980, crude oils charged to U.S. refineries have become heavier (lower °API, higher sp. gr.) and have higher sulfur contents. Both of these properties require more severe and more costly processing. In 1980, the average gravity was 33.8°API (sp. gr. 0.856) and the sulfur content, 0.89 wt%. By 2004, the gravity had decreased to 30.5°API (sp. gr. 0.874) and the sulfur content had increased to 1.42 wt% [1].

In the United States, crude oils are classified as paraffin base, naphthene base, asphalt base, or mixed base. There are some crude oils in the Far East that have up to 80% aromatic content, and these are known as aromatic-base oils. The U.S. Bureau of Mines [2,3] developed a system that classifies the crude according to two key fractions obtained in distillation: No. 1 from 482 to 527°F (250 to 275°C) at 1 atm pressure and No. 2 from 527 to 572°F (275 to 300°C) at 40 mmHg pressure. The gravity of these two fractions is used to classify crude oils into types as shown below.

	Key fractions, °API	
	No. 1	**No. 2**
Paraffin	40	30
Paraffin, intermediate	40	20–30
Intermediate, paraffin	33–40	30
Intermediate	33–40	20–30
Intermediate, naphthene	33–40	20
Naphthene, intermediate	33	20–30
Naphthene	< 33	< 20

The paraffinic and asphaltic classifications in common use are based on the properties of the residuum left from nondestructive distillation and are more descriptive to the refiner because they convey the nature of the products to be expected and the processing necessary.

3.1 CRUDE OIL PROPERTIES

Crude petroleum is very complex, and except for the low-boiling components, no attempt is made by the refiner to analyze for the pure components contained in the crude oil. Relatively simple analytical tests are run on the crude, and the results of these are used with empirical correlations to evaluate the crude oils as feedstocks for the particular refinery. Each crude is compared with the other feedstocks available and, based upon the operating cost and product realization, is assigned a value. The more useful properties are discussed.

3.1.1 API Gravity

The density of petroleum oils is expressed in the United States in terms of API gravity rather than specific gravity; it is related to specific gravity in such a fashion that an increase in API gravity corresponds to a decrease in specific gravity. The units of API gravity are °API and can be calculated from specific gravity by the following:

$$°API = \frac{141.5}{\text{specific gravity}} - 131.5 \tag{3.1}$$

In Equation 3.1, specific gravity and API gravity refer to the weight per unit volume at 60°F (15.6°C) as compared to water at 60°F. Crude oil gravity may range from less than 10°API to over 50°API, but most crudes fall in the 20 to 45°API range. API gravity always refers to the liquid sample at 60°F (15.6°C). API gravities are not linear and, therefore, cannot be averaged. For example, a gallon of 30°API gravity hydrocarbons when mixed with a gallon of 40°API hydrocarbons will not yield 2 gal of 35°API hydrocarbons, but will give 2 gal of hydrocarbons with an API gravity different from 35°API. Specific gravities can be averaged. In practice, however, API gravities are frequently averaged because the error is usually small.

3.1.2 Sulfur Content, Wt%

Sulfur content and API gravity are two properties that have had the greatest influence on the value of crude oil, although nitrogen, Total Acid Number (TAN), and metals contents are increasing in importance. The sulfur content is expressed as a percentage of sulfur by weight and varies from less than 0.1% to greater than 5%. Crudes with greater than 0.5% sulfur generally require more extensive processing than those with lower sulfur content. Although the term "sour" crude initially had reference to those crudes containing dissolved hydrogen sulfide independent of total sulfur content, it has come to mean any crude oil with a sulfur content high enough to require special processing. There is no sharp dividing line between sour and sweet crudes, but 0.5% sulfur content is frequently used as the criterion.

3.1.3 Pour Point, °F (°C)

The pour point of the crude oil, in °F or °C, is a rough indicator of the relative paraffinicity and aromaticity of the crude. The lower the pour point, the lower the paraffin content and the greater the content of aromatics.

3.1.4 Carbon Residue, Wt%

Carbon residue is determined by distillation to a coke residue in the absence of air. The carbon residue is roughly related to the asphalt content of the crude and to the quantity of the lubricating oil fraction that can be recovered. In most cases, the lower the carbon residue, the more valuable the crude. This is expressed in terms of the weight percentage of carbon residue by either the Ramsbottom (RCR) or Conradson (CCR) ASTM test procedures (D-524 or D-189).

3.1.5 Salt Content, lb/1000 bbl

If the salt content of the crude, when expressed as NaCl, is greater than 10 lb/1000 bbl, it is generally necessary to desalt the crude before processing. If the salt is not removed, severe corrosion problems may be encountered. If residua are processed catalytically, desalting is desirable at even lower salt contents of the crude. Frequently, it is economic to desalt crude oils to < 0.5 lb/1000 bbl. Although it is not possible to have an accurate conversion unit between lb/1000 bbl and parts per million (ppm) by weight because of the different densities of crude oils, 1 lb/1000 bbl is approximately 3 ppm.

3.1.6 Characterization Factors

There are several correlations between yield and the aromaticity and paraffinicity of crude oils, but the two most widely used are the UOP or Watson "characterization factor" (K_W) and the U.S. Bureau of Mines correlation index (CI).

$$K_W = \frac{T_B^{1/3}}{G} \tag{3.2}$$

$$CI = \frac{87{,}552}{T_B} + 473.7G - 456.8 \tag{3.3}$$

where

T_B = mean average boiling point, °R

G = specific gravity at 60°F

The Watson characterization factor ranges from less than 10 for highly aromatic materials to almost 15 for highly paraffinic compounds. Crude oils show a narrower range of K_W and vary from 10.5 for a highly naphthenic crude to 12.9 for a paraffinic base crude. It is an indicator of the paraffinicity of the oil.

The correlation index is useful in evaluating individual fractions from crude oils. The CI scale is based upon straight-chain paraffins having a CI value of 0 and benzene having a CI value of 100. The CI values are not quantitative, but the lower the CI value, the greater the concentrations of paraffin hydrocarbons in the fraction; and the higher the CI value, the greater the concentrations of naphthenes and aromatics [4]. It an indicator of the aromaticity of the oil.

3.1.7 Nitrogen Content, Wt%

A high nitrogen content is undesirable in crude oils because organic nitrogen compounds cause severe poisoning of catalysts used in processing and cause corrosion problems such a hydrogen blistering. Crudes containing nitrogen in amounts above 0.25% by weight require special processing to remove the nitrogen.

3.1.8 Distillation Range

The boiling range of the crude gives an indication of the quantities of the various products present. The most useful type of distillation is known as a true boiling point (TBP) distillation and generally refers to a distillation performed in equipment that accomplishes a reasonable degree of fractionation. There is no specific test procedure called a TBP distillation, but the U.S. Bureau of Mines Hempel and ASTM D-285 distillations are the tests most commonly used. Neither of these specify either the number of theoretical plates or the reflux ratio used, and as a result, there is a trend toward using the results of a 15:5 distillation (D-2892) rather than the TBP. The 15:5 distillation is carried out using 15 theoretical stages at a reflux ratio of 5:1.

The crude distillation range also has to be correlated with ASTM distillations because product specifications are generally based on the simple ASTM distillation tests D-86 and D-1160. The TBP cut point for various fractions can be approximated by use of Figure 3.1. A more detailed procedure for correlation of ASTM and TBP distillations is given in the *API Technical Data Book—Petroleum Refining*, published by the American Petroleum Institute, Washington, D.C.

3.1.9 Metals Content, ppm

The metals content of crude oils can vary from a few parts per million to more than 1000 ppm and, in spite of their relatively low concentrations, is of considerable importance [5]. Minute quantities of some of these metals (nickel, vanadium, and copper) can severely affect the activities of catalysts and result in a lower-value product distribution. Vanadium concentrations above 2 ppm in fuel oils can lead to severe corrosion to turbine blades and deterioration of refractory furnace linings and stacks [3].

Distillation concentrates the metallic constituents of crude in the residues, but some of the organometallic compounds are actually volatilized at refinery distillation temperatures and appear in the higher-boiling distillates [6].

The metallic content may be reduced by solvent extraction with propane or similar solvents as the organometallic compounds are precipitated with the asphaltenes and resins.

3.1.10 Total Acid Number

The Total Acid Number was originally used to monitor the oxidation of lubricating oils during use but now includes acidic crude oils. These crudes are so acidic they cause rapid corrosion of most crude oil fractionating columns. Highly acidic crudes require the use of expensive alloy steel equipment. The TAN is the number of mg of potassium hydroxide required to neutralize 1 g of oil.

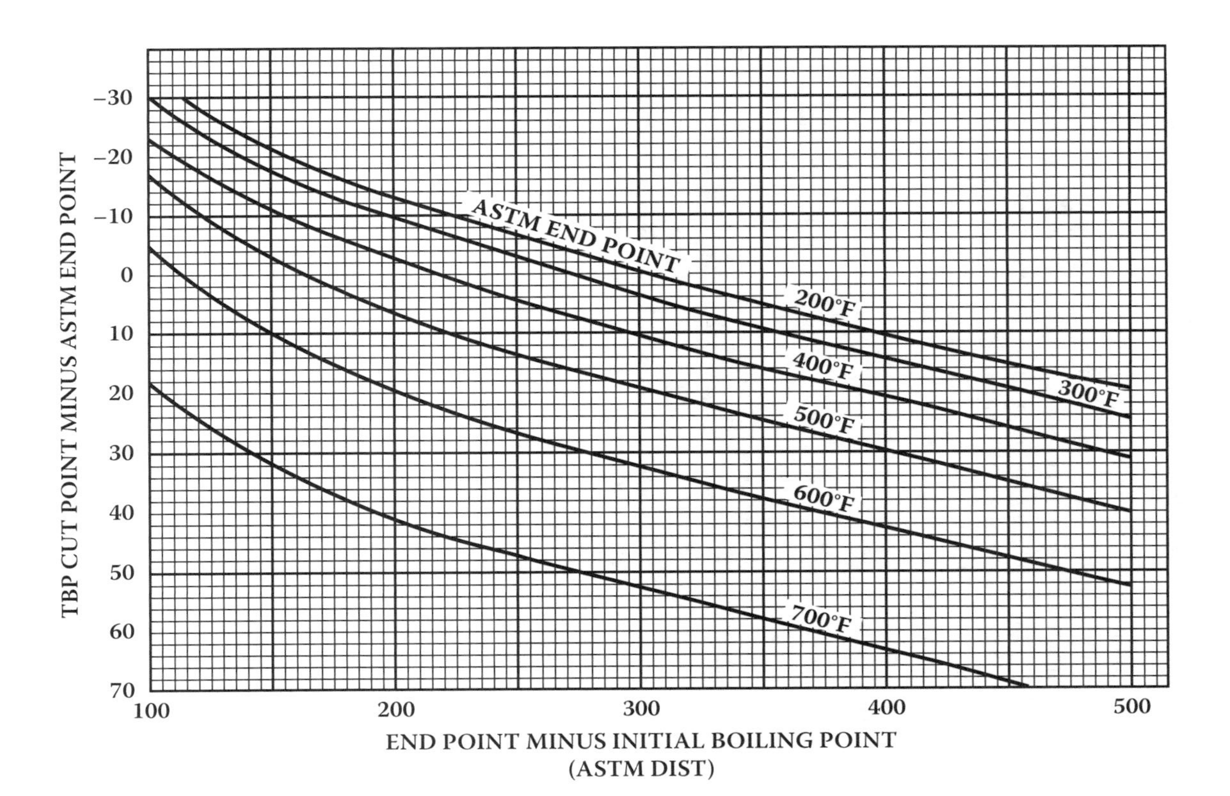

FIGURE 3.1 TBP cut point versus ASTM end point.

3.2 COMPOSITION OF PETROLEUM

Crude oils and high-boiling crude oil fractions are composed of many members of a relatively few homologous series of hydrocarbons [7]. The composition of the total mixture, in terms of elementary composition, does not vary a great deal, but small differences in composition can greatly affect the physical properties and the processing required to produce salable products. Petroleum is essentially a mixture of hydrocarbons, and even the nonhydrocarbon elements are generally present as components of complex molecules, predominantly hydrocarbon in character, but containing small quantities of oxygen, sulfur, nitrogen, vanadium, nickel, and chromium [5]. The hydrocarbons present in crude petroleum are classified into three general types: paraffins, naphthenes, and aromatics. In addition, there is a fourth type, olefins, that is formed during processing by the cracking or dehydrogenation of paraffins and naphthenes. There are no olefins in crude oils.

3.2.1 PARAFFINS

The paraffin series of hydrocarbons is characterized by the rule that the carbon atoms are connected by a single bond, and the other bonds are saturated with hydrogen atoms. The general formula for paraffins is C_nH_{2n+2}.

The simplest paraffin is methane, CH_4, followed by the homologous series of ethane; propane; normal and isobutane; and normal, iso-, and neopentane (Figure 3.2). When the number of carbon atoms in the molecule is greater than three, several hydrocarbons may exist that contain the same number of carbon and hydrogen atoms but have different structures. This is because carbon is capable not only of chain formation, but also of forming single- or double-branched chains that give rise to isomers that have significantly different properties. For example, the motor octane number of n-octane is –17 and that of isooctane (2,2,4-trimethyl pentane) is 100.

The number of possible isomers increases in geometric progression as the number of carbon atoms increases. There are 2 paraffin isomers of butane, 3 of pentane, and 17 structural isomers of octane, and by the time the number of carbon atoms has increased to 18, there are 60,533 isomers of cetane. Crude oil contains molecules with up to 70 carbon atoms, and the number of possible paraffinic hydrocarbons is very high.

3.2.2 OLEFINS

Olefins do not naturally occur in crude oils but are formed during processing. They are very similar in structure to paraffins, but at least two of the carbon atoms are joined by double bonds. The general formula is C_nC_{2n}. Olefins are generally undesirable in finished products because the double bonds are reactive and the compounds are more easily oxidized and polymerized to form gums and varnishes. In gasoline boiling-range fractions, some olefins are desirable because olefins have higher octane numbers than paraffin compounds with the same number of carbon atoms. Olefins containing five carbon atoms have high reaction rates with compounds in the

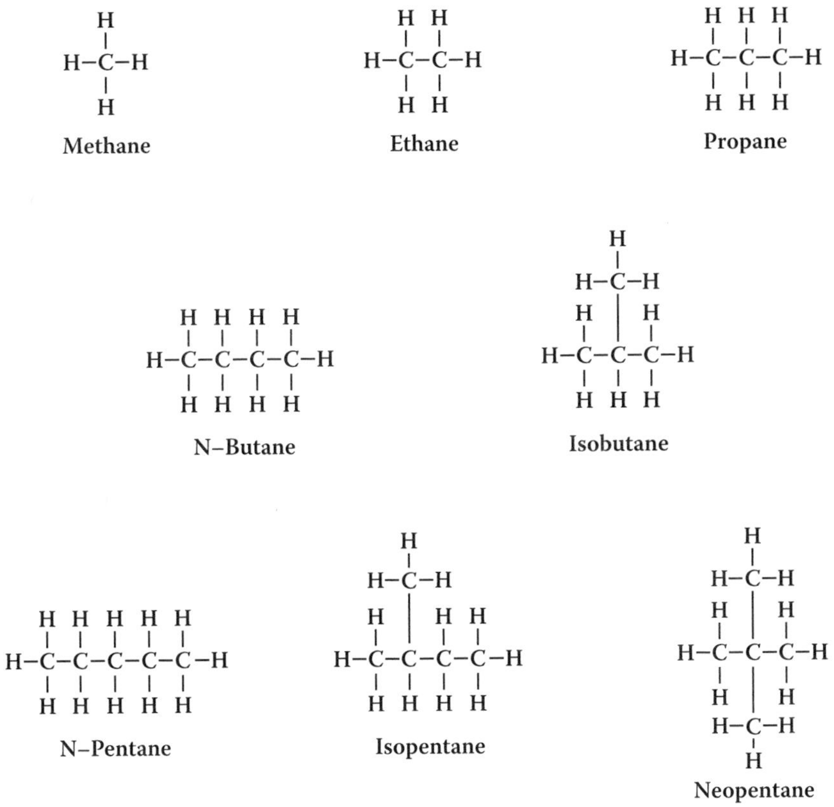

Methane

Ethane

Propane

N–Butane

Isobutane

N–Pentane

Isopentane

Neopentane

FIGURE 3.2 Paraffins in crude oil.

atmosphere that form pollutants and, even though they have high research octane numbers, are considered generally undesirable.

Some diolefins (containing two double bonds) are also formed during processing, but they react very rapidly with olefins to form high-molecular-weight polymers consisting of many simple unsaturated molecules joined together. Diolefins are very undesirable in products because they are so reactive they polymerize and form filter- and equipment-plugging compounds.

3.2.3 Naphthenes (Cycloparaffins)

Cycloparaffin hydrocarbons in which all of the available bonds of the carbon atoms are saturated with hydrogen are called naphthenes. There are many types of naphthenes present in crude oil, but, except for the lower-molecular-weight compounds such as cyclopentane and cyclohexane, they are generally not handled as individual compounds. They are classified according to boiling range and their properties determined with the help of correlation factors such as the K_W factor or CI. Some typical naphthenic compounds are shown in Figure 3.3.

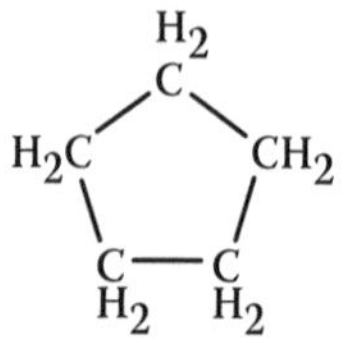

Cyclopentane 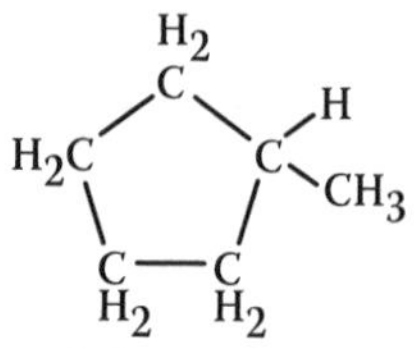 Methylcyclopentane 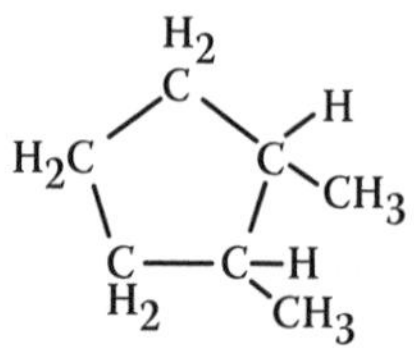 Dimethylcyclopentane

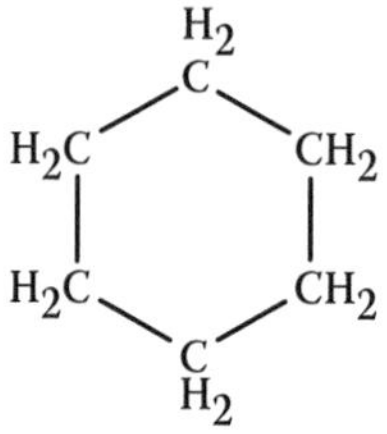

Cyclohexane 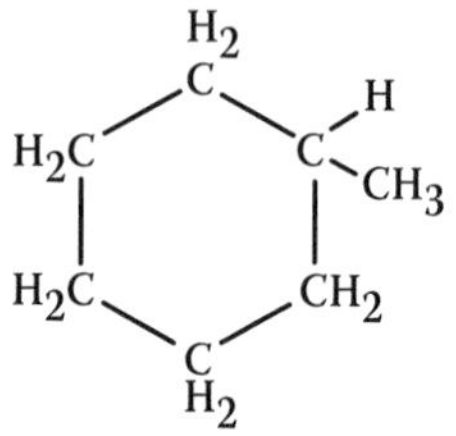 Methylcyclohexane 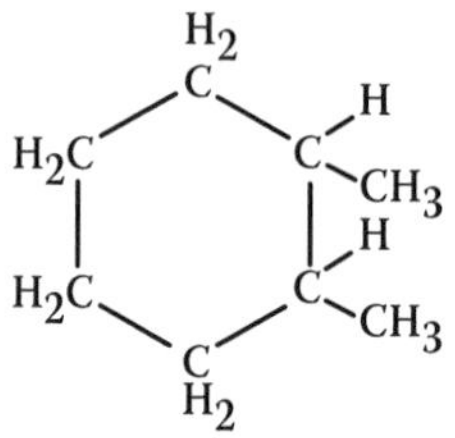 1, 2 Dimethylcyclohexane

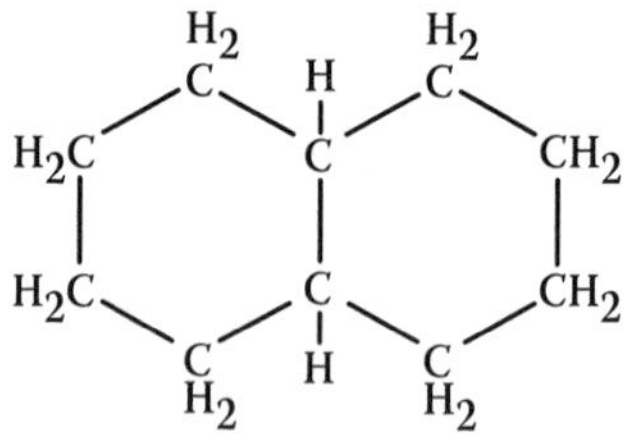

Decalin
(Decahydronaphthalene)

FIGURE 3.3 Naphthenes in crude oil.

3.2.4 Aromatics

The aromatic series of hydrocarbons is chemically and physically very different from the paraffins and cycloparaffins (naphthenes). Aromatic hydrocarbons contain a benzene ring, which is unsaturated but very stable, and frequently behave as saturated compounds. Some typical aromatic compounds are shown in Figure 3.4

The cyclic hydrocarbons, both naphthenic and aromatic, can add paraffin side chains in place of some of the hydrogen attached to the ring carbons and form a mixed structure. These mixed types have many of the chemical and physical characteristics of both of the parent compounds, but generally are classified according to the parent cyclic compound.

3.3 CRUDES SUITABLE FOR ASPHALT MANUFACTURE

It is not possible to predict with 100% accuracy whether or not a particular crude will produce specification asphalts without actually separating the asphalts from the crude and running the tests. There are, however, certain characteristics of crude oils

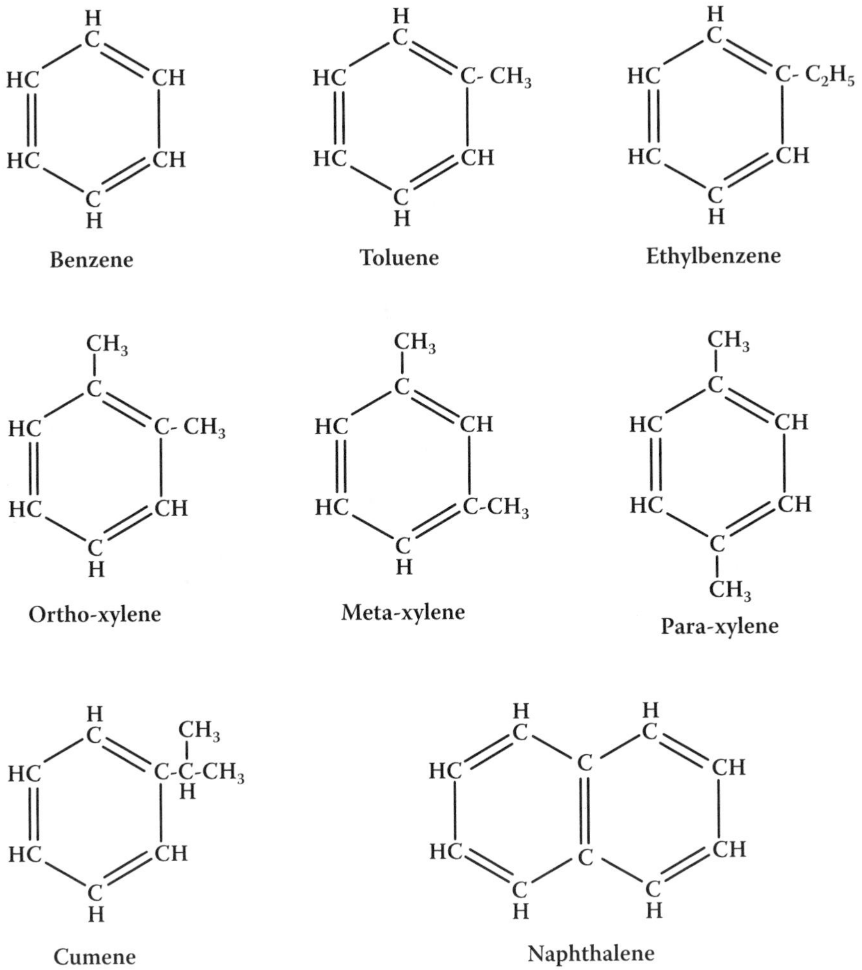

Benzene

Toluene

Ethylbenzene

Ortho-xylene

Meta-xylene

Para-xylene

Cumene

Naphthalene

FIGURE 3.4 Aromatic hydrocarbons in crude oil.

that indicate if they are possible sources of asphalt. If the crude oil contains a residue [750°F (399°C)] mean average boiling point having a Watson characterization factor of less than 11.8 and the gravity is below 35°API, it is usually suitable for asphalt manufacture [7]. If, however, the difference between the characterization factor for the 750°F and 550°F fraction is greater than 0.15, the residue may contain too much wax to meet most asphalt specifications.

3.3.1 Crude Distillation Curves

When a refining company evaluates its own crude oils to determine the most desirable processing sequence to obtain the required products, its own laboratories will provide data concerning the distillation and processing of the oil and its fractions [9]. In

CRUDE PETROLEUM ANALYSIS

Bureau of Mines Bartlesville Laboratory
Sample 53016

IDENTIFICATION

Hastings Field

Texas
Brazoria County

GENERAL CHARACTERISTICS

Gravity, specific, 0.867 Gravity, ° API, 31.7 Pour point, ° F., below 5
Sulfur, percent, 0.15 Color, brownish green
Viscosity, Saybolt Universal at 100° Nitrogen, percent,

DISTILLATION, BUREAU OF MINES ROUTINE METHOD

STAGE 1—Distillation at atmospheric pressure, 751 mm. Hg
First drop, 84 ° F.

Fraction No.	Cut temp. ° F.	Percent	Sum, percent	Sp. gr., 60/60° F.	° API, 60° F.	C. I.	Refractive index, n_D at 20° C.	Specific dispersion	S. U. visc., 100° F.	Cloud test, ° F.
1	122	0.8	0.8	0.673	78.8					
2	167	1.0	1.8	.685	75.1	15				
3	212	3.0	4.8	.725	63.7	24	1.39574	127.7		
4	257	3.4	8.2	.755	55.9	29	1.41756	128.6		
5	302	3.1	11.3	.777	50.6	32	1.42985	135.4		
6	347	3.9	15.2	.798	45.8	35	1.44192	137.8		
7	392	4.9	20.1	.817	41.7	38	1.45217	139.9		
8	437	6.8	26.9	.833	38.4	40	1.46057	140.3		
9	482	8.0	34.9	.848	35.4	41	1.46875	148.0		
10	527	10.9	45.8	.864	32.3	44	1.47679	149.8		
STAGE 2—Distillation continued at 40 mm. Hg										
11	392	7.3	53.1	0.873	30.6	45	1.48274	155.2	42	Below 5
12	437	7.8	60.9	.879	29.5	44	1.48474	156.2	50	do
13	482	6.2	67.1	.889	27.7	45	1.49058	152.7	71	do
14	527	5.7	72.8	.901	25.6	48			125	10
15	572	6.9	79.7	.916	28.0	52			280	20
Residuum		20.3	100.0	.945	18.2					

Carbon residue, Conradson: Residuum, 4.7 percent; crude, 1.0 percent.

APPROXIMATE SUMMARY

	Percent	Sp. gr.	° API	Viscosity
Light gasoline	4.8	0.708	68.4	
Total gasoline and naphtha	20.1	0.771	52.0	
Kerosine distillate	--	--	--	
Gas oil	36.9	0.858	33.4	
Nonviscous lubricating distillate	10.2	.879-.895	29.5-26.6	50-100
Medium lubricating distillate	5.8	.895-.908	26.6-24.3	100-200
Viscous lubricating distillate	6.7	.908-.924	24.3-21.6	Above 200
Residuum	20.3	0.945	18.2	
Distillation loss	0			

U. S. GOVERNMENT PRINTING OFFICE 16—57635-3

FIGURE 3.5 U.S. Bureau of Mines crude petroleum analysis.

many cases, information not readily available is desired concerning the processing qualities of crude oils. In such instances, true boiling point and gravity midpercent curves can be developed from the U.S. Bureau of Mines or similar crude petroleum analysis data sheets (Figure 3.5).

The U.S. Bureau of Mines has carried out Hempel distillations on thousands of crude oil samples from wells in all major producing fields. Although the degree of fractionation in a Hempel assay is less than that in a 15:5 distillation, the results are sufficiently similar that they can be used without correction. If desired, correction factors developed by Nelson can be applied.

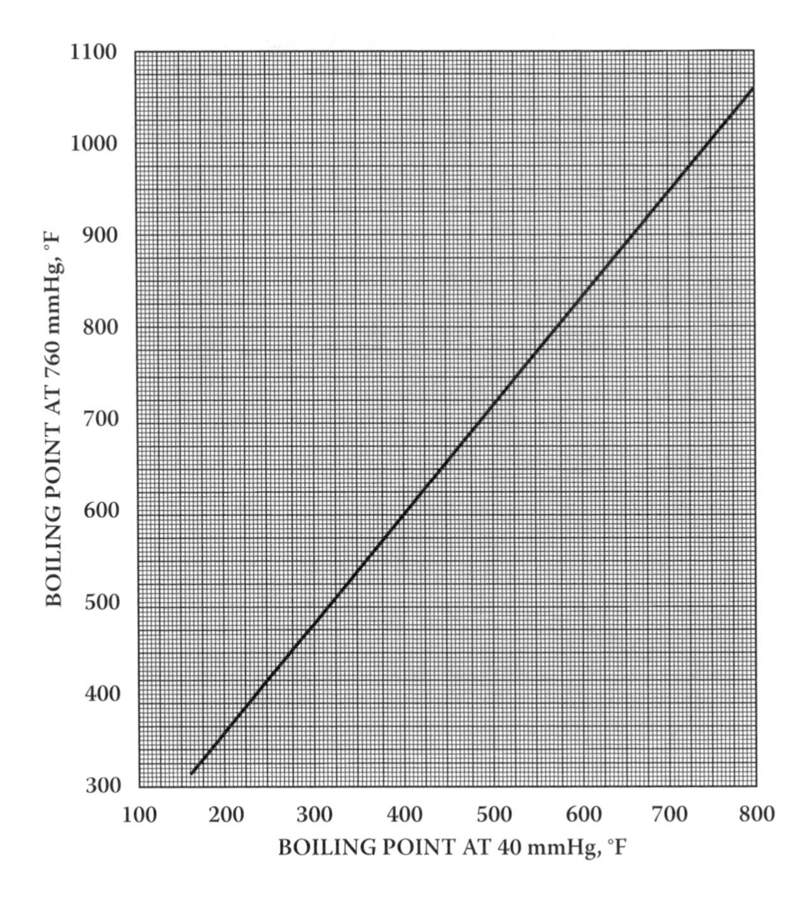

FIGURE 3.6 Boiling point at 760 mmHg versus boiling point at 40 mmHg.

The major deficiency in a Hempel assay is the lack of information concerning the low-boiling components. The materials not condensed by water-cooled condensers are reported as "distillation loss." An estimate of the composition of the butane and lighter components is frequently added to the low-boiling end of the TBP curve to compensate for the loss during distillation.

The Hempel analysis is reported in two parts: the first is the portion of the distillation performed at 1 atm pressure and up to a 527°F (275°C) end point, the second at 40 mmHg total pressure to a 572°F (300°C) end point. The portion of the distillation at reduced pressure is necessary to prevent excessive pot temperatures, which cause cracking of the crude oil.

The distillation temperatures reported in the analysis must be corrected to 760 mmHg pressure. Generally, those reported in the atmospheric distillation section need not be corrected, but if carried out at high elevations, it may also be necessary to correct these. The distillation temperatures at 40 mmHg pressure can be converted to 760 mmHg by use of charts developed by Esso Research and Engineering Co. [7]. Figure 3.6 shows the relationships between boiling temperatures at 40 mmHg and 760 mmHg pressures.

The 572°F (300°C) end point at 40 mmHg pressure corresponds to 790°F (421°C) at 760 mmHg. Refinery crude oil distillation practices take overhead streams with end points from 950 to 1050°F (510 to 566°C) at 760 mmHg. Estimates of the shape of the TBP curve above 790°F (421°C) can be obtained by plotting the distillation temperature versus percentage distilled on probability graph paper and extrapolating to 1100°F (593°C) [11] (see Figure 3.7). The data points above 790°F (421°C) can be transferred to the TBP curve.

The gravity midpercent curve is plotted on the same chart with the TBP curve. The gravity should be plotted on the average volume percentage of the fraction, as the gravity is the average of the gravities from the first to the last drops in the fraction. For narrow cuts, a straight-line relationship can be assumed and the gravity used as that of the midpercent of the fraction.

Smooth curves are drawn for both the TBP and gravity-mid-percent curves. Figure 3.8 illustrates these curves for the crude oil reported in Figure 3.5.

PROBLEMS

1. Develop a TBP and gravity midpercent curve for one of the crude oils given in Appendix C.
2. Using the TBP and gravity curves from problem 1, calculate the Watson characterization factors for the fractions having mean average boiling points of 550°F (288°C) and 750°F (399°C). Is it probable this crude oil will produce satisfactory quality asphalt?
3. Using the U.S. Bureau of Mines method for classifying crude oils from the gravities of the 482 to 527°F (250 to 275°C) and 527 to 572°F (275 to 300°C) fractions, classify the crude oil used in problem 1 according to type.

NOTES

1. Energy Information Agency, DOE/EIA-0384, 2004.
2. Dosher, J.R., *Chem. Eng. 77*(8), pp. 96–112, 1970.
3. Gruse, W.A. and Stevens, D.R., *Chemical Technology of Petroleum*, 3rd ed., McGraw-Hill Book Co., New York, 1960, p. 16.
4. Ibid., p. 13.
5. Jones, M.C.K. and Hardy, R.L., *Ind. Eng. Chem. 44,* p. 2615, 1952.
6. Lane, E.C. and Garton, E.L., U.S. Bureau of Mines, *Rept. Inves. 3279*, 1935.
7. Nelson, W.L., *Oil Gas J. 68*(44), pp. 92–96, 1970.
8. Nelson, W.L., *Petroleum Refinery Engineering*, 4th ed., McGraw-Hill Book Co., New York, 1958, p. 114.
9. Nelson, W.L., *Oil Gas J. 66*(13), pp. 125–126, 1968.
10. Maxwell, J.B. and Bonnell, L.S., Derivation and precision of a new vapor pressure correlation for petroleum hydrocarbons, presented before the Division of Petroleum Chemistry, American Chemical Society, September 1955.

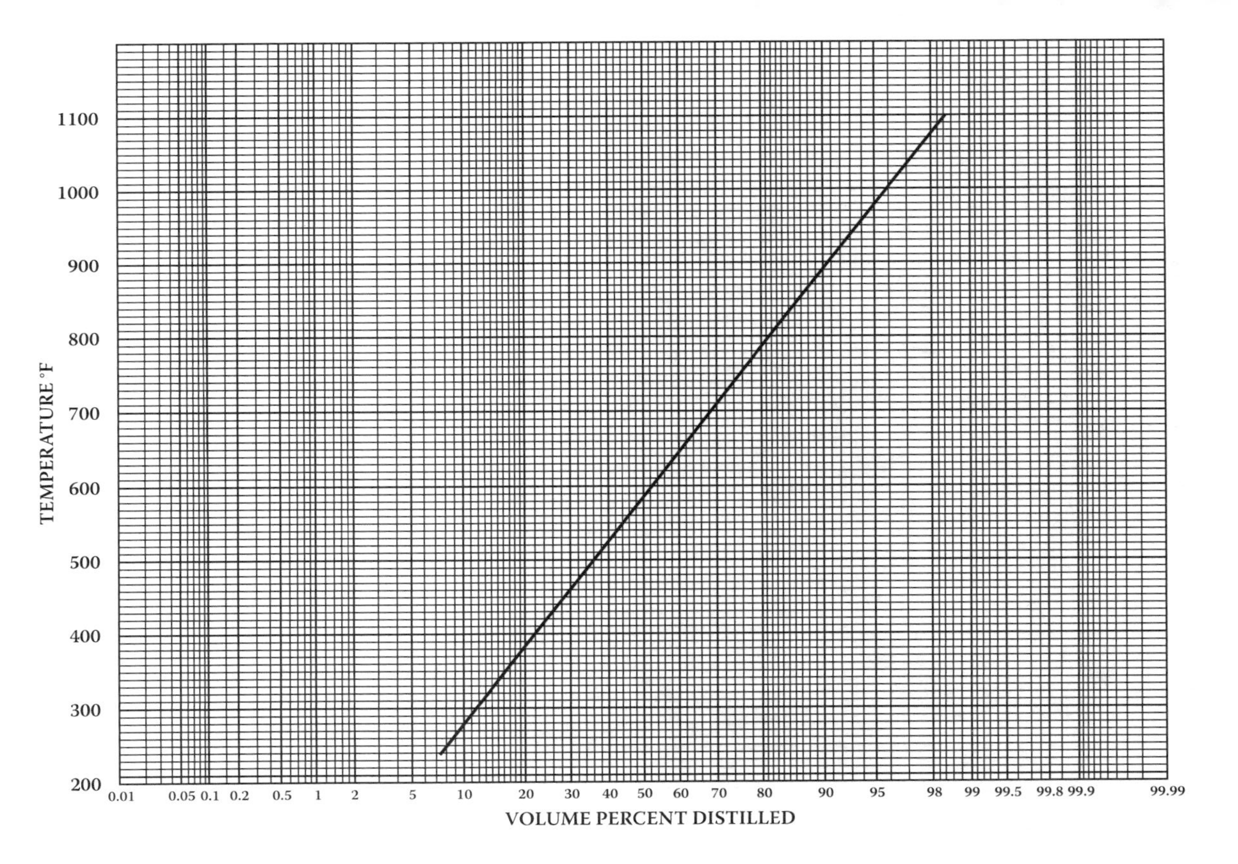

FIGURE 3.7 Crude distillation curve.

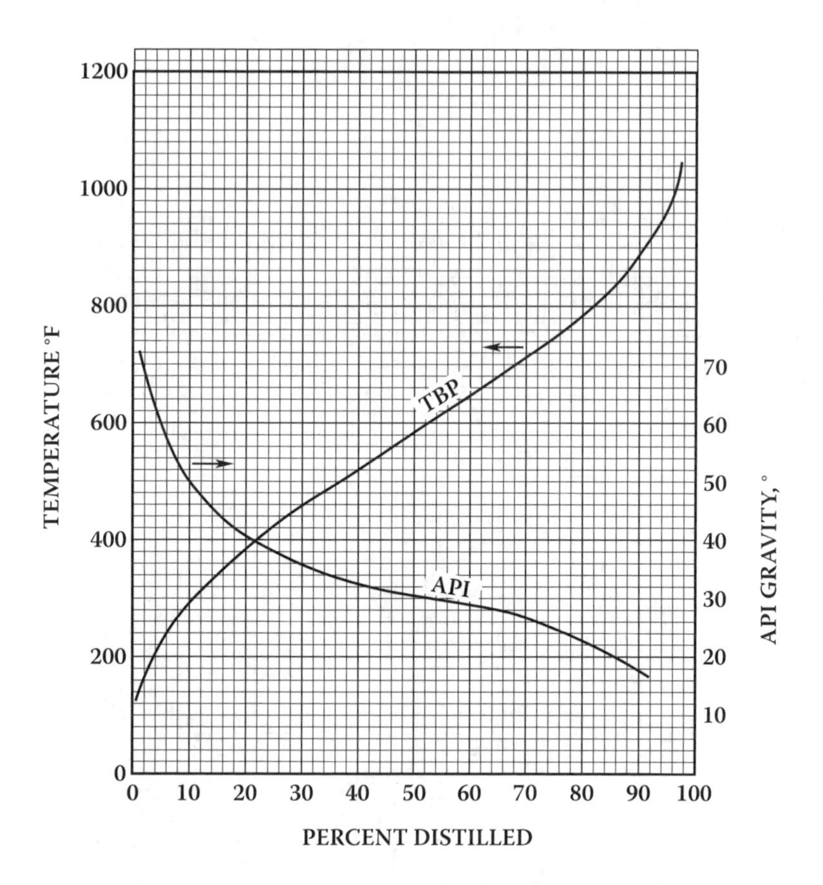

FIGURE 3.8 TBP and gravity midpercent curves. Hastings Field, TX, crude: gravity, 31.7°API; sulfur, 0.15 wt%.

ADDITIONAL READING

11. Smith, H.M., U.S. Bureau of Mines, *Tech. Paper 610,* 1940.
12. Woodle, R.A. and Chandlev, W.B., *Ind. Eng. Chem. 44*, p. 2591, 1952.

4 Crude Distillation

The crude stills are the first major processing units in the refinery. They are used to separate the crude oils by distillation into fractions according to boiling point so that each of the processing units following will have feedstocks that meet their particular specifications. Higher efficiencies and lower costs are achieved if the crude oils separation is accomplished in two steps: first by fractionating the total crude oil at essentially atmospheric pressure, then by feeding the high-boiling bottoms fraction (topped or atmospheric reduced crude) from the atmospheric still to a second fractionator operated at a high vacuum (see Photo 2, Appendix E).

The vacuum still is employed to separate the heavier portion of the crude oil into fractions, because the high temperatures necessary to vaporize the topped crude at atmospheric pressure cause thermal cracking to occur, with the resulting loss to dry gas, discoloration of the product, and equipment fouling due to coke formation. By reducing the pressure to approximately 1/20th of an atmosphere absolute, the boiling points of the hydrocarbons are reduced substantially. The boiling points are reported as though the distillation took place at 760 mm pressure.

Typical fraction cut points and boiling ranges for atmospheric and vacuum still fractions are given in Table 4.1 and Table 4.2.

Relationships among the volume-average, molal-average, and mean-average boiling points to the crude oil fractions are shown in Figure 4.1a and Figure 4.1b.

Nitrogen and sulfur contents of petroleum fractions as functions of original crude oil contents are given in Figure 4.2 through Figure 4.5.

4.1 DESALTING CRUDE OILS

If the salt content of the crude oil is greater than 10 lb/1000 bbl (expressed as NaCl), the crude requires desalting to minimize fouling and corrosion caused by salt deposition on heat transfer surfaces and acids formed by decomposition of the chloride salts. In addition, some metals in inorganic compounds dissolved in water emulsified with the crude oil, which can cause catalyst deactivation in catalytic processing units, are partially rejected in the desalting process.

The trend toward running heavier crude oils has increased the importance of efficient desalting of crudes. Until recently, the criterion for desalting crude oils was 10 lb salt/1000 bbl (expressed as NaCl) or more, but now many companies desalt all crude oils. Reduced equipment fouling and corrosion and longer catalyst life provide justification for this additional treatment. Two-stage desalting is used if the crude oil salt content is more than 20 lb/1000 bbl and, in the cases where residua are catalytically processed, there are some crudes for which three-stage desalting is used.

TABLE 4.1
Boiling Ranges of Typical Crude Oil Fractions at Atmospheric Pressure

Fraction	Boiling ranges, °F (°C)	
	ASTM	TBP
Butanes and lighter		
Light straight-run (LSR) naphtha	90–220 (32–104)	90–190 (32–88)
Heavy straight-run (HSR) naphtha	180–400 (82–204)	190–380 (88–193)
Kerosine	330–540 (166–282)	380–520 (193–271)
Light gas oil (LGO)	420–640 (216–338)	520–610 (271–321)
Atmospheric gas oil (AGO)	550–830 (288–443)	610–800 (321-427)
Vacuum gas oil (VGO)	750–1050 (399–566)	800–1050 (427–566)
Vacuum reduced crude (VRC)	1050+ (566+)	1050+ (566+)

TABLE 4.2
TBP Cut Points for Various Crude Oil Fractions

Cut	IBP °F (°C)	EP °F (°C)	Processing use
LSR gasoline cut	90 (32)	180 (82)	Min. light gasoline
	90 (32)	190 (88)	Normal LSR cut
	80 (27)	220 (104)	Max. LSR cut
HSR gasoline (naphtha)	180 (82)	380 (193)	Max. reforming cut
	190 (88)	330 (166)	Max. jet fuel opr.
	220 (104)	330 (166)	Min. reforming cut
Kerosine	330 (166)	520 (271)	Max. kerosine cut
	330 (166)	480 (249)	Max. jet-50 cut
	380 (193)	520 (271)	Max. gasoline operation
Light gas oil	420 (216)	610[a] (321)	Max. diesel fuel
	480 (249)	610[a] (321)	Max. jet fuel
	520 (271)	610[a] (321)	Max. kerosine
Heavy gas oil (HGO)	610 (321)	800 (427)	Catalytic cracker or hydrocracker feed
Vacuum gas oil	800 (427)	1050 (566)	Deasphalter or catalytic cracker feed
	800 (427)	950 (566)	Catalytic cracker or hydrocracker feed

[a] For maximum No. 2 diesel fuel production, end points as high as 650°F (343°C) can be used.
Note: In some specific locations, economics can dictate that all material between 330°F IBP and 800°F EP (166 to 427°C) be utilized as feed to a hydrocracker.

The salt in the crude is in the form of dissolved or suspended salt crystals in water emulsified with the crude oil. The basic principle is to wash the salt from the crude oil with water. Problems occur in obtaining efficient and economical water/oil mixing, water-wetting of suspended solids, and separation of the wash water from the oil. The pH, gravity, and viscosity of the crude oil, as well as the volume of wash water used per volume of crude, affect the separation ease and efficiency.

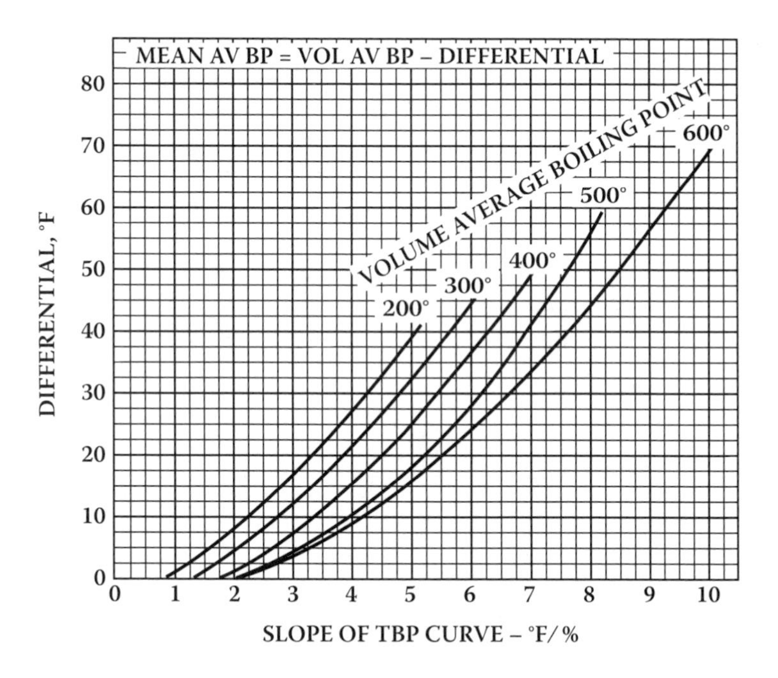

FIGURE 4.1A Mean average boiling point of petroleum fractions.

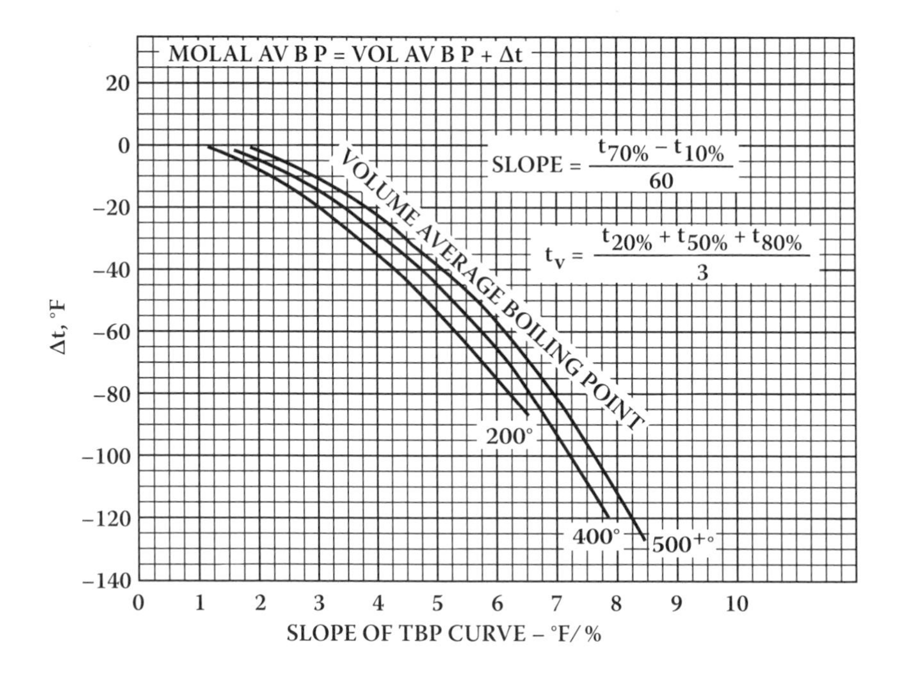

FIGURE 4.1B Molal average boiling point of petroleum fractions.

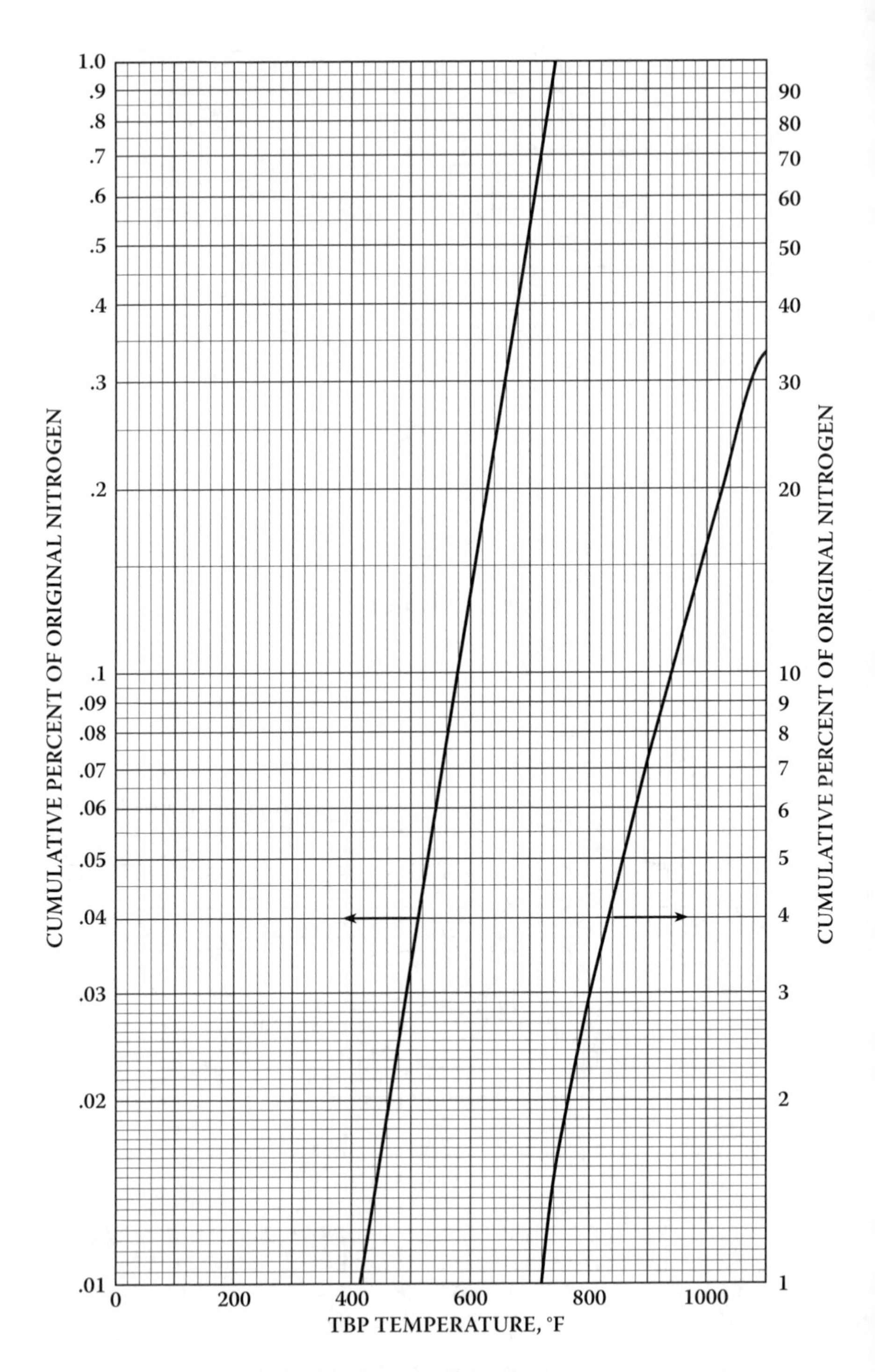

FIGURE 4.2 Nitrogen distributions in crude oil fractions.

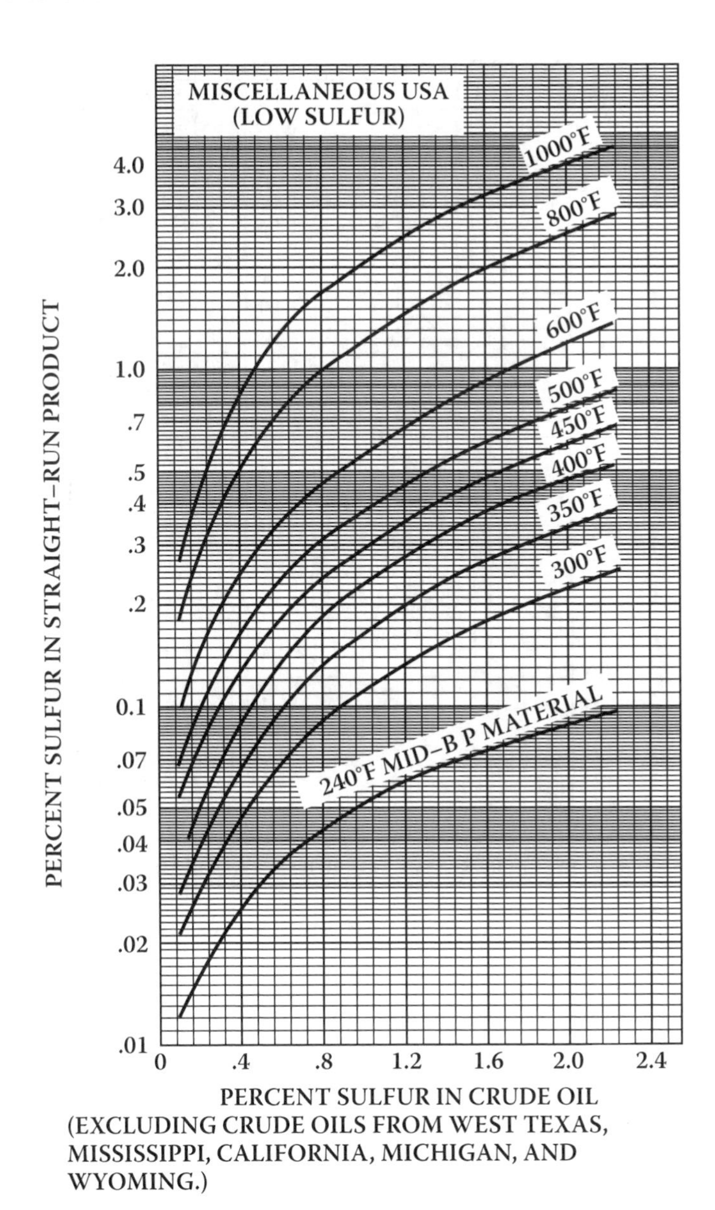

FIGURE 4.3A Sulfur content of products from miscellaneous U.S. crude oils [1].

A secondary but important function of the desalting process is the removal of suspended solids from the crude oil. These are usually very fine sand, clay, and soil particles; iron oxide and iron sulfide particles from pipelines, tanks, or tankers; and other contaminants picked up in transit or production. Total suspended solids removal should be 60% or better [2] with 80% removal of particles greater than 0.8 micron in size.

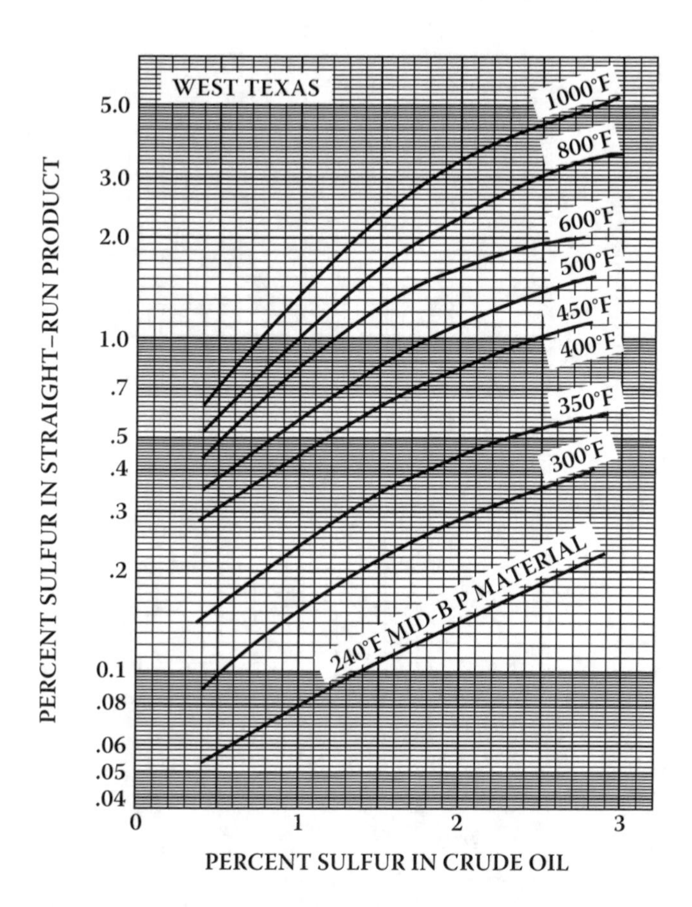

FIGURE 4.3B Sulfur content of products from West Texas crude oils [1].

Desalting is carried out by mixing the crude oil with from 3 to 10 vol% water at temperatures from 200 to 300°F (90 to 150°C). Both the ratio of the water to oil and the temperature of operation are functions of the density of the oil. Under typical operating conditions, the salts are dissolved in the wash water and the oil and water phases separated in a settling vessel either by adding chemicals to assist in breaking the emulsion or by developing a high-potential electrical field across the settling vessel to coalesce the droplets of salty water more rapidly (Figure 4.6). Either AC or DC fields may be used and potentials from 12,000 to 35,000 V are used to promote coalescence. For single-stage desalting units, 90 to 95% efficiencies are obtained, and two-stage processes achieve 99% or better efficiency.

One process uses both AC and DC fields to provide high dewatering efficiency. An AC field is applied near the oil–water interface and a DC field in the oil phase above the interface. Efficiencies of up to 99% water removal in a single stage are claimed for the dual-field process. About 90% of desalters use AC field separation only.

The dual-field electrostatic process provides efficient water separation at temperatures lower than the other processes, and as a result, higher energy efficiencies are obtained.

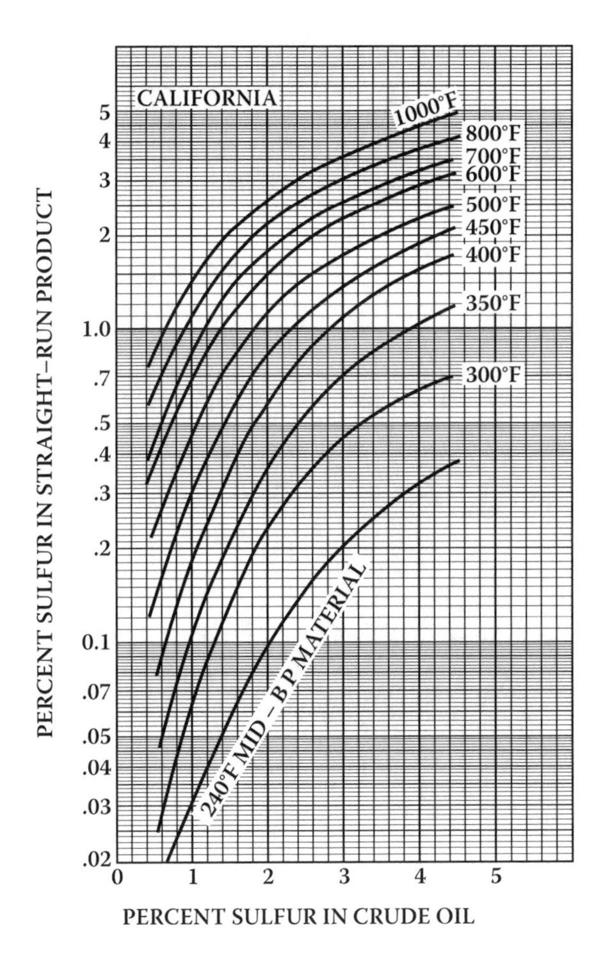

FIGURE 4.4A Sulfur content of products from California crude oils [1].

Heavy naphthenic crudes form more stable emulsions than most other crude oils, and desalters usually operate at lower efficiencies when handling them. The crude oil densities are close to the density of water and a temperature above 280°F (138°C) are needed. It is sometimes necessary to adjust the pH of the brine to obtain pH values of 7 or less in the water. If the pH of the brine exceeds 7, emulsions can be formed because of the sodium naphthenate and sodium sulfide present. For most crude oils, it is desirable to keep the pH below 8. Better dehydration is obtained in electrical desalters when they are operated in the pH range of 6 to 8, with the best dehydration obtained at a pH near 6. The pH value is controlled by using another water source or by the addition of acid to the inlet or recycled water.

Makeup water averages 4 to 5% on crude oil charge and is added to the second stage of a two-stage desalter. For very heavy crude oils (> 15°API), addition of gas oil as a diluent to the second stage is recommended to provide better separation efficiencies.

Frequently, the wash water used is obtained from the vacuum crude unit barometric condensers or other refinery sources containing phenols. The phenols are

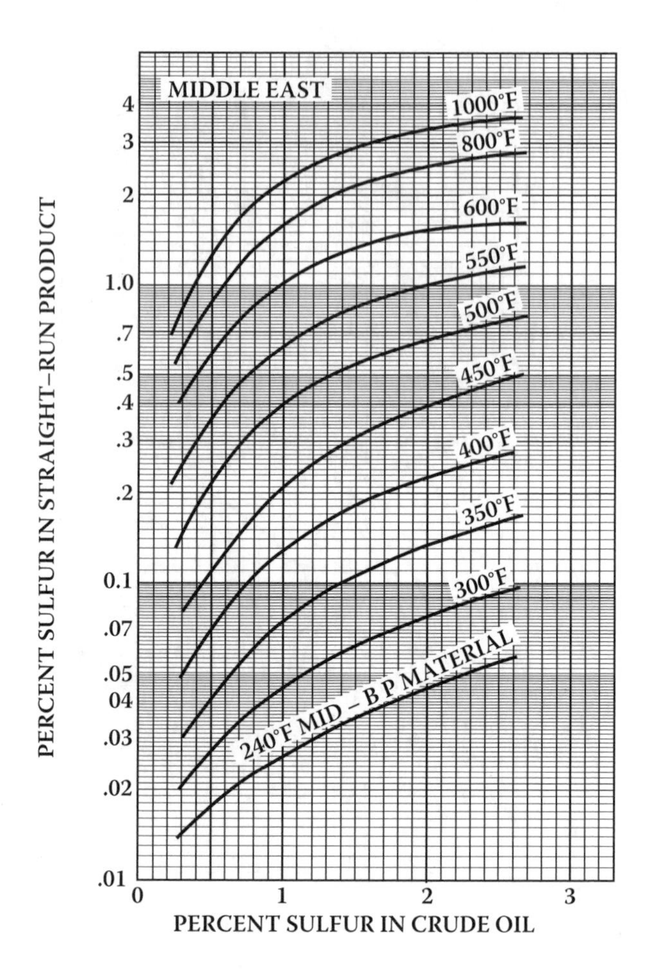

FIGURE 4.4B Sulfur content of products from Middle East crude oils [1].

preferentially soluble in the crude oil, thus reducing the phenol content of the water sent to the refinery water-handling system.

Suspended solids are one of the major causes of water-in-oil emulsions. Wetting agents are frequently added to improve the water-wetting of solids and reduce oil carry-under in desalters. Oxyalkylated alkylphenols and sulfates are the most frequently used wetting agents.

The following analytical methods are used to determine the salt content of crude oils:

1. HACH titration with mercuric nitrate after water extraction of the salt
2. Potentiometric titration after water extraction
3. Mohr titration with silver nitrate after water extraction
4. Potentiometric titration in a mixed solvent
5. Conductivity

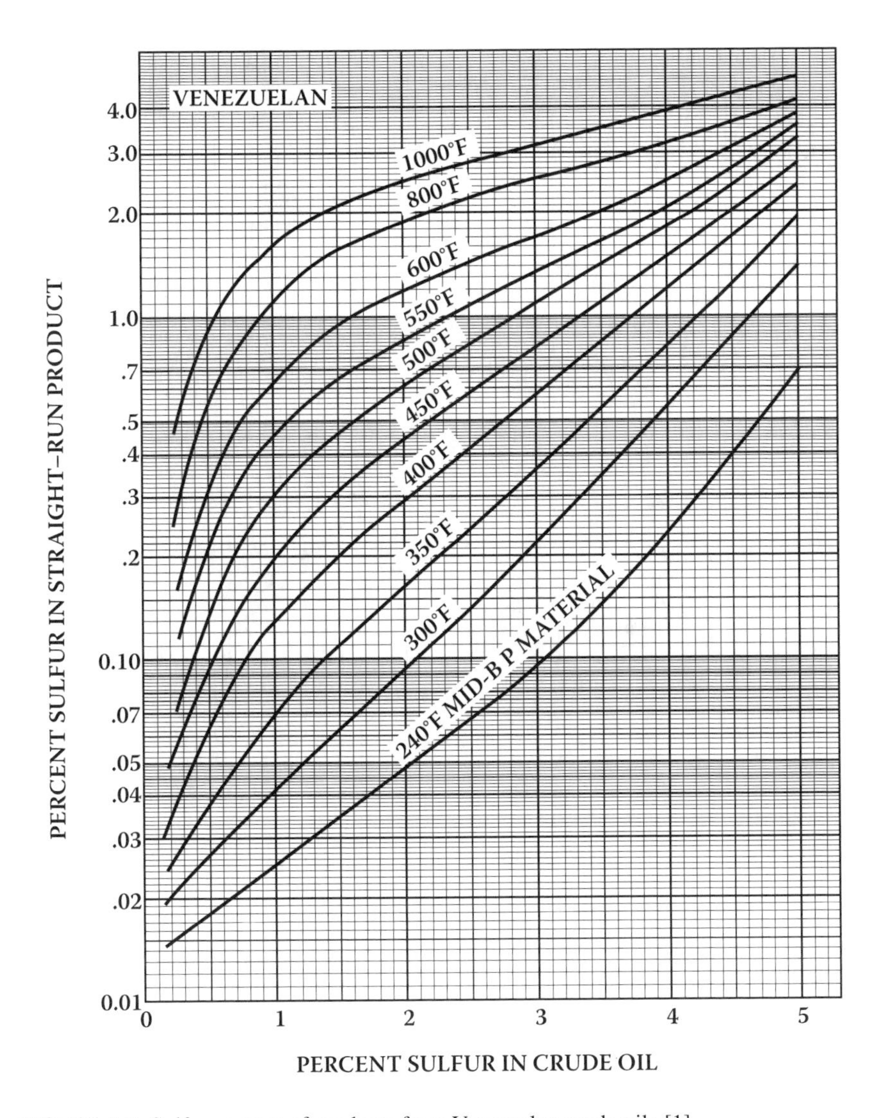

FIGURE 4.5 Sulfur content of products from Venezuelan crude oils [1].

°API	Water wash, vol%	Temp. °F (°C)
> 40	3–4	240–260 (115–25)
30–40	4–7	260–280 (125–140)
< 30	7–10	280–330 (140–150)

Although the conductivity method is the one most widely used for process control, it is probably the least accurate of these methods. Whenever used, it should be standardized for each type of crude oil processed.

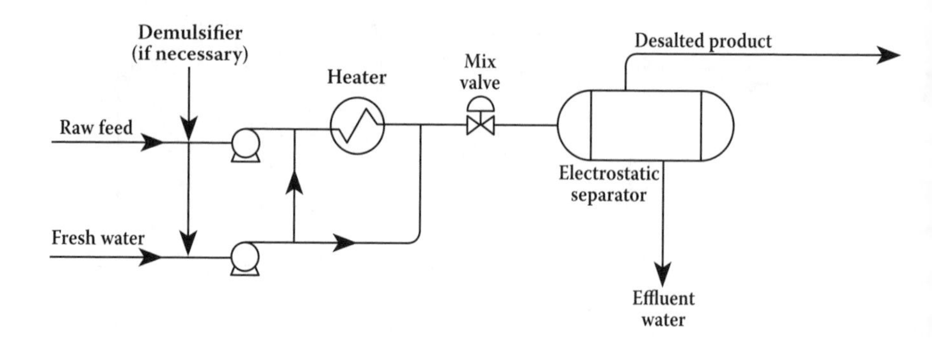

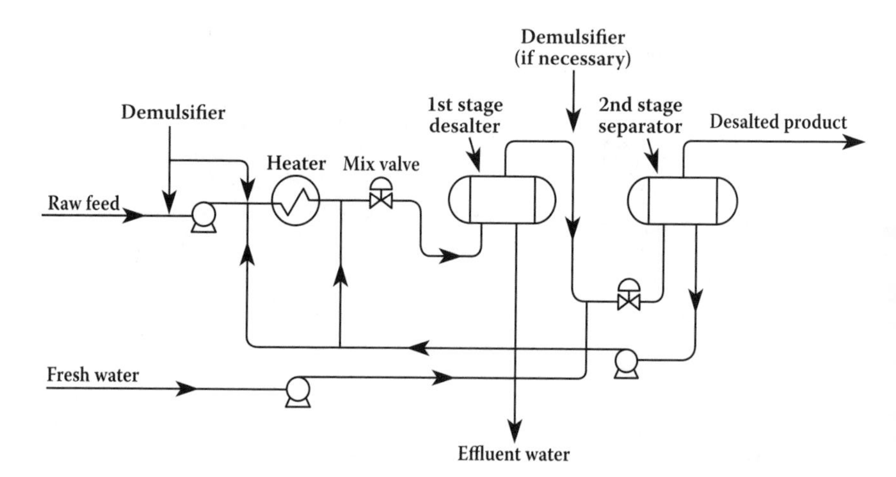

FIGURE 4.6 Single- and two-stage electrostatic desalting systems. (Flowsheets courtesy of Petrolite Corp.)

Installed costs of desalting units are shown in Figure 4.7, and utility and chemical requirements are given by Table 4.3.

4.2 ATMOSPHERIC TOPPING UNIT

After desalting, the crude oil is pumped through a series of heat exchangers and its temperature raised to about 550°F (288°C) by heat exchange with product and reflux streams [2,3]. It is then further heated to about 750°F (399°C) in a furnace (i.e., direct-fired heater or "pipe still") and charged to the flash zone of the atmospheric fractionator. The furnace discharge temperature is sufficiently high [650 to 750°F (343 to 399°C)] to cause vaporization of all products withdrawn above the flash zone plus about 10 to 20% of the bottoms product. This 10 to 20% "over-flash" allows some fractionation to occur on the trays just above the flash zone by providing internal reflux in excess of the sidestream withdrawals.

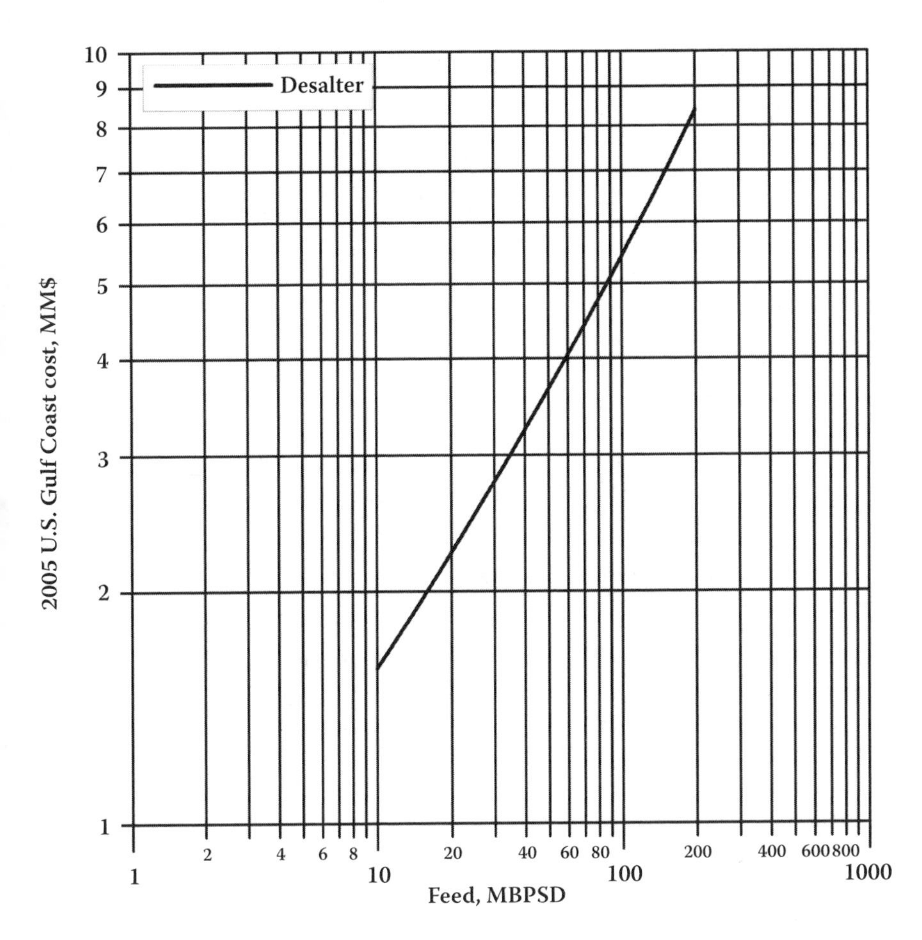

FIGURE 4.7 Crude oil desalting units investment cost: 2005 U.S. Gulf Coast (see Table 4.3).

Reflux is provided by condensing the tower overhead vapors and returning a portion of the liquid to the top of the tower, and by pump-around and pumpback streams lower in the tower. Each of the sidestream products removed from the tower decreases the amount of reflux below the point of drawoff. Maximum reflux and fractionation are obtained by removing all heat at the top of the tower, but this results in an inverted cone-type liquid loading, which requires a very large diameter at the top of the tower. To reduce the top diameter of the tower and even the liquid loading over the length of the tower as well as to improve energy efficiency, intermediate heat-removal streams are used to generate reflux below the sidestream removal points. To accomplish this, liquid is removed from the tower, cooled by a heat exchanger, and returned to the tower or, alternatively, a portion of the cooled sidestream may be returned to the tower. This cold stream condenses more of the vapors coming up the tower and thereby increases the reflux below that point.

The energy efficiency of the distillation operation is also improved by using pump-around reflux. If sufficient reflux was produced in the overhead condenser to

TABLE 4.3
Desalter Unit Cost Data

Costs included	
1. Conventional electrostatic desalting unit	
2. Water injection	
3. Caustic injection	
4. Water preheating and cooling	
Costs not included	
1. Wastewater treating and disposal	
2. Cooling water and power supply	
Utility data (per bbl feed)	
Power, kWh	0.01–0.02
Water injection, gal (m^3)	1–3 (0.004–0.012)
Demulsifier chemical, lb (kg)[a]	0.005–0.01 (0.002–0.005)
Caustic, lb (kg)	0.001–0.003 (0.005–0.0014)

[a] 2005 price approximately $1.75/lb.

Note: See Figure 4.6.

provide for all sidestream drawoffs as well as the required reflux, all of the heat energy would be exchanged at the bubble-point temperature of the overhead stream. By using pump-around reflux at lower points in the column, the heat transfer temperatures are higher, and a higher fraction of the heat energy can be recovered by preheating the feed.

Although crude towers do not normally use reboilers, several trays are generally incorporated below the flash zone, and steam is introduced below the bottom tray to strip any remaining gas oil from the liquid in the flash zone and to produce a high-flash-point bottoms. The steam reduces the partial pressure of the hydrocarbons and lowers the required vaporization temperature.

The atmospheric fractionator normally contains 30 to 50 fractionation trays. Separation of the complex mixture in crude oils is relatively easy, and generally five to eight trays are needed for each sidestream product plus the same number above and below the feed plate. Thus, a crude oil atmospheric fractionation tower with four liquid sidestream drawoffs will require from 30 to 42 trays.

The liquid sidestream withdrawn from the tower will contain lower-boiling components that lower the flash point, because the lighter products pass through the heavier products and are in equilibrium with them on every tray. These “light ends” are stripped from each sidestream in a separate small stripping tower containing four to ten trays with steam introduced under the bottom tray. The steam and stripped light ends are vented back into the vapor zone of the atmospheric fractionator above the corresponding side-draw tray (Figure 4.8).

The overhead condenser on the atmospheric tower condenses the pentane-and-heavier fraction of the vapors that passes out of the top of the tower. This is the light gasoline portion of the overhead, containing some propane and butanes and essentially all of the higher-boiling components in the tower overhead vapor. Some of

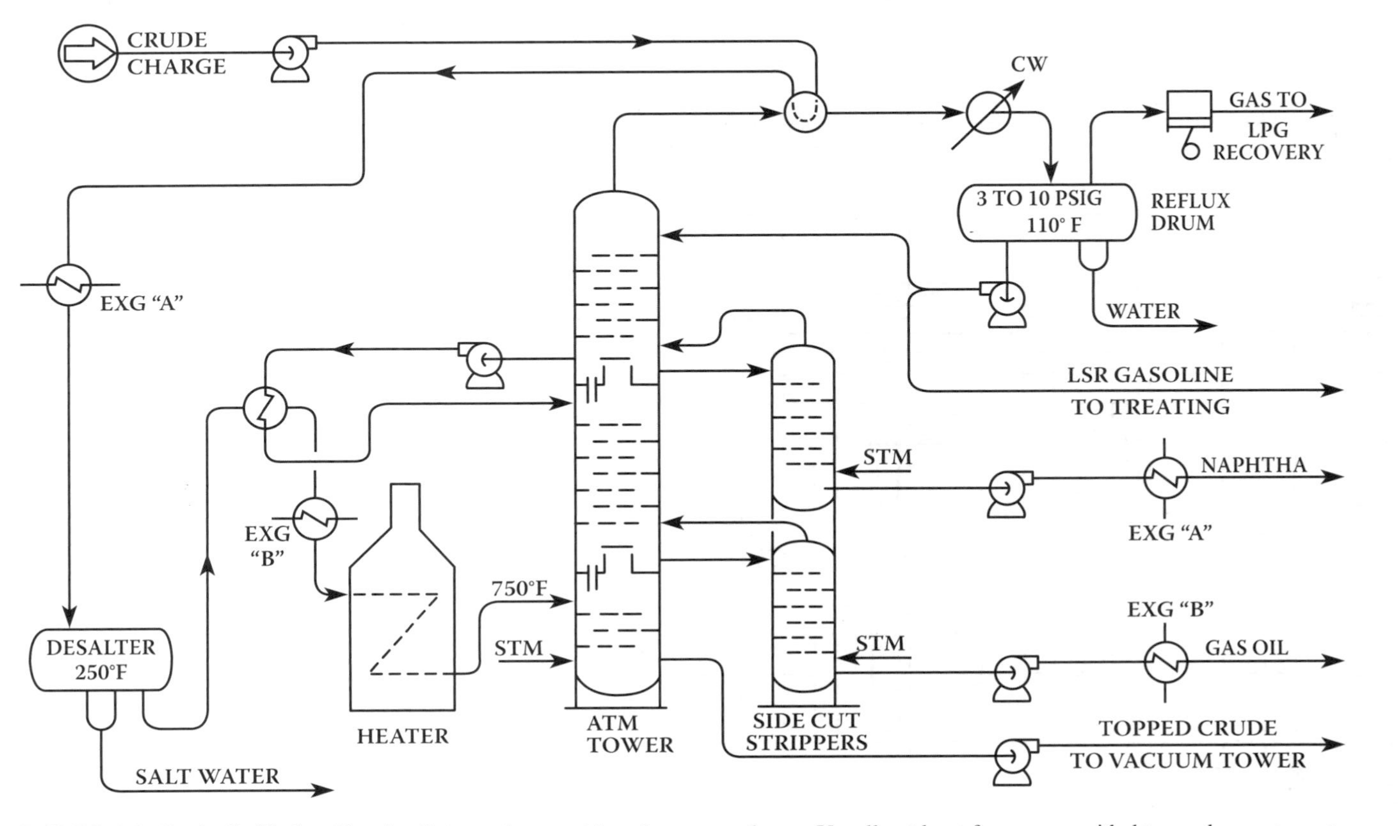

FIGURE 4.8 Crude distillation. For simplicity, only two side strippers are shown. Usually at least four are provided to produce extra cuts such as kerosine and diesel.

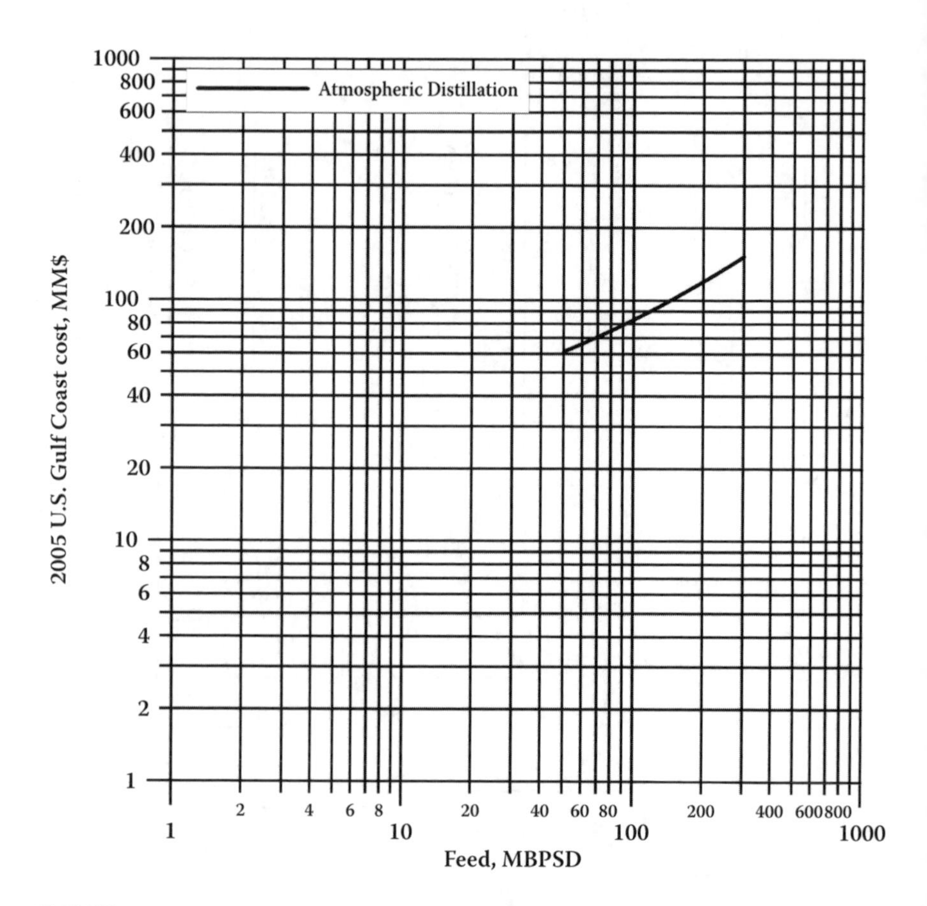

FIGURE 4.9 Atmospheric crude distillation units investment cost: 2005 U.S. Gulf Coast (see Table 4.4).

this condensate is returned to the top of the tower as reflux, and the remainder is sent to the stabilization section of the refinery gas plant, where the butanes and propane are separated from the C_5–180°F (C_5–82°C) LSR gasoline.

Installed costs of atmospheric crude distillation units are shown in Figure 4.9, and utility requirements are given by Table 4.4.

4.3 VACUUM DISTILLATION

The furnace outlet temperatures required for atmospheric pressure distillation of the heavier fractions of crude oil are so high that thermal cracking would occur, with the resultant loss of product and equipment fouling. These materials are, therefore, distilled under vacuum, because the boiling temperature decreases with a lowering of the pressure. Distillation is carried out with absolute pressures in the tower flash zone area of 25 to 40 mmHg (Figure 4.9). To improve vaporization, the effective pressure is lowered even further (to 10 mmHg or less) by the addition of steam to the furnace inlet and at the bottom of the vacuum tower. Addition of steam to the

TABLE 4.4
Atmospheric Crude Distillation Unit Cost Data

Costs included	
1. Side cuts with strippers	
2. All battery limits (BL) process facilities	
3. Sufficient heat exchange to cool top products and side cuts to ambient temperature	
4. Central control system	
Costs not included	
1. Cooling water, steam, and power supply	
2. Desalting	
3. Cooling on reduced crude (bottoms)	
4. Sour water treating and disposal	
5. Feed and product storage	
6. Naphtha stabilization	
7. Light ends recovery	
Utility data (per bbl feed)	
Steam [300 psig (2068 kPa)], lb (kg)	10.0 (4.5)
Power, kWh	0.9
Cooling water circulation, gal (m^3)[a]	150 (0.58)
Fuel, MMBtu (kJ)[b]	0.05 (52,750)

[a] 30°F (17°C) rise; about 50% of this duty can be used for BFW preheat.
[b] LHV basis, heater efficiency taken into account. (All fuel data in this text are on this same basis.)
Note: See Figure 4.9.

furnace inlet increases the furnace tube velocity and minimizes coke formation in the furnace as well as decreasing the total hydrocarbon partial pressure in the vacuum tower. The amount of stripping steam used is a function of the boiling range of the feed and the fraction vaporized, but generally ranges from 10 to 50 lb/bbl feed [4,5].

Furnace outlet temperatures are also a function of the boiling range of the feed and the fraction vaporized as well as of the feed coking characteristics. High tube velocities and steam addition minimize coke formation, and furnace outlet temperatures in the range of 730 to 850°F (388 to 454°C) are generally used [6].

Typically, the highest furnace outlet temperatures are for "dry" operation of the vacuum unit; that is, no steam is added either to the furnace inlet or to the vacuum column. The lowest furnace outlet temperatures are for "wet" operation when steam is added both to the furnace inlet and to the bottom of the vacuum tower. Intermediate temperatures are used for "damp" operation of the vacuum unit when steam is added to the furnace inlet only. For most crude oils, the furnaces can be operated from 3 to 5 years between turnarounds.

The effective pressure (total absolute pressure – partial pressure of the steam) at the flash zone determines the fraction of the feed vaporized for a given furnace outlet temperature, so it is essential to design the fractionation tower, overhead lines, and condenser to minimize the pressure drop between the vacuum-inducing device and the flash zone. A few millimeters of decrease in pressure drop will save substantially in operating costs.

The lower operating pressures cause significant increases in the volume of vapor per barrel vaporized and, as a result, the vacuum distillation columns are much larger in diameter than the atmospheric towers. It is not unusual to have vacuum towers up to 40 ft (12.5 m) in diameter.

The desired operating pressure is maintained by the use of steam ejectors and barometric condensers or vacuum pumps and surface condensers. The size and number of ejectors and condensers used are determined by the vacuum needed and the quality of vapors handled. For a flash zone pressure of 25 mmHg, three ejector stages are usually required. The first stage condenses the steam and compresses the noncondensable gases, whereas the second and third stages remove the noncondensable gases from the condensers. The vacuum produced is limited to the vapor pressure of the water used in the condensers. If colder water is supplied to the condensers, a lower absolute pressure can be obtained in the vacuum tower.

Although more costly than barometric condensers, a recent trend is the use of vacuum pumps and surface condensers in order to reduce the contamination of water with oil.

A schematic of a crude oil vacuum distillation unit is shown in Figure 4.10, installed costs are shown in Figure 4.11, and utility requirements are given by Table 4.5.

4.4 AUXILIARY EQUIPMENT

In many cases, a flash drum is installed between the feed-preheat heat exchangers and the atmospheric pipe still furnace. The lower-boiling fractions that are vaporized by heat supplied in the preheat exchangers are separated in the flash drum and flow directly to the flash zone of the fractionator. The liquid is pumped through the furnace to the tower flash zone. This results in a smaller and lower-cost furnace and lower furnace outlet temperatures for the same quantity of overhead streams produced.

A stabilizer is incorporated in the crude distillation section of some refineries instead of being placed with the refinery gas plant. The liquid condensed from the overhead vapor stream of the atmospheric pipe still contains propane and butanes, which make the vapor pressure much higher than is acceptable for gasoline blending. To remove these, the condensed liquid in excess of reflux requirements is charged to a stabilizing tower, where the vapor pressure is adjusted by removing the propane and butanes from the LSR gasoline stream. Later, in the product-blending section of the refinery, n-butane is added to the gasoline stream to provide the desired Reid vapor pressure.

4.5 CRUDE DISTILLATION UNIT PRODUCTS

In order of increasing boiling points, the main products from a typical crude distillation unit are as follows:

Fuel gas. The fuel gas consists mainly of methane and ethane. In some refineries, propane in excess of LPG requirements is also included in the fuel gas stream. This stream is also referred to as “dry gas.”

Wet gas. The wet gas stream contains propane and butanes as well as methane and ethane. The propane and butanes are separated to be used for LPG and, in the case of butanes, for gasoline blending and alkylation unit feed.

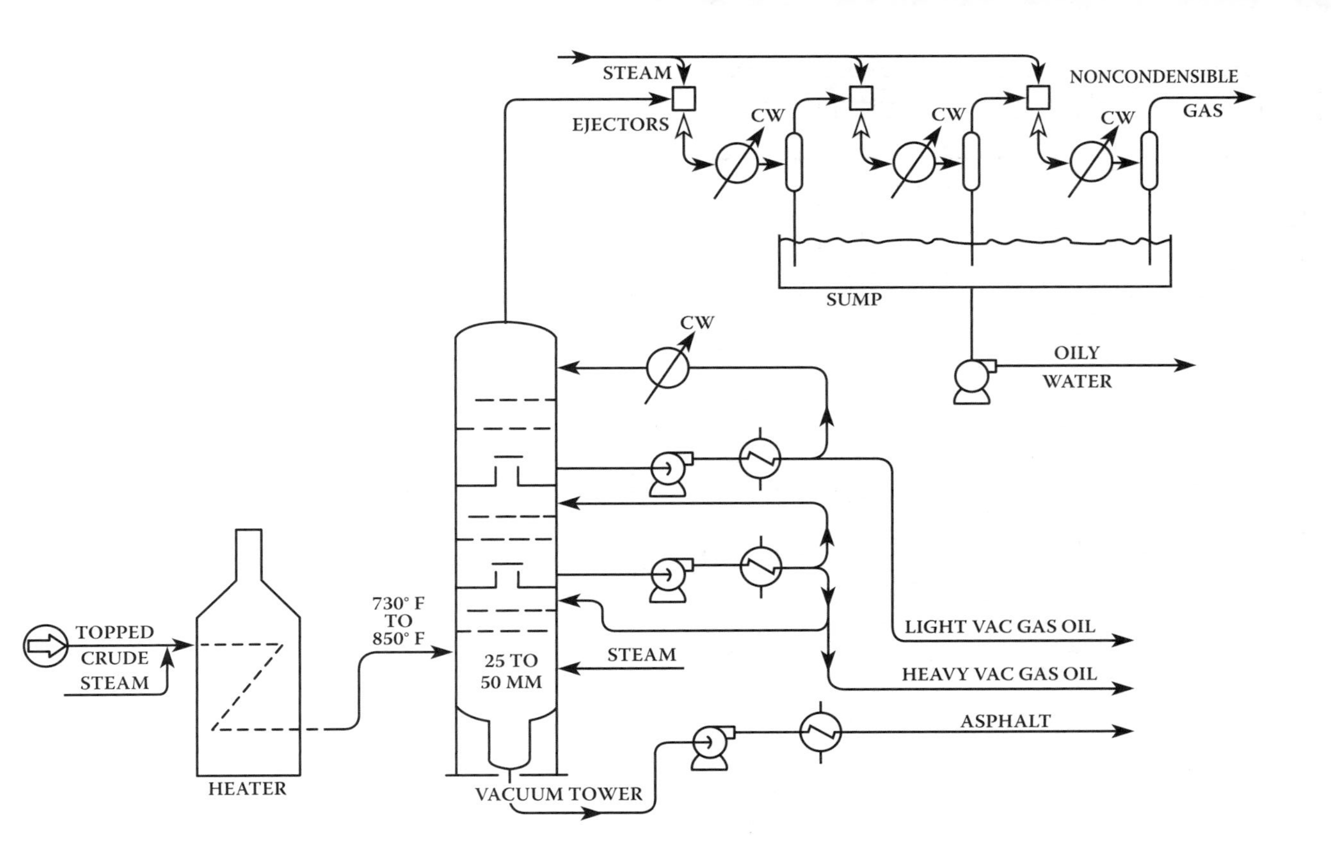

FIGURE 4.10 Vacuum distillation.

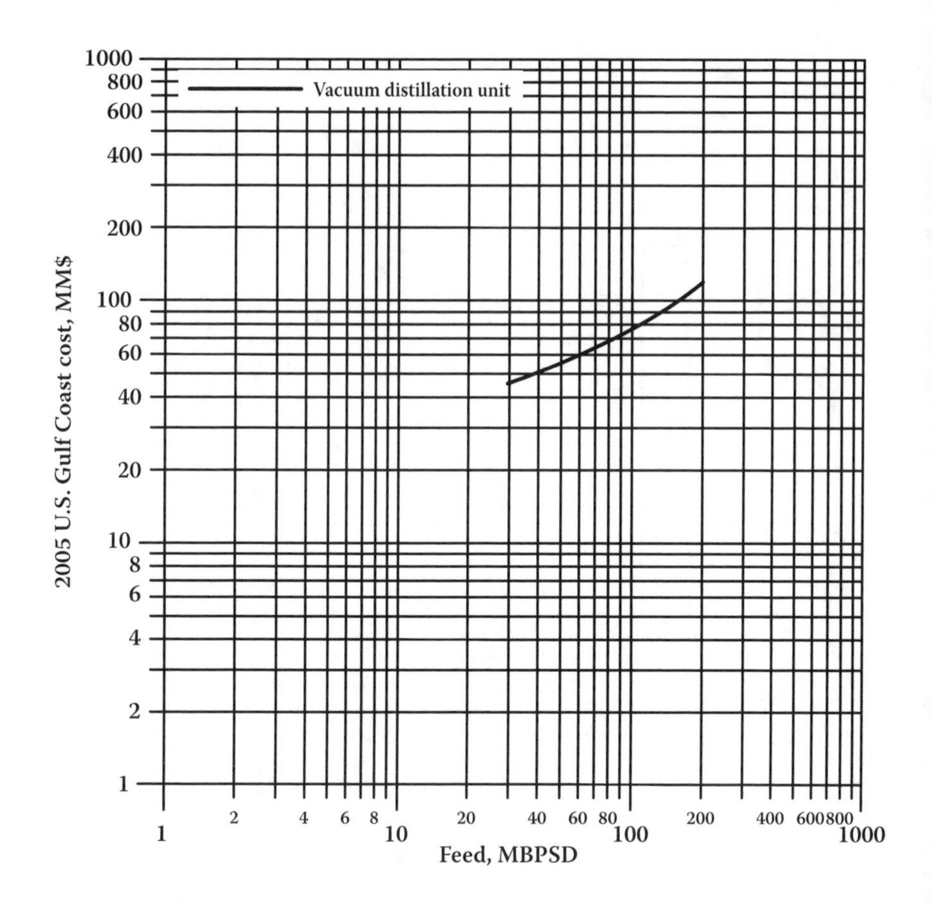

FIGURE 4.11 Vacuum distillation units investment cost: 2005 U.S. Gulf Coast (see Table 4.5).

LSR naphtha. The stabilized LSR naphtha (or LSR gasoline) stream is desulfurized and used in gasoline blending or processed in an isomerization unit to improve octane before blending into gasoline.

HSR naphtha or *HSR gasoline*. The naphtha cuts are generally used as catalytic reformer feed to produce high-octane reformate for gasoline blending and aromatics.

Gas oils. The light, atmospheric, and vacuum gas oils are processed in a hydrocracker or catalytic cracker to produce gasoline, jet, and diesel fuels. The heavier vacuum gas oils can also be used as feedstocks for lubricating oil processing units.

Residuum. The vacuum still bottoms can be processed in a visbreaker, coker, or deasphalting unit to produce heavy fuel oil or cracking or lube base stocks. For asphalt crudes, the residuum can be processed further to produce road or roofing asphalts.

4.6 CASE-STUDY PROBLEM: CRUDE UNITS

To illustrate the operation of a fuels refinery and the procedures for making a preliminary economic evaluation, a material balance will be made on each processing unit in a refinery, and operating and construction costs will be estimated. The material

TABLE 4.5
Vacuum Distillation Unit Cost Data

Costs included	
1. All facilities required for producing a clean vacuum gas oil (single cut)	
2. Three-stage jet system for operation of flash zone at 30 to 40 mmHg	
3. Coolers and exchangers to reduce VGO to ambient temperature	
Costs not included	
1. Cooling water, steam, power supply	
2. Bottoms cooling below 400°F (204°C)	
3. Feed preheat up to 670°F (354°C), ± 10 (assumes feed is direct from atmosphere crude unit)	
4. Sour water treating and disposal	
5. Feed and product storage	
6. Multiple cuts or lube oil production	
Utility data (per bbl feed)	
Steam [300 psig (2068 kPa)], lb (kg)	10.0 (4.5)
Power, kWh	0.3
Cooling water circulation, gal (m^3)[a]	150 (0.57)
Fuel, MMBtu (kJ)[b]	0.03 (31,650)

[a] 30°F (16.7°C) rise.
[b] LHV basis, heater efficiency taken into account.
Note: See Figure 4.10 and Figure 4.11.

balance for each unit is located at the end of the chapter that discusses that particular unit, and the evaluation of operating and construction costs is located at the end of Chapter 18 (Cost Estimation). An overall economic evaluation is given at the end of Chapter 19 (Economic Evaluation).

The example problem is worked using units of measurement common to U.S. refineries. Using both U.S. and international units of measurement makes it more difficult to follow the problem, and the international set of units is not standard in refinery operations throughout the world even though research and most technical publications require their use.

In order to determine the economics of processing a given crude oil or of constructing a complete refinery or individual processing units, it is necessary to make case studies for each practical processing plan and to select the one giving the best economic return. This is a very time-consuming process and can be very costly. The use of computer programs to optimize processing schemes is widespread throughout the refining industry. For the purpose of illustrating calculation methods, the process flow will not be optimized in this problem, but all units will be utilized (i.e., delayed coker, fluid catalytic cracker, and hydrocracker) even though certain ones cannot be economically justified under the assumed conditions.

In addition to the basic data contained in this book, it is necessary to use industry publications such as the *Oil and Gas Journal*, *Chemical Week*, *Oil Daily*, and *Chemical Marketing* (formerly *Oil, Paint, and Drug Reporter*) for current prices of raw materials and products. *Chemical Week*, for example, has an annual issue on plant sites that lists utility costs, wages, and taxes by location within the United States.

In many processing units, there is a change in volume between the feed and the products, so it is not possible to make a volume balance around the unit. For this reason, it is essential that a weight balance be made. Even though, in practice, the weights of individual streams are valid to only three significant figures, for purposes of calculation it is necessary to carry them out to at least the closest 100 lb. For this example, they will be carried to the nearest pound.

In order to make the balance close on each unit, it is necessary to determine one product stream by difference. Generally, the stream having the least effect is determined by difference and, in most cases, this works out to be the heaviest product.

Conversion factors (e.g., lb/hr per BPD) and properties of pure compounds are tabulated in Appendix B. Properties of products and intermediate streams are given in the chapters on products and the individual process units.

4.6.1 Statement of the Problem

Develop preliminary estimates of product yields, capital investment, operating costs, and economics of building a grassroots refinery on the West Coast to process 100,000 BPCD of North Slope Alaska crude oil.

The major income products will be motor gasoline, jet fuel, and diesel fuel, and the refinery will be operated to maximize gasoline yields within economic limits. The gasoline split will be 50/50 87 PON regular and 93 PON premium [PON = (RON + MON) / 2]. Analysis of the crude oil is given in Table 4.6. It is estimated that this crude oil will be available FOB the refinery at $40.00/bbl. Product prices will be based on the average posted prices for 2005, less the following:

1. $0.005/gal on all liquid products except fuel oil
2. $0.05/bbl on fuel oil
3. $0.50/ton on coke

Utility prices will be those reported in the 2005 Plant Sites issue of *Chemical Week* (October 2005) for the area in which the refinery will be built. Federal and state income taxes will be 45% (38% U.S. and 7% California), and land cost will be 5% of the cost of the process units, storage, steam systems, cooling water systems, and offsites.

4.6.2 General Procedure

1. From the crude distillation data given in Table 4.6, plot TBP and gravity midpercent curves. These are shown in Figure 4.12.
2. From Table 4.2, select TBP cut points of products to be made from atmospheric and vacuum pipe stills.
3. From TBP and gravity curves, determine percentages and gravities of fractions.
4. Using Tables B.1 and B.2, convert volumes to weights.
5. Determine the weight of a 1050°F (566°C) bottoms stream by difference. If the volume of a 1050°F stream is taken from the TBP curve, then gravity is calculated from weight and volume. In this case, because the gravity

TABLE 4.6
Petroleum Analysis, North Slope Alaska Crude

Gravity, °API, 26.4
Sulfur, wt%, 0.99
Viscosity, SUS at 70°F (21°C), 182.5 sec
at 100°F (38°C), 94.1 sec
Conradson carbon: crude, 5.99%; 1050+°F VRC, 14.2%
TBP Distillation

	vol% on crude		sp. gr.		Sulfur	N	Ni	V
TBP cut (°F)	Frac.	Sum	(60/60°F)	°API	(wt%)	(wppm)	(wppm)	(wppm)
C_2	0.1	0.1	0.374	—				
C_3	0.3	0.4	0.509	—				
iC_4	0.2	0.6	0.564	—				
nC_4	0.6	1.2	0.584	—				
iC_5	0.5	1.7	0.625	—				
nC_5	0.7	2.4	0.631	—				
97–178	1.7	4.1	0.697	71.6				
178–214	2.1	6.2	0.740	59.7				
214–242	2.0	8.2	0.759	55.0				
242–270	2.0	10.2	0.764	53.8				
270–296	2.0	12.2	0.781	49.6				
296–313	1.0	13.2	0.781	49.6				
313–342	2.0	15.2	0.791	47.3				
342–366	1.9	17.1	0.797	46.0				
366–395	2.0	19.1	0.806	44.0				
395–415	2.0	21.1	0.831	38.8				
415–438	2.0	23.1	0.832	38.6				
438–461	2.0	25.1	0.839	37.2				
461–479	2.0	27.1	0.848	35.4				
479–501	2.0	29.1	0.856	33.9				
501–518	2.0	31.1	0.860	33.1				
518–538	2.0	33.1	0.864	32.2				
538–557	2.0	35.1	0.867	31.8				
557–578	2.0	37.1	0.868	31.6				
578–594	2.1	39.2	0.872	30.7				
594–610	2.0	41.2	0.878	29.6				
610–632	2.0	43.2	0.887	28.0				
632–650	1.8	45.0	0.893	26.9				
650–1000	31.8	76.8	0.933	20.3	1.0	1390		
1000+	23.2	100.0	0.995	10.7	2.4	4920	56	97

Source: Note 7.

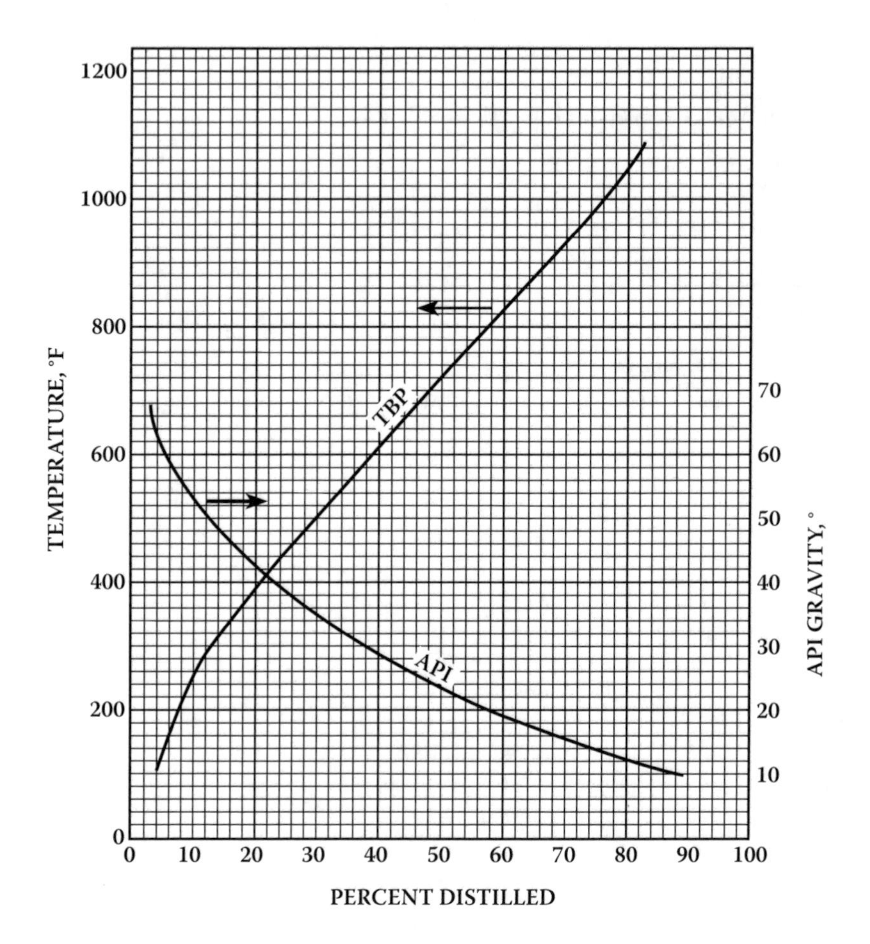

FIGURE 4.12 TBP and gravity midpercent curves, Alaska North Slope crude: gravity, 26.4 API; sulfur, 0.99 wt%. (From Note 7.)

of the 1050°F stream was determined in the laboratory and the gravity is very sensitive to small changes in weight, it was decided to use the laboratory gravity and calculate the volume. This gives a total volume recovery of 100.6%. This is reasonable because there is a negative volume change on mixing of petroleum fractions, and it is possible on some crudes to obtain liquid products [more properly, liquid-equivalent products, because C_4 and lighter products are not liquids at 60°F (16°C) and 1 atm pressure] having a total volume of up to 103% on crude. The material balance and utility requirements for the crude units are given in Table 4.7.

6. For crude oils containing significant amounts of sulfur or nitrogen, it is also necessary to make sulfur or nitrogen balances around each unit. The North Slope crude has a sufficiently high quality of sulfur to require a sulfur balance, but because there is no nitrogen shown in the analysis, a nitrogen balance is not made. It is assumed that the North Slope crude is similar to those represented in Figure 4.3.

TABLE 4.7
Crude Units Material Balance: 100,000 BPCD Alaska North Slope Crude Oil Basis

Component	vol%	BPCD	°API	(lb/BPD)	lb/h	wt%	wt% S	lb/h S
Atmospheric distillation unit								
Feed								
Crude	100.0	100,000	26.4	13.1	1,307,464	100.0	0.99	12,944
Products								
C_2–	0.1	72		5.18	372	0.03		
C_3	0.3	317		7.39	2,341	0.18		
iC_4	0.2	159		8.20	1,302	0.10		
nC_4	0.6	629		8.51	5,353	0.41		
C_5– 180°F	4.3	4,277	63.0	10.61	45,394	3.47	0.00	1
180–350	11.7	11,685	49.9	11.38	132,988	10.17	0.01	13
350–400	4.3	4,332	42.5	11.85	52,534	3.93	0.11	56
400–525	12.2	12,200	36.5	12.21	148,906	11.39	0.24	357
525–650	12.5	12,500	38.0	12.77	159,653	12.21	0.52	830
650+	57.0	54,147	15.5	14.03	759,799	58.11	1.54	11,686
Total		100,317			1,307,464	100.0		12,944
Vacuum distillation unit								
Feed								
650°F+	57.0	54,147	15.5	14.03	759,799	58.11	1.54	11,686
Products								
650–850°F	20.5	20,500	22.2	13.42	275,119	36.2	0.95	2,614
850–1050	15.6	15,644	15.5	14.03	219,518	28.9	1.51	3,315
1050+	18.0	18,003	8.7	14.73	265,162	34.9	2.17	5,757
Total		54,177			759,799	100.0		11,686

Utility requirements (per day)

	Desalter	Atm. PS	Vac. PS	Total
Steam (300 psig), Mlb	400	1,000	541	1,541
Power, MkWh	1	90	16	107
CW circ., Mgal		15,000	8,122	23,122
Fuel, MMBtu		5,000	1,624	6,624
Water inj., Mgal	200			200
Demul. Chem., lb	500			500
Caustic, Mlb	100			100

PROBLEMS

1. Using one of the crude oils in Appendix C, make TBP and gravity mid-percent curves. From these curves and the crude oil fraction specifications in Table 4.1 or Table 4.2, make a complete material balance around an atmospheric crude still. Assume a 100,000 BPCD crude oil feed rate to the atmospheric crude still. Make the balances to the nearest pound.

2. For the crude oil used in problem 1, make sulfur and nitrogen weight balances for the feed and products to the nearest pound.
3. Using the high-boiling fraction from problem 1 as feed to a vacuum pipe still, make an overall weight balance and nitrogen and sulfur balances. Assume the reduced crude stream from the vacuum pipe still has an initial TBP boiling point of 1050°F (566°C).
4. Estimate the calendar day utility and chemical requirements for the crude oil desalter, atmospheric pipe still, and vacuum pipe still of problems 1 and 3.
5. Calculate the Watson characterization factors and U.S. Bureau of Mines Correlation Indices for the crude oil and product streams from problems 1 and 3.
6. Estimate the carbon content of the reduced crude bottoms product of problem 3. Express as weight percentage Conradson carbon residue.
7. Using the information in Table 4.2, tell where each product stream from the crude still will be utilized (unit feed or product).
8. a. Calculate the Btu required per hour to heat 100,000 BPSD of the assigned crude oil from 60°F (15.6°C) to 806°F (430°C) in the atmospheric distillation unit furnace.

   b. Assuming a furnace efficiency of 73%, how many barrels of 13.0°API residual fuel oil will be burned per barrel of crude oil charged?
9. Calculate the atmospheric distillation unit furnace exit temperature needed to vaporize 30% of the assigned crude oil in the tower flash zone.
10. a. Determine the API gravity, characterization factor, and temperature of the topped crude leaving the bottom of the tower in problem 9.

    b. If the topped crude used as the hot fluid in a train of feed preheater heat exchangers is exited at 100°F (38°C), what was the temperature to which the crude oil feed to the furnace was heated? [Assume the crude inlet temperature is 60°F (15.6°C).]
11. If the crude oil leaving the preheater heat exchanger train in problem 10 was partially vaporized in a flash drum and the liquid from the flash drum heated to 752°F (400°C) in the furnace, calculate the percentage of original crude oil feed of the topped crude exiting from the bottom of the atmospheric crude tower.
12. If the assigned crude oil has an atmospheric crude unit furnace outlet temperature of 752°F (400°C) and a flash zone pressure of 1 atm gauge (101 kPa), calculate the flash zone temperature and the volume percentage vaporized.
13. The crude oil in problem 11 can be heated to 527°F (275°C) in a preheater train and flashed in a flash drum before the liquid fraction is heated to 752°F (400°C) in the crude unit furnace. If the fractionating tower flash zone pressure is 1 atm (101 kPa), what is the total volume percentage of the crude oil vaporized in the flash drum and the fractionating tower?

NOTES

1. Nelson, W.L., Fombona, G.T., and Cordero, L.J., *Proc. Fourth World Pet. Congr.*, Sec. V/A, pp. 13–23, 1955.
2. Kutler, A.A., *Petro/Chem. Eng. 41*(5), pp. 9–11, 1969.
3. Kutler, A.A., *Heat Eng. 44*(8), pp. 135–141, 1970.
4. Lee, C., *Hydrocarbon Process. 69*(5), pp. 69–71, 1990.
5. Nelson, W.L., *Petroleum Refinery Engineering*, 4th ed., McGraw-Hill Book Co., New York, 1958, p. 232.
6. Ibid, p. 228.
7. *Oil Gas J. 67*(40), p. 91, 1969; *69*(20), pp. 124–131, 1971.

5 Coking and Thermal Processes

The "bottom of the barrel" has become more of a problem for refiners because heavier crudes are being processed and the market for heavy residual fuel oils has been decreasing. Historically, the heavy residual fuel oils have been burned to produce electric power and to supply the energy needs of heavy industry, but more severe environmental restrictions have caused many of these users to switch to natural gas. Thus, when more heavy residuals are in the crude, there is more difficulty in economically disposing of them. Coking units convert heavy feedstocks into a solid coke and lower-boiling hydrocarbon products, which are suitable as feedstocks to other refinery units for conversion into higher value transportation fuels.

From a chemical reaction viewpoint, coking can be considered as a severe thermal cracking process in which one of the end products is primarily carbon (i.e., coke). Actually, the coke formed contains some volatile matter or high-boiling hydrocarbons. To eliminate essentially all volatile matter from petroleum coke, it must be calcined at approximately 1800 to 2400°F (980 to 1315°C). Minor amounts of hydrogen remain in the coke even after calcining, which gives rise to the theory held by some authors that the coke is actually a polymer.

Coking was used primarily to pretreat vacuum residuals to prepare coker gas oil streams suitable for feed to a catalytic cracker. This reduced coke formation on the cracker catalyst and thereby allowed increased cracker throughputs. This also reduced the net refinery yield of low-priced residual fuel. Added benefit was obtained by reducing the metals content of the catalytic cracker feedstocks.

In recent years, coking has also been used to prepare hydrocracker feedstocks and to produce a high-quality "needle coke" from stocks such as heavy catalytic gas oils and decanted oils from the fluid catalytic cracking unit [1,2]. Coal tar pitch is also processed in delayed coking units [3].

Delayed coking is described in Section 5.2 to Section 5.7. This is the most widely used coking process. Fluid coking and Flexicoking are described in Section 5.7 to Section 5.10. These fluid bed processes have been under development by ExxonMobil over the past 40 years and are now commercially operated in a number of refineries around the world [2,4].

5.1 TYPES, PROPERTIES, AND USES OF PETROLEUM COKE

There are several types of petroleum coke produced, depending upon the process used, operating conditions, and feedstock properties. All cokes, as produced from the coker, are called "green" cokes and contain some high molecular-weight hydrocarbons and organic molecules (they have some hydrogen in the molecules) left from incomplete carbonization reactions. These incompletely carbonized molecules are referred to as volatile materials in the coke (expressed on a moisture-free basis). Fuel-grade cokes are sold as green coke, but coke used to make anodes for aluminum production or electrodes for steel production must be calcined at temperatures from 1800 to 2400°F (980 to 1315°C) to complete the carbonization reactions and reduce the volatiles to a very low level.

Much of delayed coker coke is produced as hard, porous, irregularly shaped lumps ranging in size from 20 in. (50 cm) down to fine dust. This type of coke is called sponge coke because it looks like a black sponge.

A second form of petroleum coke being produced in increasing quantities is needle coke. Needle coke derives its name from its microscopic elongated crystalline structure. Needle coke is produced from highly aromatic feedstocks (FCC cycle oils, etc.) when a coking unit is operated at high pressures [100 psig (690 kPa)] and high recycle ratios (1:1). Needle coke is preferred over sponge coke for use in electrode manufacture because of its lower electrical resistivity and lower coefficient of thermal expansion.

Occasionally, a third type of coke is produced unintentionally. This coke is called shot coke because of the clusters of shot-sized pellets that characterize it. Its production usually occurs during operational upsets or when processing very heavy residuals such as those from some Canadian, Californian, and Venezuelan crudes. These shot clusters can grow large enough to plug the coke drum outlet (> 12 in. or 30 cm). It is also produced from some high-sulfur residuals [6]. Shot coke is undesirable because it does not have the high surface area of sponge coke nor the useful properties, characteristic of needle coke, for electrode manufacture.

The main uses of petroleum coke are as follows:

1. Fuel
2. Manufacture of anodes for electrolytic cell reduction of alumina
3. Direct use as a chemical carbon source for manufacture of elemental phosphorus, calcium carbide, and silicon carbide
4. Manufacture of electrodes for use in electric furnace production of steel, elemental phosphorus, titanium dioxide, calcium carbide, and silicon carbide
5. Manufacture of graphite

Petroleum coke characteristics and end uses by source and type are given in Table 5.1 and Table 5.2.

It is important to note that petroleum coke does not have sufficient strength to be used in blast furnaces for the production of pig iron nor is it generally acceptable for use as foundry coke. Coal-derived coke is used for these purposes. Typical analyses of petroleum cokes and specifications for anode and electrode grades are summarized in Table 5.3.

The sulfur content of petroleum coke varies with the sulfur content of the coker feedstock. It is usually in the range of 0.3 to 1.5 wt%. It can sometimes, however, be as high as 8%. The sulfur content is not significantly reduced by calcining.

TABLE 5.1
Petroleum Coke Characteristics

Process	Coke type	Characteristics
Delayed	Sponge	Spongelike appearance
		Higher surface area
		Lower contaminates level
		Higher volatile content
		Higher HGI[a] (~100 [22])
		Typical size of 0–6 in. (0–15 cm)
	Shot	Spherical appearance
		Lower surface area
		Lower volatiles
		Lower HGI[a] (< 50)
		Tends to agglomerate
	Needle	Needlelike appearance
		Low volatiles
		High carbon content
Fluid	Fluid	Low volatiles
		Higher contaminates level
		Low HGI[a] (< 40)
		Black sand-like particles
Flexicoker	Flexicoke	Highest metals level
		80% < 200 mesh

[a] Hardgrove grindability index.

Source: Note 5.

5.2 PROCESS DESCRIPTION—DELAYED COKING

This discussion relates to conventional delayed coking as shown in the flow diagram in Figure 5.1 (see also Photo 3, Appendix E).

The delayed coking process was developed to minimize refinery yields of residual fuel oil by severe thermal cracking of stocks such as vacuum residuals, aromatic gas oils, and thermal tars. In early refineries, severe thermal cracking of such stocks resulted in unwanted deposition of coke in the heaters. By gradual evolution of the art, it was found that heaters could be designed to raise residual stock temperatures above the coking point without significant coke formation in the heaters. This required high velocities (minimum retention time) in the heaters. Providing an insulated surge drum on the heater effluent allowed sufficient time for the coking to take place before subsequent processing, hence the term *delayed coking*.

Typically, furnace outlet temperatures range from 900 to 930°F (482 to 500°C). The higher the outlet temperature, the greater the tendency of producing shot coke and the shorter the time before the furnace tubes have to be decoked. Usually, furnace tubes have to be decoked every 3 to 5 months.

TABLE 5.2
Typical Coke End Uses

Application	Coke type	State	End use
Carbon source	Needle	Calcined	Electrodes
			Synthetic graphite
	Sponge	Calcined	Aluminum anodes
			TiO_2 pigments
			Carbon raiser
	Sponge	Green	Silicon carbide
			Foundries
			Coke ovens
Fuel use	Sponge	Green lump	Space heating in Europe/Japan
	Sponge	Green	Industrial boilers
	Shot	Green	Utilities
	Fluid	Green	Cogeneration
	Flexicoke	Green	Lime
			Cement

Source: Note 5.

TABLE 5.3
Typical Coke Specifications

	Sponge anodes	Needle electrodes
Calcined coke		
Moisture, wt%	< 0.5	< 0.5
Volatile matter, wt%	0.5	
Sulfur, wt%	< 3.0	< 1.5
Metals, ppm		
V	< 350	
Ni	< 300	
Si	< 150	
Fe	< 270	
Density, g/cc		
–200 Mesh RD	2.04–2.08	> 2.12
VBD	> 0.80	
CTE, $1/°C \times 10^{-7}$	< 40	< 4.0
	As produced (wt%)	**After calcining (wt%)**
Water	2–4	—
Volatile matter	7–10	2–3
Fixed carbon	85–91	95+
Ash	0.5–1.0	1–2

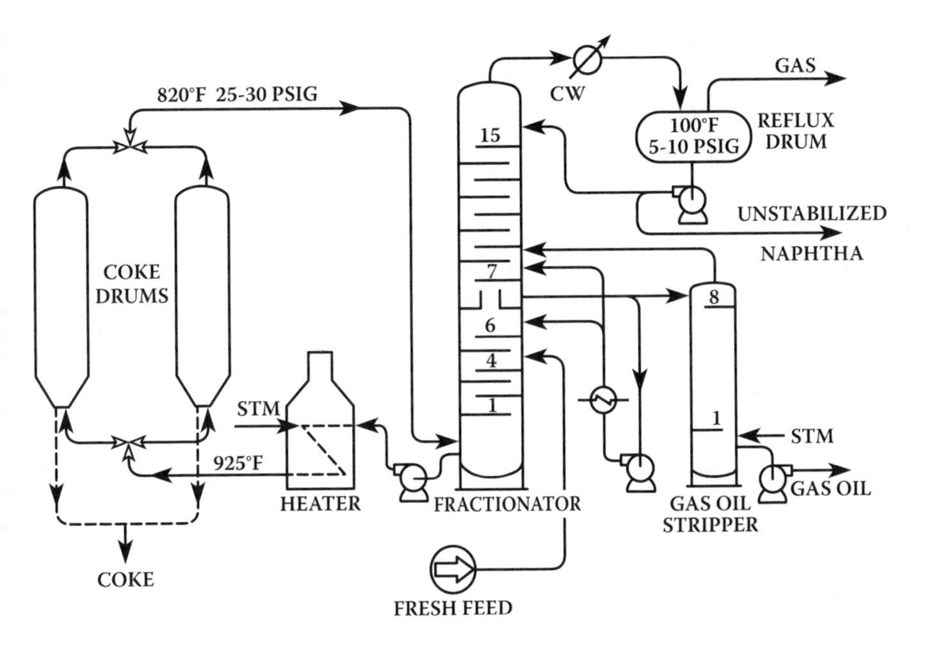

FIGURE 5.1 Delayed coking unit.

Hot fresh liquid feed is charged either directly to the furnace or to the fractionator two to four trays above the bottom vapor zone. If the feed is charged directly to the furnace, it is frequently referred to as a zero recycle operation even though a small amount of recycle is added to the fresh feed.

Charging to the fractionating column accomplishes the following:

1. The hot vapors from the coke drum are quenched by the cooler feed liquid, thus preventing any significant amount of coke formation in the fractionator and simultaneously condensing a portion of the heavy ends, which are recycled.
2. Any remaining material lighter than the desired coke drum feed is stripped (vaporized) from the fresh liquid feed.
3. The fresh feed liquid is further preheated, making the process more energy efficient.

Vapors from the top of the coke drum return to the base of the fractionator. These vapors consist of steam and the products of the thermal cracking reaction: gas, naphtha, and gas oils. The vapors flow up through the quench trays previously described. Above the fresh feed entry in the fractionator, there are usually two or three additional trays below the gas oil drawoff tray. These trays are refluxed with partially cooled gas oil in order to provide fine trim control of the gas oil end point and to minimize entrainment of any fresh feed liquid or recycle liquid into the gas oil product.

The gas oil side draw is a conventional configuration employing a six- to eight-tray stripper with steam introduced under the bottom tray for vaporization of light ends to control the initial boiling point (IBP) of the gas oil.

Steam and vaporized light ends are returned from the top of the gas oil stripper to the fractionator one or two trays above the draw tray. A pump-around reflux system is provided at the draw tray to recover heat at a high temperature level and minimize the low temperature-level heat removed by the overhead condenser. This low temperature-level heat cannot normally be recovered by heat exchange and is rejected to the atmosphere through a water cooling tower or aerial coolers.

Eight to ten trays are generally used between the gas oil draw and the naphtha draw or column top. If a naphtha side draw is employed, additional trays are required above the naphtha draw tray.

Major design criteria for coking units are described in the literature [7,8].

5.2.1 Coke Removal—Delayed Coking

When the coke drum in service is filled to a safe margin from the top, the heater effluent is switched to the empty coke drum and the full drum is isolated, steamed to remove hydrocarbon vapors, cooled by filling with water, and drained; the top and bottom headers are removed; and the coke is cut from the drum.

The ecoking operation is accomplished in some plants by a mechanical drill or reamer; [9] however, most plants use a hydraulic system. The hydraulic system is simply a number of high-pressure [2,000 to 4,500 psig (13,800 to 31,000 kPa)] water jets that are lowered into the coke bed on a rotating drill stem. A small hole [18 to 24 in. (45 to 60 cm) in diameter] called a *rat hole* is first cut all the way through the bed from top to bottom using a special jet. This is done to allow the main drill stem to enter and permit movement of coke and water through the bed.

The main bulk of coke is then cut from the drum, usually beginning at the bottom. Some operators prefer to begin at the top of the drum to avoid the chance of dropping large pieces of coke, which can trap the drill stem or cause problems in subsequent coke-handling facilities. Today, some operators use a technique referred to as "chipping" the coke out of the drum. In this technique, the cutting bit is repeatedly transferred back and forth from top to bottom as the hydraulic bit rotates, and the coke is cut from the center to the wall. This reduces cutting time, produces fewer fines, and eliminates the problem of the bit being trapped.

The coke that falls from the drum can be collected directly in railroad cars. Alternatively, it, along with the cutting water, is dumped from the bottom of the coke drum onto a concrete pad or into a sump, where the water is drained off and recycled to be used for the cutting operation. The coke is transferred by conveyor belt to storage or to a calciner. Sometimes, it is sluiced or pumped as a water slurry to a stockpile or conveyed by belt.

5.3 OPERATION—DELAYED COKING

As indicated in the paragraph describing coke removal, the coke drums are filled and emptied on a time cycle. The fractionation facilities are operated continuously. Usually, just two coke drums are provided but units having four drums are not uncommon. The following time schedule is the maximum used:

Coker operators typically will increase capacity by operating with shorter cycle times. Usual design factors will allow a 20% increase in capacity by shortening coking

Operation	Hours
Fill drum with coke	24
Switch and steam out	3
Cool	3
Drain	2
Unhead and decoke	5
Head up and test	2
Heat up	7
Spare time	2
Total	48

cycles from 24 to 20 hr [10], and moderate debottlenecking projects will allow coking cycles as low as 9 to 12 hr. Shorter cycle times usually mean a lower yield of liquid products because of higher drum and fractionating tower pressures, which may be needed to prevent high vapor velocities and fractionator and compressor overloading. Shorter cycle times can result in a shorter drum life, because of additional drum stresses due to more rapid temperature cycles. In one case, shortening the coking cycle from 21 hr to 18 hr reduced the remaining drum life by about 25% [10].

The main independent operating variables in delayed coking (Table 5.4) are the heater outlet temperature, the fractionator pressure, the temperature of the vapors rising to the gas oil drawoff tray, and the "free" carbon content of the feed as determined by the Conradson or Ramsbottom carbon tests. As would be expected, high heater outlet temperatures increase the cracking and coking reactions, thus increasing yields of gas, naphtha, and coke and decreasing the yield of gas oil. An increase in fractionator pressure has the same effect as an increase in the heater outlet temperature. This is due to the fact that more recycle is condensed in the fractionator and returned to the heater and coke drums. The temperature of the vapors rising to the gas oil drawoff tray is controlled to produce the desired gas oil end point. If this temperature is increased, more heavies will be drawn off in the gas oil, leaving less material to be recycled to the furnace.

The naphtha or gasoline fraction may be split into light and heavy cuts. After hydrotreating for sulfur removal and olefin saturation, the light cut is either isomerized to improve octane or blended directly into finished gasoline. The heavy cut is hydrotreated and reformed. A typical split of coker naphtha is as follows:

Light naphtha = 35.1 vol%, 65°API
Heavy naphtha = 64.9 vol%, 50°API

The gas oil fraction is usually split into a light and a heavy cut before further processing. The light fraction may be used as hydrocracker feed or hydrotreated and subsequently fed to an FCC. The heavy fraction may be used as heavy fuel oil or as FCC feed after hydrotreating, or sent to the vacuum distillation unit. An approximate split of the coker gas oil can be estimated from the following:

Light gas oil (LCGO) = 67.3 vol%, 30°API
Heavy gas oil (HCGO) = 32.7 vol%, ~13°API

TABLE 5.4
Relation of Operating Variables in Delayed Coking

	Independent variables							
	Heater outlet temp.		Fractionator pressure		Hat temp.[a]		Feed carbon residue[b]	
	+	–	+	–	+	–	+	–
Gas yield	+	–	+	–	–	+	–	+
Naphtha yield	+	–	+	–	–	+	+	–
Coke yield	+	–	+	–	–	+	+	–
Gas oil yield	–	+	–	+	–	+	+	–
Gas oil EP	c	c	–	+	+	–	c	c
Gas oil metals content	c	c	–	+	+	–	c	c
Coke metals content	c	c	+	–	–	+	c	c
Recycle quantity	c	c	+	–	–	+	c	c

[a] Hat temperature is the temperature of the vapors rising to the gas oil drawoff tray in the fractionator.
[b] Carbon residue is that determined by the Conradson residue test procedure (ASTM).
[c] For these items, the heater outlet temperature and the carbon residue, per se, do not have a significant independent effect.

[a] Use actual Conradson carbon when available.
[b] All °API are those for net fresh feed to coker.
Note: These yield correlations are based on the following conditions:

1. Coke drum pressure 35 to 45 psig.
2. Feed is "straight-run" residual.
3. Gas oil end point 875 to 925°F.
4. Gasoline end point 400°F.

These yield data have been developed from correlations of actual plant operating data and pilot plant data. Values calculated from these equations are sufficiently accurate for primary economic evaluation studies (see Table 5.5 for calculating yields when the Conradson carbon value is known); however, for the actual design of a specific coking unit, the yields should be determined by the pilot plant operation.

TABLE 5.6
Coke Yields, East Texas Crude Residuals

Coke wt%	=	45.76 – 1.78 × °API
Gas (C_4^-) wt%	=	11.92 – 0.16 × °API
Naphtha wt%	=	20.5 – 0.36 × °API
Gas oil wt%	=	21.82 + 2.30 × °API
Naphtha vol%	=	(131.5 + °API) 186.5 (gaso. wt%)
Gas oil vol%	=	(131.5 + °API) 155.5 (gas oil wt%)

TABLE 5.7
Coke Yields, Wilmington Crude Residuals

Coke wt%	=	39.68 – 1.60 × °API
Gas (C_4^-) wt%	=	11.27 – 0.14 × °API
Naphtha wt%	=	20.5 – 0.36 × °API
Gas oil wt%	=	28.55 + 2.10 × °API
Naphtha vol%	=	[186.5/(131.5 + °API)] (gaso. wt%)
Gas oil vol%	=	[155.5/(131.5 + °API)] (gas oil wt%)

In all cases, the weight and volume percentages given are based on the net fresh feed to the coking unit and are limited to feedstocks having gravities of less than 18°API. The yields shown will vary significantly if the coker feed is derived from material other than straight-run crude residuals. The numerical values in the equations do not represent a high degree of accuracy but are included for the purpose of establishing a complete weight balance.

Typical gas compositions and sulfur and nitrogen distributions in products produced by delayed coking of reduced crudes are given in Table 5.6 and Table 5.7. The 2005 installed costs for delayed cokers in the Gulf Coast section of the United States are given by Figure 5.2. Table 5.8 and Table 5.9 give data for typical gas composition and sulfur and nitrogen distribution for delayed coking, respectively. Table 5.10 gives utility requirements for delayed coker operation.

5.4 PROCESS DESCRIPTION—FLEXICOKING

The Flexicoking process is shown in Figure 5.3 [11].

Feed can be any heavy oil such as vacuum resid, coal tar, shale oil, or oil sand bitumen. The feed is preheated to about 600 to 700°F (315 to 370°C) and sprayed into the reactor, where it contacts a hot fluidized bed of coke. This hot coke is recycled to the reactor from the coke heater at a rate that is sufficient to maintain the rector fluid bed temperature between 950 and 1000°F (510 and 540°C). The coke recycle from the coke heater thus provides sensible heat and heat of vaporization for the feed and the endothermic heat for the cracking reactions.

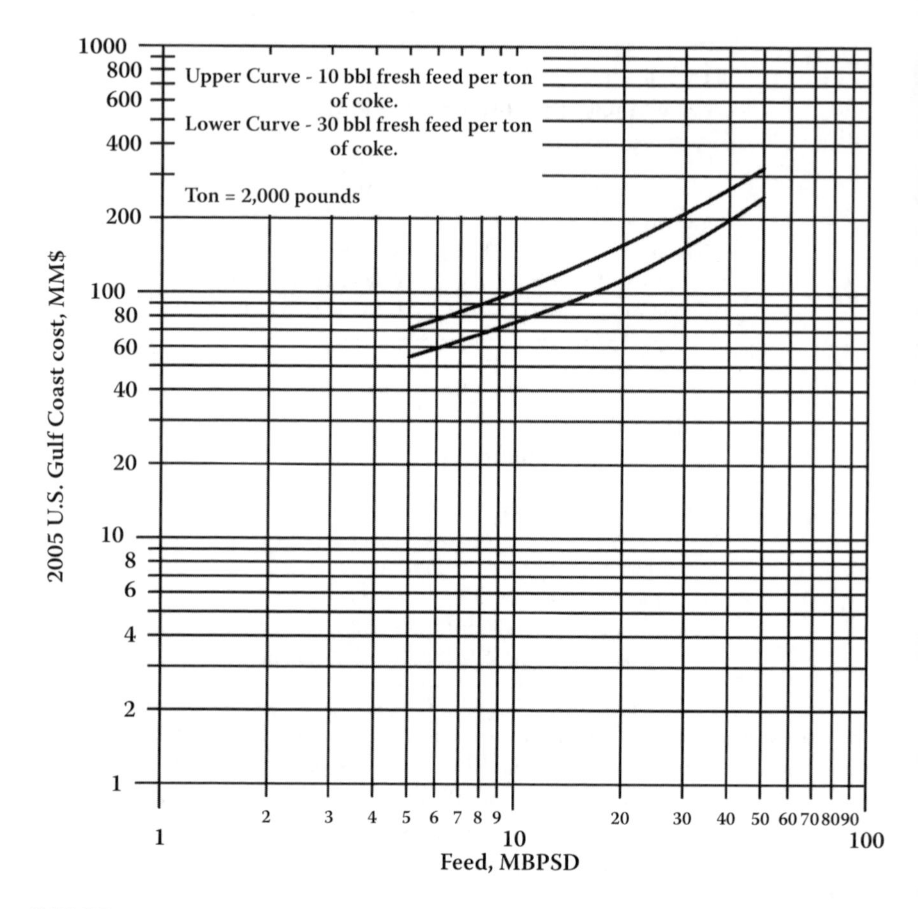

FIGURE 5.2 Delayed coking units investment cost: 2005 U.S. Gulf Coast (see Table 5.10).

The cracked vapor products pass through cyclone separators in the top of the reactor to separate most of the entrained coke particles (cyclone separators are efficient down to particles sized about 7 microns, but the efficiency falls off rapidly as the particles become smaller) and are then quenched in the scrubber vessel located at the top of the reactor. Some of the high-boiling [925+ °F (495+ °C)] cracked vapors are condensed in the scrubber and recycled to the reactor. The balance of the cracked vapors flow to the coker fractionator, where the various cuts are separated. Wash oil circulated over baffles in the scrubber provides quench cooling and also serves to reduce further the amount of entrained fine coke particles.

The coke produced by cracking is deposited as thin films on the surface of the existing coke particles in the reactor fluidized bed. The coke is stripped with steam in a baffled section at the bottom of the reactor to prevent reaction products, other than coke, from being entrained with the coke leaving the reactor. Coke flows from the reactor to the heater, where it is reheated to about 1100°F (593°C). The coke heater is also a fluidized bed, and its primary function is to transfer heat from the gasifier to the reactor.

TABLE 5.8
Typical Gas Composition for Delayed Coking (Sulfur-Free Basis)

Component	Mol%
Methane	51.4
Ethylene	1.5
Ethane	15.9
Propylene	3.1
Propane	8.2
Butylene	2.4
i-Butane	1.0
n-Butane	2.6
H_2	13.7
CO_2	0.2
Total	100.0

Note: MW = 22.12

TABLE 5.9
Sulfur and Nitrogen Distribution for Delayed Coking (Basis: Sulfur and Nitrogen in Feed to Coker)

	Sulfur (%)	Nitrogen (%)
Gas	30	—
Light naphtha	1.7	
Heavy naphtha	3.3	1
LCGO	15.4	2
HCGO	19.6	22
Coke	30	75
Total	100	100

Coke flows from the coke heater to a third fluidized bed in the gasifier, where it is reacted with air and steam to produce a fuel gas product consisting of CO, H_2, CO_2, and N_2. Sulfur in the coke is converted primarily to hydrogen sulfide (H_2S), plus a small amount of COS, and nitrogen in the coke is converted to NH_3 and N_2. This gas flows from the top of the gasifier to the bottom of the heater, where it serves to fluidize the heater bed and provide the heat needed in the reactor. The reactor heat requirement is supplied by recirculating hot coke from the gasifier to the heater.

The system can be designed and operated to gasify about 60 to 97% of the coke product in the reactor. The overall coke inventory in the system is maintained by withdrawing a stream of purge coke from the heater.

The coke gas leaving the heater is cooled in a waste-heat steam generator before passing through external cyclones and a venturi-type wet scrubber. The coke fines

TABLE 5.10
Delayed Coking Unit Cost Data

Costs included

1. Coker fractionator to produce naphtha, light gas oil, and heavy gas oil
2. Hydraulic decoking equipment
3. Coke dewatering, crushing to < 2 in. (< 5 cm), and separation of material < π in. (> 0.6 cm) from that > π in. (> 0.6 cm)
4. Three days covered storage for coke
5. Coke drums designed for 50 to 60 psig (345 to 415 kPa)
6. Blowdown condensation and purification of wastewater
7. Sufficient heat exchange to cool products to ambient temperatures

Costs not included

1. Light ends recovery facilities
2. Light ends sulfur removal
3. Product sweetening
4. Cooling water, steam, and power supply
5. Off gas compression

Utility data	
Steam, lb/t coke (kg/t)	700 (318)
Power, kWh/t coke[a] (MJ/t)	30 (108)
Cooling water, gal/bbl feed [30°F ΔT, 0°C ΔT], (m^3/m^3 feed)	70 (1.7)
Fuel, MMBtu/bbl feed[b] (MJ/m^3)	0.14 (23.5)

[a] Includes electric motor drive for hydraulic decoking pump.
[b] Based on 600°F (316°C) fresh feed. LHV basis, heater efficiency taken into account.
Note: See Figure 5.2.

collected in the venturi scrubber plus the purge coke from the heater represent the net coke yield and contain essentially all of the metal and ash components of the reactor feedstock.

After removal of entrained coke fines, the coke gas is treated for removal of hydrogen sulfide in a Stretford unit and then used for refinery fuel. The treated fuel gas has a much lower heating value than natural gas [100 to 130 Btu/scf (3.0 to 3.9 kJ/Nm^3) vs. 900 to 1000 Btu/scf (27.1 to 30.1 kJ/Nm^3)], and therefore modifications of boilers and furnaces may be required for efficient combustion of this gas.

5.5 PROCESS DESCRIPTION—FLUID COKING

Fluid coking is a simplified version of Flexicoking. In the fluid coking process, only enough of the coke is burned to satisfy the heat requirements of the reactor and the feed preheat. Typically, this is about 20 to 25% of the coke produced in the reactor. The balance of the coke is withdrawn from the burner vessel and is not gasified as it is in a Flexicoker. Therefore, only two fluid beds are used in a fluid coker—a reactor and a burner, which replaces the heater.

The primary advantage of the Flexicoker (Figure 5.3) over the more simple fluid coker is that most of the heating value of the coke product is made available

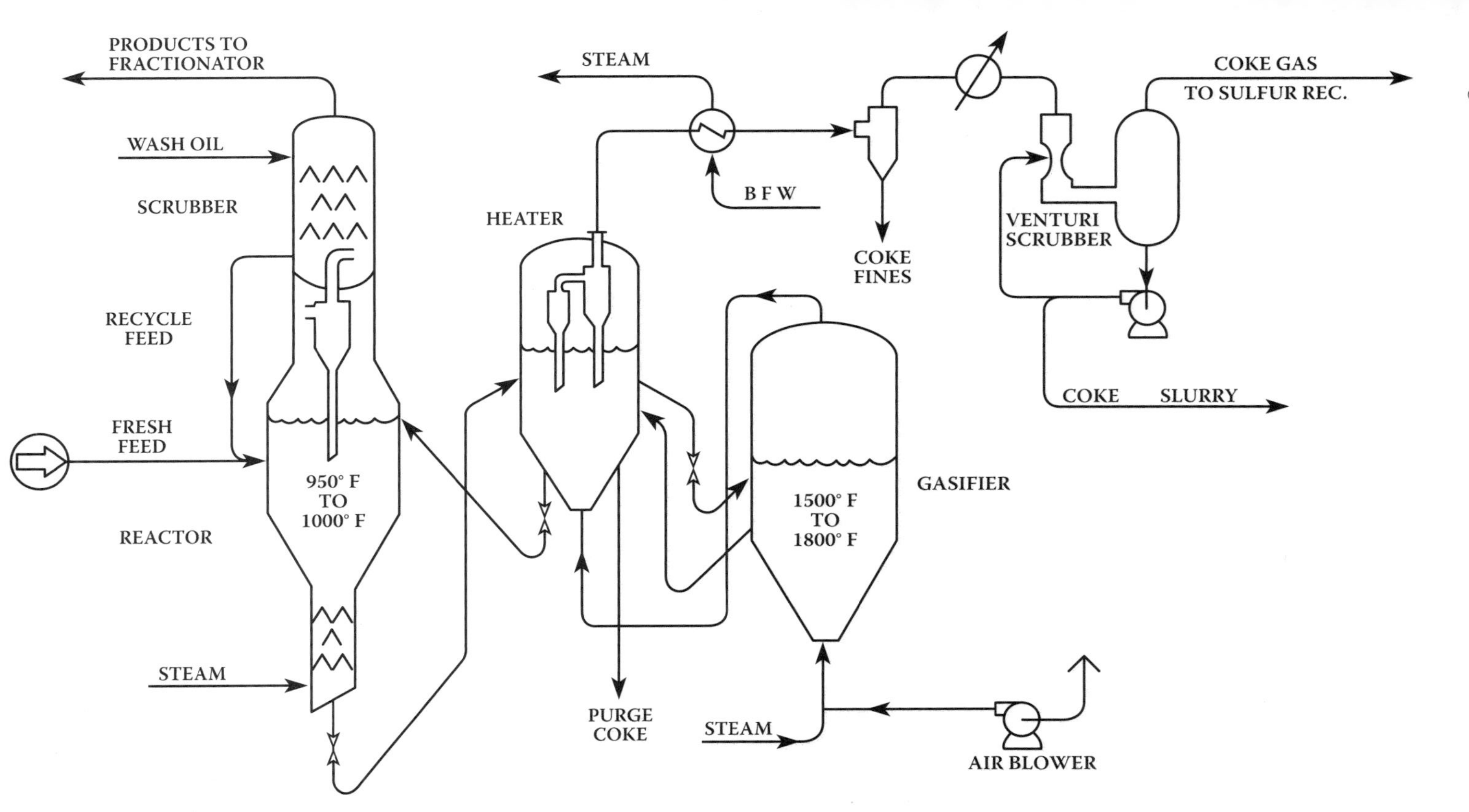

FIGURE 5.3 Simplified flow diagram for a Flexicoker.

TABLE 5.11
Comparison of Coking Yields

Feed			
Arabian medium 1050°F + Vacuum resid			
Gravity, °API	4.9		
Sulfur, wt%	5.4		
Nitrogen, wt%	0.26		
Conradson carbon, wt%	23.3		
Nickel, wt ppm	32		
Vanadium, wt ppm	86		
Iron, wt ppm	30		
Yields			
	Delayed coking	**Fluid coking**	**Flexicoking**
Recycle cut point, °F VT (°C VT)	900 (482)	975 (524)	975 (524)
Yields on Fresh Feed, wt%			
Gas	9.3	11.8	11.8
Light naphtha	2.0	1.9	1.9
Heavy naphtha	8.0	7.8	7.8
Gas oil	46.7	50.4	50.4
Gross coke	34.0	28.1	28.1
Total	100.0	100.0	100.0
Net coke	34.0	22.4	2.3
C_5^{++} Liquid	56.7	60.1	60.1

as low-sulfur fuel gas, which can be burned without an SO_2 removal system on the resulting stack gas, whereas such a system would be necessary if coke that contains 3 to 8 wt% sulfur is burned directly in a boiler. In addition, the coke gas can be used to displace liquid and gaseous hydrocarbon fuels in the refinery process heaters and does not have to be used exclusively in boilers, as is the case with fluid coke.

5.6 YIELDS FROM FLEXICOKING AND FLUID COKING

As with delayed coking, the yields from the fluidized bed coking processes can be accurately predicted only from pilot plant data from specific feeds. Typical yields for many various feeds are available from the licenser. One set of yields is shown in Table 5.11.

If specific yield data are not available, it can be assumed for preliminary estimates that the products from Flexicoking and fluid coking are the same as those from delayed coking, except for the amount of reactor coke product that is burned or gasified. Thus, the coke yield from a fluid coking unit would be about 75 to 80% of the coke yield from a delayed coker, and the yield of coke from a Flexicoker would be in the range of 2 to 40 wt% of the delayed coker yield.

The coke that is gasified in a Flexicoker produces coke gas of the following approximate composition after H_2S removal:

Component	Mol%
H_2	15
CO	20
CH_4	2
CO_2	10
N_2	53
Total	100

This composition is on a dry basis. The coke gas as actually produced from the H_2S removal step is water saturated, which means it contains typically 5 to 6 mol% water vapor.

Assuming that the coke is about 98 wt% carbon on a sulfur-free and ash-free basis, the calculated amount of coke gas produced having the above composition is 194 Mscf (5200 Nm^3) per ton of coke gasified. Combustion of this gas results in production of about 80% of the equivalent coke heating value. Recovery of sensible heat from the coke gas leaving the coke heater increases the recoverable heat to about 85% of the equivalent coke heating value.

5.7 CAPITAL COSTS AND UTILITIES FOR FLEXICOKING AND FLUID COKING

As a rough approximation, it can be assumed that the investment for a fluid coking unit is about the same as that for a delayed coking unit for a given feedstock and that a Flexicoker costs about 30 to 50% more. Operating costs for a Flexicoker are about 25 to 30% greater per barrel of feed than for a delayed coker.

The utility requirements for fluid coking are significantly higher than those for delayed coking primarily because of the energy required to circulate the solids between fluid beds. The air blower in a Flexicoker requires more power than that for a fluid coker. The process licensor should be consulted to determine reasonably accurate utility requirements.

5.8 VISBREAKING

Visbreaking is a relatively mild thermal cracking operation mainly used to reduce the viscosities and pour points of vacuum tower bottoms to meet No. 6 fuel oil specifications or to reduce the amount of cutting stock required to dilute the resid to meet these specifications. Refinery production of heavy fuel oils can be reduced from 20 to 35% and cutter stock requirements from 20 to 30% by visbreaking. The gas oil fraction produced by visbreaking is also used to increase cat cracker feedstocks and increase gasoline yields.

Long paraffinic side chains attached to aromatic rings are the primary cause of high pour points and viscosities for paraffinic base residua. Visbreaking is carried out at conditions to optimize the breaking off of these long side chains and their subsequent cracking to shorter molecules with lower viscosities and pour points.

The amount of cracking is limited, however, because if the operation is too severe, the resulting product becomes unstable and forms polymerization products during storage that cause filter plugging and sludge formation. The objective is to reduce the viscosity as much as possible without significantly affecting the fuel stability. For most feedstocks, this reduces the severity to the production of less than 10% gasoline and lighter materials.

The degree of viscosity and pour point reduction is a function of the composition of the residua feed to the visbreaker. Waxy feedstocks achieve pour point reductions from 15 to 35°F (8 to 20°C) and final viscosities from 25 to 75% of the feed. High asphaltene content in the feed reduces the conversion ratio at which a stable fuel can be made [12], which results in smaller changes in the properties. The properties of the cutter stocks used to blend with the visbreaker tars also have an effect on the severity of the visbreaker operation. Aromatic cutter stocks, such as catalytic gas oils, have a favorable effect on fuel stability and permit higher visbreaker conversion levels before reaching fuel stability limitations [4].

The molecular structures of the compounds in petroleum that have boiling points above 1000°F (538°C) are highly complex and historically have been classified arbitrarily as oils, resins, and asphaltenes according to solubility in light paraffinic hydrocarbons. The oil fraction is soluble in propane; the resin fraction is soluble (and the asphaltene fraction insoluble) in either n-pentane, hexane, n-heptane, or octane, depending upon the investigator. Usually either n-pentane or n-heptane is used. The solvent selected does have an effect on the amounts and properties of the fractions obtained, but normally little distinction is made in terminology. Chapter 7 (Catalytic Hydrocracking) and Chapter 8 (Hydroprocessing) contain a more detailed discussion of the properties of these fractions.

Many investigators believe the asphaltenes are not in solution in the oil and resins, but are very small, perhaps molecular size, solids held in suspension by the resins, and there is a definite critical ratio of resins to asphaltenes below which the asphaltenes will start to precipitate. During the cracking phase, some of the resins are cracked to lighter hydrocarbons and others are converted to asphaltenes. Both reactions affect the resin–asphaltene ratio and the resultant stability of the visbreaker tar product and serve to limit the severity of the operation.

The principal reactions [4] that occur during the visbreaking operation are

1. Cracking of the side chains attached to cycloparaffin and aromatic rings at or close to the ring so the chains are either removed or shortened to methyl groups.
2. Cracking of resins to light hydrocarbons (primarily olefins) and compounds that convert to asphaltenes.
3. At temperatures above 900°F (480°C), some cracking of naphthene rings. There is little cracking of naphthenic rings below 900°F (480°C).

The severity of the visbreaking operation can be expressed in several ways: the yield of material boiling below 330°F (166°C), the reduction in product viscosity, and the amount of standard cutter stock needed to blend the visbreaker tar to No. 6 fuel oil specifications as compared with the amount needed for the feedstock [13].

TABLE 5.12
Visbreaking Time–Temperature Relationship (Equal Conversion Conditions)

Time, min	Temperature °C	Temperature °F
1	485	905
2	470	878
4	455	850
8	440	825

Source: Note 15.

In the United States, usually the severity is expressed as the vol% product gasoline in a specified boiling range, and in Europe as the wt% yield of gas plus gasoline (product boiling below 330°F, or 166°C).

There are two types of visbreaker operations: coil and furnace cracking and soaker cracking. As in all cracking processes, the reactions are time–temperature dependent (see Table 5.12), and there is a trade-off between temperature and reaction time. Coil cracking uses higher furnace outlet temperatures [885 to 930°F (473 to 500°C)] and reaction times from 1 to 3 min, whereas soaker cracking uses lower furnace outlet temperatures [800 to 830°F (427 to 433°C)] and longer reaction times. The product yields and properties are similar, but the soaker operation with its lower furnace outlet temperatures has the advantages of lower energy consumption and longer run times before having to shut down to remove coke from the furnace tubes. Run times of 3 to 6 months are common for coil visbreakers and 6 to 18 months for soaker visbreakers. This apparent advantage for soaker visbreakers is at least partially balanced by the greater difficulty and longer time in cleaning the soaking drum [14].

Process flow diagrams are shown in Figure 5.4 and Figure 5.5. The feed is introduced into the furnace and heated to the desired temperature. In the furnace or coil cracking process, the feed is heated to cracking temperature [885 to 930°F (474 to 500°C)] and quenched as it exits the furnace with gas oil or tower bottoms to stop the cracking reaction. In the soaker cracking operation, the feed leaves the furnace between 800 and 820°F (427 and 438°C) and passes through a soaking drum, which provides the additional reaction time, before it is quenched. Pressure is an important design and operating parameter, with units being designed for pressures as high as 750 psig (5170 kPa) for liquid-phase visbreaking and as low as 100 to 300 psig (690 to 2070 kPa) for 20 to 40% vaporization at the furnace outlet [16]. Typical yields and product properties from visbreaking operations are shown in Table 5.13 and Table 5.14.

For furnace cracking, fuel consumption accounts for about 80% of the operating cost, with a net fuel consumption equivalent of 1 to 1.5 wt% on feed. Fuel requirements for soaker visbreaking are about 30 to 35% lower [15] (see Table 5.15).

Many of the properties of the products of visbreaking vary with conversion and the characteristics of the feedstocks. However, some properties, such as diesel index

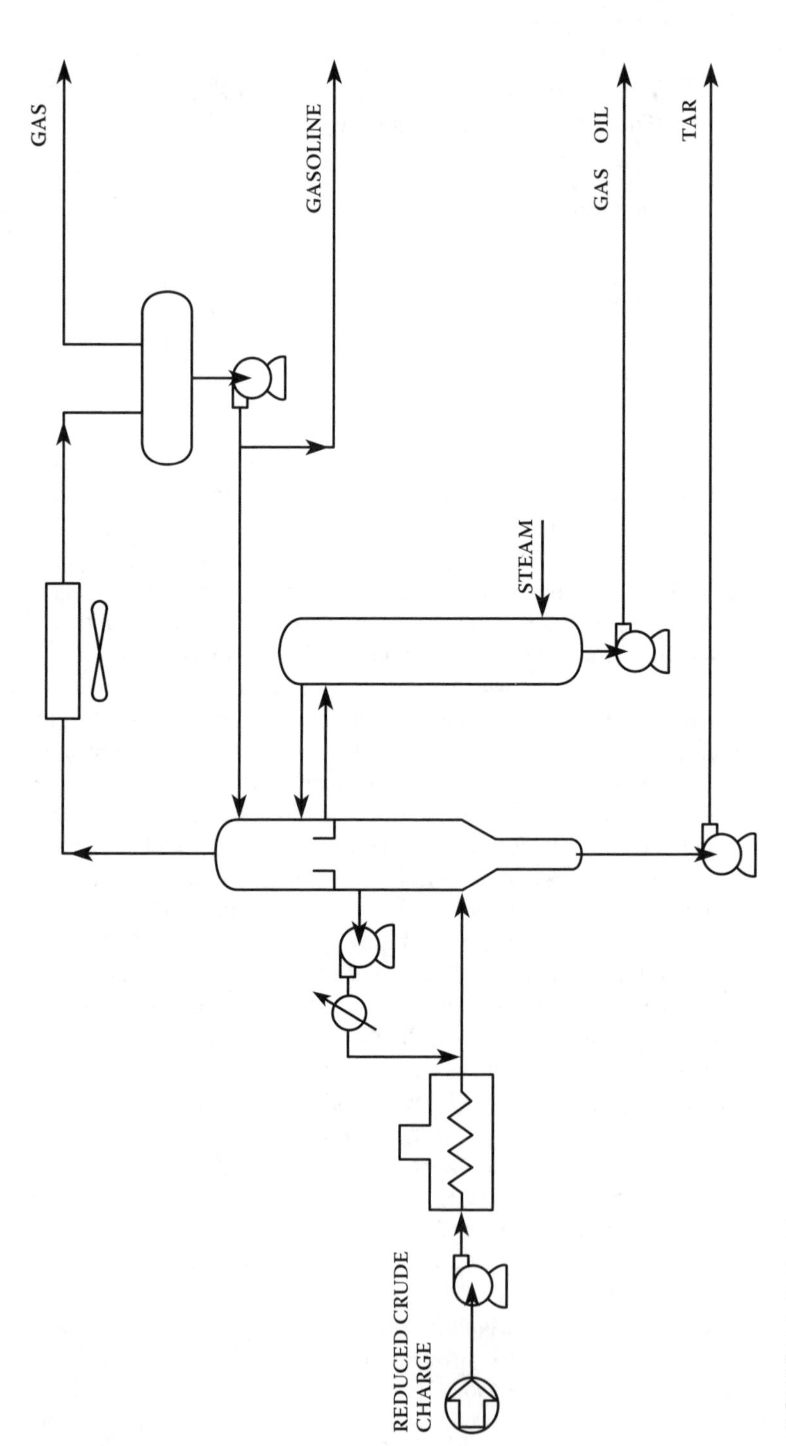

FIGURE 5.4 Coil visbreaker.

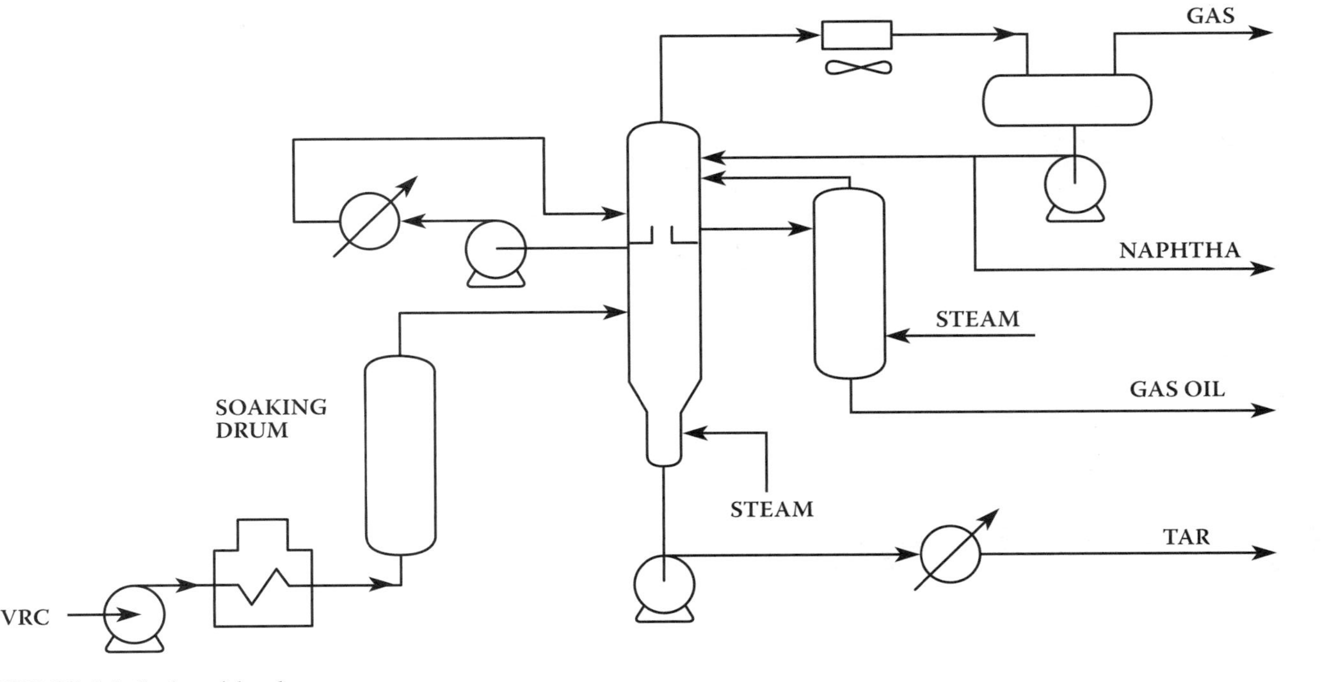

FIGURE 5.5 Soaker visbreaker.

TABLE 5.13
Visbreaking Results, Kuwait Long Resid

	Feed	Product
Yields, wt%		
Butane and lighter		2.5
C_5 – 330°F naphtha		5.9
Gas oil, 660°F EP		13.5
Tar		78.1
Product properties		
Naphtha		
°API		65.0
Sulfur, wt%		1.0
RONC		
Gas oil		
°API		32.0
Sulfur, wt%		2.5
Tar or feed		
% on crude		
°API	14.4	11.0
Sulfur, wt%	4.1	4.3
Viscosity, cSt, 50°C	720	250

Source: Reference 4.

and octane number, are more closely related to feed qualities; others, such as density and viscosity of the gas oil, are relatively independent of both conversion and feedstock characteristics [3].

5.9 CASE-STUDY PROBLEM: DELAYED COKER

See Section 4.6 for a statement of the problem and Table 4.7 for feed to delayed coker. The delayed coker material balance is calculated from the equations given in Table 5.3. The results are tabulated in Table 5.16.

Although at this time, the only feed available for the delayed coker is the vacuum tower bottoms stream, other process units in the refinery produce heavy product streams that can be either blended into heavy fuel oil or sent to the delayed coker. The market for heavy fuel oil is limited, and for this example problem, no heavy fuel oil is produced. The heavy products from the fluid catalytic cracking unit (HGO) and the alkylation unit (tar) are sent to the delayed coker for processing. These streams are included in the feed to coker in this problem.

Calculations of yields:

Coke = 1.6 (14.2) = 22.72 wt%
Gas (C_4^-) = 7.8 + 0.144 (14.2) = 9.84 wt%
Naphtha = 11.29 + 0.343 (14.2) = 16.16 wt%

TABLE 5.14
Visbreaking Results, Agha-Jari Short Resid

	Feed	Product
Yields, wt%		
Butane and lighter		2.4
C_5 – 330°F naphtha		4.6
Gas oil, 660°F EP		14.5
Tar		78.5
Product properties		
Naphtha		
°API		
Sulfur, wt%		
RONC		
Gas oil		
°API		32.2
Sulfur, wt%		
Tar or feed		
% on crude		
°API	8.2	5.5
Sulfur, wt%	—	
Viscosity, cSt, 122°F (50°C)	100,000	45,000

Source: Reference 3.

TABLE 5.15
Coil and Soaker Visbreaking

	Coil	Soaker
Furnace outlet temperature, °F (°C)	900 (480)	805 (430)
Fuel consumption, relative	1.0	0.85
Capital cost, relative	1.0	0.90

Naphtha = [186.5 / (131.5 + 7.4)] (16.16) = 21.7 vol%
Gas oil = 100.0 – (22.7 + 9.8 + 16.2) = 5.13 wt%
Gas oil = [155.5 / 138.9)] (51.3) = 57.4 vol%

Sulfur distribution is obtained from Table 5.9 and gas (C_4^-) composition from Table 5.8.

Coker gas balance:

Total gas = 28,855 lb/hr
Sulfur in gas = 1,758 lb/hr
Sulfur-free gas = 27,097 lb/hr

TABLE 5.16
Delayed Coker Material Balance: 100,000 BPCD Alaska North Slope Crude Oil Basis

Component	Vol%	BPD	°API	(lb/hr)/ BPD	lb/hr	wt% S	lb/hr S
Feed							
1050+ °F VRC	88.4	18,003	8.7	14.72	265,162	2.17	5,757
FCC HCO	11.5	2,338	–0.6	15.75	36,820	0.32	117
Alky tar	0.1	25			468		
Total	100.0	20,366	7.4	14.85	302,450	1.94	5,875
Products							
Gas (C_4^-), wt%	(22.7)				29,776	5.92	1,762
Light naphtha	7.6	1,551	65.5	10.51	16,303	0.36	59
Heavy naphtha	14.1	2,868	50.1	11.36	32,575	0.72	235
LCGO	37.3	7,587	30.0	12.78	96,967	0.93	905
HCGO	20.1	4,103	14.3	14.16	58,114	1.98	1,151
Coke, wt%	(22.7)				68,717	2.56	1,762
Total		16,110			302,450		5,875

Note: Conradson carbon, 1050+ °F RC = 14.2% (from Table 4.6)

TABLE 5.17
Coker Utility Requirement (Basis: 825 Tons Coke per Day)

Steam, Mlb/day	577
Power, MkWh/day	25
Cooling water, gpm	990
Fuel, MMBtu/day	2851

Using the composition of sulfur-free gas from Table 5.6, the total lb mol/hr is 27,097 / 22.12 = 1,225, and the sulfur-free composition is calculated as follows:

Component	Mol%	Mol/hr
C_1	51.4	650.9
$C_2^=$	1.5	19.0
C_2	15.9	201.4
$C_3^=$	3.1	39.3
C_3	8.2	103.8
$C_4^=$	2.4	30.4
i-C_4	1.0	12.7
n-C_4	2.6	32.9
H_2	13.7	173.5
CO_2	0.2	2.5
	100.0	1226.4

It is now necessary to adjust this gas composition to allow for the sulfur content. In actual operations, some of the sulfur will be combined as mercaptan molecules (R-S-H), but for preliminary calculations, it is sufficiently accurate to assume that all the sulfur in the gas fraction is combined as H_2S. Because there are 1758 lb/hr of sulfur (equivalent to 54.8 mol/hr), the free hydrogen must be reduced by 54.8 mol/hr, so the final coker gas balance is as follows:

Component	Mol/hr	MW	lb/hr	(lb/hr)BPD	BPD
C_1	650.9	16	10,415		
$C_2^=$	19.0	28	532		
C_2	201.4	30	6,041		
$C_3^=$	39.3	42	1,649	7.61	217
C_3	103.8	44	4,569	7.42	616
$C_4^=$	30.4	56	1,702	8.76	194
iC_4	12.7	58	735	8.22	89
nC_4	32.9	58	1,910	8.51	224
H_2	118.5	2	237		
CO_2	2.5	44	111		
H_2S	55.1	34	1,873		
Total	1266.4		29,773		1341

PROBLEMS

1. For the crude oil in Figure 3.5 and Figure 3.9, estimate the Conradson carbon of the 1050°F (566°C) residual crude oil fraction.
2. Estimate the coke yields for the crude oil fraction of problem 1 and make a material balance around the delayed coking unit. The 1050°F (566°C) fraction contains 0.38% sulfur by weight.
3. Using the information from problem 2, estimate the capital cost of a 7500 BPSD delayed coking unit and its utility requirements.
4. Using U.S. Bureau of Mines distillation data from Appendix C, calculate the coke yield and make a material balance from a Torrence Field, CA, crude oil residuum having an API gravity of 23.8° and a sulfur content of 1.84 wt%.
5. Estimate the capital and operating costs for a 10,000 BPSD delayed coker processing the reduced crude of problem 4. Assume four workers per shift at an average of $20.50/hr per worker.
6. Calculate the long tons per day of coke produced by charging 30,000 barrels of 1050°F (566°C) residuum from the assigned crude oil to a delayed coking unit.
7. If 95% of the hydrogen sulfide present in the coker gas product stream can be converted to elemental sulfur, how many long tons of sulfur will be produced per day when charging 30,000 barrels of 1050°F (566°C) residuum from the assigned crude oil to the delayed coker?
8. Make a material balance around the delayed coker for the change rate and the crude oil of problems 6 and 7. Also estimate the utility requirements.

9. Estimate the capital and operating costs for a 30,000 BPSD delayed coker processing the assigned reduced crude. Assume four workers per shift at an average of $20.50/hr per worker.

NOTES

1. Shea, F.L., U.S. Patent No. 2,775,549.
2. Stormont, D.H., *Oil Gas J. 67*(12), p. 75, 1969.
3. Remirez, R., *Chem. Eng.*, p. 74, February 24, 1969.
4. Rhoe, A. and de Blignieres, C., *Hydrocarbon Process. 58*(1), pp. 131–136, 1979.
5. Dymond, R.E., *Hydrocarbon Process. 70*(9), pp. 162C–162J, 1991.
6. Mochida, I., Furuno, T., Korai, Y., and Fujitsu, H., *Oil Gas J. 84*(6), pp. 51–56, 1986.
7. Mekler, V., *Refining Engineer*, p. C–7, September 1957.
8. Rose, K.E., *Hydrocarbon Process. 50*(7), p. 85, 1971.
9. Eppard, J.H., *Petrol. Refiner*, p. 98, July 1953.
10. Elliott, J.D., *Hydrocarbon Process. 71*(1), pp. 75-84, 1992.
11. Allen, D.E., et al., *Chem. Engr. Prog. 77*(12), p. 40, 1981.
12. Notarbartolo, M., Menegazzo, C., and Kuhn, J., *Hydrocarbon Process. 58*(9), pp. 144–118, 1979.
13. Hus, M., *Oil Gas J. 79*(15), pp. 109–120, 1981.
14. Allan, D.E., et al., *Chem. Engr. Prog. 79*(1), pp. 85–89, 1983.
15. Akbar, M. and Geelen, H., *Hydrocarbon Process. 60*(5), pp. 81–85, 1981.
16. Hournac, R., Kuhn, J., and Notarbartolo, M., *Hydrocarbon Process. 58*(12), pp. 97–102, 1979.

ADDITIONAL READING

17. Elliott, J.D., *Hydrocarbon Process. 71*(1), pp. 75–84 (1992).
18. Feintuch, H.M. and Negin, K.M., in *Handbook of Petroleum Refining Processes*, 2nd ed., Meyers, R.A. (Ed.), pp. 12.25–12.82, McGraw-Hill Book Co., New York, 1997.
19. Nagy, A.J. and Antalffy, L.P., *Oil Gas J. 87*(22), pp. 77–80, 1989.
20. Lieberman, N.P., *Oil Gas J. 87*(13), pp. 67–69, 1989.
21. Swain, E.J., *Oil Gas J. 89*(18), pp. 100–102, 1991.
22. Swain, E.J., *Oil Gas J. 89*(20), pp. 49–52, 1991.
23. Weisenborn, W.J., Jr., Janssen, H.R., and Hanke, T.D., *Energy Prog. 6*(4), pp. 222–225, 1986.

6 Catalytic Cracking

Catalytic cracking is the most important and widely used refinery process for converting heavy oils into more valuable gasoline and lighter products, with 10.6 MMBPD (over 1 million tons/day) of oil processed in the world [1]. In the United States today, fluid catalytic cracker (FCC) naphtha provides 35 to 45% of the blending stocks in refinery gasoline blending pools. Originally, cracking was accomplished thermally, but the catalytic process has almost completely replaced thermal cracking because more gasoline, having a higher octane and less heavy fuel oils and light gases, is produced [2]. The light gases produced by catalytic cracking contain more olefins than those produced by thermal cracking (Table 6.1).

The cracking process produces carbon (coke), which remains on the catalyst particle, covers its surface and rapidly lowers its activity. To maintain the catalyst activity at a useful level, it is necessary to regenerate the catalyst by burning off this coke with air. As a result, the catalyst is continuously moved from reactor to regenerator and back to reactor. The cracking reaction is endothermic, and the regeneration reaction exothermic. Some units are designed to use the regeneration heat to supply that needed for the reaction and to heat the feed up to reaction temperature. These are known as "heat balance" units.

Average riser reactor temperatures are in the range 900 to 1050°F (480 to 566°C), with oil feed temperatures from 500 to 800°F (260 to 425°C) and regenerator exit temperatures for the catalyst from 1200 to 1500°F (650 to 815°C).

The catalytic cracking processes in use today can all be classified as either moving-bed or fluidized bed units. There are several modifications under each of the classes depending upon the designer or builder, but within each class, the basic operation is very similar. The Thermafor catalytic cracking (TCC) process is representative of the moving-bed units, and the fluid catalytic cracker is representative of the fluidized bed units. There are very few TCC units in operation today, and the FCC unit has taken over the field. The FCC units can be classified as either bed or riser (transfer line) cracking units, depending upon where the major fraction of the cracking reaction occurs. Today, almost all of the FCC units have been reconfigured to operate with riser cracking.

The process flows of both types of processes are similar. The hot oil feed is contacted with the catalyst in either the feed riser line or the reactor. As the cracking reaction progresses, the catalyst is progressively deactivated by the formation of coke on the surface of the catalyst. The catalyst and hydrocarbon vapors are separated mechanically, and oil remaining on the catalyst is removed by steam stripping before the catalyst enters the regenerator. The oil vapors are taken overhead to a fractionation tower for separation into streams having the desired boiling ranges.

The spent catalyst flows into the regenerator and is reactivated by burning off the coke deposits with air. Regenerator temperatures are carefully controlled to

TABLE 6.1
Thermal versus Catalytic Cracking Yields on Similar Topped Crude Feed

	Thermal cracking		Catalytic cracking	
	wt%	vol%	wt%	vol%
Fresh feed	100.0	100.0	100.0	100.0
Gas	6.6		4.3	
Propane	2.1	3.7	1.3	2.2
Propylene	1.0	1.8	6.8	10.9
Isobutane	.8	1.3	2.6	4.0
n-Butane	1.9	2.9	0.9	1.4
Butylene	1.8	2.6	6.5	10.4
C_5+ gasoline	26.9	32.1	48.9	59.0
Light cycle oil	1.9	1.9	15.7	15.0
Decant oil			8.0	7.0
Residual oil	57.0	50.2		
Coke	0		5.0	
Total	100.0	96.5	100.0	109.9

prevent catalyst deactivation by overheating and to provide the desired amount of carbon burn-off. This is done by controlling the air flow to give a desired CO_2/CO ratio in the exit flue gases or the desired temperature in the regenerator. The flue gas and catalyst are separated by cyclone separators and electrostatic precipitators. The catalyst in some units is steam-stripped as it leaves the regenerator to remove adsorbed oxygen before the catalyst is contacted with the oil feed.

6.1 FLUIDIZED BED CATALYTIC CRACKING

The FCC process employs a catalyst in the form of very fine particles [average particle size about 70 micrometers (microns)], which behave as a fluid when aerated with a vapor. The fluidized catalyst is circulated continuously between the reaction zone and the regeneration zone and acts as a vehicle to transfer heat from the regenerator to the oil feed and reactor. Two basic types of FCC units in use today are the "side-by-side" type, where the reactor (separator) and regenerator are separate vessels adjacent to each other, and the Orthoflow, or stacked type, where the reactor is mounted on top of the regenerator. Typical FCC unit configurations are shown in Figure 6.1 to Figure 6.8 and Photos 6 and 7, Appendix E.

One of the most important process differences in FCC units relates to the location and control of the cracking reaction. Until about 1965, most units were designed with a discrete dense-phase fluidized catalyst bed in the reactor vessel. The units were operated so that most of the cracking occurred in the reactor bed. The extent of cracking was controlled by varying reactor bed depth (time) and temperature. Although it was recognized that cracking occurred in the riser feeding the reactor because the catalyst activity and temperature were at their highest there, no significant

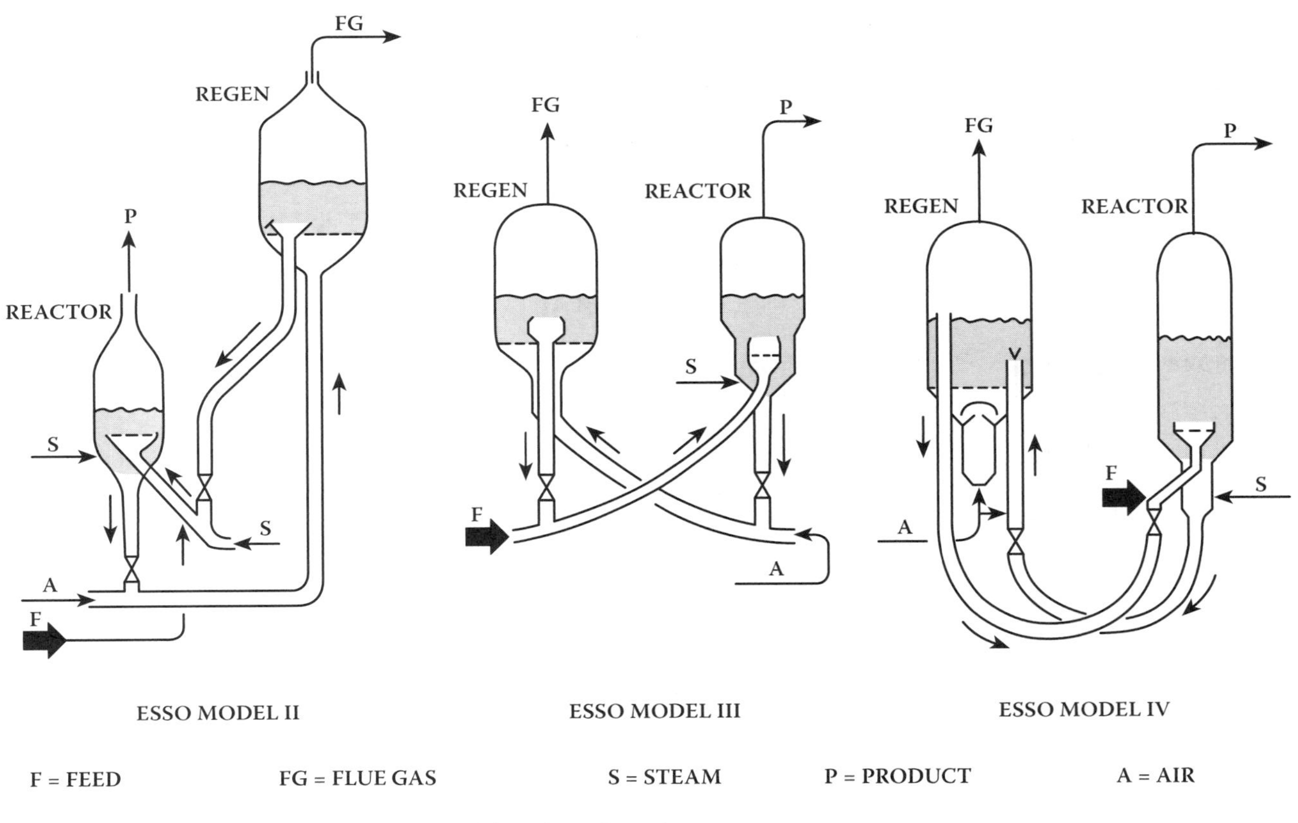

FIGURE 6.1A Older fluid catalytic cracking (FCC) unit configurations.

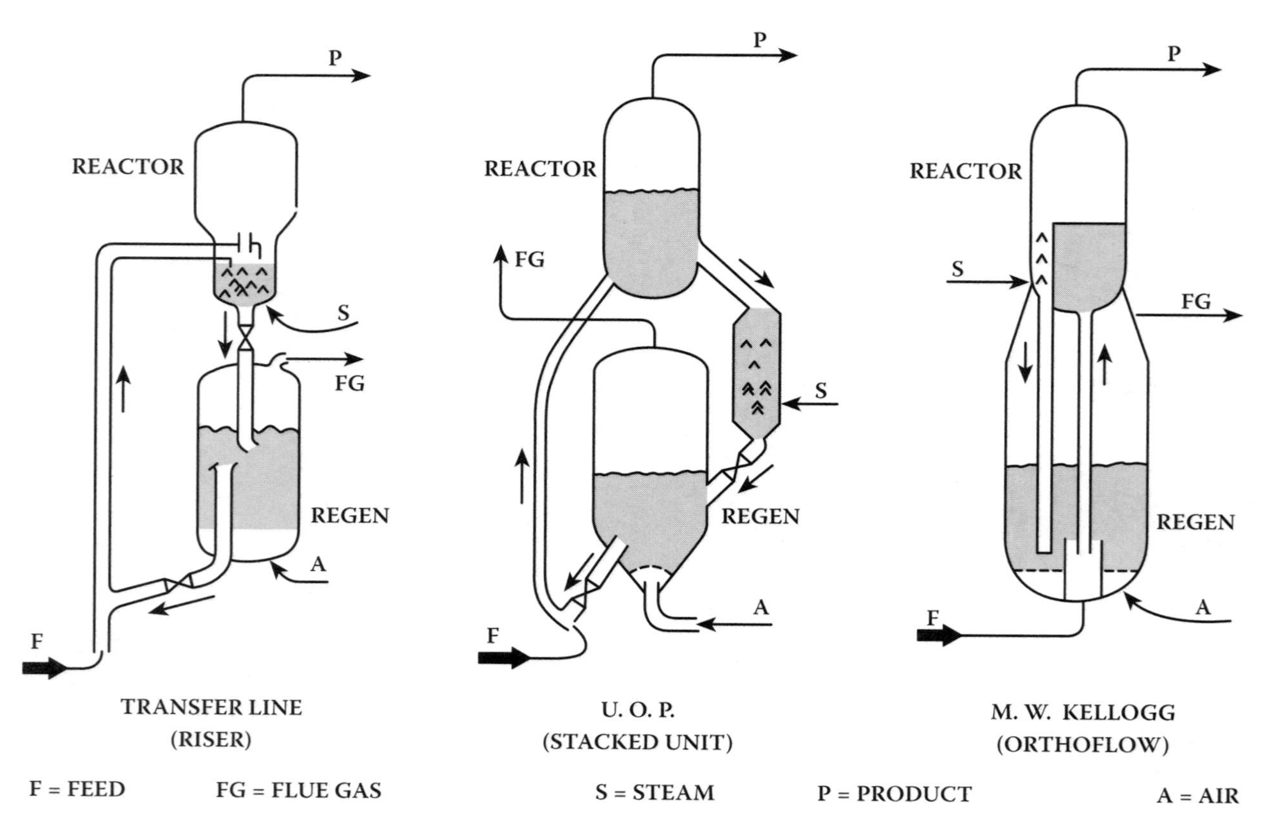

FIGURE 6.1B Older fluid catalytic cracking unit configurations.

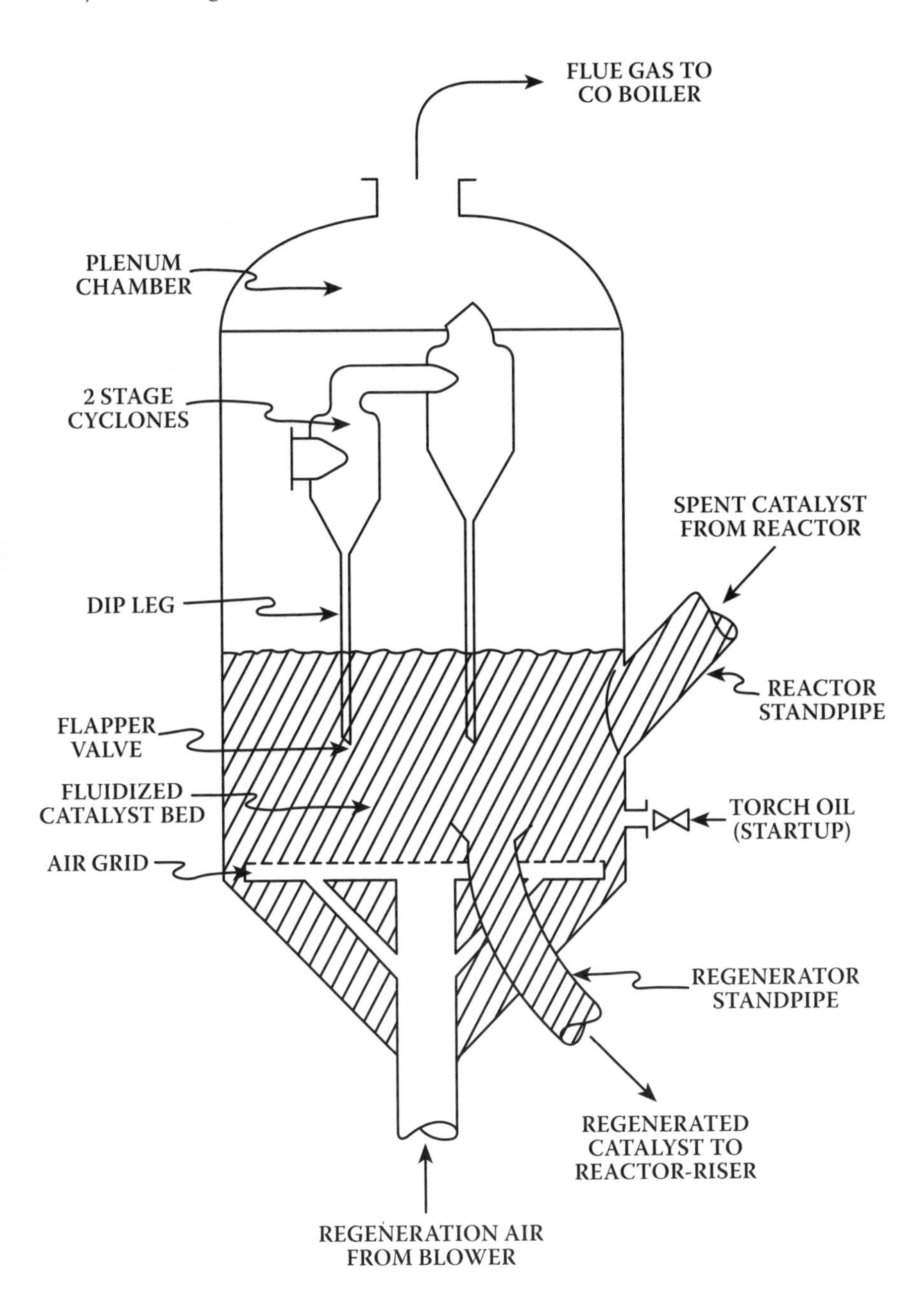

FIGURE 6.2 FCC regenerator.

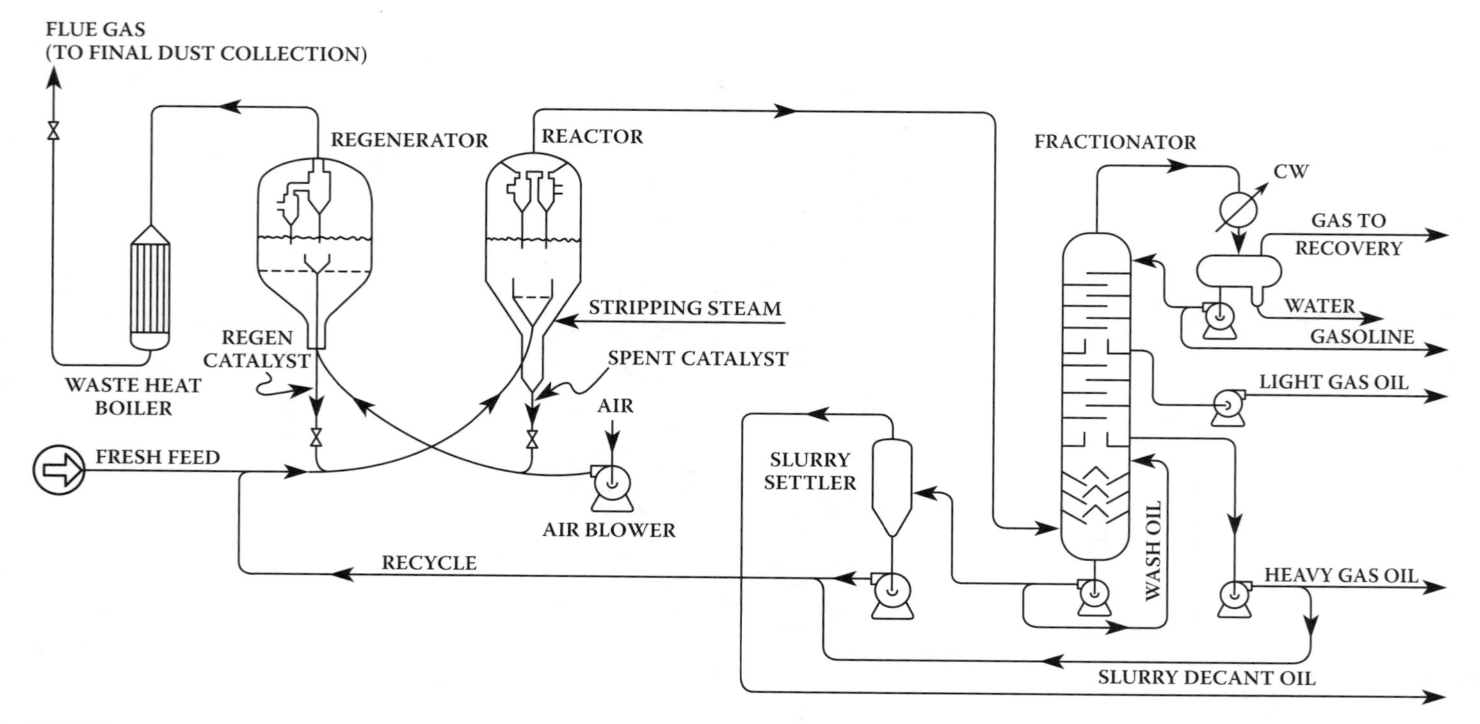

FIGURE 6.3 FCC unit, Model III.

y 16, 2012

nk you for your purchase of *Petroleum Refining: Technology and Economics, h Edition* (J.H. Gary, G.E. Handwerk, and M.J. Kaiser).

body of Table 5.5 was corrupted during the printing process. Please see below the complete table.

le 5.5
e Yields When Conradson Carbon Is Known

ke wt%	=	1.6 (wt% Conradson carbon[a])
s (C_4^-) wt%	=	7.8 + 1.44 (wt% Conradson carbon[a])
so. wt%	=	11.29 + 0.343 (wt% Conradson carbon[a])
s oil wt%	=	100 – wt% coke – wt% gas – wt% gaso.
so. vol%	=	$[186.6/(131.5 + API)]^b$ (gaso. wt%)
s oil vol%	=	$[155.5/(131.5 + API)]^b$ (gas oil wt%)

actual Conradson carbon when available.
°API are those for net fresh feed to coker.
: These yield correlations are based on the following conditions:

5. Coke drum pressure 35 to 45 psig.
6. Feed is "straight-run" residual.
7. Gas oil end point 875 to 925°F.
8. Gasoline end point 400°F.

incerely regret any inconvenience this may have caused you. Please let us y if we can be of assistance regarding this or any other title that Taylor & cis publishes.

or & Francis Group, LLC

7038/ISBN 978-0-8493-7038-0

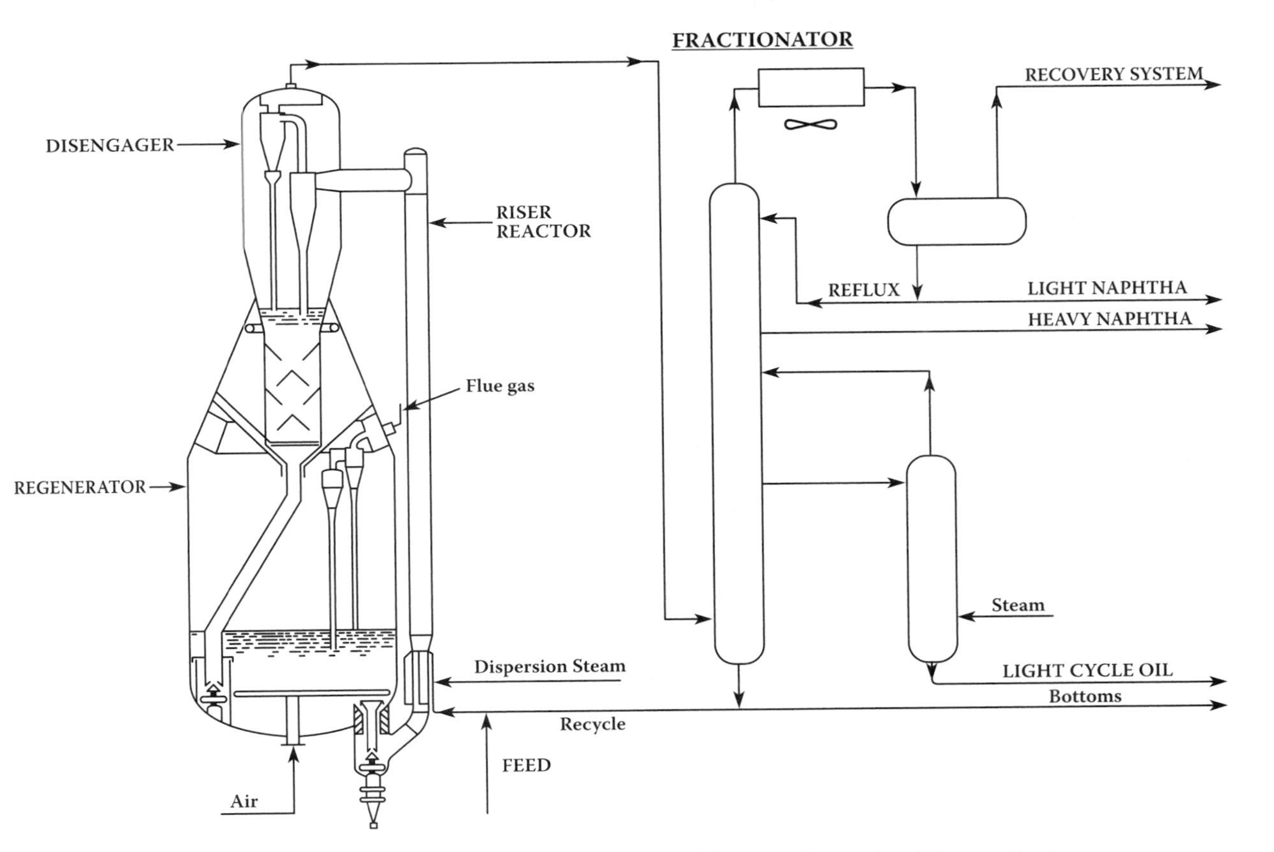

FIGURE 6.4 FCC unit, M. W. Kellogg design. (Flowsheets courtesy of Cameron International Corporation.)

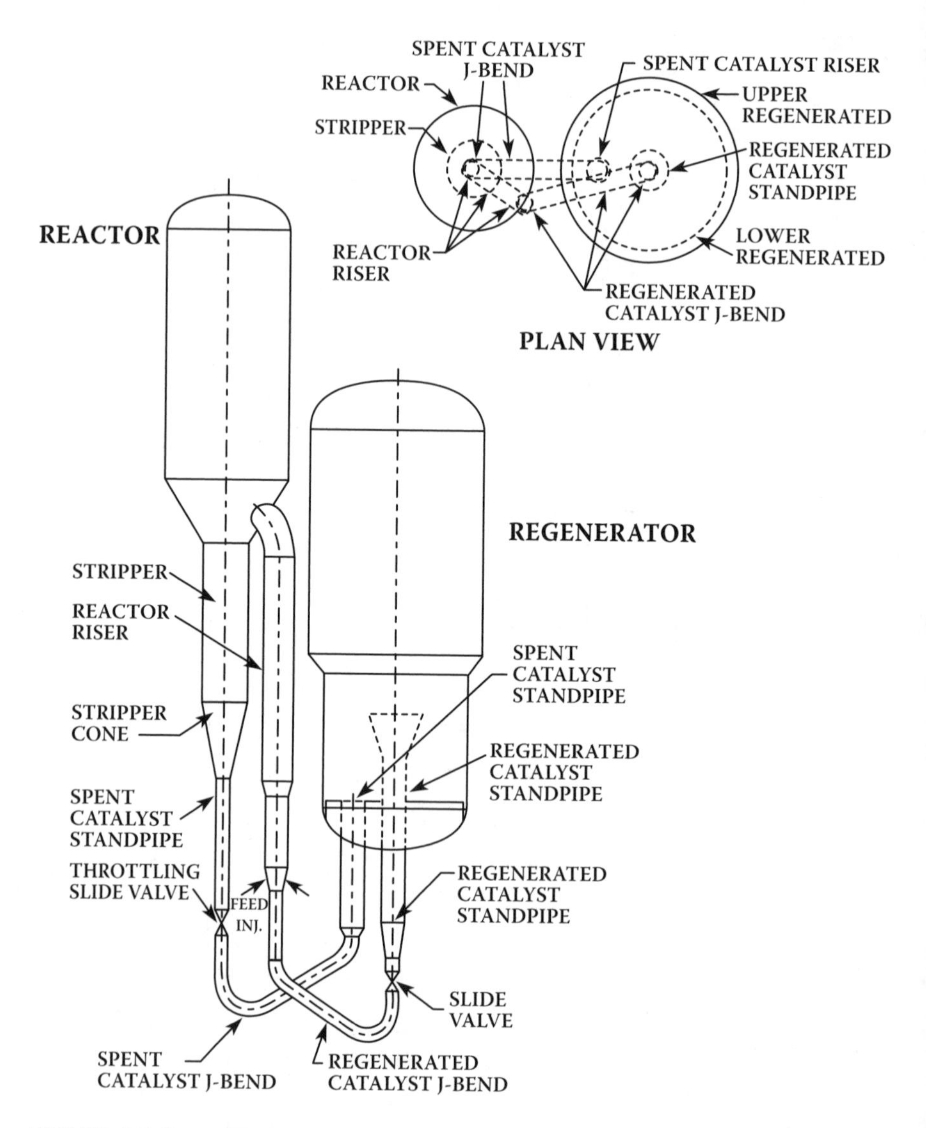

FIGURE 6.5 Exxon Flexicracking IIIR FCC unit. (Courtesy of Exxon Research and Engineering.)

attempt was made to regulate the reaction by controlling riser conditions. After the more reactive zeolite catalysts were adopted by refineries, the amount of cracking occurring in the riser (or transfer line) increased to levels requiring operational changes in existing units. As a result, recently constructed units have been designed to operate using the reactor (or its equivalent) as a separator to separate the catalyst and the hydrocarbon vapors. Control of the reaction is maintained by accelerating the regenerated catalyst to the velocity desired in the riser-reactor before introducing it into the riser and injecting the feed into the riser using spray nozzles.

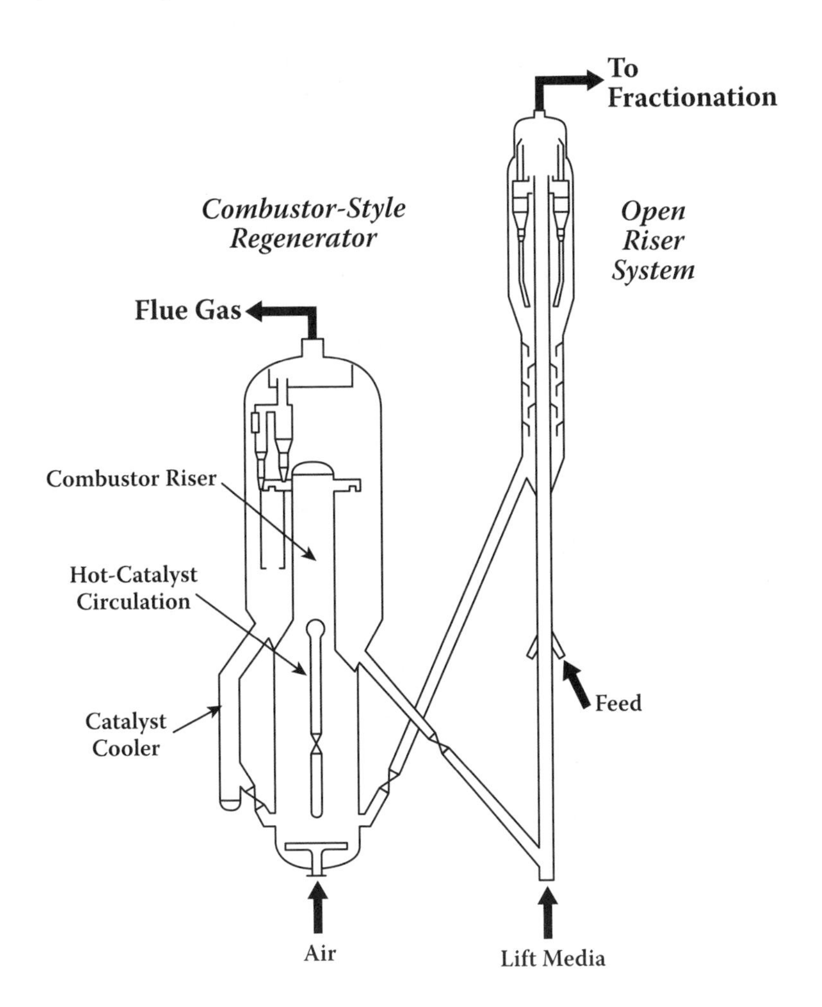

FIGURE 6.6A UOP FCC style unit. (Courtesy of UOP LLC.)

Older units have been modified to maximize and control riser cracking. Units are also operating with different combinations of feed-riser and dense-bed reactors, including feed-riser followed by dense-bed, feed-riser in parallel with dense-bed, and parallel feed-riser lines (one for fresh feed and the other for recycle) [3]. The major changes have been to take advantage of improvements in catalysts by more control over the time the catalyst is in contact with the oil and to get more efficient contact of heavy feedstocks with the catalyst particles. The results have been higher conversion levels with better selectivity (higher gasoline yields at given conversion levels), by shorter and better controlled reaction times (1 to 3 sec), closed cyclones, and improved feed distribution systems. These have been summarized by James R. Murphy [4]. Alvaro Murcia reviewed reactor–regenerator configurations from the initial commercial FCC unit through today's residue processing units [5]. Most of the designs are similar to those shown in Figure 6.1 through Figure 6.8.

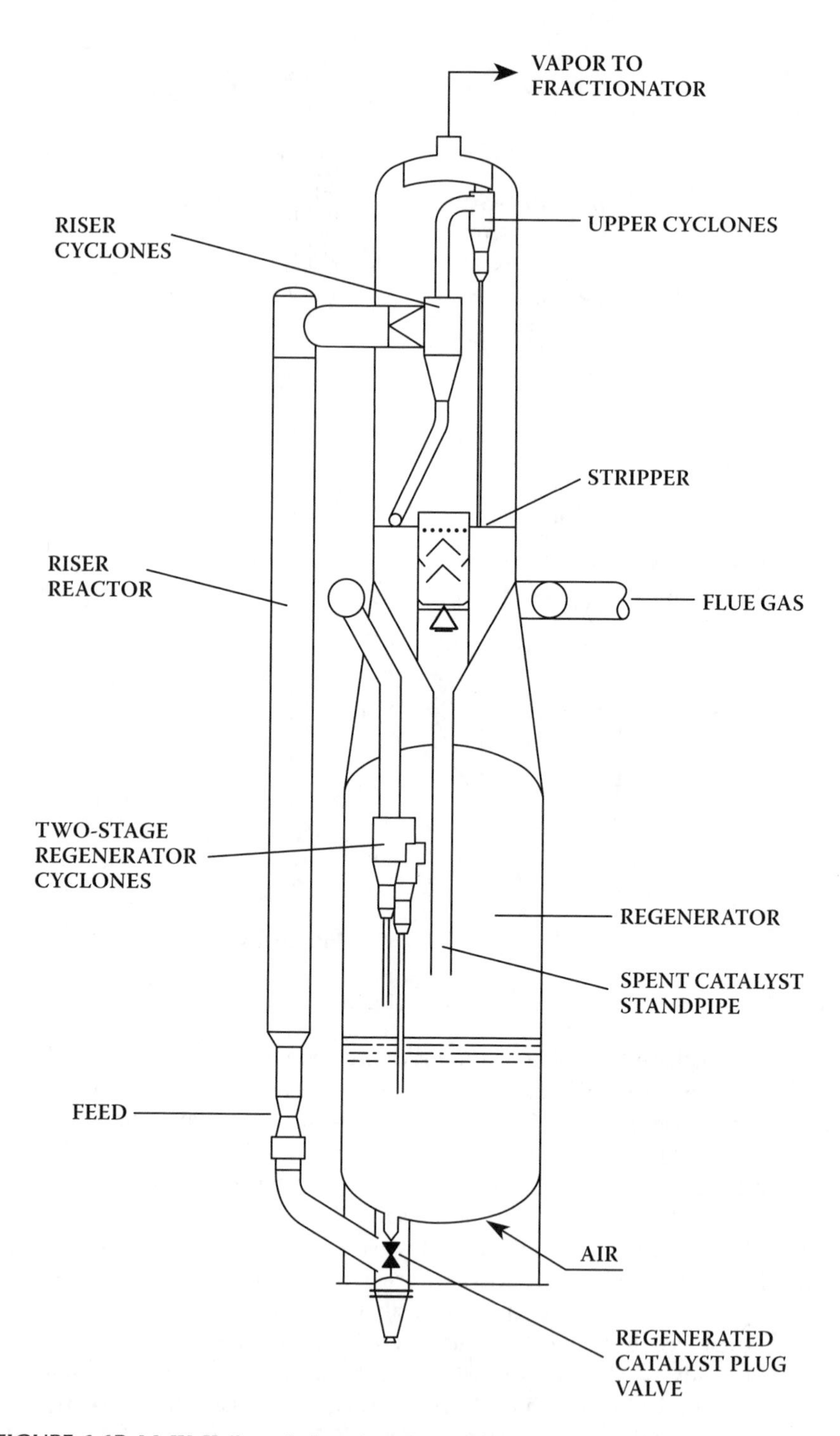

FIGURE 6.6B M. W. Kellogg design, riser FCC unit.

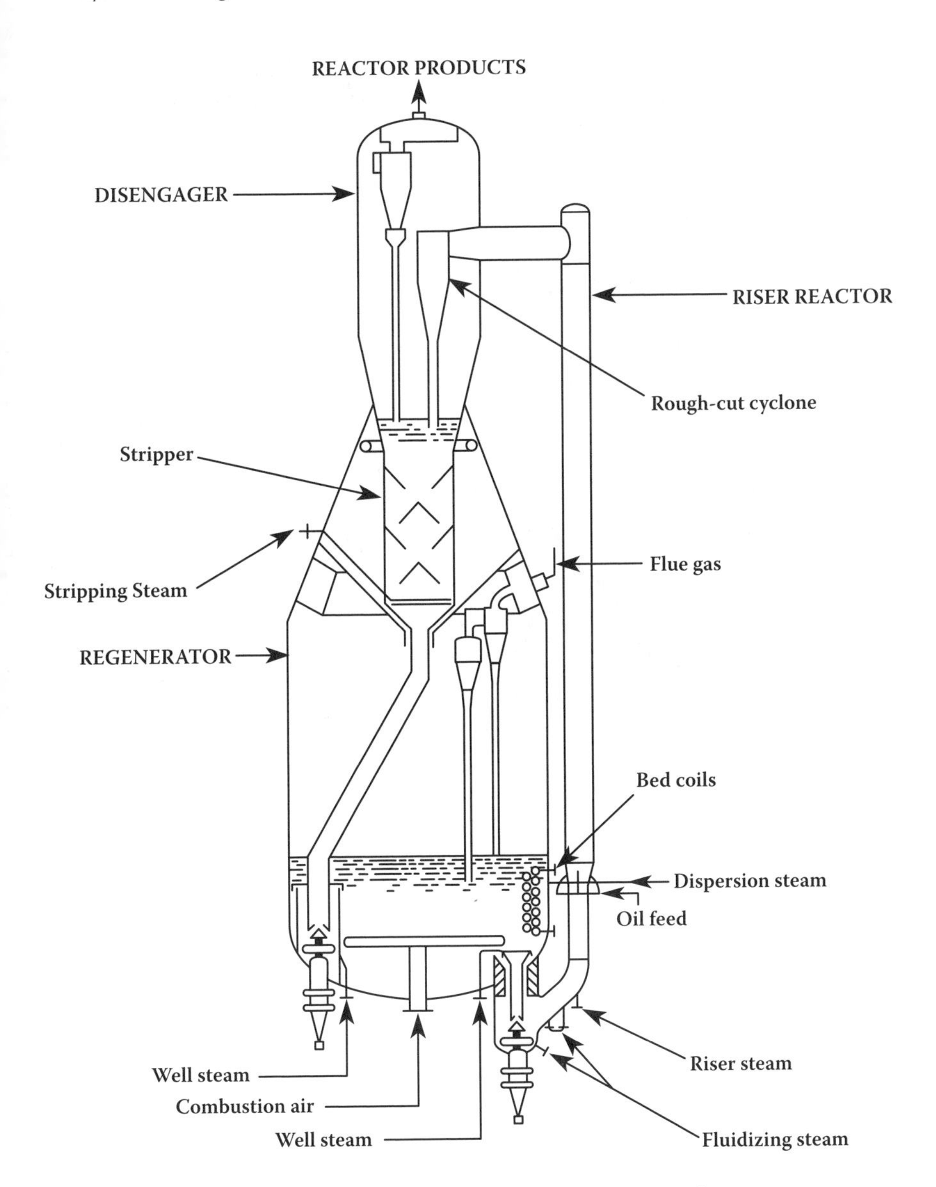

FIGURE 6.7 M. W. Kellogg resid fluid catalytic cracking (RFCC) unit.

The fresh feed and recycle streams are preheated by heat exchangers or a furnace and enter the unit at the base of the feed riser, where they are mixed with the hot generated catalyst. The heat from the catalyst vaporizes the feed and brings it up to the desired reaction temperature. The mixture of catalyst and hydrocarbon vapor travels up the riser into the separator. The cracking reactions start when the feed contacts the hot catalyst in the riser and continues until the oil vapors are separated from the catalyst in the reactor-separator. The hydrocarbon vapors are sent to the synthetic crude fractionator for separation into liquid and gaseous products.

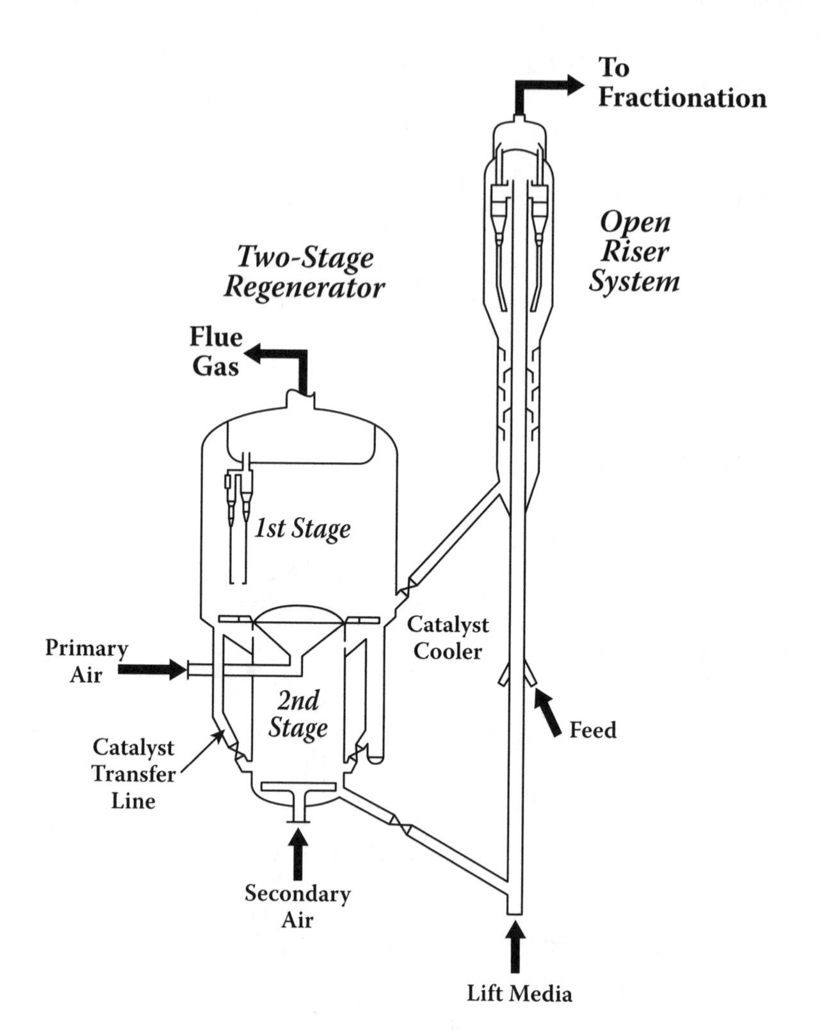

FIGURE 6.8 UOP RCC reduced crude cracking unit. (Courtesy of UOP LLC.)

The catalyst leaving the reactor is called spent catalyst and contains hydrocarbons adsorbed on its internal and external surfaces as well as the coke deposited by the cracking. Some of the adsorbed hydrocarbons are removed by steam stripping before the catalyst enters the regenerator. In the regenerator, coke is burned from the catalyst with air. The regenerator temperature and coke burn-off are controlled by varying the air flow rate. The heat of combustion raises the catalyst temperature to 1150 to 1550°F (620 to 845°C), and most of this heat is transferred by the catalyst to the oil feed in the feed riser. The regenerated catalyst contains from 0.01 to 0.4 wt% residual coke, depending upon the type of combustion (burning to CO or CO_2) in the regenerator.

The regenerator can be designed and operated to burn the coke on the catalyst to either a mixture of carbon monoxide and carbon dioxide or completely to carbon dioxide. Older units were designed to burn to carbon monoxide to minimize blower capital and operating costs, because only about half as much air had to be compressed to burn to carbon monoxide rather than to carbon dioxide. Newer units are designed and operated to burn the coke to carbon dioxide in the regenerator, because they can burn to a much lower residual carbon level on the regenerated catalyst. This gives a more reactive and selective catalyst in the riser, and a better product distribution results at the same equilibrium catalyst activity and conversion level.

For units burning to carbon monoxide, the flue gas leaving the regenerator contains a large quantity of carbon monoxide, which is burned to carbon dioxide in a CO furnace (waste heat boiler) to recover the available fuel energy. The hot gases can be used to generate steam or to power expansion turbines to compress the regeneration air and generate electric power. A schematic diagram of a typical FCC reactor and regenerator is shown in Figure 6.2, a flow diagram for a Model III FCC unit is given in Figure 6.3, and a Kellogg FCC unit is shown in Figure 6.4.

6.2 NEW DESIGNS FOR FLUIDIZED BED CATALYTIC CRACKING UNITS

Zeolite catalysts have a much higher cracking activity than amorphous catalysts, and shorter reaction times are required to prevent overcracking of gasoline to gas and coke. This has resulted in units that have a catalyst-oil separator in place of the fluidized bed reactor to achieve maximum gasoline yields at a given conversion level. Many of the newer units are designed to incorporate up to 25% reduced crude in the FCC unit feed. Schematic drawings of Exxon Research and Engineering's Flexicracking IIIR design are shown in Figure 6.5 [6,7]. These units incorporate a high height-to-diameter ratio, a lower regenerator section for more efficient single-stage regeneration, and an offset side-by-side or stacked vessel design. UOP-designed units utilize high-velocity, low-inventory regenerators (Figure 6.6a), and Kellogg-designed units use plug valves to minimize erosion by catalyst containing streams (Figure 6.6b). All units are now designed for both complete combustion to CO_2 and CO combustion control.

The large differential value between residual fuel and other catalytic cracking feedstocks has caused refiners to blend atmospheric and vacuum tower bottoms into the FCC feed. Residual feedstocks have orders of magnitude higher metals contents (especially nickel and vanadium) and greater coke forming potential (Ramsbottom and Conradson carbon values) than distillate feeds. These contaminants reduce catalyst activity, promote coke and hydrogen formation, and decrease gasoline yield. It has been shown that catalyst activity loss due to metals is caused primarily by vanadium deposition, and increased coke and hydrogen formation is due to nickel deposited on the catalyst [6,8]. The high coke laydown creates problems because of the increased coke-burning requirement, with the resulting increased air or oxygen demand, higher regenerator temperatures, and greater heat removal. For feeds containing up to 15 ppm Ni + V, passivation of nickel catalyst activity can be achieved by the addition of organic antimony or bismuth compounds to the feed [8], with reductions in coke and hydrogen yields up to 15% and 45%, respectively. For feeds containing > 15 ppm metals and having Conradson carbon contents above 3%, hydroprocessing of the FCC feed may be required to remove metals, reduce the carbon forming potential, and increase the yield of gasoline and middle distillates by increasing the hydrogen content of the feed.

Designs of FCC units have been modified for higher carbon burning, heat removal, and gas product rates by larger equipment sizing and introduction of steam generation coils. Kellogg's resid fluid catalytic cracking (RFCC) unit, formerly called the heavy oil cracking (HOC) FCC unit, developed with Phillips Petroleum Co., is shown in Figure 6.7 [9].

6.3 CRACKING REACTIONS

The products formed in catalytic cracking are the result of both primary and secondary reactions [10]. Primary reactions are designed as those involving the initial carbon–carbon bond scission and the immediate neutralization of the carbonium ion [11]. The primary reactions can be represented as follows:

$$\begin{aligned} \text{Paraffin} &\rightarrow \text{Paraffin} + \text{olefin} \\ \text{Alkyl naphthene} &\rightarrow \text{Naphthene} + \text{olefin} \\ \text{Alkyl aromatic} &\rightarrow \text{Aromatic} + \text{olefin} \end{aligned}$$

Thomas [12] suggested the mechanism that carbonium ions are formed initially by a small amount of thermal cracking of n-paraffins to form olefins. These olefins add a proton from the catalyst to form large carbonium ions, which decompose according to the beta rule (carbon–carbon bond scission takes place at the carbon in the position beta to the carbonium ions and olefins) to form small carbonium ions and olefins. The small carbonium ions propagate the chain reaction by transferring a hydrogen ion from an n-paraffin to form a small paraffin molecule and a new large carbonium ion [3,13]. As an example of a typical n-paraffin hydrocarbon cracking reaction, we may look at the borrowing sequence for n-octane (where $R = CH_3CH_2CH_2CH_2CH_2-$).

Step 1: Mild thermal cracking initiation reaction.

$$nC_8H_{18} \rightarrow CH_4 + R - CH = CH_2$$

Step 2: Proton shift.

$$R{-}CH{=}CH_2 + H_2O + \left[\begin{array}{c} O \\ | \\ Al - O - Si \\ | \\ O \end{array}\right]$$

$$\rightarrow R{-}\overset{+}{C}H{-}CH_3 + [HO{-}Al{-}O{-}Si]^-$$

Step 3: Beta scission.

$$R{-}\overset{+}{C}H{-}CH_3 \rightarrow CH_3CH_2 = CH_2 + \overset{+}{C}H_2CH_2CH_2CH_3$$

Step 4: Rearrangement toward a more stable structure. The order of carbonium ion stability is tertiary > secondary > primary.

$$\overset{+}{C}H_2CH_2CH_2CH_3 \leftrightarrow CH_3\overset{+}{C}HCH_2CH_3$$

$$\leftrightarrow CH_3{-}\underset{}{\overset{CH_3}{\overset{|}{C}}}H{-}CH_2 \leftrightarrow CH_3{-}\underset{+}{\overset{CH_3}{\overset{|}{C}}}{-}CH_3$$

Step 5: Hydrogen ion transfer.

$$CH_3{-}\underset{+}{\overset{CH_3}{\overset{|}{C}}}{-}CH_3 + C_8H_{18} \rightarrow i\text{-}C_4H_{10} + CH_3\overset{+}{C}HCH_2R$$

Thus, another large carbonium ion is formed, and the chain is ready to repeat itself.

Even though the basic mechanism is essentially the same, the manner and extent of response to catalytic cracking differs greatly among the various hydrocarbon types.

6.4 CRACKING OF PARAFFINS

The catalytic cracking of paraffins is characterized by high production of C_3 and C_4 hydrocarbons in the cracked gases, reaction rates and products determined by size and structure of paraffins, and isomerization to branched structures and aromatic hydrocarbons formation, resulting from secondary reactions involving olefins [14]. In respect to reaction rates, the effect of the catalyst is more pronounced as the number of carbon atoms in the molecule increases, but the effect is not appreciable until the number of carbon atoms is at least six.

The cracking rate is also influenced by the structure of the molecule, with those containing tertiary carbon atoms cracking most readily, whereas quaternary carbon atoms are most resistant. Compounds containing both types of carbon atoms tend to neutralize each other on a one-to-one basis. For example, 2,2,4-trimethylpentane (one tertiary and one quaternary) cracks only slightly faster than n-octane, whereas 2,2,4,6,6-pentamethylheptane (one tertiary and two quaternary) cracks at a slower rate than does n-dodecane.

6.5 OLEFIN CRACKING

The catalytic cracking rates of olefinic hydrocarbons are much higher than those of the corresponding paraffins. The main reactions are [13]

1. Carbon–carbon bond scissions
2. Isomerization
3. Polymerization
4. Saturation, aromatization, and carbon formation

Olefin isomerization followed by saturation and aromatization are responsible for the high octane number and lead susceptibility of catalytically cracked gasolines. The higher velocity of hydrogen transfer reactions for branched olefins results in ratios of iso- to normal paraffins higher than the equilibrium ratios of the parent olefins. In addition, naphthenes act as hydrogen donors in transfer reactions with olefins to yield isoparaffins and aromatics.

6.6 CRACKING OF NAPHTHENIC HYDROCARBONS

The most important cracking reaction of naphthenes in the presence of silica-alumina is dehydrogenation to aromatics. There is also carbon–carbon bond scission in both the ring and attached side chains, but at temperatures below 1000°F (540°C), the dehydrogenation reaction is considerably greater. Dehydrogenation is very extensive for C_9 and larger naphthenes, and a high-octane gasoline results. The nonring liquid products and cracked gases resulting from naphthenic hydrocarbon cracking are more saturated than those resulting from cracking paraffins.

6.7 AROMATIC HYDROCARBON CRACKING

Aromatic hydrocarbons with alkyl groups containing less than three carbon atoms are not very reactive. The predominant reaction for aromatics with long alkyl chains is the clean splitting off of side chains without breaking the ring. The carbon–carbon bond ruptured is that adjacent to the ring, and benzene compounds containing alkyl groups can be cracked with nearly quantitative recovery of benzene [13].

6.8 CRACKING CATALYSTS

Commercial cracking catalysts can be divided into three classes: (1) acid-treated natural aluminosilicates, (2) amorphous synthetic silica-alumina combinations, and (3) crystalline synthetic silica-alumina catalysts called zeolites or molecular sieves [16]. Most catalysts used in commercial units today are either class 3 or mixtures of classes 2 and 3 catalysts [17] (see Table 6.2 and Table 6.3). The advantages of the zeolite catalysts over the natural and synthetic amorphous catalysts are

TABLE 6.2
Comparison of Amorphous and Zeolite Catalysts

	Amorphous	Zeolite
Coke, wt%	4	4
Conversion, vol%	55	65
C_5+ gasoline, vol%	38	51
C_3– gas, wt%	7	6
C_4, vol%	17	16

Source: Reference 15.

TABLE 6.3
Y Zeolite Types in Catalyst

Type Y zeolite	Major attributes
Rare earth exchanged Y (REY)	Highest gasoline yield
Ultrastable Y (USY)	Highest octane, low H-transfer, best coke selectivity (resid)
Rare earth exchanged USY (RE USY)	Gasoline-octane balance, better coke selectivity

Source: Reference 18.

1. Higher activity
2. Higher gasoline yields at a given conversion
3. Production of gasolines containing a larger percentage of paraffinic and aromatic hydrocarbons
4. Lower coke yield (and therefore usually a larger throughput at a given conversion level)
5. Increased isobutene production
6. Ability to go to higher conversions per pass without overcracking

The high activity of zeolitic cracking catalyst permits short residence time cracking and has resulted in most cracking units being adapted to riser cracking operations [17]. Here, the adverse effects of carbon deposits on catalyst activity and selectivity are minimized because of the negligible amount of catalyst back-mixing in the riser. In addition, separate risers can be used for cracking the recycle stream and the fresh feed so that each can be cracked at their own optimum conditions.

The catalytic effects of zeolitic catalysts can be achieved with only 10 to 25% of the circulating catalyst as zeolites [19] and the remainder amorphous silica-alumina cracking catalyst. Amorphous catalysts have higher attrition resistance and are less costly than zeolitic catalysts. Most commercial catalysts contain approximately 15 to 25% zeolites and thus obtain the benefits of the higher activity and gasoline selectivity of the zeolites and the lower costs and makeup rates of the amorphous catalysts. Lower attribution rates also greatly improve particulate emission rates.

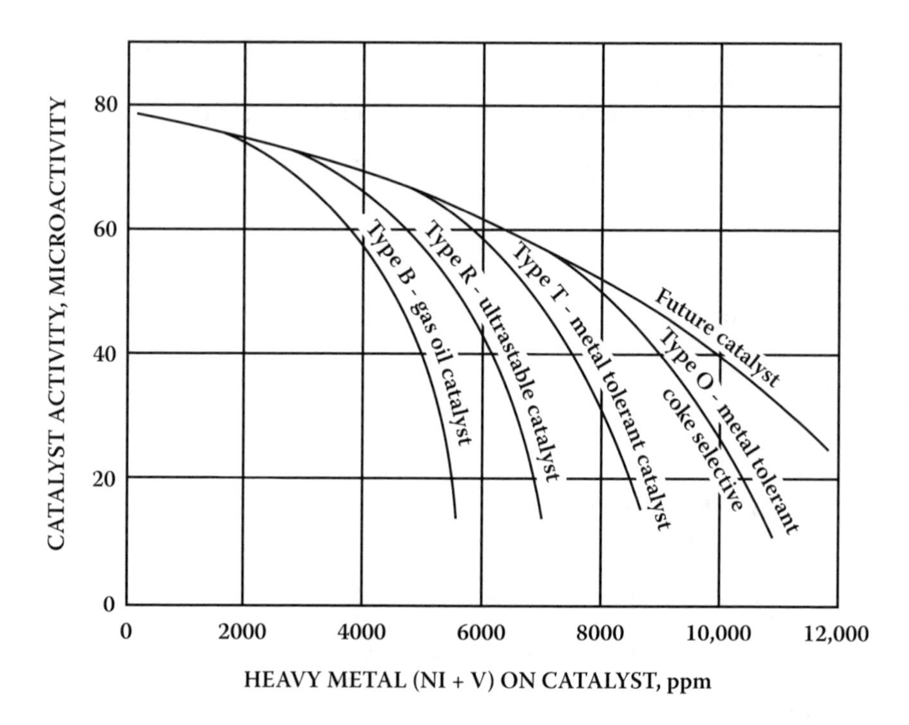

FIGURE 6.9A Catalyst activity loss from heavy metals. (From Note 21.)

Basic nitrogen compounds, iron, nickel, vanadium, and copper in the oil act as poisons to cracking catalysts [16]. The nitrogen reacts with the acid centers on the catalyst and lowers the catalyst activity. The metals deposit and accumulate on the catalyst and cause a reduction in throughput by increasing coke formation and decreasing the amount of coke burn-off per unit of air by catalyzing coke combustion to CO_2 rather than to CO.

It is generally accepted that nickel has about four times as great an effect on catalyst activity and selectivity as vanadium, and some companies use a factor of 4Ni + V (expressed in ppm) to correlate the effects of metals loading, whereas others use Ni + V/4 or Ni + V/5 [11,14,20].

Although the deposition of nickel and vanadium reduces catalyst activity by occupying active catalytic sites, the major effects are to promote the formation of gas and coke and reduce the gasoline yield at a given conversion level. Tolen [21] and others [17] have discussed the effects of nickel and vanadium deposits on equilibrium catalysts (Figure 6.9a and Figure 6.9b). Metals loadings on equilibrium catalysts are now as high as 10,000 ppm, and in 1998, the mean was over 3,000 ppm of nickel plus vanadium [27]. The effects of nickel can be partially offset by the addition of passivators, such as antimony and barium compounds, to the feed, and catalysts containing tin, barium and strontium titanates, and magnesium oxide have been developed that act as metal traps for vanadium [23]. Metals removal processes can also be used to reactivate the catalyst by cycling a slipstream through a metals

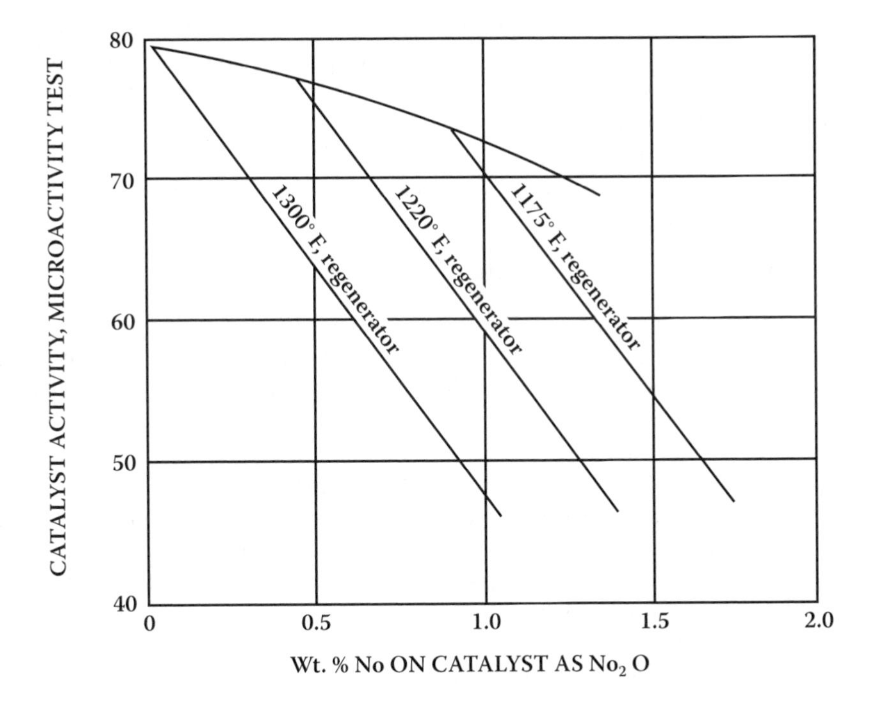

FIGURE 6.9B Catalyst activity loss from sodium contamination. (From Note 21.)

removal system (Demet). This permits the equilibrium catalyst metals concentrations to be controlled at the level at which the fresh catalyst required to maintain activity and selectivity equals catalyst losses [24].

A range of catalysts is available from catalyst manufacturing companies that are compounded to give high gasoline octanes (0.5 to 1.5 RON improvement), resist deactivation due to sulfur or metals in the feed, and transfer sulfur from the regenerator to the reactor (to reduce SO_x in the regenerator flue gas) [19]. For each operation and feed, economic or environmental considerations affect the catalyst used.

The average microactivity of FCC catalyst in North and South America in 2004 was between 67 and 70, whereas world activity averaged between 65 and 70. In the United States, approximately two thirds of the FCC units supply propylene to the chemicals market and supply about 50% of the U.S. propylene demand. Additives are available that increase the percentage of propylene in the butane and lighter gases from the FCC unit. By operating at higher conversion levels, the butylenes and propylene production (as well as total butane and lighter gas yields) can be increased at the expense of naphtha. The naphtha has one to two higher blending octane numbers because the additives crack the low-octane straight-chain and lightly branched paraffins and olefins to propylenes and butylenes [20].

The zeolite content of catalysts ranges from 15 to 40 wt% [18].

Catalysts for the processing of resids in specially designed FCC units have to be designed with a range of pore size distributions to handle the large molecules (> 30 Å) present and also smaller pores to give higher activities. Avidan [18] summarizes the need for high matrix activity to increase bottoms conversion along with a distribution of large liquid-catching pores (> 100 Å) with lower activity to control coke and gas make, meso pores (30 to 100 Å) with higher activity, and small pores (< 20 Å) with very high activity. The meso pores are most effective for reducing bottoms yields for aromatic and naphthenic feedstocks and small pores for paraffinic feedstocks.

6.9 FCC FEED PRETREATING

The trend toward low sulfur and nitrogen contents in gasolines and diesel fuels requires that either the FCC unit feed or products be treated to reduce sulfur and nitrogen. Treating feed to the FCC unit offers the advantages that the sulfur and nitrogen in the gasoline and diesel fuel products are reduced, and by adding hydrogen to the feed, naphtha and LCO yields are increased without lowering the olefins content and octanes of the naphtha fraction [25]. For refineries that do not hydrotreat the FCC feed or naphtha products, over 95% of the sulfur in the gasoline blending pool is from the FCC naphtha.

The hydrotreating unit can be operated in several ways: as a hydrodesulfurization (HDS) unit, a mild hydrocracking (MHC) unit, or a partial-conversion hydrocracking unit. In all cases, the product sulfur content has to be less than 135 wppm to produce a refinery gasoline blending pool with less than 50 wppm sulfur and less than 85 wppm to produce a refinery gasoline blending pool of less than 30 wppm.

If operated as an HDS or MHC unit, the operating pressure is in the 800 to 1000 psig (55 to 70 barg) range. Higher operating pressures [1400 to 1600 psig (95 to 110 barg)] are necessary to provide the aromatic ring hydrogenation necessary for partial-conversion hydrocracking (Figure 6.10) [26]. The partial-conversion hydrocracking unit also produces distillates with 50 cetane indices (CIs), less than 15 wppm sulfur, and smoke points of 15 to 19 mm. The capital costs of the units designed for 1000 and 1500 psig (70 and 102 barg) are from 15 to 110% higher than the HDS unit designed for 900 psig (60 barg) [43].

Using MHC or partial-conversion hydrocracking instead of hydrocracking offers greater flexibility by increasing the yield of higher-value products such as diesel fuel and jet fuel. MHC, because of its lower operating pressure and less hydrogenation of aromatic rings, typically produces diesel fuels with cetane indices from 39 to 42 and kerosine smoke points well below the 19 mm smoke point required for jet fuels. Partial-conversion hydrocracking, operating at 1400 to 1600 psig (95 to 110 barg) and 40% conversion to diesel or diesel plus naphtha, produces diesel with cetane indices near 50 and kerosine smoke points close to 19 [26].

6.10 PROCESS VARIABLES

In addition to the nature of the charge stock, the major operating variables affecting the conversion and product distribution are the cracking temperature, catalyst/oil ratio, space velocity, catalyst type and activity, and recycle ratio. For a better understanding of the process, several terms should be defined:

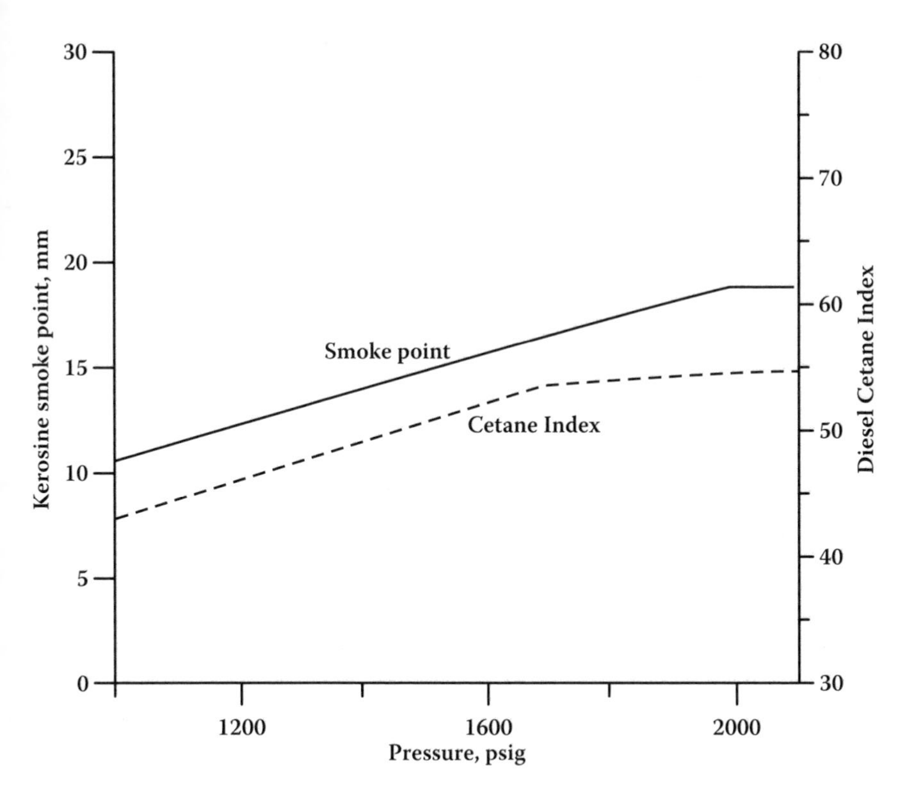

FIGURE 6.10 Effects of pressure on smoke point and cetane index. (From Note 26.)

Activity: Ability to crack a gas oil to lower boiling fractions.
Catalyst/oil ratio (C/O):= lb catalyst/lb feed.
Conversion: 100 (volume of feed – volume of cycle stock) / volume of feed.
Cycle stock: Portion of catalytic cracker effluent not converted to naphtha and lighter products [generally the material boiling above 430°F (220°C)].
Efficiency: (% gasoline) × conversion.
Recycle ratio: Volume recycle / volume fresh feed.
Selectivity: The ratio of the yield of desirable products to the yield of undesirable products (coke and gas).
Space velocity: Space velocity may be defined on either a volume (LHSV) or a weight (WHSV) basis. In a fluidized bed reactor, the LHSV has little meaning because it is difficult to establish the volume of the bed. The weight of the catalyst in the reactor can be easily determined or calculated from the residence time and C/O ratio.
LHSV: Liquid hour space velocity in volume feed / (volume catalyst) (hr).
WHSV: (Weight hour space velocity in lb feed) / (lb catalyst) (hr). If t is the catalyst residence time in hours, then WHSV = 1 / (t)(C/O).

Within the limits of normal operations, increasing reaction temperature, catalyst/oil ratio, catalyst activity, and contact time results in an increase in conversion, whereas a decrease in space velocity increases conversion. It should be noted that an increase in conversion does not necessarily mean an increase in gasoline yield, as an increase in temperature above a certain level can increase conversion, coke and gas yields, and octane number of the gasoline but decrease gasoline yield [5,14]. In many FCC units, conversion and capacity are limited by the regenerator coke-burning ability. This limitation can be due to either air compression limitations or the afterburning temperatures in the last stage regenerator cyclones. In either case, FCC units are generally operated at the maximum practical regenerator temperature with the reactor temperature and throughput ratio selected to minimize the secondary cracking of gasoline to gas and coke. With the trend to heavier feedstocks, the carbon-forming potential of catalytic cracker feeds is increasing, and some units limited in carbon-burning ability because of limited blower capacity are adding oxygen to the air to the regenerator to overcome this limitation. Oxygen contents of the gases to the regenerator are being increased to 24 to 30% by volume and are limited by regenerator temperature capacity and heat removal capacity [4,27].

In fluidized bed units, the reactor pressure is generally limited to 15 to 20 psig by the design for the unit and is, therefore, not widely used as an operating variable. Increasing pressure increases coke yield and the degree of saturation of the gasoline but decreases the gasoline octane. It has little effect on the conversion.

The initial catalyst charge to an FCC unit using riser cracking is about 3 to 5 tons of catalyst per 1000 BPSD charge rate. Catalyst circulation rate is approximately 1 ton/min per MBPD charge rate. Typical operations of these units are given in Table 6.4.

6.11 HEAT RECOVERY

Fuel and energy costs are a major fraction of the direct costs of refining crude oils, and as a result of large increases in crude oil and natural gas prices, there is a great incentive to conserve fuel by the efficient utilization of the energy in the off-gases from the catalytic cracker regenerator. Temperature control in the regenerator is easier if the carbon on the catalyst is burned to carbon dioxide rather than carbon monoxide, but much more heat is evolved and regenerator temperature limits many be exceeded. A better yield structure with lower coke laydown and higher gasoline yield is obtained at a given conversion level when burning to carbon dioxide to obtain a lower residual carbon on the catalyst. In either case, the hot gases are at high temperatures [1100 to 1250°F (595 to 675°C) in the former case and 1250 to 1500°F (675 to 815°C) in the latter] and at pressures of 15 to 25 psig (103 to 172 kPa). Many catalytic crackers include waste heat boilers, which recover the sensible heat by steam generation, and others use power recovery turbines to generate electric power or compress the air used in the catalytic cracker regenerator. Some refineries recover the heat of combustion of the carbon monoxide in the flue gas by installing CO-burning waste heat boilers in place of those utilizing only the sensible heat of the gases. An even higher rate of energy recovery can be achieved by using a power recovery turbine prior to the CO or waste heat boiler, although when regenerator pressures are less than 15 psig, power recovery turbines usually are not economic.

TABLE 6.4
Comparison of Fluid, Thermafor, and Houdry Catalytic Cracking Units

	FCC	TCC	HCC
Reactor space velocity	1.1–13.4[a]	1–3[b]	1.5–4[b]
C/O	5–16[c]	2–7[d]	3–7[d]
Recycle/fresh feed, vol	0–0.5	0–0.5	0–0.5
Catalyst requirement, lb/bbl feed	0.15–0.25	0.06–0.13	0.06–0.13
Cat. crclt. rate, ton cat./bbl total feed	0.9–1.5	0.4–0.6	0.4–0.6
On-stream efficiency, %	96–98		
Reactor temp., °F	885–950[e]	840–950	875–950
°C	475–510	450–510	470–510
Regenerator temp., °F	1200–1500	1100–1200	1100–1200
°C	650–815	595–650	595–650
Reactor pressure, psig (barg)	8–30[e] (0.54–2.0)	8–12 (0.54–0.8)	9–10 (0.6–0.8)
Regenerator pressure, psig	15–30 (1–2)		
Turndown ratio			2:1
Gasoline octane, clear			
RON	92–99	88–94	88–94
MON	80–85		

[a] lb/hr/lb.
[b] v/hr/v.
[c] wt.
[d] vol.
[e] One company has operated at 990°F and 40 psig to produce a 98 RON (clear) gasoline with a C_3-650°F liquid yield of 120 vol% on feed (once-through); there was approximately 90% yield of the C_5-650°F product.

Assuming a regenerator flue gas discharge pressure of 20 psig (138 kPa) and 1000°F (538°C), the available horsepower per pound per second of gas flow for the various schemes is

	hp	kW
Waste heat boiler only	45	34
Power recovery only	78	58
Power recovery plus waste heat boiler	106	79
CO-burning waste heat boiler	145	108
Power recovery plus CO-burning waste heat boiler	206	154

6.12 YIELD ESTIMATION

Correlations have been developed to estimate the product quality and yields from the catalytic cracking of virgin gas oils. The correlations are very useful for esti-

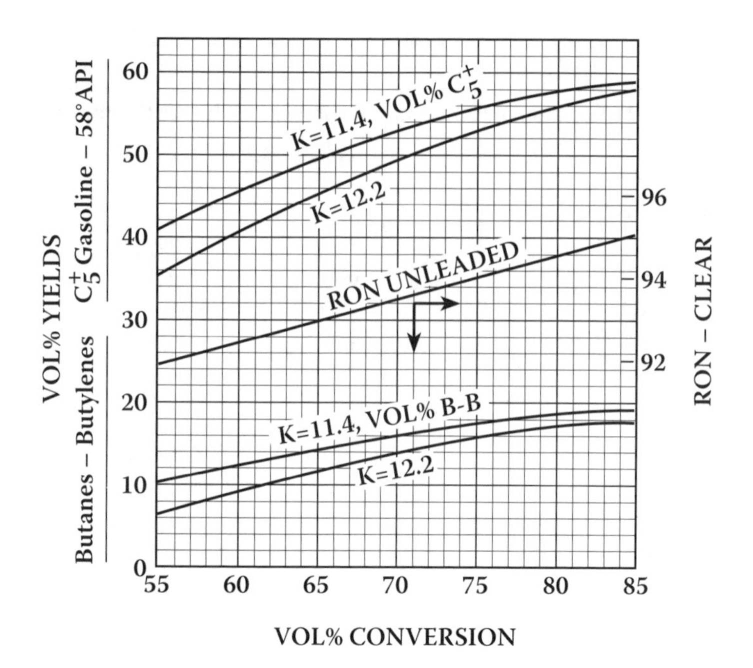

FIGURE 6.11 Catalytic cracking yields. Silica-alumina catalyst (butanes, butylenes, C_5^+ gasoline). The butane-butylene fraction typically contains about 40 vol% isobutane, 12 vol% n-butane, and 48 vol% butylenes.

mating typical yields for preliminary studies and to determine yield trends when changes are made in conversion levels. The yield structure is very dependent upon catalyst type. Figure 6.17 through Figure 6.27 are for the zeolitic or molecular-sieve silica-alumina catalyst. If the feedstocks contain sulfur, Figure 6.28 can be used to estimate the distribution of sulfur in the product streams.

It is necessary to make a weight balance in order to obtain product distribution and properties as the cycle gas oil yields are obtained by difference. When using a molecular-sieve catalyst, the procedure is as follows (gravities shown on curves are those of the fresh feed):

1. Calculate the weight of feed.
2. Determine yields and weights of all products except gasoline from Figure 6.17 through Figure 6.19.
3. Use Figure 6.20 and Figure 6.21 to obtain the split of LPG components and adjust to total LPG as shown by Figure 6.19.
4. Determine the weight of TCGO by use of Figure 6.26, then gasoline by difference.
5. Using Figure 6.26 to estimate gasoline gravity, calculate the volume of gasoline produced, and check against the yield estimated by Figure 6.22.
6. From Figure 6.23 and Figure 6.24 and the feed rate, calculate the volume of heavy cycle gas oil (HGO).

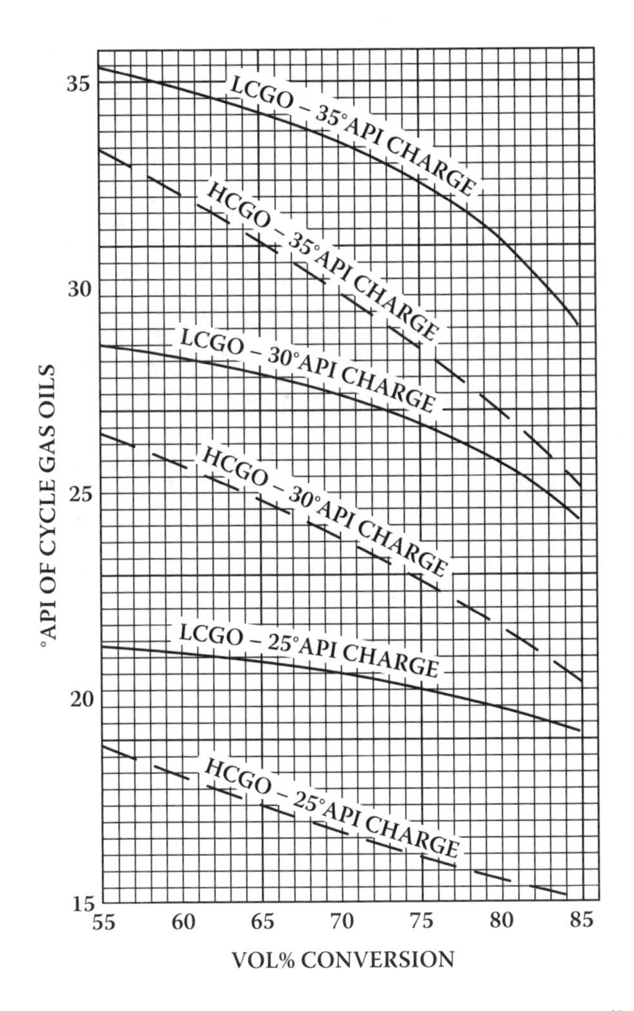

FIGURE 6.12 Catalytic cracking yields. Silica-alumina catalyst (cycle gas oils).

7. Subtract the volume of HGO from the volume of total cycle gas oil (100 – conversion) to obtain volume yield of light cycle gas oil (LGO).
8. Use Figure 6.26 to estimate HGO gravity and calculate the weight yield of HGO.
9. Subtract the weight yield of HGO from the weight of total cycle gas oil to obtain weight yield of LGO.
10. Divide the weight yield of LGO by the volume yield of LGO (step 7) to obtain the density of LGO. Use tables to find API gravity.

Product yields and properties are obtained in a similar manner when a silica-alumina catalyst is used in the reactor, except the gravity of the LGO can be obtained from Figure 6.10 and the yield of HGO from Figure 6.12. It is necessary to find the weight yield of LGO by difference.

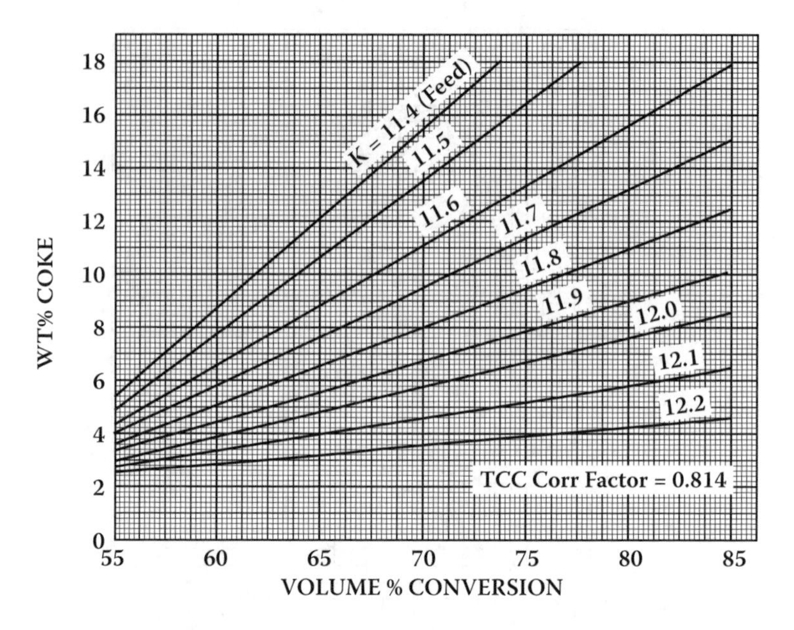

FIGURE 6.13 Catalytic cracking yields. Silica-alumina catalyst (coke).

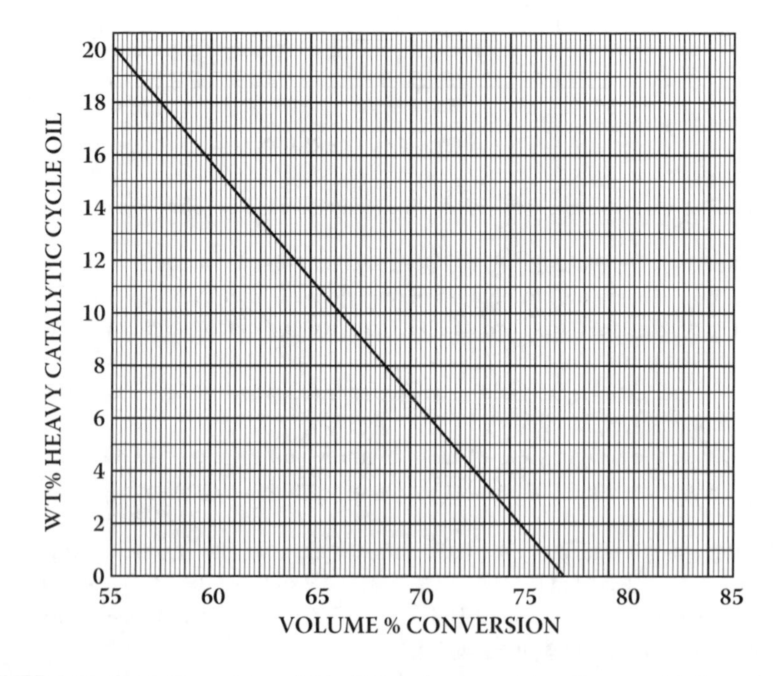

FIGURE 6.14 Catalytic cracking yields. Silica-alumina catalyst (heavy catalytic cycle oil).

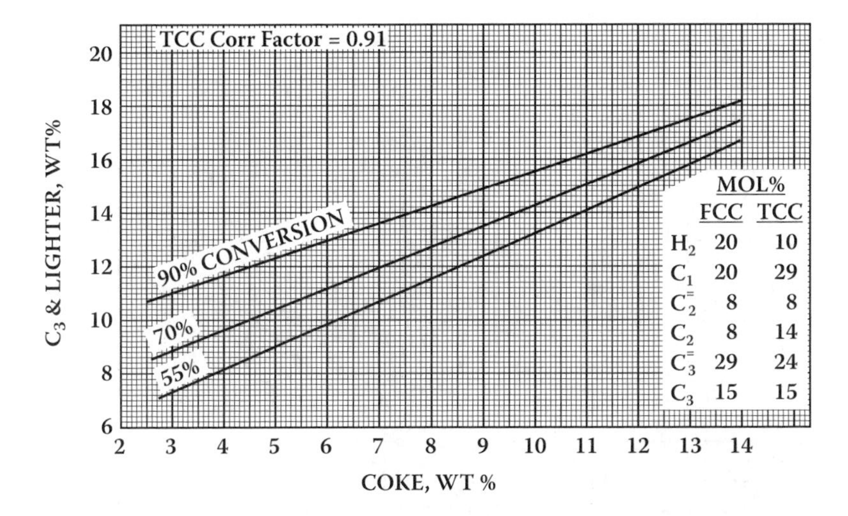

FIGURE 6.15 Catalytic cracking yields. Silica-alumina catalyst (C_3 and lighter).

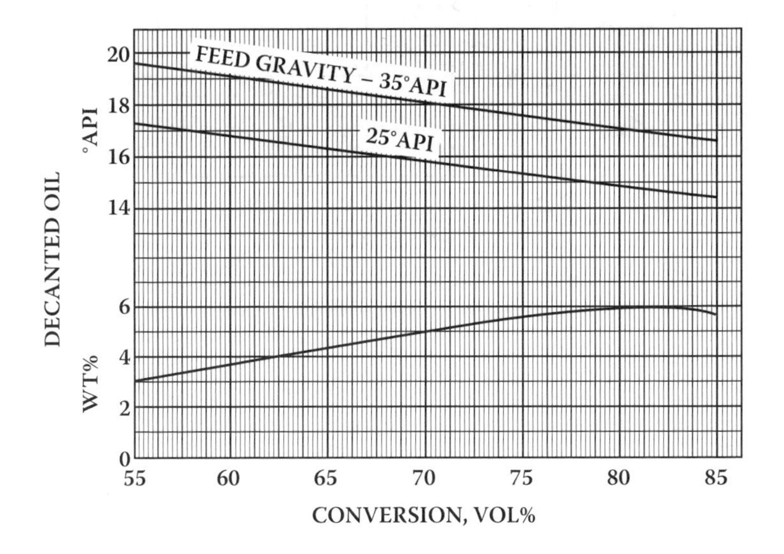

FIGURE 6.16 Catalytic cracking yields. Silica-alumina catalyst (decanted oil).

6.13 CAPITAL AND OPERATING COSTS

Capital construction and operating costs for a fluid catalytic cracking unit can be estimated using Figure 6.27 and its accompanying descriptive material (Table 6.5). Multiplying factors can be found in Chapter 18 (Cost Estimation).

6.14 CASE-STUDY PROBLEM: CATALYTIC CRACKER

The choice of feedstocks for a catalytic cracking unit should be based on an economic evaluation of alternative uses of the hydrocarbon streams. This is especially important

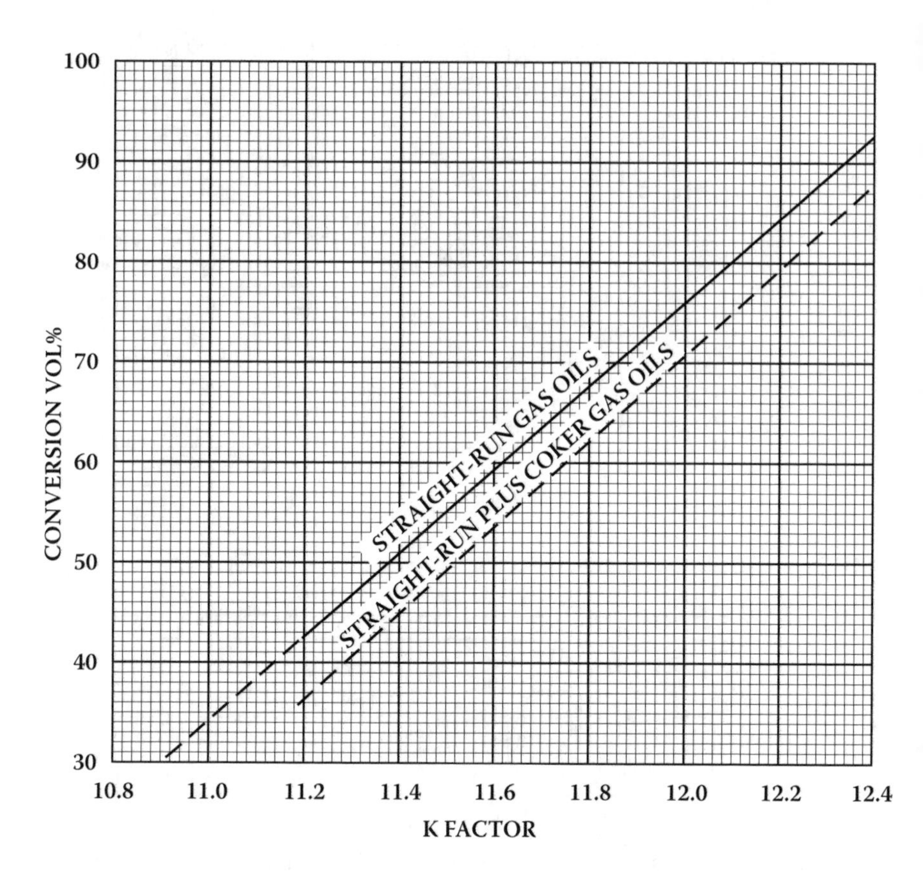

FIGURE 6.17 Effect of feed composition on conversion at constant operating conditions.

in the case of a refinery having both catalytic cracking and hydrocracking units. For this example, both the heavy gas oil from the atmospheric pipe still and the vacuum gas oil from the vacuum unit can be used as FCC feedstocks, although it is quite possible the vacuum gas oil might give a better return if used as hydrocracker feed.

The selection of a 75% conversion level was made because of the high-boiling characteristics of the feed. Checks should be made at higher and lower conversion levels to determine the level of conversion giving the best economic return.

The gravity of the combined charge stock is obtained by dividing the total pounds of charge by the volume. If the gravity is to be used for the calculation of a characterization factor (K_W) for yield prediction, the weight of sulfur and nitrogen should be subtracted from the weight of feed before dividing by the volume. If this is not done, the characterization factor will not be representative of the types of hydrocarbon compounds present in the feedstocks.

The yields for the various product streams are obtained from Figure 6.15 through Figure 6.21 for zeolitic catalyst. The sulfur distribution in the products is estimated from Figure 6.22. In the case of the C_4 and lighter stream shown in Figure 6.22, it is assumed the sulfur is in the form of hydrogen sulfide (H_2S).

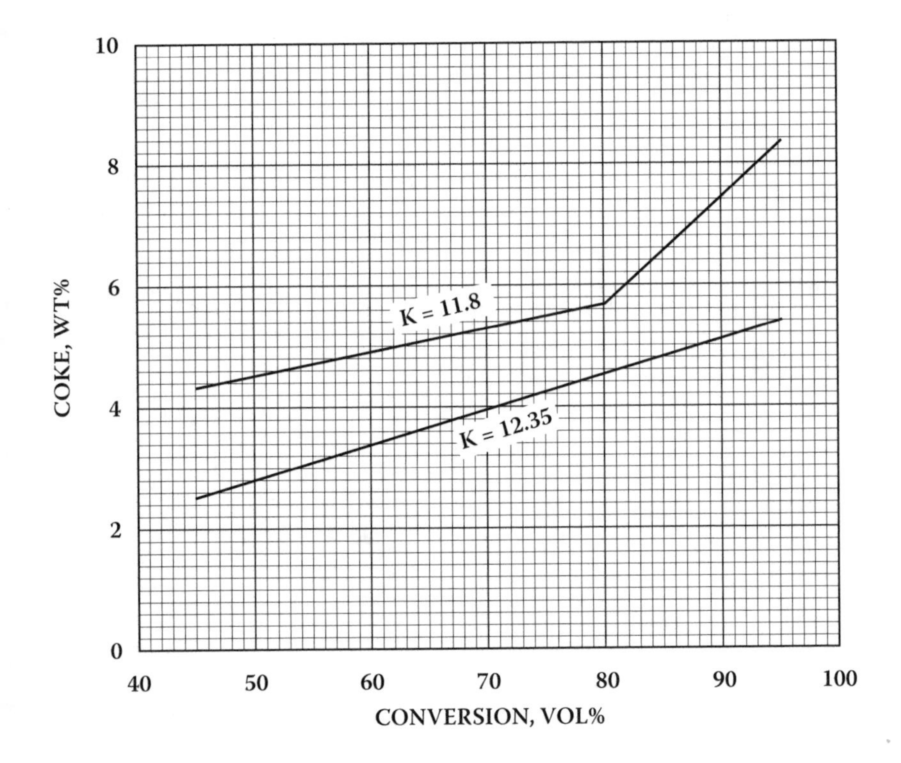

FIGURE 6.18 Catalytic cracking yields. Zeolite catalyst (coke).

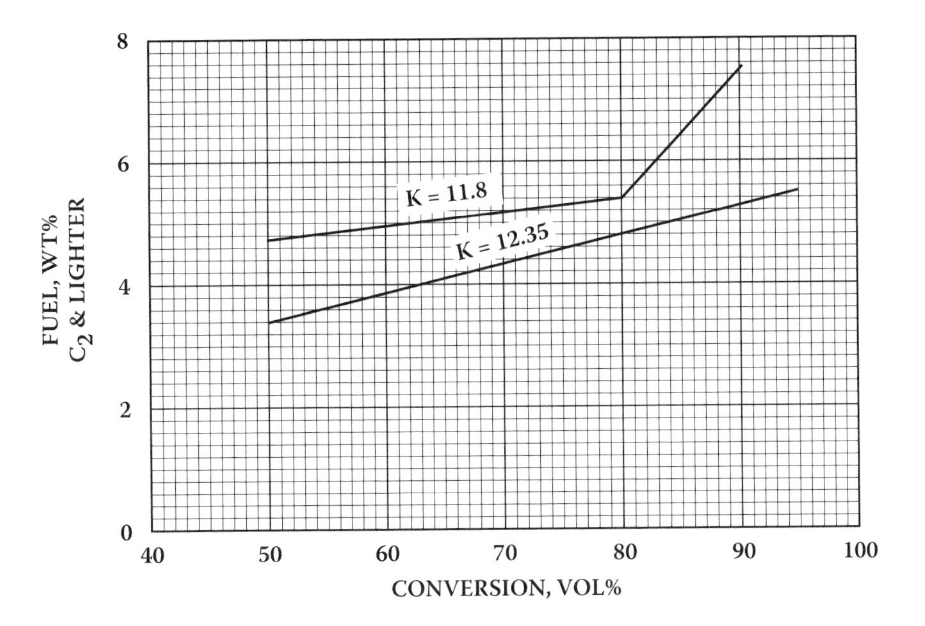

FIGURE 6.19 Catalytic cracking yields. Zeolite catalyst (fuel gas).

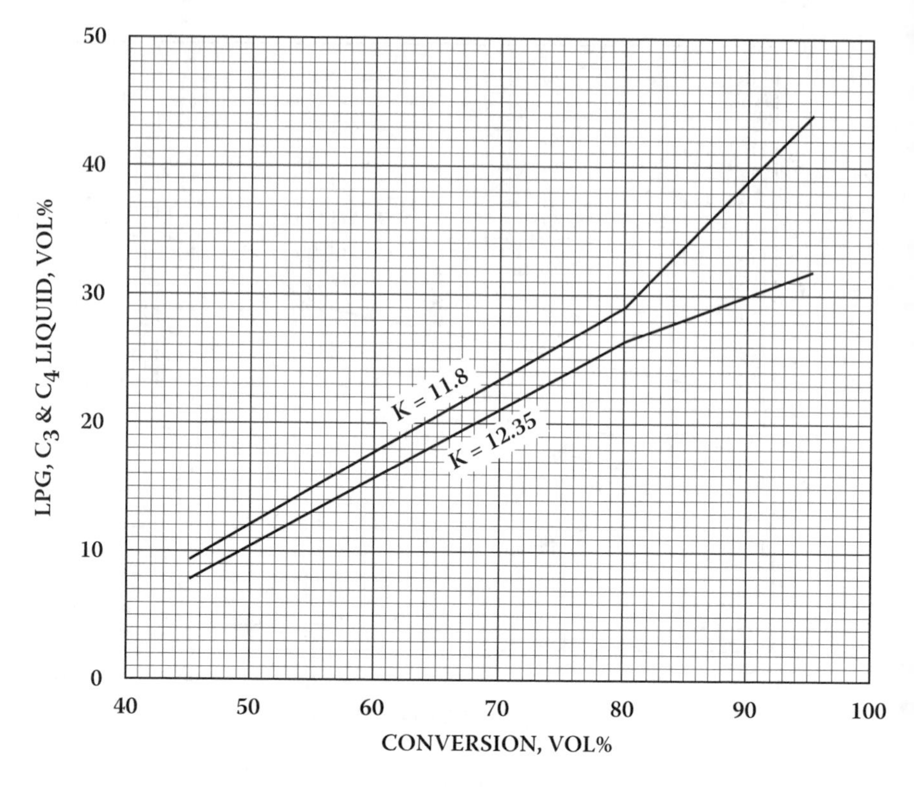

FIGURE 6.20 Catalytic cracking yields. Zeolite catalyst (C_3 and C_4).

The weight of the total cycle gas oil (TGO) product stream is determined by difference, the gravity from Figure 6.21, and the volume by dividing the weight by the gravity. This volume is then used to check the conversion. In this example, the conversion obtained is 75.2% versus the 75% used as the base. This is a satisfactory check. If it does not check within ±1%, some other method of determining product yields may better fit the characteristics of the feed.

The material balance is shown in Table 6.6.

Utility requirements are estimated from Table 6.5 and are listed in Table 6.7.

PROBLEMS

1. For a 27.0°API catalytic cracker feedstock with a boiling range of 650 to 900°F (343 to 482°C) and a sulfur content of 1.2% by weight, make an overall weight and volume material balance for 10,000 BPD feed rate when operating at a 65% conversion level and a once-through operation with zeolite catalyst.
2. If the catalytic gas oil produced in problem 1 is recycled to extinction (no products heavier than kerosine or No. 1 fuel oil are withdrawn from the unit), how much will gasoline production be increased? Coke laydown?

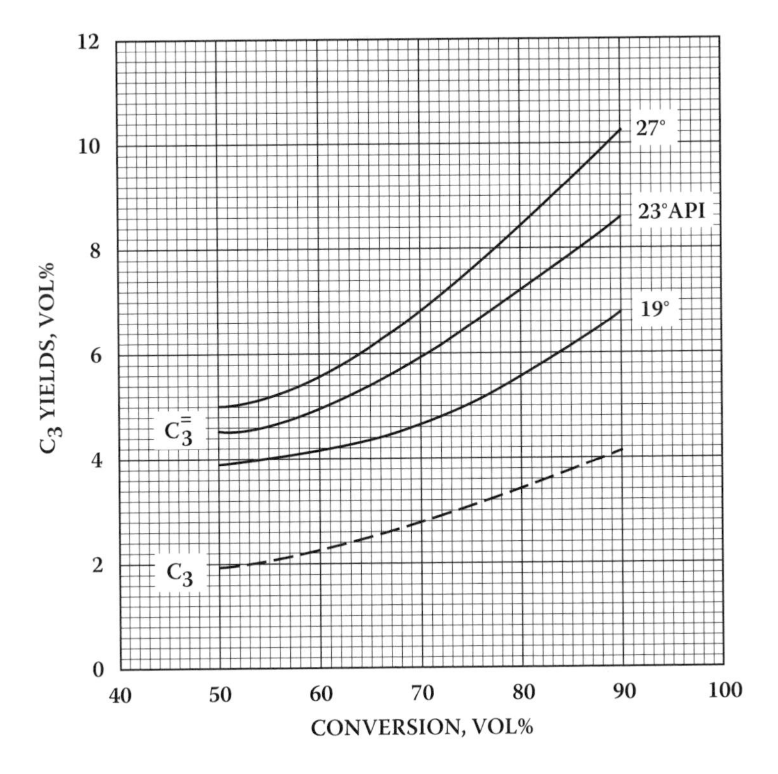

FIGURE 6.21 Catalytic cracking yields. Zeolite catalyst (C_3 ratios).

3. Estimate the direct operating cost, including royalty, per barrel of feed for a 20,000 BPD catalytic cracking unit if labor costs are $900/day, electric power is $0.05/kWh, steam is $3.15/Mlb, fuel is $2.25/MMBtu, and zeolite catalyst is $3000/ton.
4. Using today's construction costs for a 20,000 BPD catalytic cracking unit and double-rate declining-balance depreciation, estimate the cost per barrel of feed added by depreciation costs. Assume a 20-year life and a salvage value equal to dismantling costs. Calculate for the first and fifth years of operation. Compare with depreciation costs per barrel when straight-line depreciation is used.
5. For the following conditions, calculate the (a) wt% hydrogen in coke, (b) coke yield, and (c) catalyst-to-oil ratio.

Carbon on spent catalyst:	1.50 wt%
Carbon on regenerated catalyst:	0.80 wt%
Air from blower:	155,000 lb/hr
Hydrocarbon feed to reactor:	295,000 lb/hr

6. Estimate the cost, including royalty, per gallon of gasoline boiling range product for a 20,000 BPSD catalytic cracking unit if labor costs are

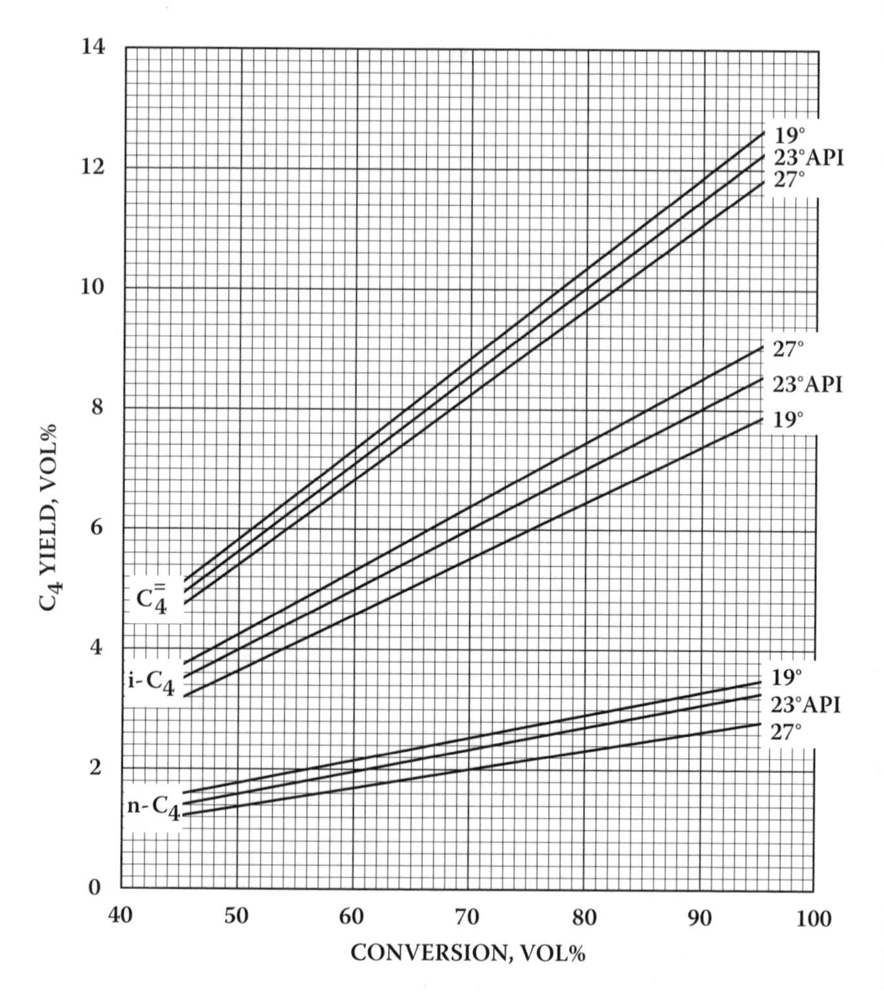

FIGURE 6.22 Catalytic cracking yields. Zeolite catalyst (C_4 ratios).

Flue gas analysis (Orsat), vol%:	
CO	12.0
CO_2	6.0
O_2	0.7
N_2	81.3
	100.0

$1150/day, electric power is $0.06/kWh, steam is $2.25/Mlb, fuel is $2.49/MMBtu, zeolite catalyst is $3000/ton, and depreciation is $625,000/yr. The on-stream factor is 96.5%. The C_5^+ gasoline yield is 58% at 85% conversion level.

7. For a 24.0°API catalytic cracker feedstock with a boiling range of 617 to 950°F (325 to 510°C) and a sulfur content of 0.45% by weight, make an

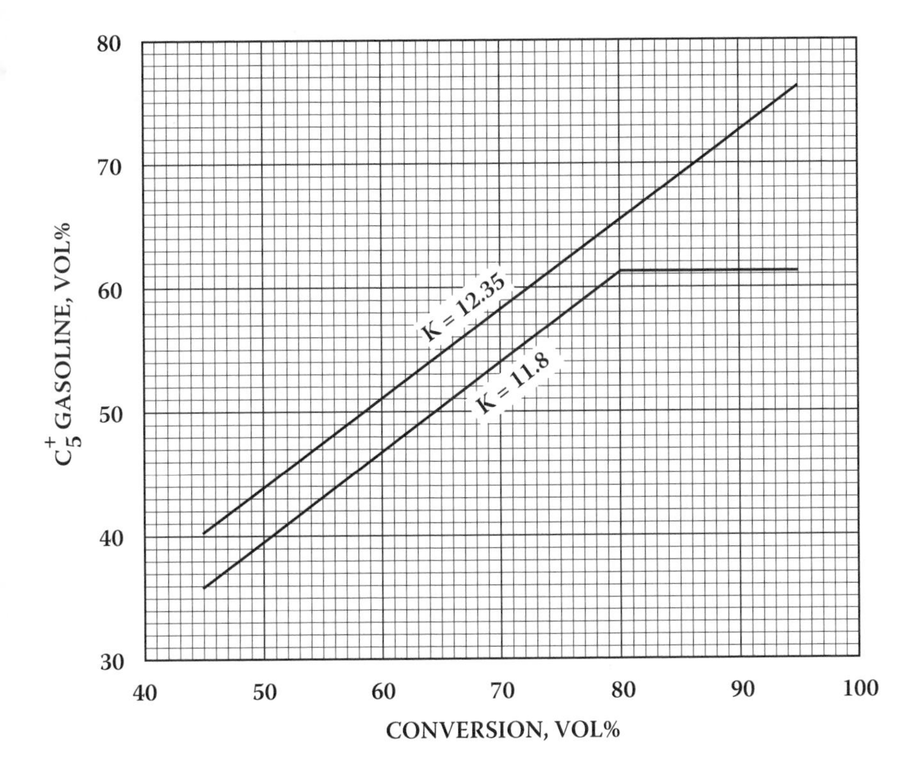

FIGURE 6.23 Catalytic cracking yields. Zeolite catalyst (C_5^+ gasoline).

overall weight and volume material balance for a 25,000 BPSD feed rate when operating at an 85% conversion level and a once-through operation with a zeolite catalyst.

8. For the feedstock and conditions in problem 7, compare the relative cost per barrel of C_5^+ gasoline produced at 75% conversion with that at 85% conversion assuming constant direct operating costs. Use *Oil and Gas Journal* values for other fuels.
9. For a 23.0°API catalytic cracker feedstock with a boiling range of 600 to 900°F (315 to 482°C) and containing 1.5 wt% sulfur, make an overall weight and volume material balance for a 50,000 BPCD feed rate when operating at 70% conversion using a zeolite catalyst.

NOTES

1. O'Connor, P., et al., *Hydrocarbon Process. 70*(11), pp. 76–84, 1991.
2. Nelson, W.L., *Petroleum Refinery Engineering*, 4th ed., pp. 759–810, McGraw-Hill Book Co., New York, 1958.
3. Ritter, R.E., Rheaume, L., Welch, W.A., and Magee, J.S., *Oil Gas J. 79*(27), pp. 103–106, 1981.

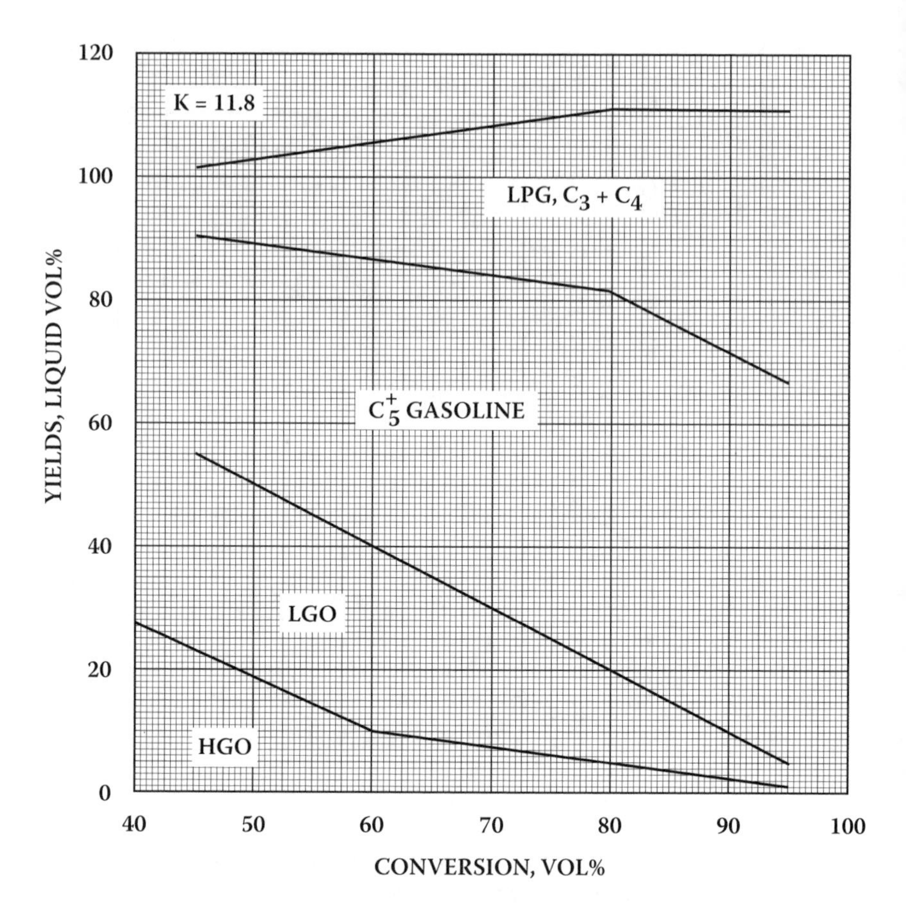

FIGURE 6.24 Catalytic cracking yields. Zeolite catalyst (heavy gas oil, feed K = 11.8).

4. Murphy, J.R., Personal communication, 1991.
5. Murcia, A.G., *Oil Gas J. 90*(10), pp. 68–71, 1992.
6. Blazek, J.J., *Davison Catalagram*, 63, pp. 2–10, 1981.
7. *Hydrocarbon Process. 61*(9), pp. 155–158, 1982.
8. Dale, G.H. and McKay, D.L., *Hydrocarbon Process. 56*(9), pp. 97–9, 1979.
9. Sloan, H.D., *Hydrocarbon Process. 70*(11), pp. 99–102, 1991.
10. Hatch, L.F., *Hydrocarbon Process. 48*(2), pp. 77–88, 1969.
11. Mitchell, B.R., *Ind. Eng. Chem. Prod. Res. Dev. 19*, p. 209, 1980.
12. Thomas, C.L., *Ind. Eng. Chem. 41*, p. 2564, 1949.
13. Gruse, W.A. and Stevens, D.R., *Chemical Technology of Petroleum*, 3rd ed., pp. 375–387, McGraw-Hill Book Co., New York, 1960.
14. Habib, E.T., et al., *Ind. Eng. Chem. Prod. Res. Dev. 16*, p. 291, 1977.
15. Avidan, A.A., Edwards, M., and Owen, H., *Oil Gas J. 88*(2), pp. 33–58, 1990.
16. Anon., Fluid Catalytic Cracking with Molecular Sieve Catalysts, *Petro/Chem. Eng.*, pp. 12–15, May 1969.
17. Hemler, C.L., Strother, C.W., McKay, B.E., and Myers, G.D., *Catalytic Conversion of Residual Stocks*, 1979 NPRA Annual Meeting, San Antonio, TX, March 25–27, 1979.

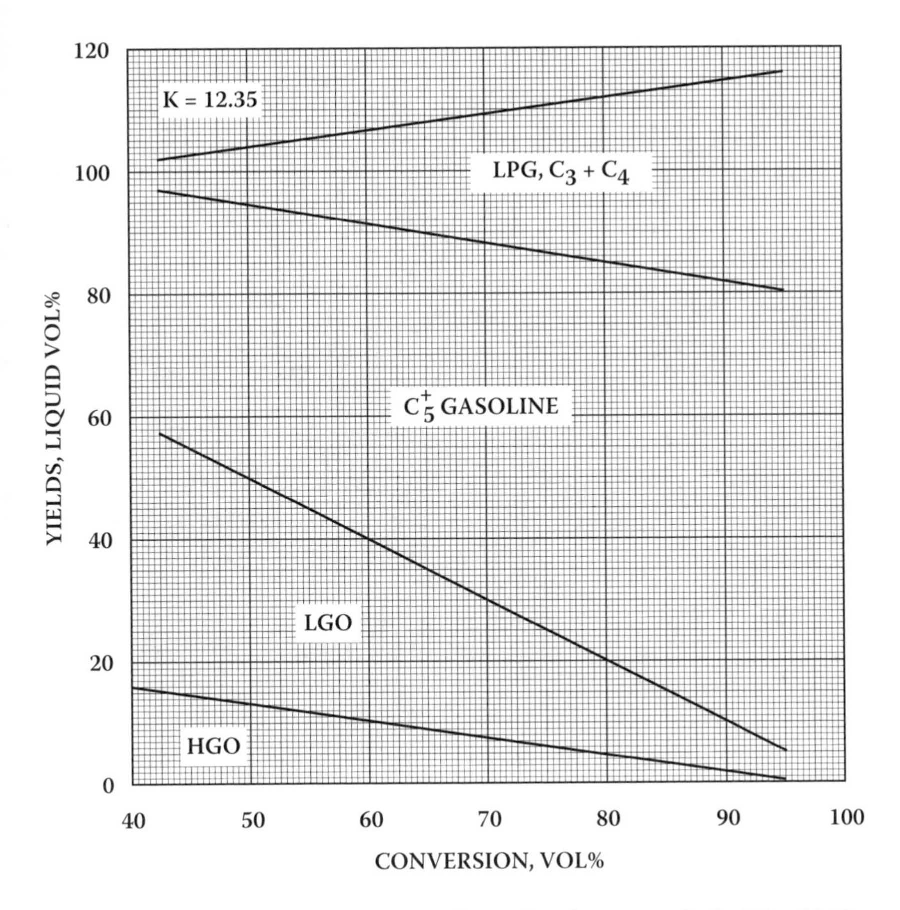

FIGURE 6.25 Catalytic cracking yields. Zeolite catalyst (heavy gas oil, feed K = 12.35).

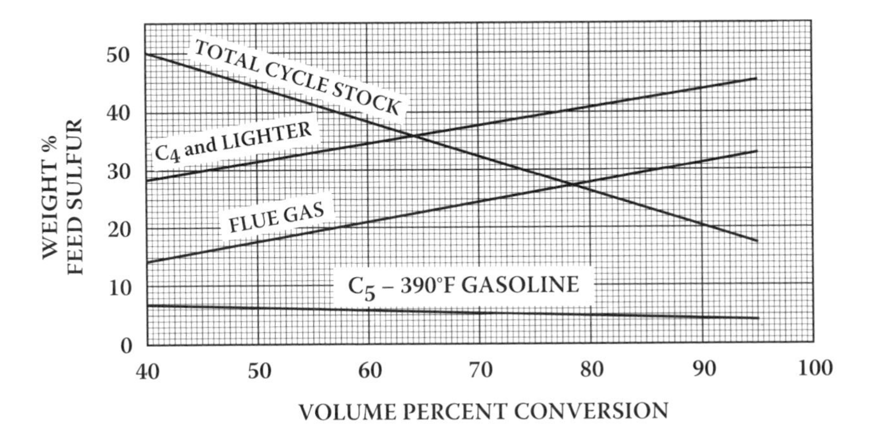

FIGURE 6.26 Distribution of sulfur in catalytic cracking products. (From Note 10.)

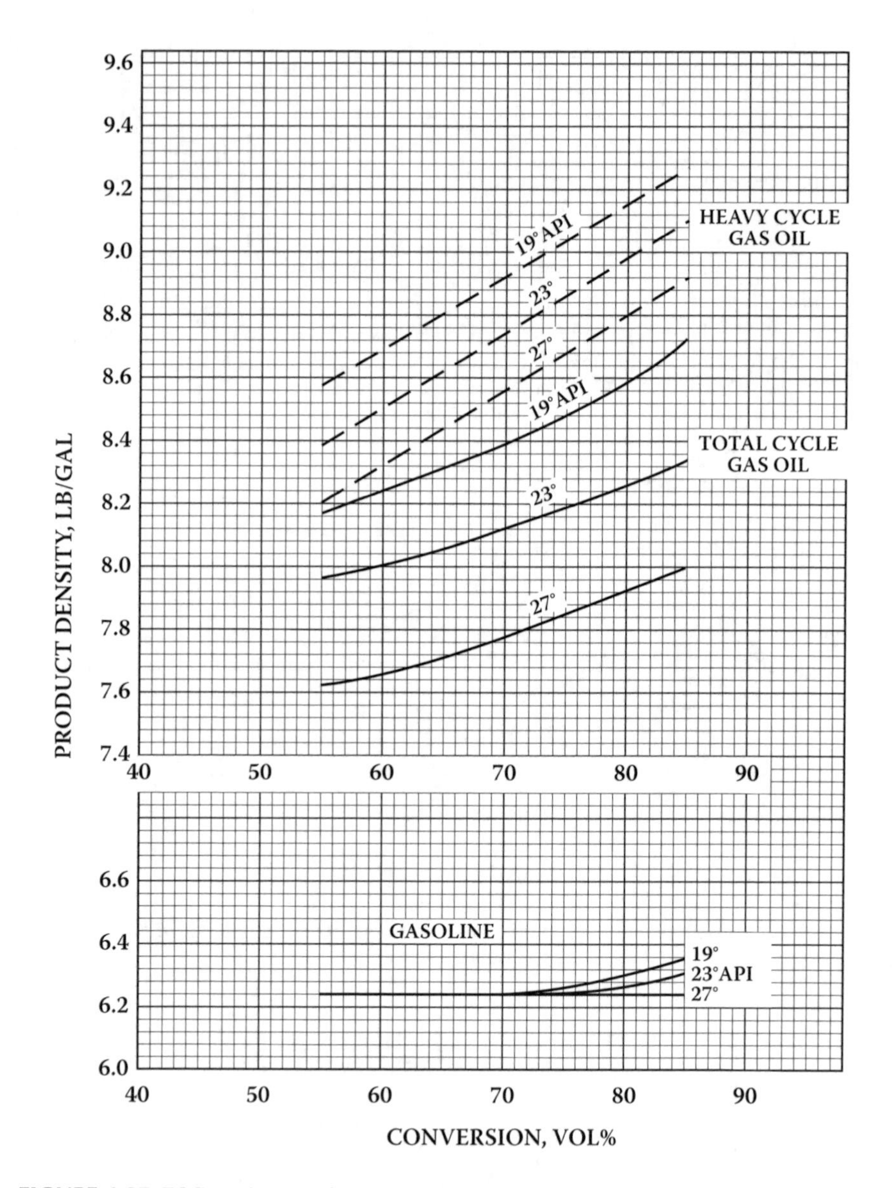

FIGURE 6.27 FCC product gravity. Zeolite catalyst.

18. Avidan, A.A., *Oil Gas J. 90*(10), pp. 59–67, 1992.
19. Knowlton, H.W., Beck, R.R., and Melnyk, J.J., *Oil Gas J. 68*(45), pp. 57–61, 1970.
20. Davey, S.W., Haley, J.T., and Zhao, X., *Davison Catalagram,* No. 86, pp. 1–12, 1998.
21. Tolen, D.F., *Oil Gas J. 79*(13), pp. 90–93, 1981.
22. Smith, M.T., *Davison Catalagram*, 86, pp. 13–22, 1998.
23. Macerato, F. and Anderson, S., *Oil Gas J. 79*(9), pp. 101–106, 1981.
24. Elvin, F.J. and Pavel, S.K., *Oil Gas J. 89*(30), pp. 94–98, 1991.
25. Danzinger, F., Groeneveld, L.R., Tracy, W.J., and Macris, A., *Oil Gas J. 97*(18), pp. 97–106, 1999.

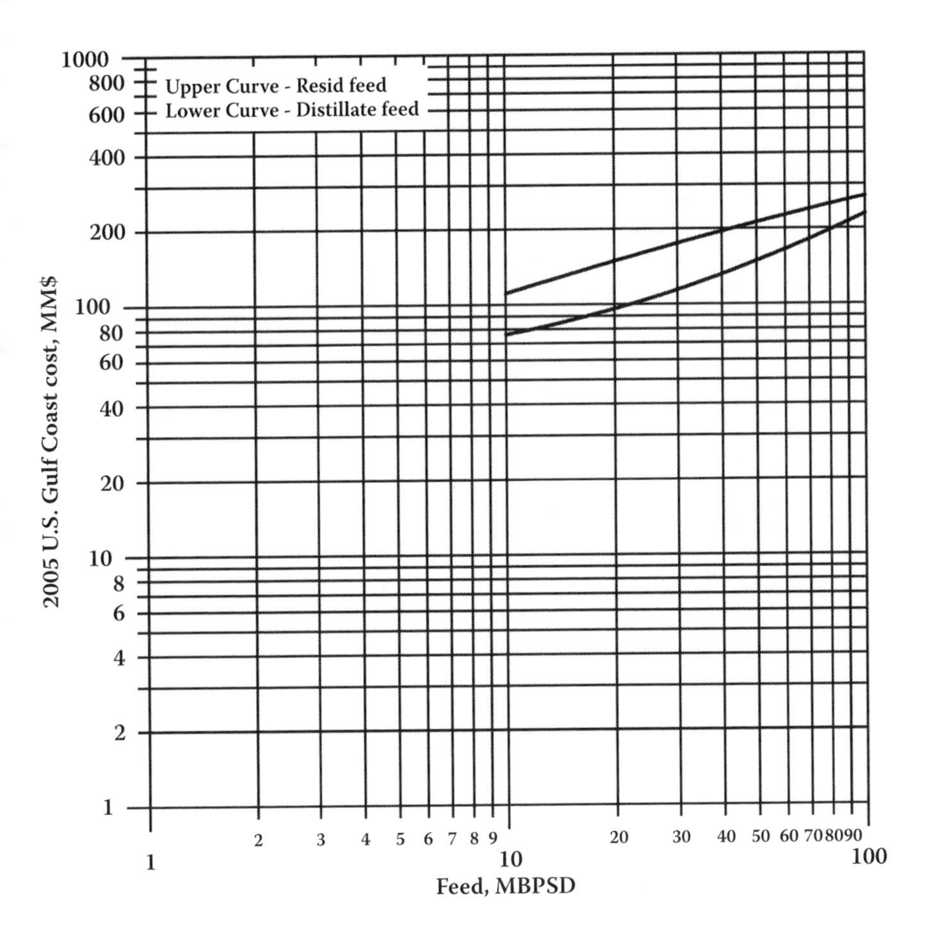

FIGURE 6.28 Fluid catalytic cracking units investment cost: 2005 U.S. Gulf Coast (see Table 6.5).

26. Shorey, S.W., Lomas, D.A., and Keesom, W.H., *Hydrocarbon Process.* *78*(11), pp. 43–51, 1999.
27. Murphy, J.R., *Oil Gas J.* *90*(10), pp. 49–58, 1992.
28. Fowle, M.J. and Bent, R.D., *Petrol. Refiner.* *26*(11), pp. 719–727, 1947.

ADDITIONAL READING

29. Castiglioni, B.P., *Hydrocarbon Process.* *62*(2), pp. 35–38, 1983.
30. Dean, R.R., Mauleon, J.L., and Letsch, W.S., *Oil Gas J.* *80*(40), pp. 75–80, *80*(41), pp. 168–176, 1982.

TABLE 6.5
Fluid Catalytic Cracking Unit Cost Data

Costs included	
1. Product fractionation	
2. Gas compression of concentration for recovery of 95% and fractionation of C_4s and 80% of C_3	
3. Complete reactor-regenerator section	
4. Sufficient heat exchange to cool products to ambient temperature	
5. Central control system	
Costs not included	
1. Feed fractionation	
2. Off-gas and product treating	
3. Cooling water, steam, and power supply	
4. Initial catalyst charge	
Royalty	
Running royalty about $0.10/bbl	
Paid-up royalty about $110/BPD	
Utility data (per bbl feed)	
Steam, lb[a]	
Power, kWh[b]	6.0
Cooling water, gal (30FΔt)	500
Fuel (LHV), MMBtu	0.1
Catalyst Replacement, $	0.25–0.80

[a] Waste heat steam production usually is in excess of consumption by approximately 30 lb of steam per barrel of fresh feed.

[b] Includes electric drive for air blower and off-gas compressor.

Note: See Figure 6.2.

TABLE 6.6
FCC Material Balance: 100,000 BPCD Alaska North Slope Crude Oil Basis (Severity, 75% Conversion; Zeolite Catalyst)

Component	vol%	BPD	°API	(lb/h)/ BPD	lb/h	wt% S	lb/h S
Feed							
650–850°F (343–455°C)	56.7	20,090	23.2	13.35	268,103	0.10	261
850–1050°F (455–566°C)	43.3	15,331	16.5	13.95	213,857	0.29	630
kW = 11.67	100.0	35,421	20.0	13.61	481,960	0.18	891
Product							
Coke, wt%	5.80				27,954	0.83	232[a]
C_2 and lighter, wt%	5.50				26,508	1.30	344
$C_3^=$	5.30	1,877		7.61	14,286		
C_3	2.90	1,027		7.42	7,591		
$C_4^=$	9.50	3,365		8.76	29,477		
iC_4	6.10	2,161		8.22	17,718		
nC_4	2.70	956		8.51	8,139		
C_5^+ naphtha	56.79	20,117	57.2	10.94	(220,115)	0.02	40
TGO	(25.0)	(8,855)	8.9	14.70	130,172	0.21	275
LGO	18.4	(6,517)	12.6	(14.32)	(93,352)	0.17	158
HGO	6.6	2,338	–0.6	15.75	36,820	0.32	117
Total		38,359			481,960		891

[a] In flue gas as SO_2.

TABLE 6.7
FCC Catalyst and Utility Requirements

Power, MkWh/d	213
Cooling water, gpm	17,711
Fuel, MMBtu/d	3,542
Catalyst, $/d	5,313
Steam, Mlb/d	(1063)

Atomspheric & Vacuum Gas Oils → Cat Cracker
Cycle oil & Coker Distillates → Hydrocracking Cata
Atmospheric Resid → Hydroprocessing
Vaccuum Resid → Heavy oil cracking / coker
Heavy Naphtha → Platformer

Resid processing

↓ catalytic
ARC
- hydroprocessing (moving bed)
- hydrocracking (expanded bed) } technically still hydroprocessing
- solvent extraction

↘ non-catalytic
VRC
coking

Hydrotreating: reduce sulfur, saturate olefins, reduce N, does not change BP much

Hydrocracking: primary purpose is to reduce b.p.

Hydroprocessing: significant sulfur removal and b.p reduction

Hydroprocessing: operates on residuals
Hydrocracking: operates on distillates

7 Catalytic Hydrocracking

Although hydrogenation is one of the oldest catalytic processes used in refining petroleum, only since the 1960s has catalytic hydrocracking developed to any great extent in this country. This interest in the use of hydrocracking has been caused by several factors, including (1) the demand for petroleum products has shifted to high demand for gasoline, diesel, and jet fuel compared with the usages of other products, (2) by-product hydrogen at low cost and in large amounts has become available from catalytic reforming operations, and (3) environmental concerns limiting sulfur and aromatic compound concentrations in motor fuels have increased.

The hydrocracking process was commercially developed by I. G. Farben Industrie in 1927 for converting lignite into gasoline and was brought to this country by Esso Research and Engineering Co. in the early 1930s for use in upgrading petroleum feedstocks and products, but the first modern distillate hydrocracker was put into commercial operation by Chevron in 1958 [1]. Improved catalysts have been developed that permit operations at lower pressures than the earlier units, and the demand for high-octane unleaded gasolines, jet fuels, and low aromatics and very low sulfur-content diesel fuels has promoted the conversion of higher-boiling petroleum materials to gasoline, jet fuels, and low-sulfur, low-aromatics diesel fuels.

Product balance is of major importance to any petroleum refiner. There are a number of things that can be done to balance the products made with the demand, but there are relatively few operations that offer the versatility of catalytic hydrocracking. Some of the advantages of hydrocracking are

1. Better balance of gasoline and distillate production
2. Greater gasoline boiling-range naphtha yields
3. Improved gasoline pool octane quality and sensitivity
4. Production of relatively high amounts of isobutane in the butane fraction
5. Supplementing of fluid catalytic cracking to upgrade heavy cracking stocks, aromatics, cycle oils, and coker oils to gasoline, jet fuels, and diesel

In a modern refinery, catalytic cracking and hydrocracking work as a team. The catalytic cracker takes the more easily cracked paraffinic atmospheric and vacuum gas oils as charge stocks, whereas the hydrocracker uses more aromatic cycle oils and coker distillates as feed [2]. These streams are very refractory and resist catalytic cracking, whereas the higher pressures and hydrogen atmosphere make them relatively easy to hydrocrack. The new zeolite cracking catalysts help improve the gasoline yields and octanes from catalytic crackers as well as reduce the cycle stock and gas make. However, the cycle oil still represents a difficult fraction to crack catalytically to extinction. One alternative is to use the cycle stock as a component for fuel oil blending, but this is limited, as it is a relatively poor-burning stock and burns with a smoky flame. For this reason, a limit is placed on the percentage that

TABLE 7.1
Typical Hydrocracker Feedstocks

Feed	Products
Kerosine	Naphtha
Straight-run diesel	Naphtha or jet fuel
Atmospheric gas oil	Naphtha, jet fuel, or diesel
Vacuum gas oil	Naphtha, jet fuel, diesel, lube oil
FCC LCO	Naphtha
FCC HCO	Naphtha or distillates
Coker LCGO	Naphtha or distillates
Coker HCGO	Naphtha or distillates

Source: Reference 3 and Reference 6.

can be blended into distillate fuel oils. The cycle oils that result from cracking operations with zeolite catalysts tend to be highly aromatic and, therefore, make satisfactory feedstocks for hydrocracking. Vacuum and coker gas oils are also used as hydrocracker feed, but the end points are lower than those for FCC feedstocks.

Sometimes, diesel boiling range material is included in hydrocracker feed to make jet and motor gasoline products. Both straight-run and FCC LCO can be used, and in some cases, 100% LCO is used. In cases where 100% LCO is the feed, there is a follow-up with a high-pressure hydrotreater to reduce the aromatic content and increase the smoke point to meet specifications. When the feed contains large amounts of LCO, the major effects are increased heat release and lower smoke point of the jet fuel product [3].

In addition to middle distillates and cycle oils used as feed for hydrocracking units, it is also possible to process residual fuel oils and reduced crude by hydrocracking. This usually requires a different technology, and for the purposes of our discussion, the hydrocracking operation is broken into two general types of processes: those that operate on distilled feed (hydrocracking) and those that process residual materials (hydroprocessing). These processes are similar, and some licensed processes have been adapted to operate on both types of feedstocks. There are major differences, however, between the two processes in regard to the type of catalyst and operating conditions. During the design stages of the hydrocracker, the process can be tailored to convert heavy residue into lighter oils or to change straight-run naphthas into liquefied petroleum gases. This is difficult to do after the unit is built, as the processing of residual oil requires special consideration with respect to such factors as asphaltenes, ash, and metal contents of the feedstocks [4,5]. Hydroprocessing is discussed in Chapter 8, and typical feedstocks and products for hydrocrackers are given in Table 7.1.

7.1 HYDROCRACKING REACTIONS

Although there are hundreds of simultaneous chemical reactions occurring in hydrocracking, it is the general opinion that the mechanism of hydrocracking is that of catalytic cracking with hydrogenation superimposed (see Figure 7.1). Catalytic

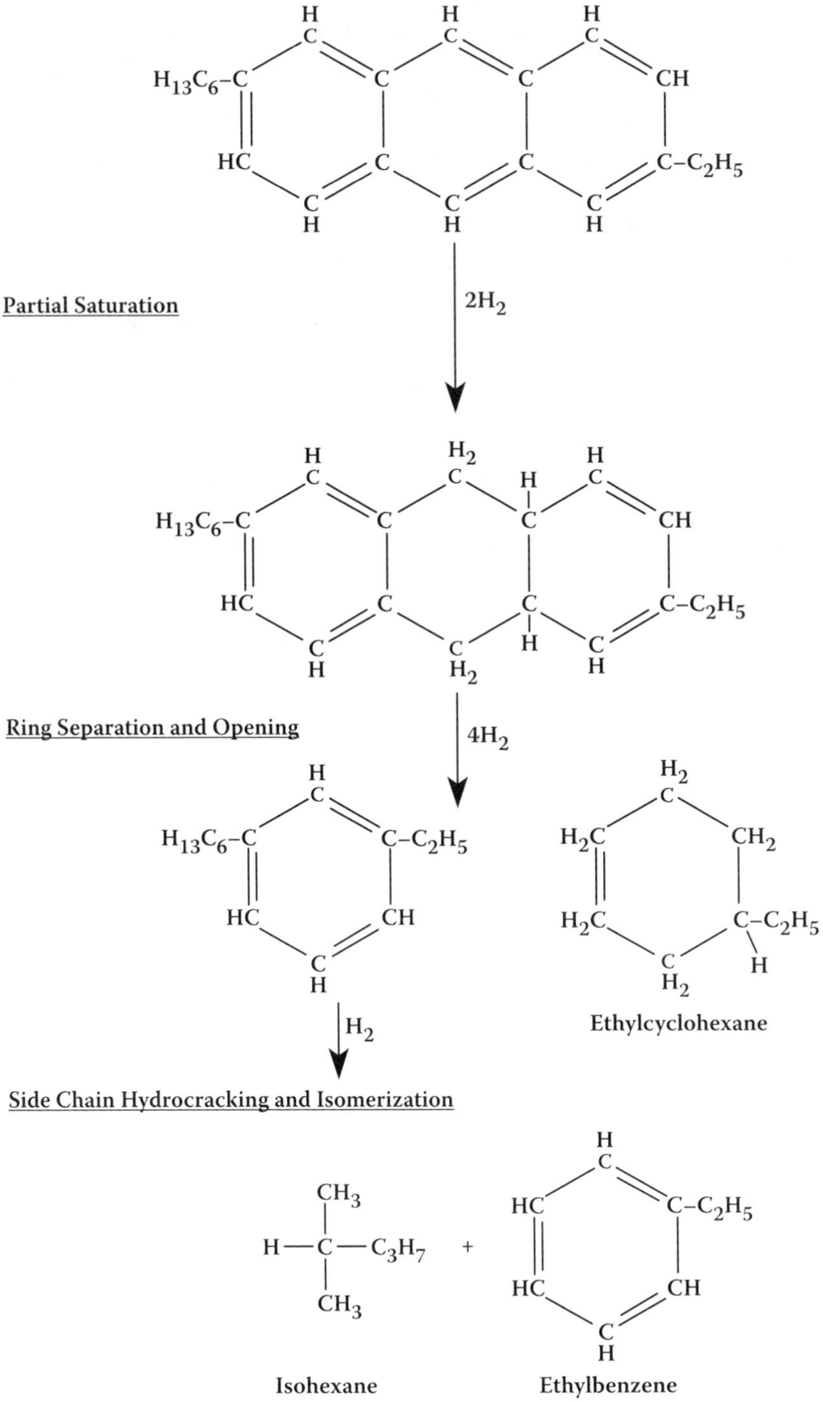

Partial Saturation

$2H_2$

Ring Separation and Opening

$4H_2$

Ethylcyclohexane

H_2

Side Chain Hydrocracking and Isomerization

Isohexane

Ethylbenzene

FIGURE 7.1 Typical hydrocracking reactions.

cracking is the scission of a carbon–carbon single bond, and hydrogenation is the addition of hydrogen to a carbon–carbon double bond. An example of the scission of a carbon–carbon single bond followed by hydrogenation is the following:

This shows that cracking and hydrogenation are complementary, for cracking provides olefins for hydrogenation, whereas hydrogenation in turn provides heat for cracking. The cracking reaction is endothermic, and the hydrogenation reaction is exothermic. The overall reaction provides an excess of heat because the amount of heat released by the exothermic hydrogenation reaction is much greater than the amount of heat consumed by the endothermic cracking reactions. This surplus of heat causes the reactor temperature to increase and accelerate the reaction rate. This is controlled by injecting cold hydrogen as a quench into the reactors to absorb the excess heat of reaction.

Another reaction that illustrates the complementary operation of the hydrogenation and cracking reactions is the initial hydrogenation of a condensed aromatic compound to a cycloparaffin. This allows subsequent ring cracking to proceed to a greater extent and thus converts a low-value component of catalytic cycle oils to a useful product.

Isomerization is another reaction type that occurs in hydrocracking and accompanies the cracking reaction. The olefinic products formed are rapidly hydrogenated, thus maintaining a high concentration of high-octane isoparaffins and preventing the reverse reaction back to straight-chain molecules. An interesting point in connection with the hydrocracking of these compounds is the relatively small amounts of propane and lighter materials that are produced as compared with normal cracking processes. The volumetric yield of liquid products can be as high as 125% of the feed, because the hydrogenated products have a higher API gravity (lower specific gravity) than the feed.

Hydrocracking reactions are normally carried out at average catalyst temperatures between 550 and 750°F (290 and 400°C) and at reactor pressures between 1200 and 2200 psig (8,275 and 15,200 kPa). The circulation of large quantities of hydrogen with the feedstock prevents excessive catalyst fouling and permits long runs without catalyst regeneration. Careful preparation of the feed is also necessary in order to remove catalyst poisons and to give long catalyst life. Frequently, the feedstock is hydrotreated to remove sulfur and nitrogen compounds as well as metals before it is sent to the first hydrocracking stage, or sometimes, the first reactor in the reactor train (called a guard reactor) can be used for this purpose.

7.2 FEED PREPARATION

Hydrocracking catalyst is susceptible to poisoning by metallic salts, oxygen, organic nitrogen compounds, and sulfur in the feedstocks. The feedstock is hydrotreated to saturate the olefins and remove sulfur, nitrogen, and oxygen compounds. Molecules containing metals are cracked, and the metals are retained on the catalyst. The nitrogen and sulfur compounds are removed by conversion to ammonia and hydrogen sulfide. Although organic nitrogen compounds are thought to act as permanent poisons to the catalyst, the ammonia produced by reaction of the organic nitrogen compounds with hydrogen does not affect the catalyst permanently. For some types of hydrocracking catalysts, the presence of hydrogen sulfide in low concentrations acts to inhibit the saturation of aromatic rings. This is a beneficial effect when maximizing gasoline production, as it conserves hydrogen and produces a higher octane product.

In the hydrotreater, a number of hydrogenation reactions, such as olefin saturation and aromatic ring saturation, take place, but cracking is almost insignificant at the operating conditions used. The exothermic heats of the desulfurization and denitrogenation reactions are high [about 65 to 75 Btu/scf of hydrogen consumed (2400 to 2800 kJ/std m^3)]. If the nitrogen and sulfur contents of the feedstock are high, this effect contributes appreciably to the total heat of reaction.

Another reaction contributing to high heat release in the hydrotreating process is the saturation of olefins, as the heat of reaction for olefin saturation is about 140 Btu/scf (42 kJ/Nm^3) of hydrogen consumed. For cracked feedstocks, the olefins content is very high, and olefin saturation is responsible for a large portion of the total heat of reaction. For virgin stocks, however, the olefin content is negligible, and this is not an important contribution to the heat of reaction. The overall heat of reaction for most hydrotreating reactors used for the preparation of hydrocracker feed is approximately 25,000 to 35,000 Btu per barrel (166 to 232 MJ/m^3) feed.

In addition to the removal of nitrogen and sulfur compounds and metals, it is also necessary to reduce the water content of the feed streams to less than 25 ppm because, at the temperatures required for hydrocracking, steam causes the crystalline structure of the catalyst to collapse and the dispersed rare-earth atoms to agglomerate. Water removal is accomplished by passing the feed stream through a silica gel or molecular-sieve dryer. Exceptions to this are the Unicracking and GOFining processes, which can tolerate water contents as high as 400 to 500 ppm; it is necessary only to remove free water from the feed.

On the average, the hydrogen treating processing requires approximately 150 to 300 ft^3 of hydrogen per barrel of feed (27 to 54 m^3 of hydrogen per m^3 of feed).

7.3 THE HYDROCRACKING PROCESS

There are a number of hydrocracking processes available for licensing, and some of these are given in Table 7.2. With the exception of the H-Oil, LC-Fining, HYCON,

TABLE 7.2
Hydrocracking Processes Available for License

Process	Company	Note
Isomax	Chevron and UOP LLC	7
Unicracking	UOP LLC	
GOFining	ExxonMobil Research and Engineering	
Ultracracking	BP	8
Shell	Shell Development Co.	
BASF-IFP hydrocracking	Badische Anilin and Soda Fabrik, and Institute Francais Petrole	
Unibon	UOP LLC	

and OCR processes, all hydrocracking and hydroprocessing processes in use today are fixed-bed catalytic processes with liquid downflow. The hydrocracking process may require either one or two stages, depending upon the process and the feedstocks used. The process flows of most of the fixed-bed processes are similar, and the GOFining process will be described as a typical fixed-bed hydrocracking process.

The GOfining process is a fixed-bed regenerative process employing a molecular-sieve catalyst impregnated with a rare-earth metal. The process employs either single-stage or two-stage hydrocracking with typical operating conditions ranging from 660 to 785°F (350 to 420°C) and from 1000 to 2000 psig (6900 to 13,800 kPa). The temperature and pressure vary with the age of the catalyst, the product desired, and the properties of the feedstock.

The decision to use a single- or two-stage system depends upon the size of the unit and the product desired. For most feedstocks, the use of a single stage will permit the total conversion of the feed material to gasoline and lighter products by recycling the heavier material back to the reactor. The process flow for a two-stage reactor is shown in Figure 7.2. If only one stage is used, the process flow is the same as that of the first stage of the two-stage plant, except the fractionation tower bottoms is recycled to the reactor feed.

The fresh feed is mixed with makeup hydrogen and recycle gas (high in hydrogen content) and passed through a heater to the first reactor. If the feed has not been hydrotreated, there is a guard reactor before the first hydrocracking reactor. The guard reactor usually has modified hydrotreating catalyst such as cobalt-molybdenum on silica-alumina to convert organic sulfur and nitrogen compounds to hydrogen sulfide, ammonia, and hydrocarbons to protect the precious metals catalyst in the following reactors. The hydrocracking reactor(s) is operated at a sufficiently high temperature to convert 40 to 50 vol% of the reactor effluent to material boiling below 400°F (205°C). The reactor effluent goes through heat exchangers to a high-pressure separator, where the hydrogen-rich gases are separated and recycled to the first stage for mixing both makeup hydrogen and fresh feed. The liquid product from the separator is sent to a distillation column, where the C_4 and lighter gases are taken off overhead, and the light and heavy naphtha, jet fuel, and diesel fuel boiling range streams are removed as liquid sidestreams. The fractionator bottoms are used as feed to the second-stage reactor system. The unit can be operated to produce all gasoline and lighter products or to maximize jet fuel or diesel fuel products (see Figure 7.3).

The bottoms stream from the fractionator is mixed with recycle hydrogen from the second stage and sent through a furnace to the second-stage reactor. Here, the temperature is maintained to bring the total conversion of the unconverted oil from the first-stage and second-stage recycle to 50 to 70 vol% per pass. The second-stage product is combined with the first-stage product prior to fractionation.

Both the first- and second-stage reactors contain several beds of catalysts. The major reason for having separate beds is to provide locations for injecting cold recycled hydrogen into the reactors for temperature control. In addition, redistribution of the feed and hydrogen between the beds helps to maintain a more uniform utilization of the catalyst.

When operating hydrocrackers for total conversion of distillate feeds to gasoline, the butane-and-heavier liquid yields are generally from 120 to 125 vol% of fresh feed.

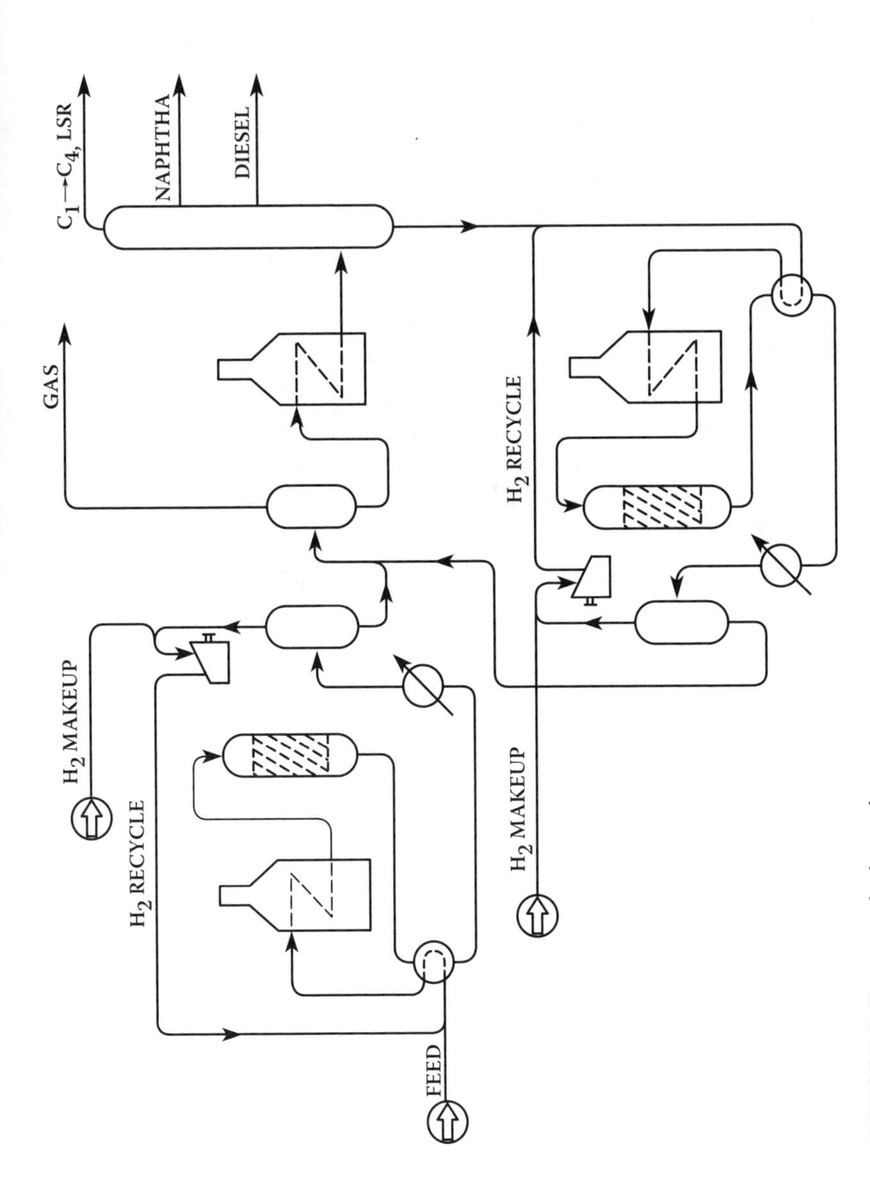

FIGURE 7.2 Two-stage hydrocracker.

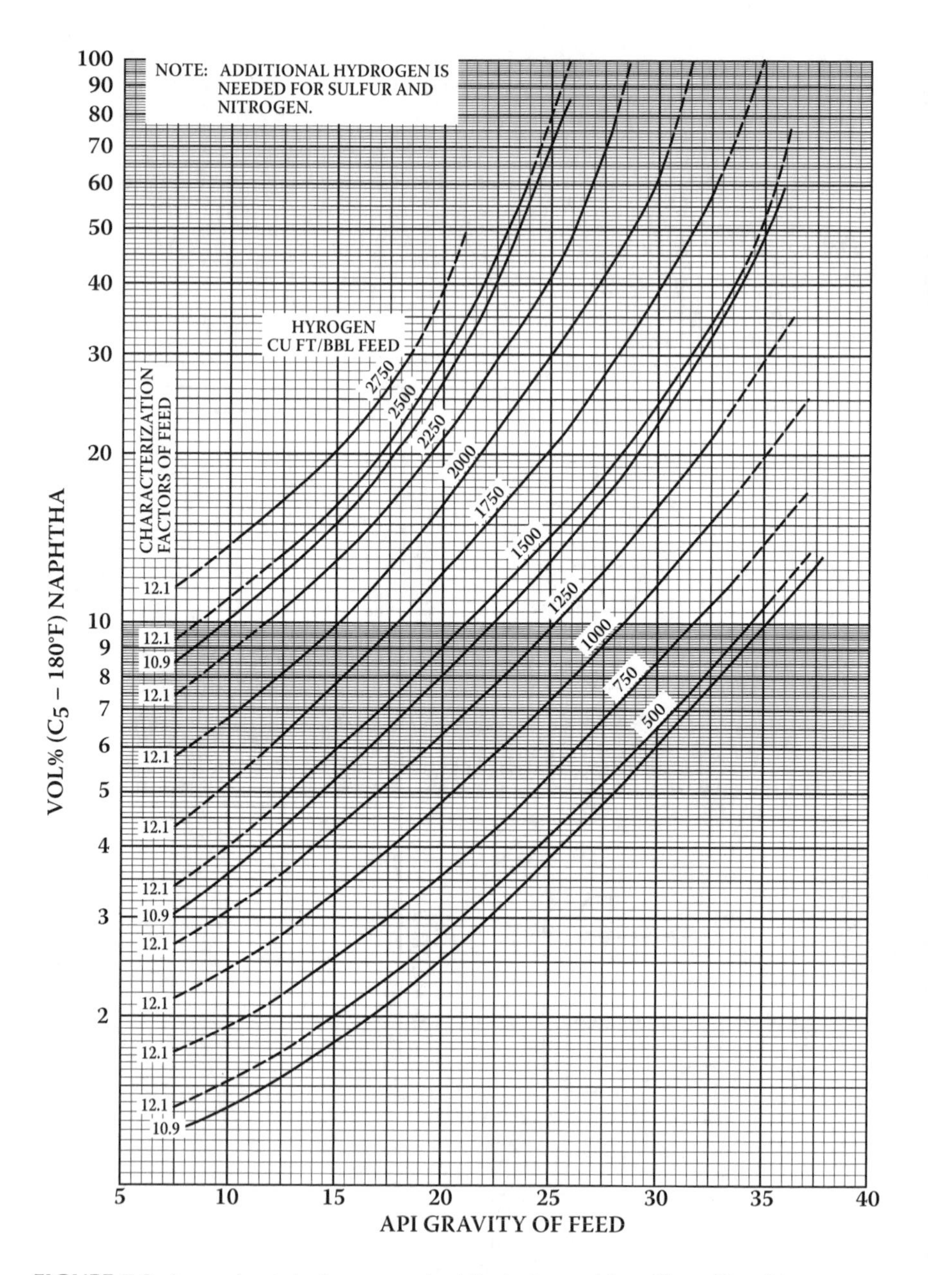

FIGURE 7.3 Approximate hydrogen required for hydrocracking. (From Note 9.)

7.4 HYDROCRACKING CATALYST

There are a number of hydrocracking catalysts available, and the actual composition is tailored to the process, feed material, and products desired. Most of the hydrocracking catalysts consist of a crystalline mixture of silica-alumina with a small uniformly distributed amount of rare earths contained within the crystalline lattice. The silica-alumina portion of the catalyst provides cracking activity, whereas the rare-earth metals promote hydrogenation. Catalyst activity decreases with use, and reactor temperatures are raised during a run to increase reaction rate and maintain conversion. The catalyst selectivity also changes with age, and more gas is made and less naphtha produced as the catalyst temperature is raised to maintain conversion. With typical feedstocks, it will take from 2 to 4 years for catalyst activity to decrease from the accumulation of coke and other deposits to a level that will require regeneration. Regeneration is accomplished by burning off the catalyst deposits, and catalyst activity is restored to close to its original level. The catalyst can undergo several regenerations before it is necessary to replace it.

Almost all hydrocracking catalysts use silica-alumina as the cracking base, but the rare-earth metals vary according to the manufacturer. Those in most common use are platinum, palladium, tungsten, and nickel.

7.5 PROCESS VARIABLES

The severity of the hydrocracking reaction is measured by the degree of conversion of the feed to lighter products. Conversion is defined as the volume percentage of the feed that disappears to form products boiling below the desired product end point. In order to compare operation severities, it is necessary to equate conversions to the same product end point. A given percentage conversion at a low product end point represents a more severe operation than does the same percentage conversion at a higher product end point.

The primary reaction variables are reactor temperature and pressure, space velocity, hydrogen consumption, nitrogen content of the feed, and hydrogen sulfide content of the gases. The effects of these are as follows:

7.5.1 Reactor Temperature

Reactor temperature is the primary means of conversion control. At normal reactor conditions, a 20°F (10°C) increase in temperature almost doubles the reaction rate, but does not affect the conversion level as much, because a portion of the reaction involves material that has already been converted to materials boiling below the desired product end point. As the run progresses, it is necessary to raise the average temperature about 0.1 to 0.2°F per day to compensate for the loss in catalyst activity.

7.5.2 Reactor Pressure

The primary effect of reactor pressure is in its effect on the partial pressures of hydrogen and ammonia. An increase in total pressure increases the partial pressures of both hydrogen and ammonia. Conversion increases with increasing hydrogen partial pressure and decreases with increasing ammonia partial pressure. The hydrogen effect is greater, however, and the net effect of raising total pressure is to increase conversion.

7.5.3 Space Velocity

The volumetric space velocity is the ratio of liquid flow rate, in barrels per hour, to catalyst volume, in barrels. The catalyst volume is constant; therefore, the space velocity varies directly with feed rate. As the feed rate increases, the time of catalyst contact for each barrel of feed is decreased and conversion is lowered. In order to maintain conversion at the proper level when the feed rate is increased, it is necessary to increase the temperature.

7.5.4 Nitrogen Content

The organic nitrogen content of the feed is of great importance, as the hydrocracking catalyst is deactivated by contact with organic nitrogen compounds. An increase in organic nitrogen content of the feed causes a decrease in conversion.

7.5.5 Hydrogen Sulfide

At low concentrations, the presence of hydrogen sulfide acts as a catalyst to inhibit the saturation of aromatic rings. This conserves hydrogen and produces a product with a higher octane number because the aromatic naphtha has a higher octane than does its naphthenic counterpart. However, hydrocracking in the presence of a small amount of hydrogen sulfide normally produces a very low smoke-point jet fuel. At high hydrogen sulfide levels, corrosion of the equipment becomes important, and the cracking activity of the catalyst is also affected adversely.

7.5.6 Heavy Polynuclear Aromatics

Heavy polynuclear aromatics (HPNA) are formed in small amounts from hydrocracking reactions and, when the fractionator bottoms are recycled, can build up to concentrations that cause fouling of heat exchanger surfaces and equipment. Steps such as reducing the feed end point or removal of a drag stream may be necessary to control this problem [10].

7.6 HYDROCRACKING YIELDS

Yields from the hydrocracker are functions of the type of crude oil, previous processing operations, type and activity of catalyst used, and operating conditions. Representative yields for straight-run and cracked feedstocks are given in Table 7.3.

For the purpose of the example problem, the yields for hydrocracking to produce gasoline as the primary product can be calculated from charts and equations developed by Nelson [8, 10, 11]. The data needed to start the calculations are the Watson characterization factor (K_W) of the feed and the hydrogen consumption in scf/bbl feed. With this information proceed as follows:

1. Use Figure 7.4 to determine the vol% (C_5-180°F) naphtha.
2. With Figure 7.5, use the vol% (C_5-180°F) naphtha and feed K_W to obtain vol% (180 to 400°F) naphtha.

TABLE 7.3
Hydrocracking Yields

	Naphtha	Jet	Diesel
Yields on Coker Gas Oil and FCC Decanted Oil (vol% on feed)			
Butanes	17	8	5
C_5-180°F	32	15	9
180–380°F	81	24	20
Jet or diesel	—	74	84
Total	130	121	118
Yields on Coker Gas Oil and FCC Decant Oil (vol% on feed)			
Butanes		5.2	
C_5-185°F		8.8	
180–435°F		31.8	
435–650°F, diesel		33.8	
650°F+, gas oil		35.0	
Total		114.6	

Source: Note 6.

3. Calculate the liquid vol% butanes formed using:

$$\text{LV\% iC}_4 = 0.377\ [\text{LV\% }(\text{C}_5 - 180°\text{F})]$$

$$\text{LV\% nC}_4 = 0.186\ [\text{LV\% }(\text{C}_5 - 180°\text{F})]$$

4. Calculate the wt% of propane and lighter using:

$$\text{wt\% C}_3 \text{ and lighter} = 1.0 + 0.09\ [\text{LV\% }(\text{C}_5 - 180°\text{F})].$$

It is necessary to make both weight and hydrogen balances on the unit. The gravities of the product streams can be calculated using the K_W factors of the product streams obtained from Figure 7.5 and average midboiling points of 131, 281, and 460°F for the C_5-180°F naphtha, 180 to 400°F naphtha, and 400+ °F streams, respectively. The weight of the 400+ °F stream is obtained by difference. The chemical hydrogen consumed should be included with the total weight of the feed.

Hydrogen contents of the streams can be estimated using the wt% hydrogen for each stream, except the heavy hydrocrackate (180 to 400°F), obtained from Figure 7.6 [12]. The heavy hydrocrackate is highly naphthenic and contains from 13.3 to 14.5 wt% hydrogen (avg. 13.9%). Assume hydrogen loss by solution in products of 1 lb/bbl feed (range 0.8 to 1.3 lb H_2/bbl feed) [9].

It should be noted that if the yield of C_5-180°F naphtha is greater than 25 to 30 vol%, the yield of heavy hydrocrackate (180 to 400°F naphtha) will be determined from a curve having a negative slope. This indicates an economically unattractive situation because heavy hydrocrackate is being cracked to lighter materials. A less severe operation should be used.

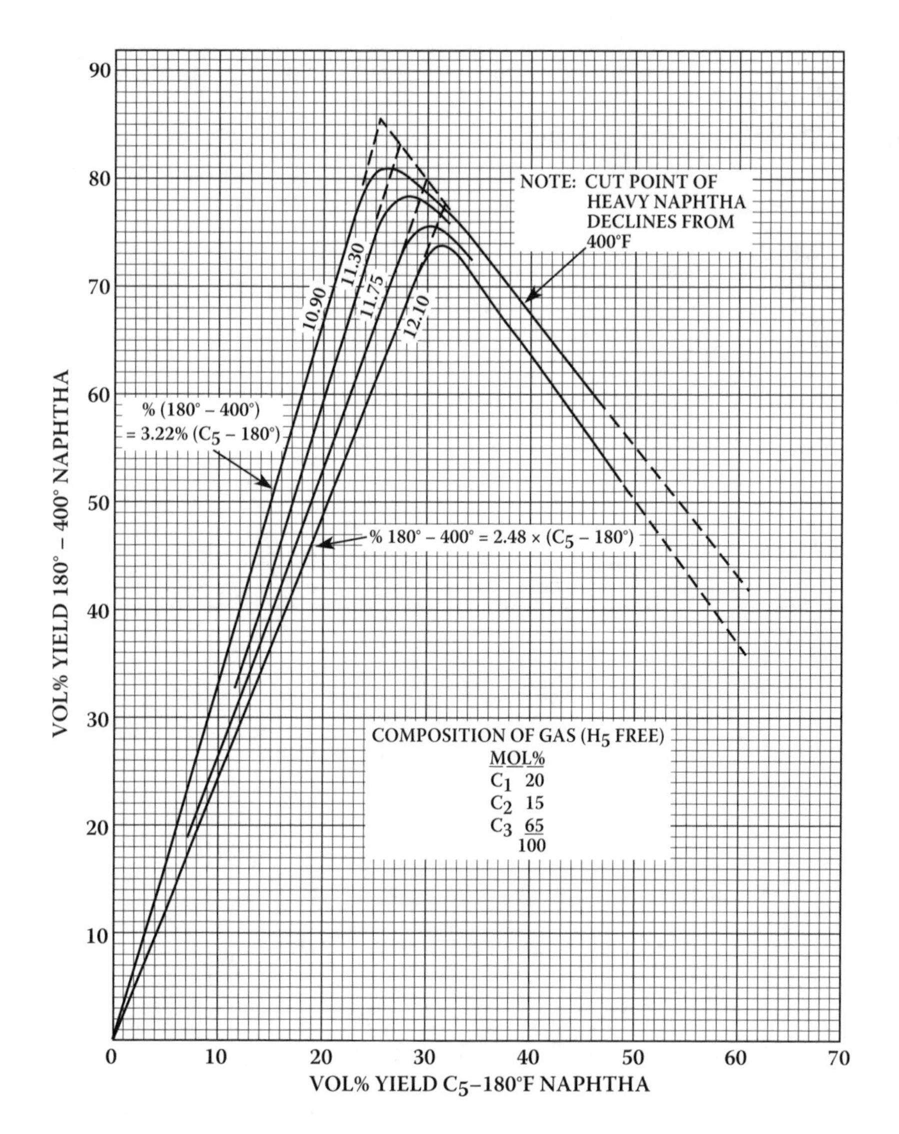

FIGURE 7.4 Relationship between yields of C_5-180°F and 180–400°F hydrocrackates. (From Note 9.)

The composition of the C_3 and lighter stream will vary depending upon feedstock properties and operating conditions. For the purpose of preliminary studies, the following composition can be assumed:

	mol%	Wt%
C_1	20	8.8
C_2	15	12.4
C_3	65	78.8
Total	100	100.0

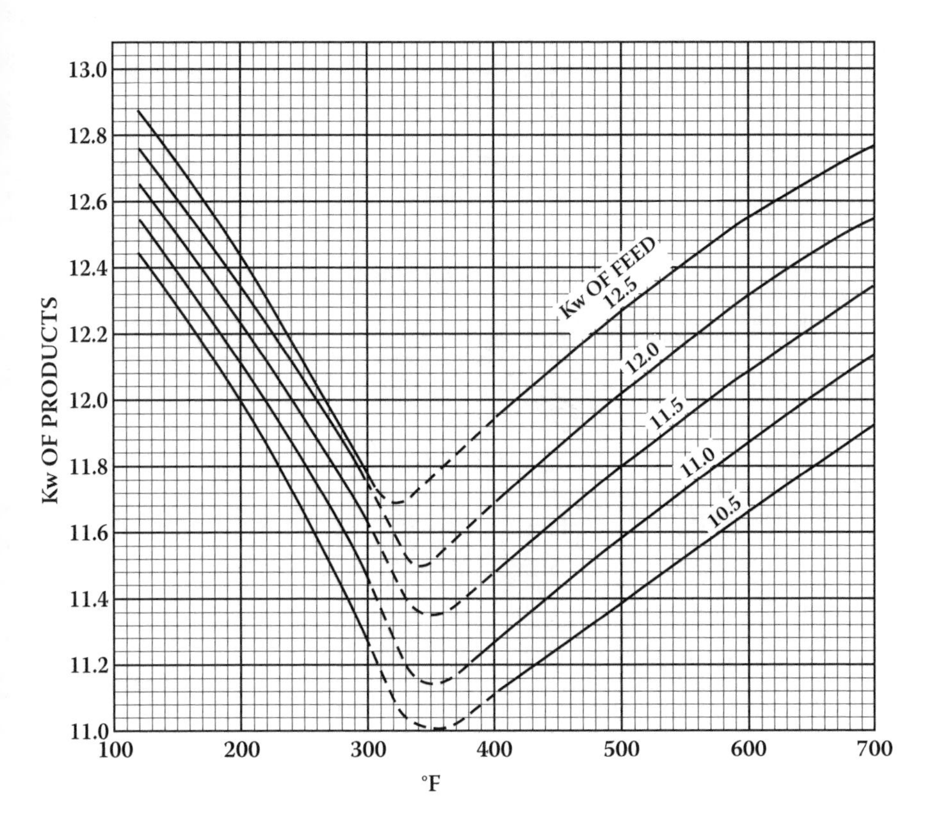

FIGURE 7.5 Characterization factor of hydrocracker products. (From Note 11.)

These values were determined by averaging the compositions obtained when processing 13 feedstocks ranging from virgin gas oil to coker and fluid catalytic cracker gas oil. Within the precision of the data, the same average composition was found for the C_3 and lighter streams when operating to obtain all gasoline or maximum jet fuel liquid products.

7.7 INVESTMENT AND OPERATING COSTS

Capital investment costs for catalytic hydrocracking units can be estimated from Figure 7.7. Table 7.4 lists the items included in the investment cost obtained from Figure 7.7 and also the utility requirements for operation.

7.8 MODES OF HYDROCRACKER OPERATION

Full-conversion hydrocracking operations are very expensive in terms of both original capital cost and direct operating cost because of the high pressures at which the units operate [1800 to 2500 psig (120 to 170 barg, 12.4 to 35 MPa)]. As a result, units designed to operate at lower pressures are being used to obtain some of the benefits of hydrocracking at lower costs. These units are called mild hydrocracking

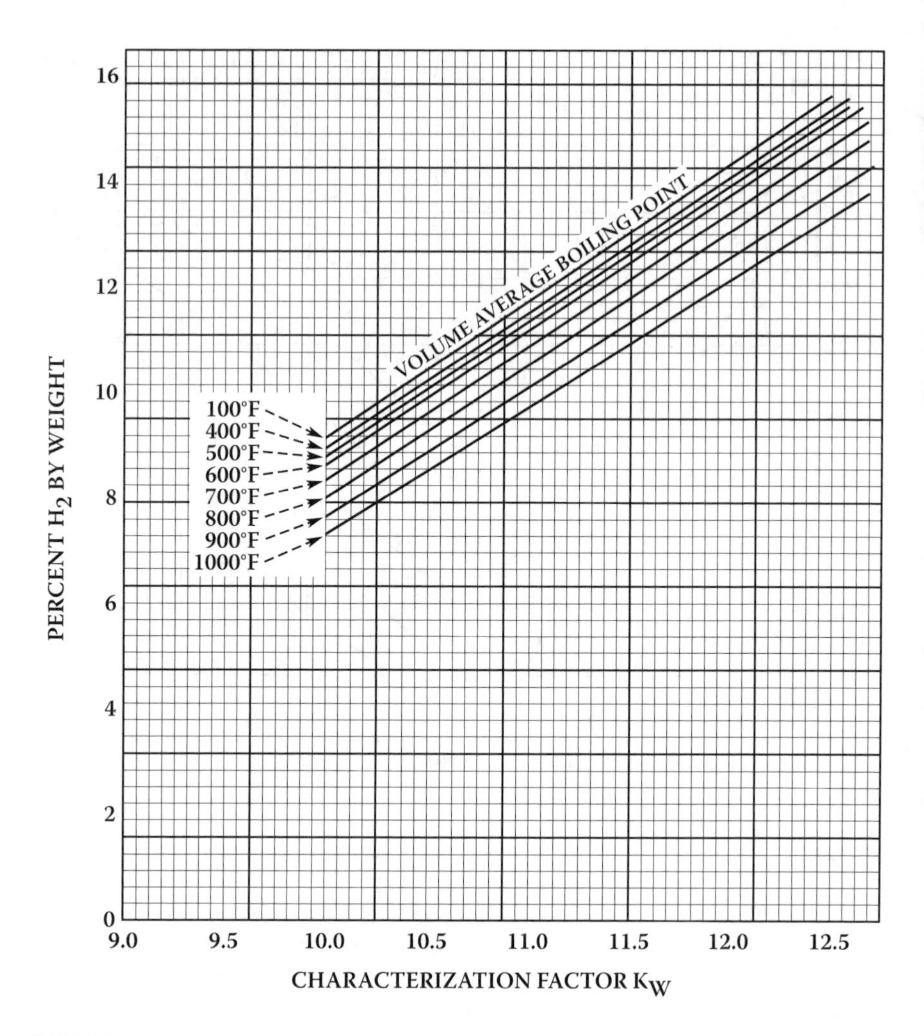

FIGURE 7.6 Hydrogen content of hydrocarbons. (From Note 13.)

(MHC) units or moderate pressure hydrocracking (MPHC) units. UOP refers to the MPHC units as partial-conversion units [14]. The MHC units operate at pressures in the range of 800 to 1200 psig (55 to 82 barg, 5.5–8.3 MPa), and the MPHC units at pressures in the range of 1400 to 1500 psig (95 to 102 barg, 9.7–10.3 MPA).

The MHC units are frequently used to reduce the sulfur and nitrogen contents of fluid catalytic cracking unit feedstocks while also increasing the amount of middle distillate fuel (diesel) blending stocks. Yields of up to 30% full-range diesel fuel with cetane indices between 30 and 40 can be produced from virgin vacuum gas oil feedstocks [15].

The MPHC units, operating at higher pressures and costing from 1.5 to 1.8 times the cost of a MHC unit, can produce from 35 to 40% full-range diesel blending stocks with cetane indices of 45 to 50. They can also be operated to produce 10 to 20% kerosine blending stocks with smoke points of 15 to 20 [14].

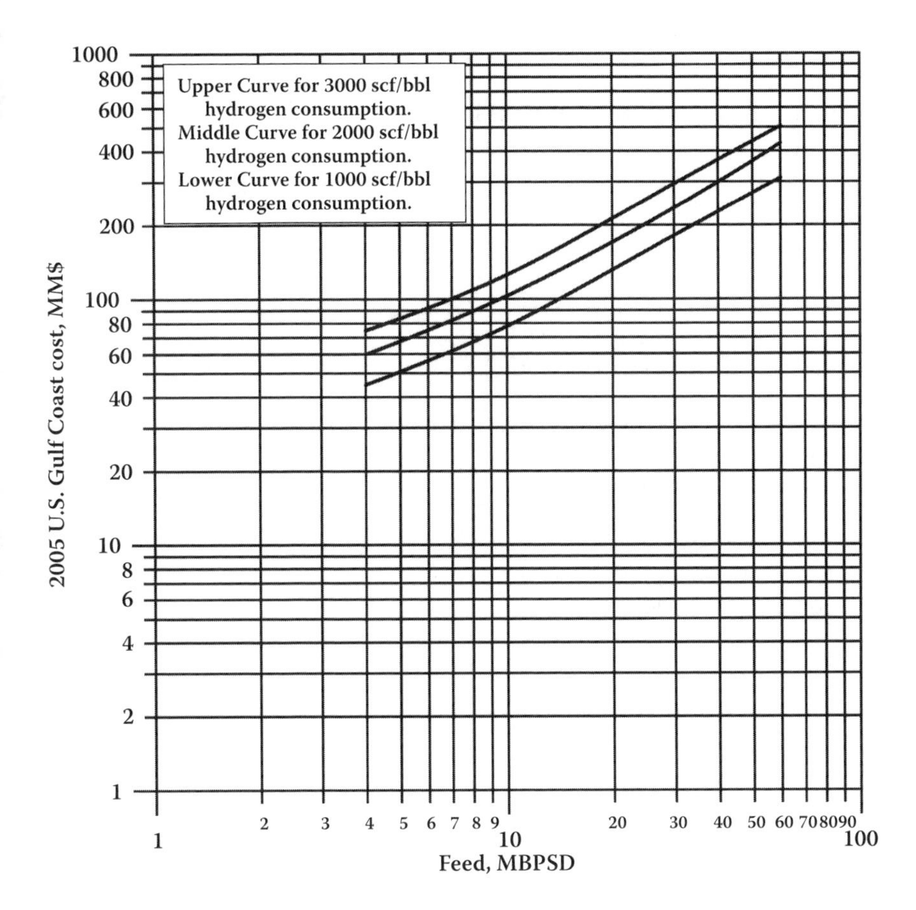

FIGURE 7.7 Catalytic hydrocracking unit investment cost: 2005 U.S. Gulf Coast (see Table 7.4).

The MHC and MPHC units offer refineries the flexibility of producing higher yields of high-quality middle distillate products with capital and operating costs from 50 to 80% of those of a full-conversion hydrocracker.

7.9 CASE-STUDY PROBLEM: HYDROCRACKER

Table 7.5 shows the hydrocracker material balance. Hydrocracker catalyst and utility requirements are shown in Table 7.6.

PROBLEMS

1. A hydrocracker feedstock has a boiling range of 650 to 920°F (343 to 495°C) and an API gravity of 23.7°, and contains 1.7 wt% sulfur. If the hydrocracking hydrogen consumption is 1500 scf/bbl of feed and the feed rate is 7500 BPSD, determine (a) total hydrogen consumption, (b) barrels of gasoline, and (c) barrels of jet fuel produced per day.

TABLE 7.4
Catalytic Hydrocracking Unit Cost Data

Costs included			
1. Stabilization of gasoline			
2. Fractionation into two products			
3. Complete preheat, reaction, and hydrogen circulation facilities			
4. Hydrogen sulfide removal from hydrogen recycle			
5. Sufficient heat exchange to cool products to ambient temperature			
6. Central control system			
7. Electric motor-driven hydrogen recycle compressors			
Costs not included			
1. Initial catalyst charge, approximately $175/BPD of feed			
2. Hydrogen generation and supply facilities			
3. Spare hydrogen recycle compressors			
4. Recovery of butanes, propane, etc., from gas			
5. Feed fractionation			
6. Conversion of hydrogen sulfide to sulfur			
7. Cooling water, steam, and power supply			
8. Paid-up royalty			
Royalty			
Running royalty is about $0.15 to $0.25/bbl.			
Paid-up royalty is about $150 to $250/BPD.			
Utility data (per bbl feed)			
Hydrogen consumption, scf	1000	2000	3000
Steam, lb	50	75	100
Power, kWh[a]	8	13	18
Cooling water, gal crclt. (30°FΔt)	300	450	600
Fuel (LHV), MMBtu	0.1	0.2	0.3
Catalyst replacement, $	0.08	0.16	0.32

[a] Includes electric drive for hydrogen recycle compressors.

Note: See Figure 7.8.

2. For the feed of problem 1, calculate the feed rate in barrels per day needed to produce sufficient isobutene for an alkylation unit producing 3500 BPD of alkylate. Assume the hydrocracking hydrogen consumption is 1750 scf/bbl of feed, and 0.65 bbl of isobutene is needed to produce 1 bbl of alkylate.
3. For the feedstock and under the conditions of problem 1, estimate the characterization factors of the gasoline and jet fuel fractions produced by the hydrocracker.
4. Make an overall material balance, including a hydrogen balance for a 10,000 BPSD hydrocracker with a 26.5°API feedstock having a characterization factor of 12.1 and containing 0.7% sulfur, 0.3% nitrogen, and 0.15% oxygen by weight. The hydrocracking hydrogen consumption is 2000 scf/bbl of feed.

TABLE 7.5
Hydrocracker Material Balance: 100,000 BPCD, Alaska North Slope Crude Oil Basis (10.621 BPCD Fresh Feed; Severity, 2000 scf H_2/bbl)

Component	vol%	BPD	°API	(lb/hr)/BPD	lb/hr	wt% S	lb/hr S	Kw
Feed								
Coker GO	38.6	4,103	14.3	14.16	58,114	1.98	1,151	10.56
FCC LCO	61.4	6,517	12.6	14.32	93,352	0.17	158	10.44
Combined	100.0	10,621	13.3	14.26	151,456		1,309	
Hydrogen, scfb	2,000				4,671			
Total					156,137			
Products								
H_2S, wt%	0.89				1,391		1,309	
C_3 and lighter, wt%	1.72				1,603			
iC_4	2.89	307		8.22	2,523			
nC_4	1.43	151		8.51	1,289			
C_5-180°F	7.7	817	77.1	9.90	8,096			12.37
180–400°F	28.12	2,937	47.1	11.56	34,519			11.42
400–520°	79.5	8,441	32.2	12.61	106,716			11.25
Total	120.9	12,841			156,137			

	wt% H_2	lb/hr	lb/hr H_2
Hydrogen balance out			
H_2S		1,391	82
C_3 and lighter	20.0	1,603	321
C_4	17.2	3,805	654
C_5-180°F	15.3	8,053	1,231
180–400°F	13.9	34,519	4,798
400–520°F	12.1	106,439	12,870
H_2 in soln. (1 lb/bbl feed)		443	
Total			20,399
Hydrogen balance in			
Coker GO	10.0	58,114	5,791
FCC LCGO	9.9	93,352	9,235
Hydrogen	100.0	4,671	4,671
Total			19,697
Added H_2[a]			702
Total			20,399

[a] Increase H_2 consumption by 702 lb/hr to give 2301 scfb H_2 used.

5. Calculate direct operating costs, excluding labor, per barrel of feed for a 10,000 BPSD hydrocracker that has a total hydrogen consumption of 1780 scf/bbl of feed, and hydrogen has a value of $3.00/Mscf.
6. Calculate the total hydrogen consumption for hydrocracking 10,000 BPSD of 24.0°API, 617 to 950°F (325 to 510°C) boiling range feedstock containing 0.45% sulfur, 0.18% nitrogen, and 0.11% oxygen by weight to a total gasoline liquid product.

TABLE 7.6
Hydrocracker Catalyst and Utility Requirements

Steam, Mlb/day	797
Power, MkWh/day	138
Cooling water, Mgpn	3.3
Fuel, MMBtu/day	2124
Catalyst, $/day	1062

7. Using the cost figures in problem 6 of Chapter 6 and assuming labor costs are equal on the two units, compare the costs per barrel for gasoline product for hydrocracking and fluid catalytic cracking the feedstocks in problems 6 and 7 of Chapter 6. The cost of hydrogen is $3.00/Mscf.
8. Calculate the hydrogen consumption and jet fuel production for the feedstock in problem 6 if the hydrocracker were operated to maximize jet fuel production (45°API fraction). How much gasoline would be produced?
9. Using the cost figures in problem 6 of Chapter 6 and product prices in a current issue of the *Oil and Gas Journal*, estimate the hydrocracking processing costs per barrel of liquid product and the value added per barrel of product.

NOTES

1. Bridge, A.B., Personal communication, 1986.
2. Murphy, J.R., Smith, M.R., and Viens, C.H., *Oil Gas J. 68*(23), pp. 108–112, 1970.
3. Tajbl, D.G., *Petroleum Refining Handbook*, Myers, R.A. (Ed.), McGraw-Hill Book Co., New York, 1986, Sections 2.35–2.42.
4. *Hydrocarbon Process. 49*(9), pp. 167–173, 1970.
5. *Petro/Chem. Eng. 41*(5), pp. 30–52, 1969.
6. Baral, W.J. and Miller, J.R., *Hydrocracking—a Rouse to Superior Distillate Products from Heavy Oil*, Kellogg Symposium on Heavy Oil Upgrading, Nice, France, September 16, 1982.
7. Watkins, C.H. and Jacobs, W.L., *Hydrocarbon Process. 45*(5), pp. 159–164, 1966.
8. Nelson, W.L., *Oil Gas J. 65*(26), pp. 84–85, 1967.
9. Pedersen, B.S., Elkwall, G.R., Gruia, A.J., Humbach, M.J., Kauff, D.A., and Meurling, P.J., UOP Technology Conference, 1990.
10. Nelson, W.L., *Oil Gas J. 71*(44), p. 108, 1973.
11. Nelson, W.L., *Oil Gas J. 69*(9), pp. 64–65, 1971.
12. Hougen, D.A. and Watson, K.M., *Chemical Process Principles*, Vol. 1, John Wiley & Sons, New York, 1943, p. 333.
13. Hougan, D.A. and Watson, K.M., *Industrial Chemical Calculations*, John Wiley & Sons, New York, 1938.
14. Hunter, M.G., Pappal, D.A., and Pesek, C.L., *Handbook of Petroleum Refining Processes*, 2nd ed., Meyers, R.A. (Ed.), McGraw-Hill Book Co., New York, 1997, Sections 7.3–7.20.
15. Shorey, S.W., Lomas, D.A., and Keesom, W.H., *Hydrocarbon Process. 78*(11), pp. 43–51, 1999.

ADDITIONAL READING

16. Ward, J.W., Hansford, R.C., Reichle, A.D., and Sosnowski, J., *Oil Gas J. 71*(22), pp. 69–73 (1973).
17. Nelson, W.L., *Oil Gas J. 65*(26), 84 (1967).
18. VanDriesen, R.P., *Altering Today's Refinery,* Energy Bureau Conference on Petroleum Refining, Houston, September 28, 1981.
19. Bridge, A.G., *Handbook of Petroleum Refining Processes*, 2nd ed., Meyers, R.A. (Ed.), McGraw-Hill Book Co., New York, 1997, Sections 7.21–7.40.
20. Ibid., Sections 7.41–7.49.

8 Hydroprocessing and Resid Processing

The term *resid* refers to the bottom of the barrel and is usually the atmospheric tower bottoms (atmospheric reduced crude, or ARC) with an initial boiling point of 650°F (343°C) or vacuum tower bottoms (vacuum reduced crude, or VRC) with an initial boiling point of 1050°F (566°C). In either case, these streams contain higher concentrations of sulfur, nitrogen, and metals than does the crude oil from which they were obtained, and hydrogen/carbon ratios in the molecules are much lower. These concentrations are much higher in the case of VRC.

In recent years, the density and sulfur contents of crude oils charged to U.S. refineries have been increasing, and consequently a higher percentage of the crude oils are in the 1050+°F (566+°C) boiling range. In the past, this resid has been sold as asphalt (if the qualities of the crude permit) or as heavy fuel oil (No. 6 or bunker fuel oil). Stricter environmental emission standards have made it much more difficult and costly to use these heavy oils for fuels, and more of these must be converted in the refinery to feedstocks for refining processes that will convert them to transportation fuel blending stocks.

High carbon-forming potentials of resids, caused by the low hydrogen/carbon ratios in the molecules and as indicated by the Conradson and Ramsbottom carbon residues, cause rapid catalyst deactivation and high catalyst costs, and the nickel and vanadium in the resids act as catalysts for reactions creating carbon and light gaseous hydrocarbons. As a result, catalytic processes for converting resids usually use ARC for their feedstocks, and VRC feedstocks are usually processed by noncatalytic processes. The processes most commonly used for processing ARC are reduced crude catalytic cracking units and hydroprocessing units. Thermal cracking processes such as delayed coking and Flexicoking or solvent extraction are processes used for VRC feedstocks.

8.1 COMPOSITION OF VACUUM TOWER BOTTOMS

Vacuum tower bottoms are complex mixtures of high molecular-weight and high boiling-point materials containing thousands of hydrocarbon and organic compounds. All of the bad processing features of refinery feedstocks are present in the bottoms streams in greater concentrations than in any of the distillate feedstocks [1]. Because they are so complex, it has been difficult to express the compositions in ways meaningful to processing operations.

Several investigators have reported the results of studies using solubility techniques to separate vacuum resids into fractions whose properties can be related to processing techniques and results [2,3]. Liquid propane is used to extract the oil fraction from vacuum tower bottoms, and liquid n-pentane, n-hexane, or n-heptane is then used to extract the resin fraction from the residue left from the propane extraction. The material insoluble in either the propane or the higher hydrocarbons is termed the asphaltene fraction.

Hensley and associates from Amoco Research summarized the properties of the fractions in a report given at a meeting of the American Institute of Chemical Engineers [2], from which the following description is derived:

The asphaltene fraction has a very low hydrogen-to-carbon ratio and consists of highly condensed ring compounds with predominating molecular weights, as determined by solution techniques, in the 5,000 to 10,000 range (mass spectrometer techniques estimate molecular weights about an order of magnitude lower). The molecule is built up of sheets of these highly condensed ring structures held together by valence bonds between hetro atoms such as sulfur, oxygen, and metals. Support to this structure is given by work reported by Bonné [4]. An asphaltene molecule contains three to five unit sheets consisting of condensed aromatic and naphthenic rings with paraffinic side chains. These sheets are held together by hetro atoms such as sulfur or nitrogen or polymethylene bridges, thio-ether bonds, and vanadium and nickel complexes. Separation into unit sheets is accompanied by sulfur and vanadium removal. A significant feature of the asphaltene fraction is that 80 to 90% of the metals in the crude (nickel and vanadium) are contained in this material. Apparently, 25 to 35% of these metals are held in porphryin structures and the remainder in some undetermined type of organic structure. The asphaltene fraction contains a higher content of sulfur and nitrogen than does the vacuum resid and also contains higher concentrations of carbon-forming compounds (as shown by Conradson and Ramsbottom carbon residues). A hypothetical asphaltene molecule structure developed by Bridge and coworkers at Chevron is depicted in Figure 8.1. Vacuum resids from crude oils examined by Hensley using extraction by n-heptane contained from 2 to 15 wt% asphaltenes. Results on similar crude oils by Rossi and coworkers from Chevron using n-pentane as the solvent gave asphaltene contents up to 25 wt% [3].

The resin fraction contains certain condensed-ring aromatics but also has a substantial amount of paraffinic structures and serves as a solvent for the asphaltenes. Average molecular weights from 600 to 5000 were determined by solvent techniques. Sulfur concentrations are approximately the same as the vacuum resids from which they are derived, so there is not a significant concentration of sulfur in this fraction. The resins contain 10 to 20 wt% of the metals in the crude, so the oil fraction is almost free of metals contamination.

The oil fraction is highly paraffinic, usually contains no metals, and has lower sulfur and nitrogen contents than the vacuum resid.

8.2 PROCESSING OPTIONS

Although there have been improvements in existing processes for converting resids to more salable products or reducing the quantity of material difficult to sell, the main processes used have been available for many years, but now the economics have changed. The processes are classified as catalytic or noncatalytic. The catalytic

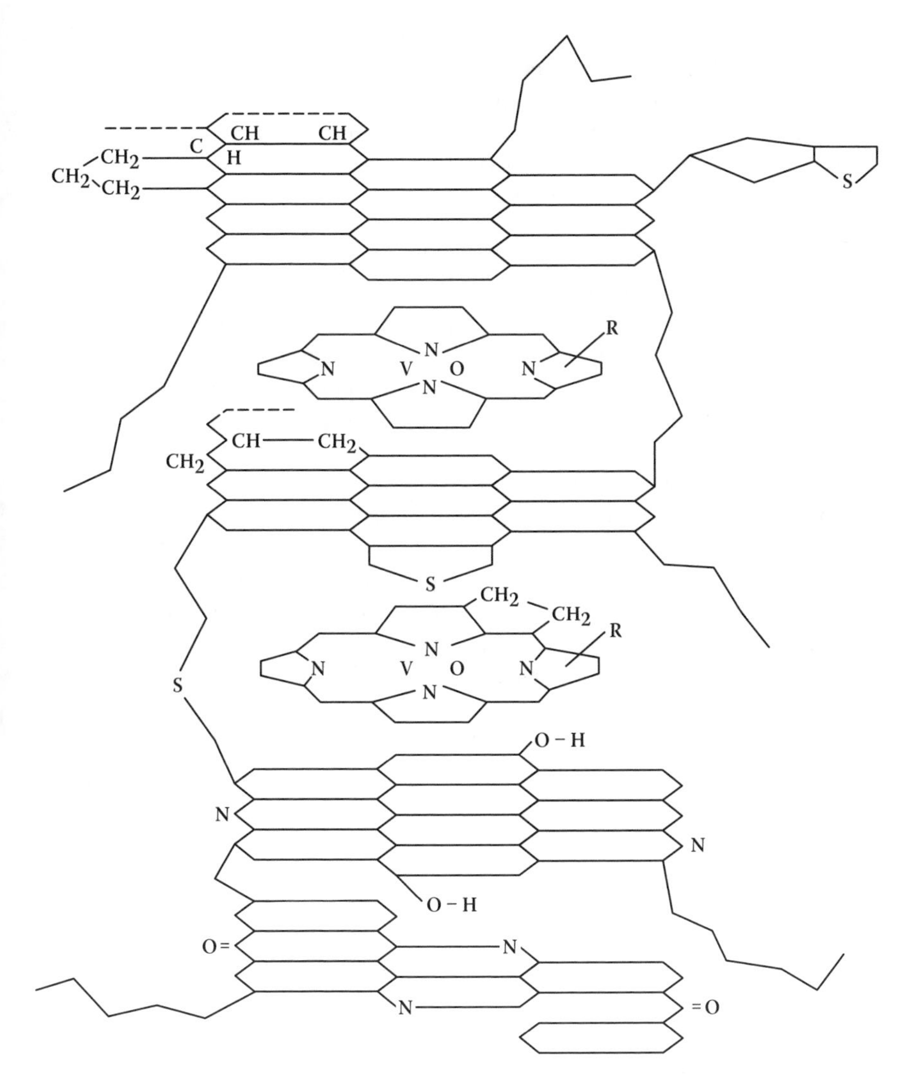

FIGURE 8.1 Hypothetical asphaltene molecule structure. (From Reference 5.)

processes normally use atmospheric reduced crude as the feedstock and include fixed-bed hydroprocessing, ebullated- or expanded-bed hydroprocessing, moving-bed hydroprocessing, and reduced crude fluid catalytic cracking. The noncatalytic processes typically use vacuum reduced crude as the feedstock and include solvent extraction, delayed coking, and Flexicoking.

8.3 HYDROPROCESSING

The term *hydroprocessing* is used to denote those processes used to reduce the boiling range of the feedstock as well as to remove substantial amounts of impurities such as metals, sulfur, nitrogen, and high carbon-forming compounds. As presently practiced, fixed-bed processes—such as Residfining, ARDS, VRDS, and RESID HDS, and the ebullated- or expanded-bed H-Oil and LC-fining—fall into this category rather than hydrocracking. In hydroprocessing processes, feed conversion levels of 25 to 65% can be attained. Other names applied to this operation are hydroconversion, hydrorefining, and resid HDS.

In U.S. refineries, hydroprocessing units are used to prepare residual stream feedstocks for cracking and coking units. Although vacuum resids can be used as feedstocks, most units are atmospheric resids as feeds because the lower viscosities and impurity levels give better overall operations and greater impurity reductions in the 1050+ °F (566+ °C) fractions. Typically, the heavy naphtha fraction of the products will be catalytically reformed to improve octanes, the atmospheric gas oil fraction hydrotreated to reduce aromatic content and improve cetane number, the vacuum gas oil fraction used as conventional FCC unit feed, and the vacuum tower bottoms sent to a heavy oil cracker or coker.

With the exception of H-Oil, LC-fining, HYCON, and Onstream Catalyst Replacement (OCR), the processes have fixed-bed reactors and usually require the units to be shut down to have the catalyst changed when catalyst activity declines below the accepted level of activity. A typical fixed-bed process flow diagram is shown in Figure 8.2. Ebullated- or expanded-bed processes have similar overall flow diagrams in terms of sequence of operations.

All units operate at very high pressures, above 2000 psig (1.38 MPa) and usually near 3000 psig (20.7 MPa), and low space velocities of 0.2 to 0.5 v/hr/v. The low space velocities and high pressures limit charge rates to 30,000 to 40,000 BPSD (4760 to 6360 M^3/SD) per train of reactors. Typically, each train will have a guard reactor to reduce the metals contents and carbon forming potential of the feed, followed by three to four hydroprocessing reactors in series. The guard reactor's catalyst is a large pore-size (150 to 200 Å) silica-alumina catalyst with a low-level loading of hydrogenation metals such as cobalt and molybdenum. The catalysts in the other reactors are tailor-made for the feedstocks and conversion levels desired and may contain catalysts with a range of pore and particle sizes as well as different catalytic metal loadings and types (e.g., cobalt and molybdenum or nickel and molybdenum). Typical pore sizes will be in the 80 to 100 Å range.

The process flow, as shown in Figure 8.2 [3,6], is very similar to that of a conventional hydrocracking unit except for the amine absorption unit to remove hydrogen sulfide from the recycle hydrogen stream and the guard reactor to protect the catalyst in the reactor train. The heavy crude oil feed to the atmospheric distillation unit is desalted in a two- or three-stage desalting unit to remove as much of the organic salts and suspended solids as possible, because these will be concentrated in the resid. The atmospheric resids are filtered before being fed to the hydroprocessing unit to remove solids greater than 25 Å in size, mixed with recycle hydrogen, heated to reaction temperature, and charged into the top of the guard reactor. Suspended solids in the feed will be deposited in the top section of the guard reactor, and most of the metals will be deposited on the catalyst. There is a substantial reduction in the Conradson and Ramsbottom carbons in the guard reactor, and the feed to the following reactors is low in metals and carbon-forming precursors.

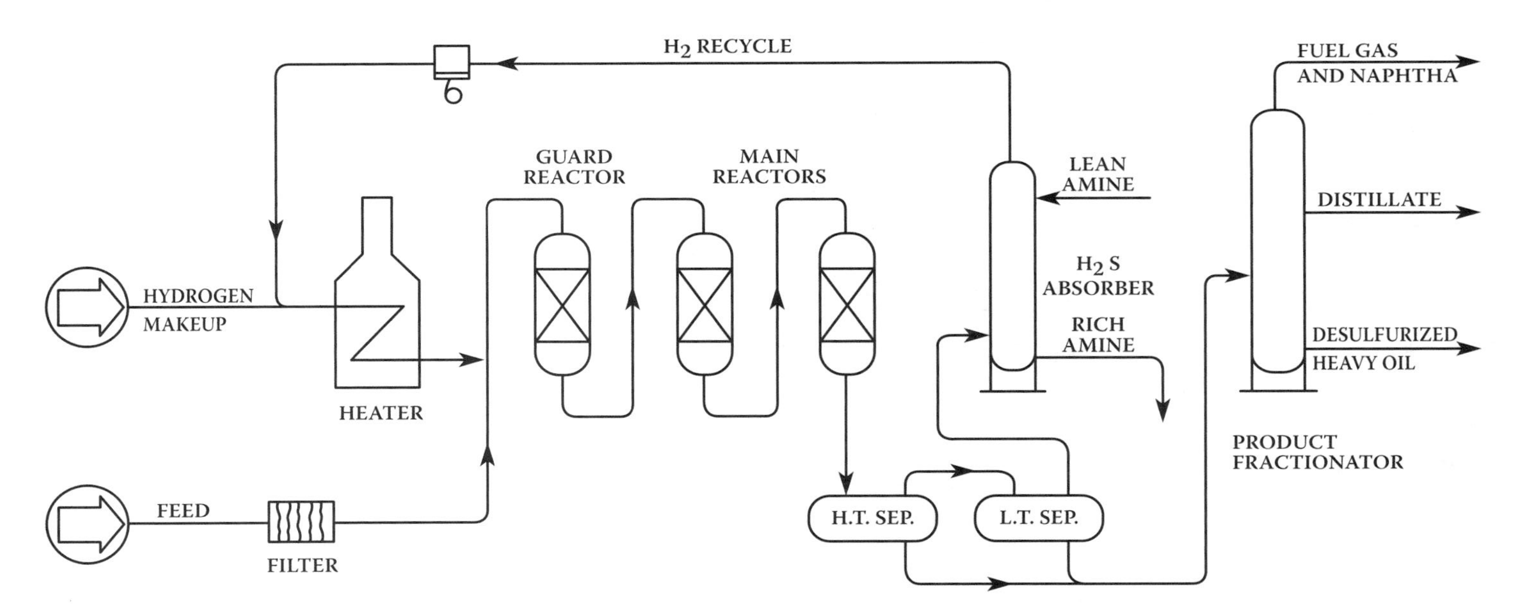

FIGURE 8.2 ExxonMobil RESIDfining hydroprocessing unit. (Courtesy ExxonMobil Research and Engineering.)

TABLE 8.1
Results from Hydroprocessing Jobo Crude

	Feed	Product
Gravity, °API	8.5	22.7
Sulfur, wt%	4.0	0.8
Nickel, ppm	89	5
Vanadium, ppm	440	19
Residue, wt%	13.8	2.8

Source: Note 1.

The three or four reactors following the guard reactor are operated to remove sulfur and nitrogen and to crack the 1050+ °F (566+ °C) material to lower-boiling compounds. Recycle hydrogen is separated and the hydrocarbon liquid stream fractionated in atmospheric and vacuum distillation columns. Results of hydroprocessing a heavy Venezuelan crude oil (Jobo) are shown in Table 8.1.

8.4 EXPANDED-BED HYDROCRACKING PROCESSES

There are two expanded- or ebullated-bed processes available for license today. These are the H-Oil and LC-fining processes, which were developed by Hydrocarbon Research Inc. (HRI) and Cities Service and C-E Lummus. The LC-fining and H-Oil processes are designed to process heavy feeds such as atmospheric tower bottoms or vacuum reduced crude and use catalysts with metals removal, hydrotreating, and cracking activities. A simplified process flow diagram for the LC-fining process is shown in Figure 8.3a and Figure 8.3b (see also Photo 9, Appendix E).

The terms *ebullated bed* and *expanded bed* are names given by HRI and C-E Lummus to a fluidized bed-type operation that utilizes a mixture of liquids and gases to expand the catalyst bed rather than just gases (definition of fluidized bed). Both HRI and Lummus use similar technologies but offer different mechanical designs [2,7,8].

The preheated feed, recycle, and makeup hydrogen are charged to the first reactor of the unit. The liquid passes upward through the catalyst, which is maintained as an ebullient bed. The first-stage reactor effluent is sent to the second-stage reactor for additional conversion. The product from the last reactor passes through a heat exchanger to a high-pressure separator, where the recycle gas is removed. The liquid from the high-pressure separator is sent to a low-pressure flash drum to remove additional gases. The liquid stream at low pressure then goes to a rectification column for separation into products. The operating pressure for an H-Oil unit is a function of feed boiling point, with operating pressures up to 3000 psig (21 MPa) used when charging vacuum tower residuum. The operating temperature is a function of charge stock and conversion but is normally in the range of 800 to 850°F (425 to 455°C).

One of the main advantages of the ebullated-bed reactor process is the ability to add and remove catalyst during operation. This permits operators to regenerate

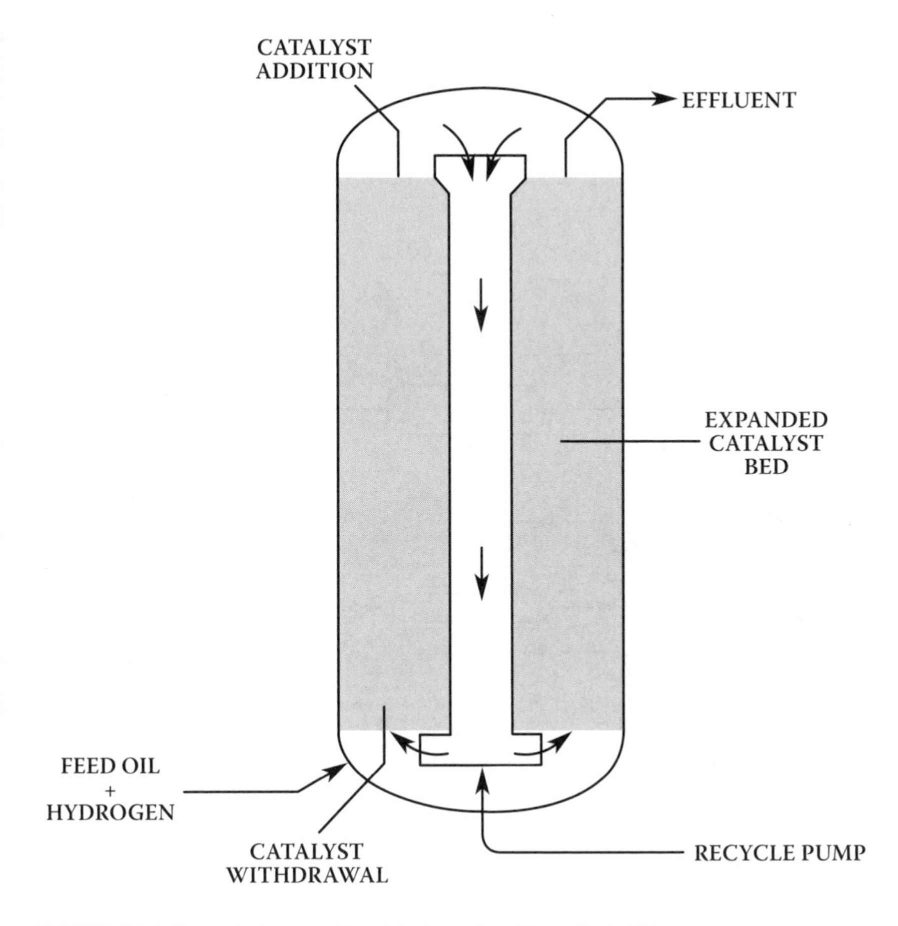

FIGURE 8.3A Expanded- or ebullated-bed reactor. (From Note 2.)

catalyst while remaining on-stream and to maintain catalyst activity by either regeneration or the addition of fresh catalyst. Because the unit operates from start-of-run to end-of-run with an equilibrium activity catalyst, with a constant quality feedstock and constant operating conditions, the product yields and quality will also be constant. This significantly improves refinery operation and efficiency.

Another advantage to the ebullated-bed reactor system is that small solid particles are flushed out of the reactor and do not contribute to plugging or increase in pressure drop through the reactor.

It is necessary to recycle effluent from each of the reactor's catalyst beds into the feed to that reactor in order to have sufficiently high velocities to keep the catalyst bed expanded, to minimize channeling, to control the reaction rates, and to keep heat released by the exothermic hydrogenation reactions to a safe level. This back-mixing dilutes the concentrations of the reactants and slows the rates of reactions as compared with fixed-bed (plug-flow) reactors. Based on work reported by Shell [9], ebullated-bed reactors require up to three times as much catalyst per barrel of feed to obtain the same conversion level as fixed-bed reactors.

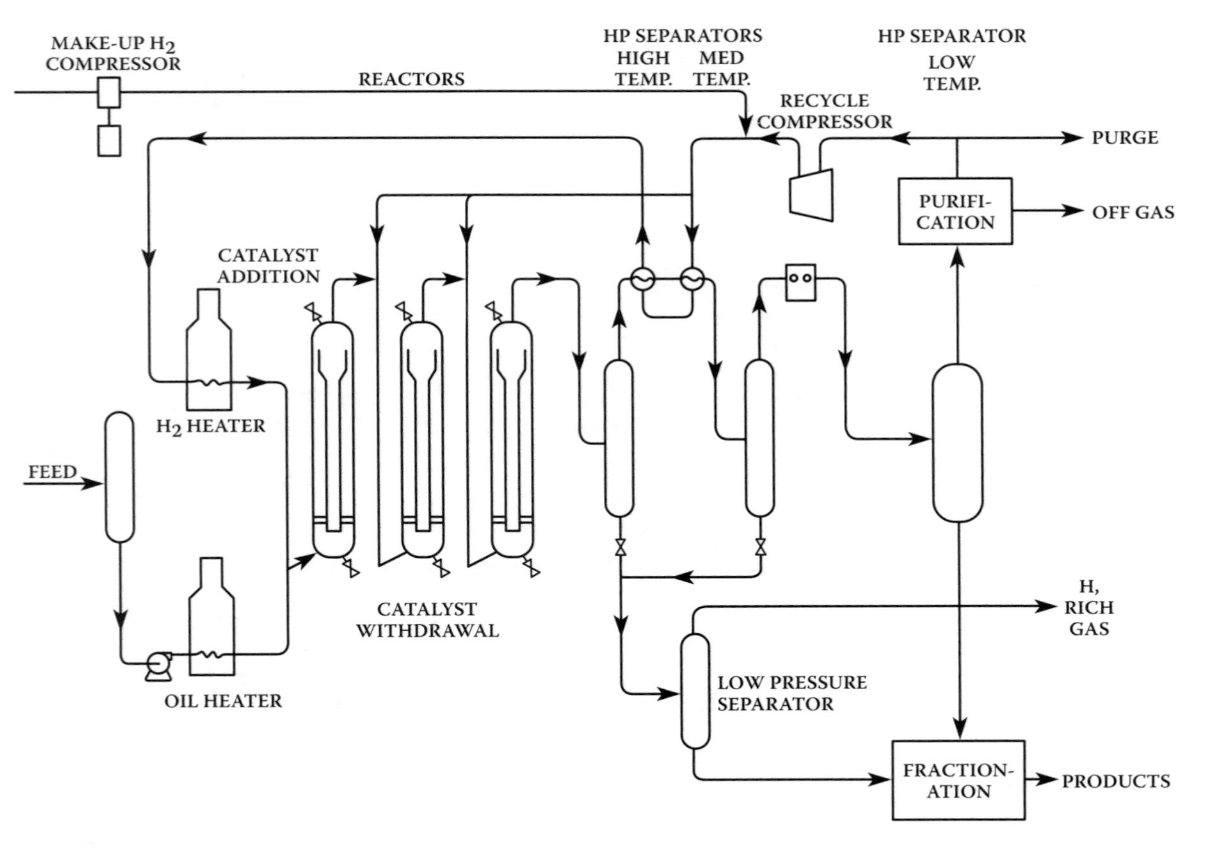

FIGURE 8.3B LC-fining expanded-bed hydroprocessing unit. (From Reference 2.)

TABLE 8.2
LC-fining Yields

	Long resid	Short resid
Feedstocks		
Gravity, °API	15.7	10.7
Sulfur, wt%	2.7	3.2
RCR, wt%	9.4	—
Metals		
Vanadium, ppm	110	147
Nickel, ppm	27	35
IBP-1050°F (IBP-566°C), vol%	55.0	33.2
1050+ °F (566+ °C), vol%	45.0	66.8
Yields		
Gas, C_3^-, scf/bbl (Nm^3/m^3)	350 (62)	590 (105)
Naphtha, C_4-400°F (C_4-294°C), vol%	17.8	13.1
Gravity, °API	61.0	64.0
Sulfur, wt%	<0.1	0.04
Kerosine		
400–500°F (294–260°C), vol%	10.2	8.9
Gravity, °API	38.8	37.2
Sulfur, wt%	0.1	0.1
AGO, 500–600°F (260–315°C), vol%	19.7	14.3
Gravity, °API	31.8	30.1
Sulfur, wt%	0.25	0.18
VGO, 650–1050°F (343–566°C), vol%	37.1	36.1
Gravity, °API	22.4	23.0
Sulfur, wt%	0.6	0.6
Pitch, 1050+ °F (566+ °C), vol%	20.0	32.7
Gravity, °API	7.7	7.0
Sulfur, wt%	1.3	2.3
H_2 consumption, scf/bbl (Nm^3/m^3)	985 (175)	1310 (233)
Catalyst consumption, lb/bbl (kg/m^3)	0.15 (0.43)	0.12 (0.34)

Source: Note 2.

Typical product yields from LC-fining cracking are given in Table 8.2.

8.5 MOVING-BED HYDROPROCESSORS

Shell and Chevron have developed technology that combines the advantages of fixed-bed and ebullated-bed hydroprocessing. These systems use reactors designed for catalyst flow by gravity from top to bottom with mechanisms designed to allow spent catalyst to be removed continuously or periodically from the bottom and fresh catalyst added to the top. This permits a low-activity high-metals catalyst to be removed from the reactor and replaced with fresh catalyst without taking the unit off-stream. Lower catalyst consumption rates are required than with the ebullated-bed systems because, in the ebullated-bed system, equilibrium-activity and metals-loaded catalyst is removed rather than the lowest activity-spent catalyst. As there is no recycling of product from the reactor outlets to the reactor inlet, the reactors operate in a plug-flow condition, and reaction rates are the same as in a fixed-bed operation. Shell technology is known as the HYCON process and the Chevron process is called Onstream Catalyst Replacement (OCR).

8.6 SOLVENT EXTRACTION

Solvent extraction technology is used to extract up to two thirds of the vacuum reduced crude to be used as a good quality feed for a fluid catalytic cracking unit to convert into gasoline, diesel fuel, and home heating oil blending stocks. There are a number of processes licensed but the two most common are the DEMEX process licensed by UOP and the ROSE process licensed by Kerr-McGee. Both technologies use light hydrocarbons (propane to pentanes) as the solvents and use subcritical extraction but use supercritical techniques to recover the solvents. A simplified process flow diagram of the UOP DEMEX process is shown in Figure 8.4.

Light hydrocarbons have reverse solubility properties; that is, as temperature increases, the solubility of higher molecular-weight hydrocarbons decreases. Also, paraffinic hydrocarbons have higher solubilities than aromatic hydrocarbons. A temperature can be selected at which all of the paraffins go into solution along with the desired percentage of the resin fraction. The higher molecular-weight resins precipitate along with the asphaltenes. The extract is then separated from the precipitated raffinate fraction and stripped of the solvent by increasing the temperature to just above the critical temperature of the solvent. At the critical temperature, the oil-plus-resin portion separates from the solvent, and the solvent can be recovered without having to supply latent heat of vaporization. This reduces energy requirements by 20 to 30% as compared with solvent recovery by evaporation.

The hydrocarbon solvent used is feedstock-dependent. As the molecular weight of the solvent increases (propane to pentane), the amount of solvent needed for a given amount of material extracted decreases, but the selectivity of the solvent also decreases. Therefore, the choice of solvent is an economic one because, for a given recovery of FCC unit feedstock from a particular resid, propane will give a better quality extract but will use more solvent. Solvent recovery costs will be greater than if the higher molecular-weight solvent is used because more solvent has to be recovered. The higher molecular-weight solvents give lower solvent recovery costs but, for a given feedstock and yield, give a lower quality extract and have higher capital costs because the critical pressure of the solvent increases with molecular weight and a higher equipment design pressure must be used.

As 80 to 90% of the metals in the crude oil are in the asphaltenes, and most of the remaining metals are in the resin fraction, a good-quality FCC unit feedstock can be obtained. As shown in Figure 8.5, the quality of the extract decreases as the percentage extracted increases. Typically, extract is limited to 50 to 65% of the VRC.

The asphaltene fraction is a very hard asphalt (0.1 penetration) and is usually blended into asphalt or residual fuels in order to dispose of it.

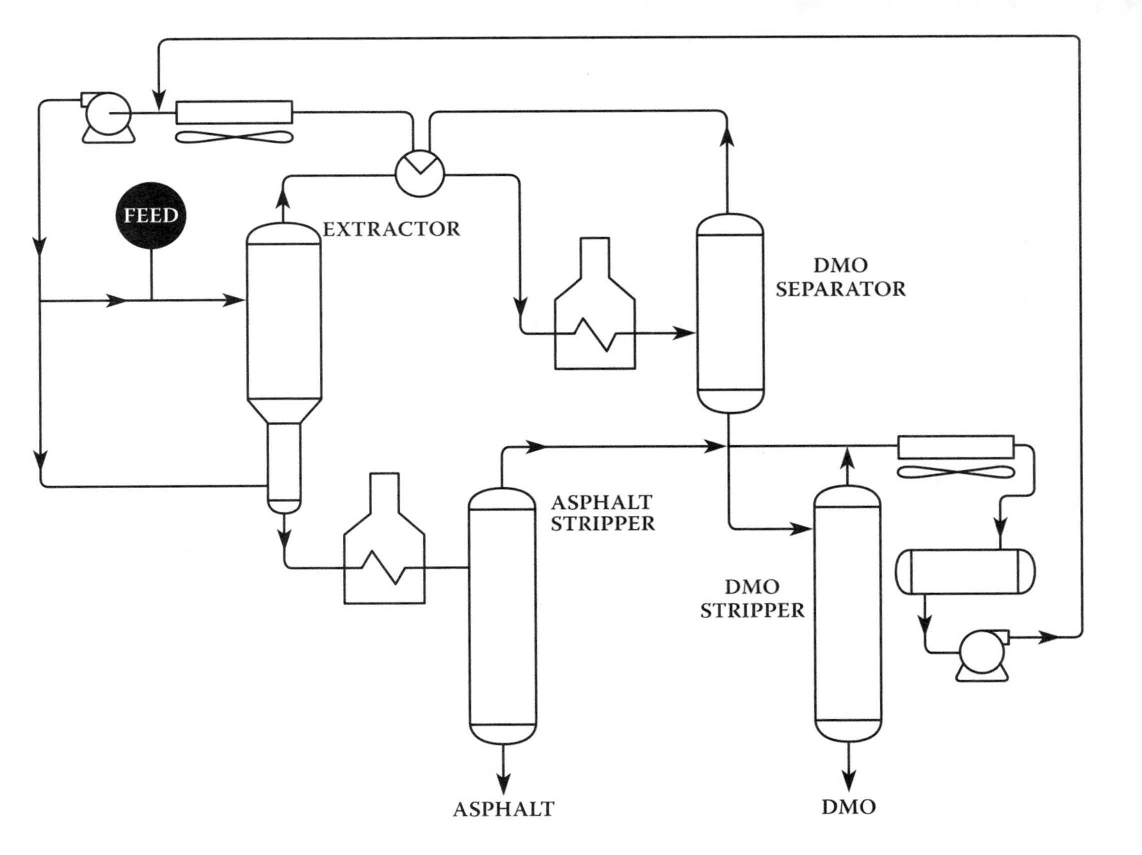

FIGURE 8.4 UOP DEMEX solvent extraction unit flow. (Courtesy of UOP LLC.)

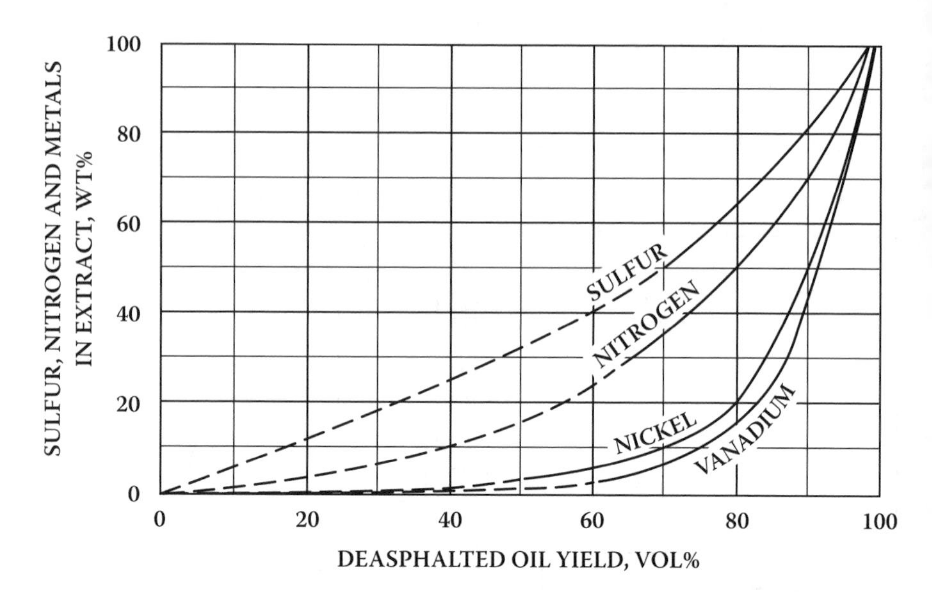

FIGURE 8.5 Extract quality varies with extract quantity. (Courtesy of UOP LLC.)

8.7 SUMMARY OF RESID PROCESSING OPERATIONS

The selection of the most economic method of processing the portion of crude oil boiling above 1050°F (566°C) is a very complex matter because political, environmental, and technical issues are all involved. Unlike decisions driven by the market, criteria that are difficult to evaluate from an economic viewpoint must also be included. As a result, the "best" decision for one refinery may be the "worst" for another refinery even in the same company. A summary of the advantages and disadvantages of the processes illustrates the complexity of the matter.

Thermal processes (delayed coking and Flexicoking) have the advantage that the vacuum reduced crude is eliminated, so there is no residual fuel for disposal, and most of the VRC is converted to lower-boiling hydrocarbon fractions suitable for feedstocks to other processing units to convert them into transportation fuels. However, for high-sulfur crude oils, delayed coking produces a fuel-grade coke of high sulfur content. This coke may be very difficult to sell. The alternative is to hydroprocess the feed to the coker to reduce the coker feed sulfur level and make a low-sulfur coke. This can add a great deal to the cost.

Flexicoking is more costly than delayed coking, both from capital and operating cost viewpoints, but has the advantage of converting the coke to a low heating-value fuel gas to supply refinery energy needs and elemental sulfur, for which there is now a market. A disadvantage is that the fuel gas produced is more than the typical refinery can use, and the energy required for compression does not permit it to be transported very far. It can be used for cogeneration purposes or sold to nearby users.

Hydroprocessing reduces the sulfur and metal contents of the VRC and improves the hydrogen/carbon ratio of the products by adding hydrogen, but the products are very aromatic and may require a severe hydrotreating operation to obtain satisfactory middle distillate stocks. Crude oils with higher sulfur and metal levels will also have high catalyst replacement costs.

Solvent extraction recovers 55 to 70% of the VRC for FCC or hydrocracker feedstocks to be converted into transportation fuel blending stocks, but the asphaltene fraction can be difficult to process or sell.

NOTES

1. Elvin, J., NPRA Annual Meeting, 1983.
2. Hensley, A.L. and Quick, L.M., A.I.Ch.E. Meeting, Philadelphia, June 9–12, 1980.
3. Rossi, W.J., Deighton, B.S., and MacDonald, A.J., *Hydrocarbon Process. 56*(5), pp. 103–109, 1977.
4. Bonné, R.L.C., Ph. D. Thesis, Department of Chemical Engineering, University of Amsterdam, 1992.
5. Bridge, A.G., Personal communication, 1986.
6. Exxon Research and Engineering Co., *RESIDfining Information,* Florham Park, NJ, May 1982.
7. Ragsdale, R., Ewy, G.L., and Wisdom, L.I., NPRA Annual Meeting, 1990.
8. Ragsdale, R. and Wisdom, L.I., A.I.Ch.E. Meeting, Houston, April pp. 7–127.1, 1991.
9. Oelderik, J.M., Sie, S.T., and Bode, D., *Applied Catalysis*, *47*, pp. 1–24, 1989.

ADDITIONAL READING

10. Nelson, W.L., *Oil Gas J. 65*(26), 84, 1967.
11. Van Driesen, R.P., *Altering Today's Refinery,* Energy Bureau Conference on Petroleum Refining, Houston, September 28, 1981.
12. Ward, J.W., Hansford, R.C., Reichle, A.D., and Sosnowski, J., *Oil Gas J. 71*(22), pp. 69–73, 1973.

9 Hydrotreating

The terms *hydrotreating*, *hydroprocessing*, *hydrocracking*, and *hydrodesulfurization* are used rather loosely in the industry because, in the hydrodesulfurization and hydrocracking processes, cracking and desulfurization operations occur simultaneously, and it is relative as to which predominates. In this text, hydrotreating refers to a relatively mild operation whose primary purpose is to saturate olefins or reduce the sulfur or nitrogen content (and not to change the boiling range) of the feed. Hydrocracking refers to processes whose primary purpose is to reduce the boiling range and in which most of the feed is converted to products with boiling ranges lower than that of the feed. Hydrotreating and hydrocracking set the two ends to the spectrum, and those processes with a substantial amount of sulfur or nitrogen removal and a significant change in boiling range of the products versus the feed are called hydroprocessing in this text.

Hydrotreating is a process to catalytically stabilize petroleum products by converting olefins to paraffins or remove objectionable elements from products or feedstocks by reacting them with hydrogen. Stabilization usually involves converting unsaturated hydrocarbons such as olefins and gum-forming unstable diolefins to paraffins. Objectionable elements removed by hydrotreating include sulfur, nitrogen, oxygen, halides, and trace metals. Hydrotreating is applied to a wide range of feedstocks, from naphtha to reduced crude. When the process is employed specifically for sulfur removal, it is usually called hydrodesulfurization. To meet environmental objectives, it also may be necessary to hydrogenate aromatic rings to reduce aromatic content by converting aromatics to paraffins.

Although there are about 30 hydrotreating processes available for licensing [1], most of them have essentially the same process flow for a given application. Figure 9.1 illustrates a typical hydrotreating unit.

The oil feed is mixed with hydrogen-rich gas either before or after it is preheated to the proper reactor inlet temperature. Most hydrotreating reactions are carried out below 800°F (427°C) to minimize cracking, and the feed is usually heated to between 500 and 800°F (260 and 427°C). The oil feed combined with the hydrogen-rich gas enters the top of the fixed-bed reactor. In the presence of the metal-oxide catalyst, the hydrogen reacts with the oil to produce hydrogen sulfide, ammonia, saturated hydrocarbons, and free metals. The metals remain on the surface of the catalyst, and the other products leave the reactor with the oil–hydrogen stream. The reactor effluent is cooled before separating the oil from the hydrogen-rich gas. The oil is stripped of any remaining hydrogen sulfide and light ends in a stripper. The gas may be treated to remove hydrogen sulfide and ammonia, then recycled to the reactor.

9.1 HYDROTREATING CATALYSTS

Catalysts developed for hydrotreating include tungsten and molybdenum sulfides on alumina. These metals are considered the hydrogenating catalysts but their properties

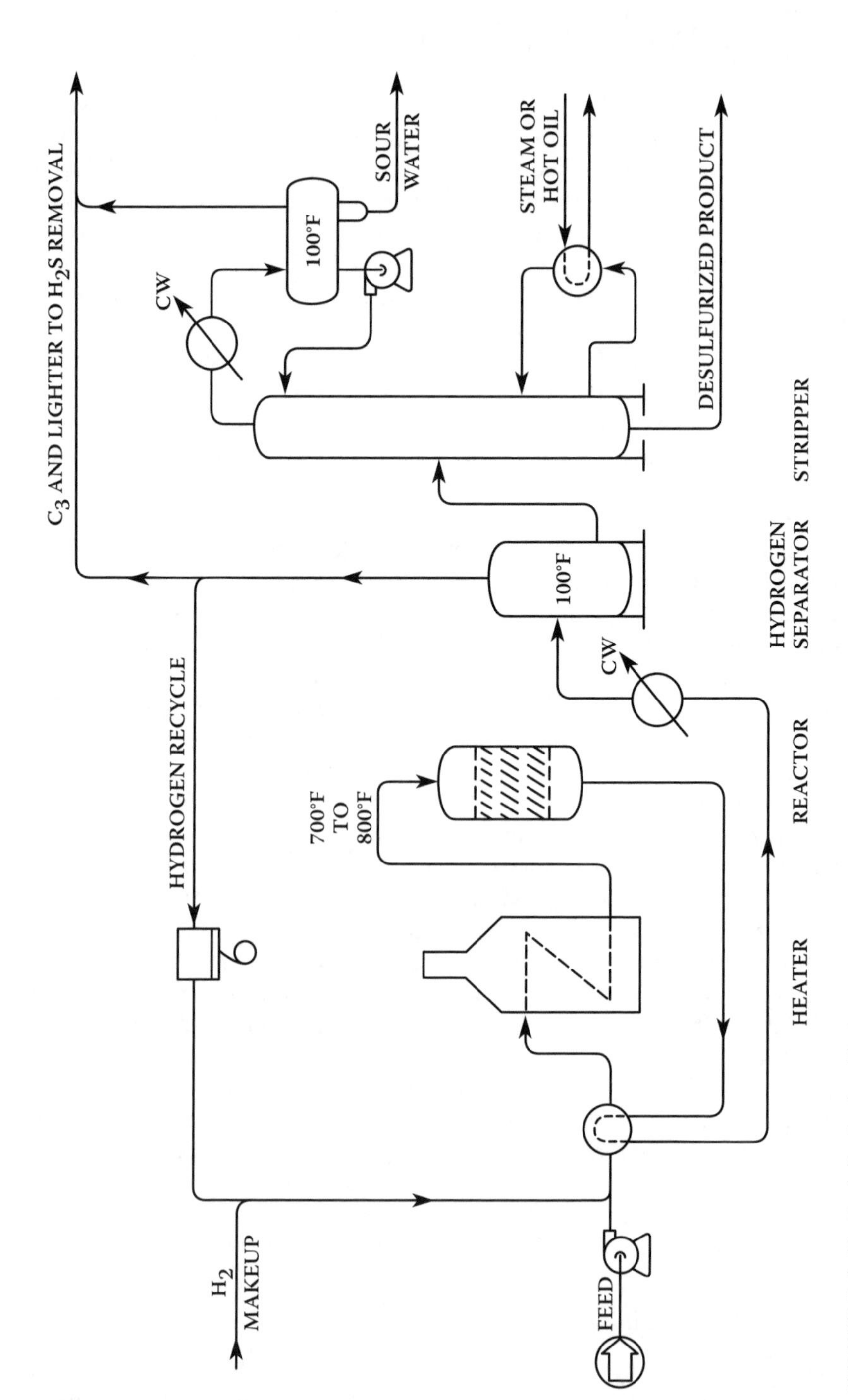

FIGURE 9.1 Catalytic hydrodesulfurizer.

are modified by adding either cobalt or nickel sulfides. Nickel sulfide, nickel thiomolybdate, tungsten and nickel sulfides, and vanadium oxide are also hydrogenation catalysts [2]. Cobalt and molybdenum sulfide on alumina catalysts are in most general use today because they have proven to be highly selective, easy to regenerate, and resistant to poisons. Usually, when purchased the metals are in the oxide state, and they must be activated by converting the hydrogenation metals from the oxide to the sulfide form.

The most economic catalysts for sulfur removal contain cobalt and molybdenum sulfides (CoMo) on alumina supports. If, however, the removal of nitrogen is a significant consideration, catalysts composed of nickel-cobalt-molybdenum or nickel-molybdenum (NiMo) compounds supported on alumina are more efficient. Nitrogen is usually more difficult to remove from hydrocarbon streams than sulfur, and any treatment that reduces excess nitrogen concentration to a satisfactory level usually will effectively remove excess sulfur. Nickel-containing catalysts generally require activation by presulfiding with carbon disulfide, mercaptans, or dimethl sulfide before bringing up to reaction temperature. However, some refiners activate cobalt-molybdenum catalysts by injecting the sulfiding chemical into the oil feed during start-up [3]. The sulfiding reaction is highly exothermic, and care must be taken to prevent excessive temperatures during activation to prevent permanent catalyst deactivation.

Cobalt-molybdenum catalysts are selective for sulfur removal, and nickel-molybdenum catalysts are selective for nitrogen removal, although both catalysts will remove both sulfur and nitrogen [4]. Nickel-molybdenum catalysts have a higher hydrogenation activity than cobalt-molybdenum, which results, at the same operating conditions, in a greater saturation of aromatic rings. Simply stated, if sulfur reduction is the primary objective, then a cobalt-molybdenum catalyst will reduce the sulfur a given amount at less severe operating conditions with a lower hydrogen consumption than a nickel-molybdenum catalyst. If nitrogen reduction or aromatic ring saturation is desired, nickel-molybdenum catalyst is the preferred catalyst. Actually, nickel-tungsten is most effective for nitrogen removal and saturation of aromatic rings to reduce aromatic content but is much more expensive than nickel-molybdenum and is, therefore, seldom used in fuels refineries.

The ability to adjust pore size to concentrate pores around a particular diameter has a great impact on the hydrotreating activity both at start-of-run (SOR) and as the catalyst ages. Reactions taking place in the hydrotreating of gas oils [400 to 1050°F (200 to 566°C)] generally require a minimum pore size to overcome most diffusional restrictions. Pores that are larger than necessary lend little to improving diffusional characteristics, and as the pore diameters of the catalyst increase, the surface area decreases (at constant pore volume). Activity generally decreases with surface area, and loss in pore volume occurs in the smallest diameter pores first. Highest activity retention is maintained if the pore volume is concentrated in a very narrow range of pore diameters.

At the hydrotreating severity to reduce sulfur in LCO to 0.05 wt%, the performance of high-activity NiMo and CoMo catalysts appears to be equivalent.

Catalyst consumption varies from 0.001 to 0.007 lb/bbl (0.003 to 0.02 kg/m^3) feed depending upon the severity of operation and the gravity and metals content of the feed.

9.2 AROMATICS REDUCTION

Hydrogen partial pressure is the most important parameter controlling aromatic saturation. Depending on the type of feedstock, the required hydrogen partial pressure to reduce aromatic content to 10 vol% may vary as much as 40%. Several investigators [2,4] have shown that at liquid hourly space velocities (LHSVs) of 2.0, aromatics in diesel fuel blending stocks can be reduced to < 10 vol% only at pressures of 1500 psig (10.4 MPa) or greater.

Wilson, Fisher, and Kriz [5] demonstrated that at 715°F (380°C), LHSVs of 0.75 to 2.0 h^{-1}, and 2450 psig (16.9 MPa), even the most difficult hydrodearomatizations can be achieved to levels of less than 10 vol% aromatics with nickel-tungsten on gamma-alumina catalysts.

Hydrogenation is an exothermic reaction, and equilibrium yields are favored by low temperatures. Reaction rates increase with temperature, and hydrogenation of aromatic ring compounds is a compromise between using low reactor temperatures to achieve maximum reduction of aromatic content and a high temperature to give high reaction rates and a minimum amount of catalyst charge per barrel of feed. Maximum aromatic reduction is achieved between 700 and 750°F (370 and 400°C) [usually between 705 and 725°F (375 and 385°C)] because of the interrelation between thermodynamic equilibrium and reaction rates. Relationships between reaction temperature and pressure are shown in Figure 9.2 [6]. For a given pressure, the optimum temperature is a function of the types of aromatic compounds in the feed and space velocity.

High-pressure single-stage hydrotreating of only the front end [400 to 550°F (205 to 288°C)] of an LCO reduces hydrogen consumption and extends catalyst life. Usually, this fraction originally contains about 11.1 wt% mono-aromatics and 17.5 wt% di-aromatics. Hydrogenation at 1200 psig (8.2 MPa) reduces the di-aromatic content to 0.4 wt% and increases the mono-aromatic content to 18.3 wt%. Saturation of the final aromatic ring is difficult because of the resonance stabilization of the mono-aromatic ring. Hydrogenation at 1500 psig (10.3 MPa) is required to reduce the aromatic content to 10 wt%, but only about one third as much hydrogen is required as compared to reducing the aromatic content of the full range [400 to 650°F (205 to 345°C)] LCO. This is because the back end of the LCO contains only di- and tri-aromatics and the front end contains almost all of the mono-aromatics, about one third of the di-aromatics, and none of the tri-aromatics in the LCO.

Hydrotreating the feed to the FCC unit reduces the sulfur contents of the FCC products but also increases the aromatic content of the LCO (probably because the percentage of mono-aromatic compounds in the feed is increased). Hydrotreating the FCC feed also makes it more difficult to reduce the aromatics content of the LCO to < 20 vol%.

9.3 REACTIONS

The main hydrotreating reaction is that of desulfurization, but many others take place to a degree proportional to the severity of the operation. Typical reactions are

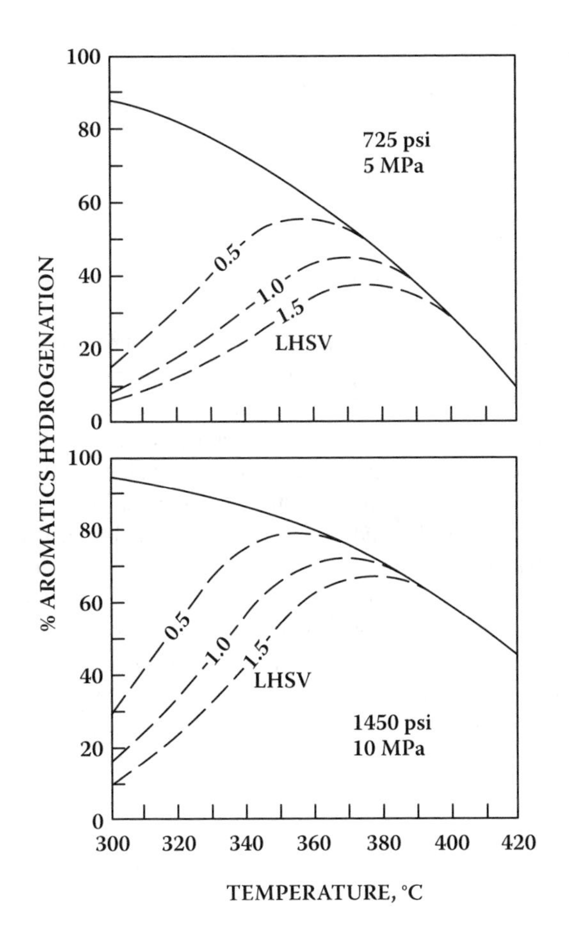

FIGURE 9.2 Kinetic rate and thermodynamic equilibrium effects on aromatics reduction. (From Note 7.)

1. Desulfurization
   a. Mercaptans: $RSH + H_2 \rightarrow RH + H_2S$
   b. Sulfides: $R_2S + 2H_2 \rightarrow 2RH + H_2S$
   c. Disulfides: $(RS)_2 + 3H_2 \rightarrow 2RH + 2H_2S$
   d. Thiophenes:

$$\text{HC=CH-CH=CH-S (thiophene ring)} + 4H_2 \rightarrow C_4H_{10} + H_2S$$

2. Denitrogenation
   a. Pyrrole: $C_4H_4NH + 4H_2 \rightarrow C_4H_{10} + NH_3$
   b. Pyridine: $C_5H_5N + 5H_2 \rightarrow C_5H_{12} + NH_3$

3. Deoxidation
   a. Phenol: $C_6H_5OH + H_2 \rightarrow C_6H_6 + H_2O$
   b. Peroxides: $C_7H_{13}OOH + 3H_2 \rightarrow C_7H_{16} + 2H_2O$
4. Dehalogenation
   Chlorides: $RCl + H_2 \rightarrow RH + HCl$.
5. Hydrogenation:
   Pentene: $C_5H_{10} + H_2 \rightarrow C_5H_{12}$
6. Hydrocracking: $C_{10}H_{22} + H_2 \rightarrow C_4H_{10} + C_6H_{14}$

The ease of desulfurization is dependent upon the type of compound. Lower-boiling compounds are desulfurized more easily than higher-boiling ones. The difficulty of sulfur removal increases in this order: paraffins, naphthenes, aromatics [4].

Nitrogen removal requires more severe operating conditions than does desulfurization. For middle distillate fractions from crude oils containing high concentrations of nitrogen compounds, more efficient nitrogen reduction is achieved by using a catalyst charge of 90% nickel-molybdenum and 10% nickel-tungsten [8].

Hydrogen consumption is about 70 scf/bbl (12.5 Nm^3/m^3) of feed per percentage sulfur, about 320 scf/bbl (57 Nm^3/m^3) oil feed per percentage nitrogen, and 180 scf/bbl (32 Nm^3/m^3) per percentage oxygen removed. Hydrogen consumption for olefin and aromatics reduction can be estimated from the stoichiometric amounts required. If operating conditions are severe enough that an appreciable amount of cracking occurs, hydrogen consumption increases rapidly. It is important to note that actual hydrogen makeup requirements are from two to ten times the amount of stoichiometric hydrogen required. This is due to the solubility loss in the oil leaving the reactor effluent separator and the saturation of olefins produced by cracking reactions.

All reactions are exothermic, and depending on the specific conditions, a temperature rise through the reactor of 5 to 20°F (3 to 11°C) is usually observed.

9.4 PROCESS VARIABLES

The principal operating variables are temperature, hydrogen partial pressure, and space velocity.

Increasing temperature and hydrogen partial pressure increases sulfur and nitrogen removal and hydrogen consumption. Increasing pressure also increases hydrogen saturation and reduces coke formation. Increasing space velocity reduces conversion, hydrogen consumption, and coke formation. Although increasing temperature improves sulfur and nitrogen removal, excessive temperatures must be avoided because of the increased coke formation. Typical ranges of process variables in hydrotreating operations are [1]

Temperature	520–645°F	270–340°C
Pressure	100–3,000 psig	690–20,700 kPag
Hydrogen, per unit of feed		
Recycle	2,000 scf/bbl	360 Nm^3/m^3
Consumption	200–800 scf/bbl	36–142 Nm^3/m^3
Space velocity (LHSV)	1.5–8.0	

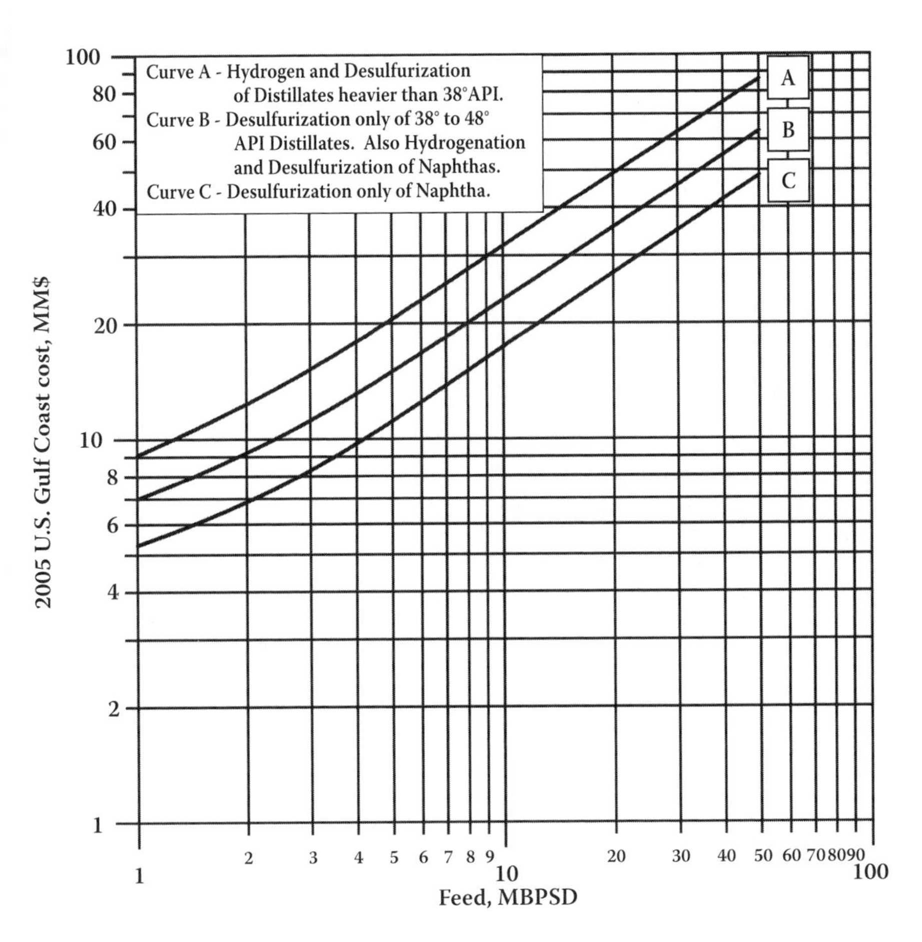

FIGURE 9.3 Catalytic desulfurization and hydrogenation unit investment cost: 2005 U.S. Gulf Coast (see Table 9.1).

9.5 CONSTRUCTION AND OPERATING COSTS

Hydrogen consumption for sulfur, nitrogen, and oxygen removal can be estimated from a monograph prepared by Nelson [10] or from Section 9.2. Assume hydrogen loss by solution in products is approximately 1 lb per barrel of feed [190 scf/bbl (34 Nm^3/m^3)].

Construction and operating costs can be estimated from Figure 9.3 and Table 9.1.

9.6 CASE-STUDY PROBLEM: HYDROTREATERS

U.S Environmental Protection Agency standards for motor fuels require that sulfur, olefin, and aromatic contents be less than specified values. Hydrotreating is used to reduce the concentrations in finished products by processing feedstocks for units producing motor fuel blending components or by processing blending stocks to lower

TABLE 9.1
Catalytic Desulfurization and Hydrogenation Unit Cost Data

Costs included
1. Product fractionation.
2. Complete preheat, reaction, and hydrogen circulation facilities.
3. Sufficient heat exchange to cool products to ambient temperature.
4. Central control system.
5. Initial catalyst charge.

Costs not included
1. Feed fractionation.
2. Makeup hydrogen generation.
3. Sulfur recovery from off-gas.
4. Cooling water, system, and power supply.

Royalty[a]

Running royalty is about \$0.03/bbl (\$0.19/m^3).

Paid-up royalty is about \$30/BPD (\$189/m^3/d).

Utility data (per unit feed)

	"A"	"B"	"C"
Steam, lb/bbl	10	8	6
kg/m^3	0.7	0.6	0.4
Power, kWh	6.0	3.0	2.0
Cooling water, gal crclt. (30°FΔt)	500	400	300
m^3 crclt. (17°CΔt)	1.9	1.5	1.1
Fuel (LHV), MMBtu	0.2	0.15	0.1
MMJ	211	158	105
Hydrogen makeup, scf	400–800	150–400	100–150
m^3	10–20	4–10	3–4.2
Catalyst replacement, \$/bbl	0.06	0.05	0.03
\$/m^3	0.4	0.3	0.2

[a] Royalties apply to gas oils and resids. Naphtha and kerosine are royalty-free.

Note: See Figure 9.3.

concentrations of the specific components (Table 9.2). Without hydrotreating, the sulfur contents of gasolines and diesel fuels from North Slope Alaskan crude exceed EPA specification. By hydrotreating the fluid catalytic cracking unit feedstocks and the jet and diesel fuel blending stocks, the low sulfur specifications for gasolines, diesel, and jet fuel can be met. Two hydrotreaters will be used: one for FCC feedstocks (Table 9.4 and Table 9.5) and the other for diesel and jet fuel blending stocks (Table 9.3).

There is very little in the literature on yields from hydrotreating, but these operations are relatively mild, and product yields from 95 to 98 vol% on feed (same boiling range as feed) can be expected with an increase of 1°API in the gravity of the heavier products. Light product (C_4-) yield distribution is assumed to be similar to that obtained from hydrocracking. Maples has published a compilation of literature

TABLE 9.2
Medium Distillate Hydrotreater Material Balance

Component	vol%	BPD	°API	lb/h/BPD	lb/h	wt% S	lb/h S
Feed							
AGO	39.0	12,500	30.0	12.77	159,653	0.52	830
LCGO	23.5	7,587	30.0	12.78	96,967	0.93	905
Atm JET	37.8	12,200	37.5	12.21	148,906	0.24	357
H_2, scf/bbl		500			3,524		
Total	100.0	32,287	32.3	12.67	409,076	0.51	2092
Products							
H_2S					2,013		1895
C_2-					1,385		
C_3	1.0	337		7.42	2,500		
iC_4	1.9	634		8.22	5,210		
nC_4	1.0	315		8.51	2,677		
Jet	36.3	11,956	38.5	12.14	145,194	0.05	73
Diesel	59.8	19,686	31.0	12.70	250,096	0.05	125
Total	100.0	32,927			409,076		2092

TABLE 9.3
Medium Distillate Hydrotreater Utility Requirements

Utility	Per bbl feed	Per day	Per min
Steam, Mlb	0.01	323	
Power, MkWh	0.006	194	
Cooling water, Mgal circulated	0.5	16,144	11,21
Fuel, MMBtu	0.2	6,457	
H_2, Mscf	0.5	16,144	
Catalyst repl, $	0.02	646	

data [11] that indicate a hydrotreating severity of 400 scf of hydrogen per barrel (71 Nm^3/m^3) of feed is adequate for each of these hydrotreaters, and a solution loss of 100 scf per barrel (18 Nm^3/m^3) of feed is assumed.

PROBLEMS

1. Estimate the hydrogen consumption required to completely remove the sulfur from a hydrotreater feedstock and to reduce the nitrogen content of the product to 15 ppm by weight. The 48.5°API naphtha feed to the unit contains 0.62% sulfur, 0.15% nitrogen, and 0.09% oxygen by weight.
2. The hydrogen required for hydrotreating in a refinery is usually obtained from catalytic reforming operations. Calculate the minimum barrels per day of reformer feed required to provide 120% of the hydrogen necessary

TABLE 9.4
FCC Hydrotreater Material Balance

Component	vol%	BPD	°API	lb/h/BPD	lb/h	wt% S	lb/h S
Feed							
LCGO	56.7	20,500	22.2	13.42	275,119	0.95	2614
HVGO	43.3	15,644	15.5	14.03	219,518	1.51	3357
H_2, scf/bbl		500			3,974		
Total	100.0	36,144	19.3	13.69	494,637	1.20	5928
Products							
H_2S					5,352		5037
C_2-					1,193		
C_3	0.5	199		7.42	1,476		
iC_4	1.0	374		8.22	3,076		
nC_4	0.5	186		8.51	1,580		
HT LVGO	55.5	20,090	23.2	13.35	268,103	0.10	261
HT HVGO	42.4	15,331	16.5	13.95	213,857	0.29	630
Total	100.0	36,180			494,637		5928

TABLE 9.5
FCC Feed HT Utility Requirements

Utility	Per bbl feed	Per day	Per min
Steam, Mlb	0.01	361	
Power, MkWh	0.006	217	
Cooling water, Mgal circulated	0.5	18,072	12.55
Fuel, MMBtu	0.2	7,729	
H_2, Mscf	0.5	18,072	
Catalyst repl, $	0.02	723	

to completely desulfurize 1000 bbl/day (159 m^3/d) of naphtha having the following properties:

°API	55.0
S, as mercaptans (RSH), wt%	0.5
S, as sulfides (R_2S), wt%	0.5
S, total, wt%	1.0

The reformer feed has the following properties:

Component	vol%	gal/lb mol
C_6 naphthenes	20.0	13.1
C_7 naphthenes	10.0	15.4
C_8 naphthenes	10.0	18.3
Paraffins	50.0	18.5
Aromatics	10.0	13.0

Make the following simplifying assumptions:

1. Conversion of naphthenes is 90% for each type.
2. C_6 naphthenes convert only to benzene, C_7 naphthenes convert only to toluene, and C_8 naphthenes convert only to xylene.
3. Aromatics and paraffins in the reformer feed do not react.

Express the actual hydrogen consumption in the hydrotreater and the hydrogen yield from the reformer as scf/bbl of feed.

NOTES

1. Asim, M.Y., et al., NPRA Annual Meeting, March 25–27, 1990.
2. Baron, K., Miller, R.E., Tang, A., and Palmer, L., NPRA Annual Meeting, March 22–24, 1992.
3. Carbrera, C.N., *Process Refining Handbook*, Myers, R.A. (Ed.), McGraw-Hill Book Co., New York, 1987, Sections 6.15–6.25.
4. Gialella, R.M., Andrews, J.W., Cosyns, J., and Heinrich, G., NPRA Annual Meeting, March 22–24, 1992.
5. Wilson, M.F., Fisher, I.P., and Kriz, J.F., *Catalysis 95*, pp. 155–166, 1985.
6. Yui, S.M. and Sanford, E.C., API Meeting, Kansas City, MO, May 13–16, 1985.
7. Yoes, J.R. and Asim, M.Y., *Oil Gas J. 85*(19), pp. 54–58, 1987.
8. Malik, T., NPRA Q&A Meeting, October 14–16, 1992.
9. Gruse, W.A. and Stevens, D.R., *Chemical Technology of Petroleum*, 3rd ed., McGraw-Hill-Hill Book Co., New York, 1960, pp. 117–121, pp. 306–309.
10. Nelson, W.L., *Oil Gas J.* 69(9), p. 64, 1971.
11. Maples, R.E., *Petroleum Refinery Process Economics,* PennWell Publishing Co., Tulsa, OK, 1993, pp. 185–198.

10 Catalytic Reforming and Isomerization

The demand of today's automobiles for unleaded high-octane gasolines has stimulated the use of catalytic reforming. Catalytic reformate furnishes approximately 30 to 40% of the U.S. gasoline requirements but, with the implementation of restrictions on the aromatic contents of gasolines, this quantity can be expected to decrease as the EPA and California Air Resources Board (CARB) continue to restrict automobile emissions standards.

In catalytic reforming, the change in the boiling point of the stock passed through the unit is relatively small, as the hydrocarbon molecular structures are rearranged to form higher-octane aromatics with only a minor amount of cracking. Thus, catalytic reforming primarily increases the octane of motor gasoline rather than increasing its yield; in fact, there is a decrease in yield because of hydrocracking reactions that take place in the reforming operation and the higher densities of the aromatic compounds containing the same number of carbon atoms as the paraffins in the feed. Hydrocarbons boiling above 400°F (204°C) are easily hydrocracked and cause an excessive coke laydown on the catalyst.

The typical feedstocks to catalytic reformers are heavy straight-run (HSR) naphthas [180 to 375°F (82 to 190°C)] and heavy hydrocracker naphthas. These are composed of the four major hydrocarbon groups: paraffins, olefins, naphthenes, and aromatics (PONA). Typical feedstock and reformer products have the following PONA analyses (vol %):

Component	Feed	Product
Paraffins	30–70	30–50
Olefins	0–2	0–2
Naphthenes	20–60	0–3
Aromatics	7–20	45–60

The paraffins and naphthenes undergo two types of reactions in being converted to higher octane components: cyclization and isomerization. The ease and probability of either of these occurring increases with the number of carbon atoms in the molecules, and this is one of the reasons that only the naphthas containing seven or more carbon atoms are used for reformer feed. The LSR naphtha [C_5-180°F (C_5-82°C)] is largely composed of pentanes and hexanes; the pentanes do not contain enough carbon atoms to make an aromatic ring, and the hexanes can be converted into benzene. Because the benzene content of gasolines is limited by the EPA, the hexane content of the reformer feed is minimized. Hydrocarbons boiling above 400°F (204°C) are easily hydrocracked and cause an excessive carbon laydown on the catalyst.

10.1 REACTIONS

As in any series of complex chemical reactions, reactions occur that produce undesirable products in addition to those desired. Reaction conditions have to be chosen that favor the desired reactions and inhibit the undesired ones. Desirable reactions in a catalytic reformer all lead to the formation of aromatics and isoparaffins, as follows:

1. Paraffins are isomerized and to some extent converted to naphthenes. The naphthenes are subsequently converted to aromatics.
2. Olefins are saturated to form paraffins, which then react as in (1).
3. Naphthenes are converted to aromatics.
4. Aromatics are left essentially unchanged.

Reactions leading to the formation of undesirable products include

1. Dealkylation of side chains on naphthenes and aromatics to form butane and lighter paraffins
2. Cracking of paraffins and naphthenes to form butane and lighter paraffins

As the catalyst ages, it is necessary to change the process operating conditions to maintain the reaction severity and to suppress undesired reactions (Table 10.1).

There are four major reactions that take place during reforming. They are (1) dehydrogenation of naphthenes to aromatics, (2) dehydrocyclization of paraffins to aromatics, (3) isomerization, and (4) hydrocracking. The first two of these reactions involve dehydrogenation and will be discussed together.

10.1.1 DEHYDROGENATION REACTIONS

The dehydrogenation reactions are highly endothermic and cause a decrease in temperature as the reaction progresses; dehydrogenation reactions have the highest reaction rates of the reforming reactions, which necessitates the use of the interheaters between catalyst beds to keep the mixture at sufficiently high temperatures for the reactions to proceed at practical rates (see Figure 10.1).

The major dehydrogenation reactions are

1. Dehydrogenation of alkylcyclohexanes to aromatics:

$$\text{Methylcyclohexane} \longrightarrow \text{Toluene} + 3\ H_2$$

Methylcyclohexane **Toluene**

TABLE 10.1
Some Basic Relationships in Catalytic Reforming

Reaction	Reaction rate	Heat effect	Effect of high pressure	Effect of high temperature	Effect of high space velocity	Effect on hydrogen production	Effect on RVP	Effect on density	Effect on volumetric yield	Effect on octane
Hydrocracking	Slowest	Exothermic	Aids	Aids	Hinders	Absorbs	Increase	Decrease	Varies	Increase
Isomerization	Rapid	Mildly exothermic	None	Aids	Hinders	None	Increase	Slight decrease	Slight increase	Increase
Cyclization	Slow	Mildly exothermic	Hinders	Aids	Hinders	Evolves	Decrease	Increase	Decrease	Increase
Naphthene isomerization	Rapid	Mildly exothermic	None	Aids	Hinders	None	Decrease	Slight increase	Slight increase	Slight decrease
Naphthene dehydrogenation	Very fast	Quite endothermic	Hinders	Aids	Hinders	Evolves	Decrease	Increase	Decrease	Increase

Source: Note 1.

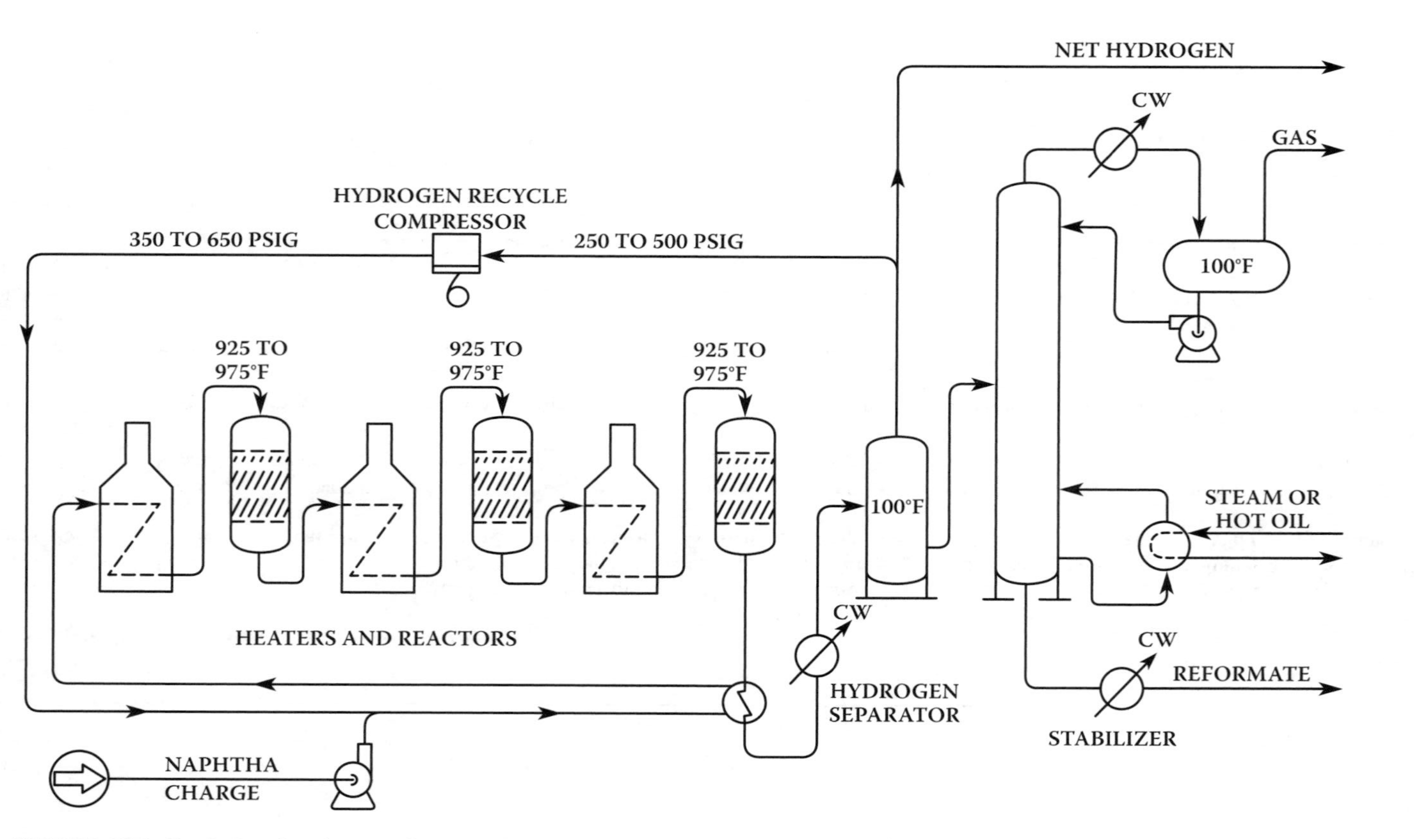

FIGURE 10.1 Catalytic reforming, semiregenerative process.

2. Dehydroisomerization of alkylcyclopentanes to aromatics:

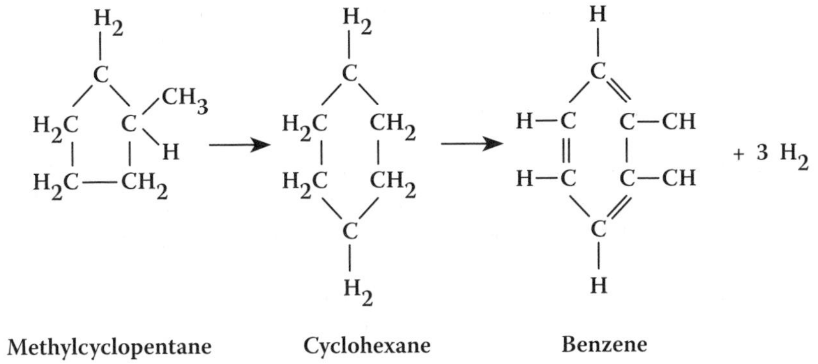

```
      H2                 H2                  H
      |                  |                   |
      C                  C                   C
    /   \   CH3        /   \               /   \\
 H2C     C/          H2C    CH2         H-C     C-CH
  |      | \H   --->  |      |    --->    ||     |        + 3 H2
 H2C --- CH2         H2C    CH2         H-C     C-CH
                       \   /               \   //
                         C                   C
                         |                   |
                         H2                  H
```

Methylcyclopentane **Cyclohexane** **Benzene**

3. Dehydrocyclization of paraffins to aromatics:

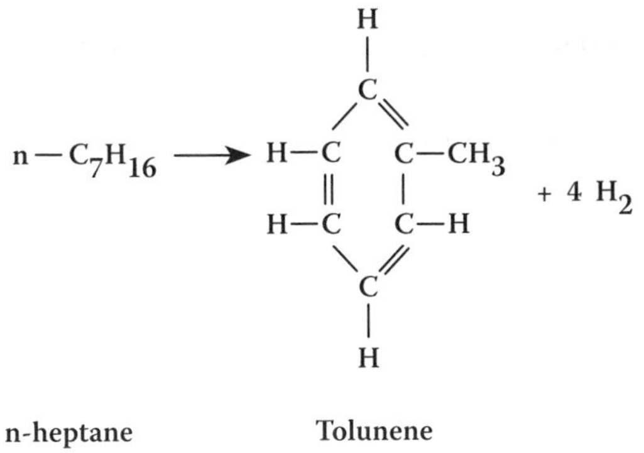

```
                      H
                      |
                      C
                    /   \\
 n-C7H16  --->   H-C     C-CH3
                   ||     |          + 4 H2
                 H-C     C-H
                    \   //
                      C
                      |
                      H
```

n-heptane **Tolunene**

The dehydrogenation of cyclohexane derivatives is a much faster reaction than either the dehydroisomerization of alkylcyclopentanes or the dehydrocyclization of paraffins; however, all three reactions take place simultaneously and are necessary to obtain the aromatic concentration needed in the reformate product to give the octane improvement needed.

Aromatics have a higher liquid density than paraffins or naphthenes with the same number of carbon atoms, so 1 volume of paraffins produces only 0.77 volume of aromatics, and 1 volume of naphthenes about 0.87 volume. In addition, conversion to aromatics increases the gasoline end point, because the boiling points of aromatics are higher than the boiling points of paraffins and naphthenes with the corresponding number of carbons.

The yield of aromatics is increased by

1. High temperature (increases reaction rate but adversely affects chemical equilibrium)
2. Low pressure (shifts chemical equilibrium "to the right")

3. Low space velocity (promotes approach to equilibrium)
4. Low hydrogen-to-hydrocarbon mole ratios (shifts chemical equilibrium "to the right"; however, a sufficient hydrogen partial pressure must be maintained to avoid excessive coke formation)

10.1.2 Isomerization Reactions

Isomerization of paraffins and cyclopentanes usually results in a lower octane product than does conversion to aromatics. However, there is a substantial increase over that of the un-isomerized materials. These are fairly rapid reactions with small heat effects.

1. Isomerization of normal paraffins to isoparaffins:

$$CH_3-CH_2-CH_2-CH_2-CH_3 \rightarrow CH_3-\underset{\displaystyle |}{\overset{\displaystyle CH_3}{CH}}-CH_2-CH_3$$

n-pentane **Isopentane**

2. Isomerization of alkylcyclopentanes to cyclohexanes, plus subsequent conversion to benzene:

$$\text{Methylcyclopentane} \longrightarrow \text{Cyclohexane} \longrightarrow \text{Benzene} + 3\,H_2$$

Methylcyclopentane
91 RON

Cyclohexane
83 RON

Benzene
> 100 RON

Isomerization yield is increased by

1. High temperature (which increases reaction rate)
2. Low space velocity (which increases reaction time)
3. Low pressure

There is no isomerization effect due to the hydrogen-to-hydrocarbon mole ratios, but high hydrogen-to-hydrocarbon ratios reduce the hydrocarbon partial pressure and thus favor the formation of isomers.

10.1.3 Hydrocracking Reactions

The hydrocracking reactions are exothermic and result in the production of lighter liquid and gas products. They are relatively slow reactions, and therefore most of the hydrocracking occurs in the last section of the reactor. The major hydrocracking reactions involve the cracking and saturation of paraffins.

```
                     H   H   H   H   H               H   H   H   H
                     |   |   |   |   |               |   |   |   |
C10H22 + H  →   H—C—C—C—C—C—H  +  H—C—C—C—C—H
                     |   |   |   |   |               |   |   |   |
                     H   H   |   H   H               H   H   H   H
                         H—C—H
                             |
                             H

      Decane              Isohexane                n-butane
```

The concentration of paraffins in the charge stock determines the extent of the hydrocracking reaction, but the relative fraction of isomers produced in any molecular weight group is independent of the charge stock.

Hydrocracking yields are increased by

1. High temperature
2. High pressure
3. Low space velocity

In order to obtain high product quality and yields, it is necessary to carefully control the hydrocracking and aromatization reactions. Reactor temperatures are carefully monitored to observe the extent of each of these reactions.

Low-pressure reforming is generally used for aromatics production, and the following generalizations hold for feedstocks in the 155 to 345°F (68 to 175°C) TBP boiling range:

1. On a mole basis, naphthene conversion to aromatics is about 98%, with the number of carbon atoms in the precursor being retained in the product as follows:

   Methylcyclopentane produces benzene.
   Cyclohexane produces benzene.
   Dimethylcyclopentane produces toluene.
   Dimethylcyclohexane produces xylene.
   Cycloheptane produces toluene.
   Methylcycloheptane produces xylene.

2. For paraffins, the following number of moles of aromatics are produced from 1 mole of paraffins having the indicated number of carbon atoms:

   1 mole P_6 yields 0.05 moles A_6.
   1 mole P_7 yields 0.10 moles A_7.
   1 mole P_8 yields 0.25 moles A_8.
   1 mole P_9 yields 0.45 moles A_9.
   1 mole P_{10} yields 0.45 moles A_{10}.

10.2 FEED PREPARATION

The active material in most catalytic reforming catalysts is platinum. Certain metals, hydrogen sulfide, ammonia, and organic nitrogen and sulfur compounds will deactivate the catalyst [2]. Feed pretreating, in the form of hydrotreating, is usually employed to remove these materials. The hydrotreater employs a cobalt-molybdenum catalyst to convert organic sulfur and nitrogen compounds to hydrogen sulfide and ammonia, which then are removed from the system with the unreacted hydrogen. Any metals in the feed are retained by the hydrotreater catalyst. Hydrogen needed for the hydrotreater is obtained from the catalytic reformer. If the boiling range of the charge stock must be changed, the feed is redistilled before being charged to the catalytic reformer.

10.3 CATALYTIC REFORMING PROCESSES

There are several major reforming processes in use today. These include the Platforming™ process licensed by UOP LLC (Photo 4, Appendix E), Powerforming (ExxonMobil) (Photo 5, Appendix E), Ultraforming (BP), Catalytic Reforming (Engelhard), Magnaforming (BP), Reforming (Institut Francais du Petrole [IFP]), and Rheniforming (Chevron). There are several other processes in use at some refineries but these are limited to a few installations and are not of general interest.

Reforming processes are classified as continuous, cyclic, or semiregenerative, depending upon the frequency of catalyst regeneration. The equipment for the continuous process is designed to permit the removal and regeneration or replacement of catalyst during normal operation. As a result, the catalyst can be regenerated continuously and maintained at a high activity. As increased coke laydown and thermodynamic equilibrium yields of reformate are both favored by low-pressure operation, the ability to maintain high catalyst activities and selectivities by continuous catalyst regeneration (CCR) is the major advantage of the continuous type of unit. This advantage has to be evaluated with respect to the higher capital costs and possible lower operating costs due to lower hydrogen recycle rates and pressures needed to keep coke laydown at an acceptable level.

The semiregenerative unit is at the other end of the spectrum and has the advantage of minimum capital costs (a CCR unit without the regeneration section does not cost much more than the semiregenerative unit and permits the replacement of catalyst while the unit is still on-stream). Regeneration requires the unit to be taken off-stream. Depending upon the severity of operation, regeneration is required at intervals of 3 to 24 months. High hydrogen recycle rates and operating pressures are utilized to minimize coke laydown and consequent loss of catalyst activity.

The cyclic process is a compromise between these extremes and is characterized by having a swing reactor, in addition to those on-stream, in which the catalyst can be regenerated without shutting the unit down. When the activity of the catalyst in one of the on-stream reactors drops below the desired level, this reactor is isolated from the system and replaced by the swing reactor containing freshly regenerated catalyst. The catalyst in the replaced reactor is then regenerated by admitting hot gas containing about 0.5% oxygen into the reactor to burn the carbon off the catalyst.

After regeneration and reactivation of the catalyst, it is used to replace the next reactor needing regeneration.

The reforming process can be obtained as a continuous or semiregenerative operation and other processes as either continuous, cyclic, or semiregenerative. The semiregenerative reforming process is typical of fixed-bed reactor reforming operations and will be described here.

The semiregenerative process is shown in the simplified process flow diagram given in Figure 10.1 [3]. The pretreated feed and recycle hydrogen are heated to 925 to 975°F (498 to 524°C) before entering the first reactor. In the first reactor, the major reaction is the dehydrogenation of naphthenes to aromatics, and as this is strongly endothermic, a large drop in temperature occurs. To maintain the reaction rate, the gases are reheated before being passed over the catalyst in the second reactor. As the charge proceeds through the reactors, the reaction rates decrease and the reactors become larger, and the reheat needed becomes less. Usually three or four reactors are sufficient to provide the desired degree of reaction, and heaters are needed before each reactor to bring the mixture up to reaction temperature. In practice, either separate heaters can be used or one heater can contain several separate coils. A typical gas composition leaving each of the reactors in a four-reactor system, with an HSR naphtha feed of 180 to 380°F; severity, 99 RON; and pressure, 163 psi (1124 kPa), is as follows [4]:

	Feed	1	2	3
P	60	59	50	29
N	29	8	6	5
A	11	33	45	87
Total	100	100	96	87

Note: Reactor outlet, mol/100 mol feed.

The reaction mixture from the last reactor is cooled, and the liquid products condensed. The hydrogen-rich gases are separated from the liquid phase in a drum separator, and the liquid from the separator is sent to a fractionator to be debutanized.

The hydrogen-rich gas stream is split into a hydrogen recycle stream and a net hydrogen by-product, which is used in hydrotreating or hydrocracking operations or as fuel.

The reformer operating pressure and the hydrogen/feed ratio are compromises among obtaining maximum yields, long operating times between regeneration, and stable operation. It is usually necessary to operate at pressures from 50 to 350 psig (345 to 2415 kPa) and at hydrogen charge ratios of 3 to 8 mol H_2/mol feed [2800 to 7600 scf/bbl (500–1350 Nm^3/m^3)]. Liquid hourly space velocities in the area of 1 to 3 are in general use.

The original reforming process is classified as a semiregenerative type because catalyst regeneration is infrequent and runs of 6 to 24 months between regenerations are common. In the cyclic processes, regeneration is typically performed on a 24- or 48-hour cycle, and a spare reactor is provided so that regeneration can be accomplished while the unit is still on-stream. Because of these extra facilities, the cyclic processes are more expensive but offer the advantages of lower pressure operation and higher yields of reformate at the same severity.

The continuous catalyst regeneration unit moves the catalyst between the reactor and regenerator and permits the catalyst to be regenerated and returned to the reactor while the unit is operating. The catalyst flows by gravity through the reactor. It is then picked up in a nitrogen stream and carried to the top of the regeneration unit. As the catalyst flows through the regenerator, the coke is burned from the catalyst using a nitrogen stream containing a small amount of oxygen. The oxygen content is carefully regulated to prevent the catalyst from overheating and becoming permanently deactivated. After regeneration, it is carried in a hydrogen stream to the top of the reactor to begin its journey through the cycle again. From the time a catalyst particle enters the top of the reactor until it goes through the cycle and is returned to the top of the reactor, a period of 5 to 7 days is required.

10.3.1 Example Problem

Calculate the length of time between regeneration of catalyst in a reformer operating at the following conditions:

Liquid hourly space velocity (LHSV) = 3.0 v/hr/v
Feed rate = 5000 BPSD (795 m^3/d)
Feed gravity = 55.0° API
Catalyst bulk density = 50 lb/ft^3 (802 kg/m^3)
Hydrogen-to-feed ratio = 8000 scf/bbl (1424 Nm^3/m^3)
Number of reactors = 3

The catalyst deactivates after processing 90 barrels of feed per pound of catalyst. If the catalyst bed is 6 ft deep in each reactor, what are the reactor inside diameters? Assume an equal volume of catalyst in each reactor.

10.3.1.1 Solution

Time between regenerations:

5000 BPD = 1170 ft^3/hr (795 m^3/d)
Total catalyst = 1170 / 3 = 390 ft^3 (m^3)
(390 ft^3) (50 lb/ft^3) = 19,500 lb (8,845 kg)
(19,500 lb) (90 bbl/lb) / 5,000 bbl/day = 351 days

Inside diameter:

Volume of catalyst per reactor = 390 / 3 = 130 ft^3 (3.7 m^3)
Inside area = (130 ft^3) / (6 ft) = 21.67 ft^2 (2 m^2)
Inside diameter = 5.25 ft (1.6 m)

The continuous catalytic reforming process unit can consist of either a stacked design (UOP LLC), in which the reactors are stacked on top of each other, or side-by-side reactors (IFP). In both cases, the sequence of flow of the reactants is similar to that shown for the semiregenerative system in Figure 10.1.

In the stacked design (CCR Platforming Unit), freshly regenerated catalyst is introduced in the top of the upper reactor between two concentric perforated cylinders (made from Johnson screens) and flows by gravity from top to bottom. The reactants are introduced on the outside of the outer cylinder and flow radially through the catalyst to the center of the inner cylinder (see Figure 10.2). Partially aged catalyst is removed from the bottom of the lowest reactor and sent to an external regenerator, where the carbon is burned from the catalyst, and the catalyst is reduced and acidified before being returned to the upper reactor [12].

The IFP system is similar to that of UOP LLC except that the catalyst removed from all of the reactors, except the last, is transported by hydrogen lifts to the top of the next reactor in the series [6]. The catalyst removed from the last reactor is conveyed by nitrogen to the regenerator section.

10.4 REFORMING CATALYST

All of the reforming catalyst in general use today contains platinum supported on an alumina base. In most cases, rhenium is combined with platinum to form a more stable catalyst that permits operation at lower pressures. Platinum is thought to serve as a catalytic site for hydrogenation and dehydrogenation reactions, and chlorinated alumina provides an acid site for isomerization, cyclization, and hydrocracking reactions [2]. Reforming catalyst activity is a function of surface area, pore volume, and active platinum and chlorine content. Catalyst activity is reduced during operation by coke deposition and chloride loss. In a high-pressure process, up to 200 bbl of charge can be processed per pound of catalyst (64 m^3/kg) before regeneration is needed. The activity of the catalyst can be restored by high-temperature oxidation of the carbon followed by chlorination. This type of process is referred to as semiregenerative and is able to operate for 6- to 24-month periods between regenerations. The activity of the catalyst decreases during the on-stream period, and the reaction temperature is increased as the catalyst ages to maintain the desired operating severity. Normally, the catalyst can be regenerated *in situ* at least three times before it has to be replaced and returned to the manufacturer for reclamation.

Catalyst for fixed-bed reactors is extruded into cylinders 1/32 to 1/16 in. (0.8 to 1.6 mm) diameter with lengths about 3/16 in. (5 mm). The catalyst for continuous units is spherical, with diameters approximately 1/32 to 1/16 in. (0.8 to 1.6 mm).

10.5 REACTOR DESIGN

Fixed-bed reactors used for semiregenerative and cyclic catalytic reforming vary in size and mechanical details, but all have basic features as shown in Figure 10.3. Very similar reactors are used for hydrotreating, isomerization, and hydrocracking.

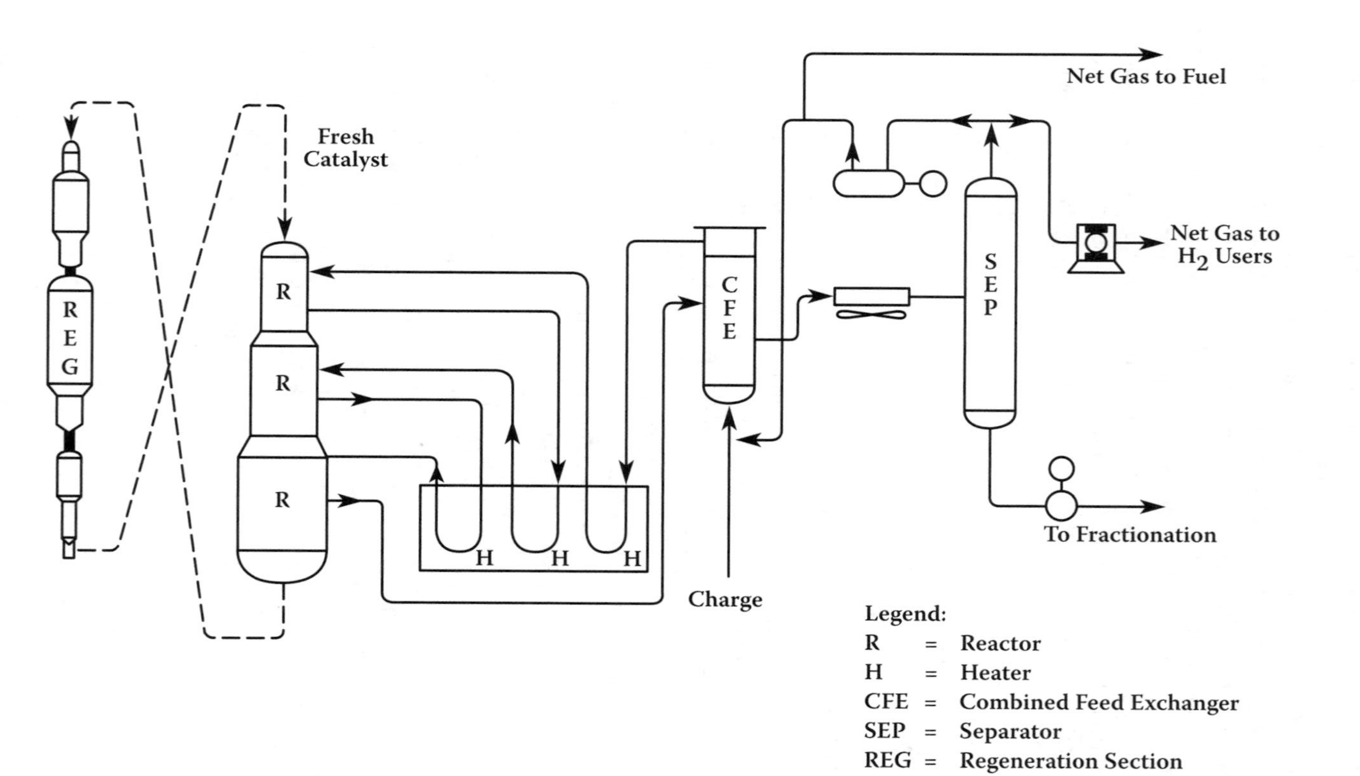

FIGURE 10.2 Continuous catalyst regeneration (CCR Platforming™) catalytic reformer. (Courtesy of UOP LLC.)

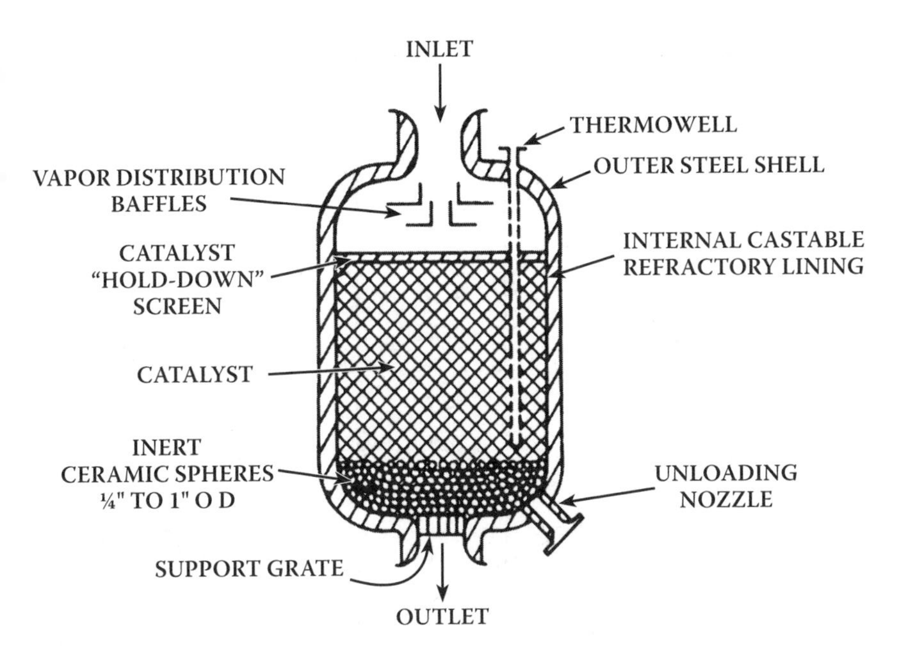

FIGURE 10.3 Typical fixed-bed downflow catalytic reactor.

The reactors have an internal refractory lining that is provided to insulate the shell from the high reaction temperatures and thus reduce the required metal thickness. Metal parts exposed to the high-temperature hydrogen atmosphere are constructed from steel containing at least 5% chromium and 0.5% molybdenum to resist hydrogen embrittlement. Proper distribution of the inlet vapor is necessary to make maximum use of the available catalyst volume. Some reactor designs provide for radial vapor flow rather than the simpler straight-through type shown here. The important feature of vapor distribution is to provide maximum contact time with minimum pressure drop.

Temperature measurement at a minimum of three elevations in the catalyst bed is considered essential to determine catalyst activity and as an aid in coke burn-off operations.

The catalyst pellets are generally supported on a bed of ceramic spheres about 12 to 16 in. (30 to 40 cm) deep. The spheres vary in size from about 1 in. (25 mm) on the bottom to about 0.35 in. (9 mm) on the top.

10.6 YIELDS AND COSTS

Catalytic reforming yields can be estimated using Figure 10.4 through Figure 10.7. These simplified yield correlations are approximations only and are not specific for any catalyst, operating parameters, or process configuration. Actual yields are functions of reactor pressure, catalyst type and activity, and feed quality. Capital and

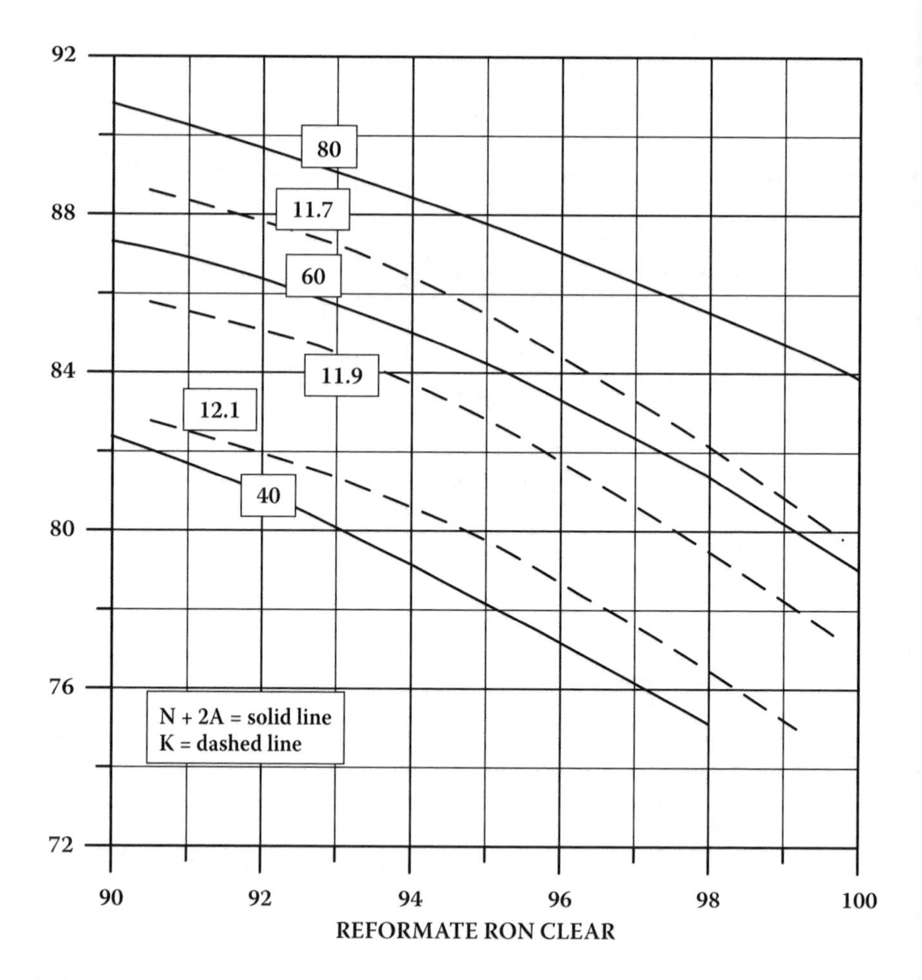

FIGURE 10.4 Catalytic reforming yield correlations.

operating costs can be obtained from Figure 10.8 and its accompanying descriptive material (Table 10.2).

10.7 ISOMERIZATION

The octane numbers of the LSR naphtha [C_5-180°F (C_5-82°C)] can be improved by the use of an isomerization process to convert normal paraffins to their isomers. This results in significant octane increases, as n-pentane has an unleaded (clear) RON of 61.7 and isopentane has a rating of 92.3. In once-through isomerization where the normal and iso compounds come essentially to thermodynamic equilibrium, the unleaded RON of LSR naphtha can be increased from 70 to about 82 to 84. If the normal components are recycled, the resulting research octane numbers will be about 87 to 93 RONC.

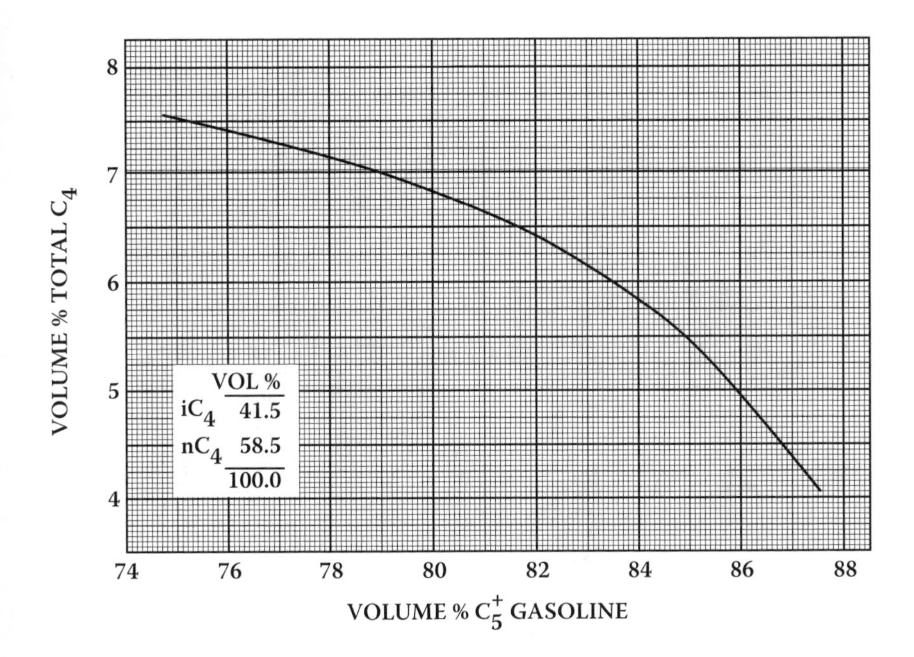

FIGURE 10.5 Catalytic reforming yield correlations.

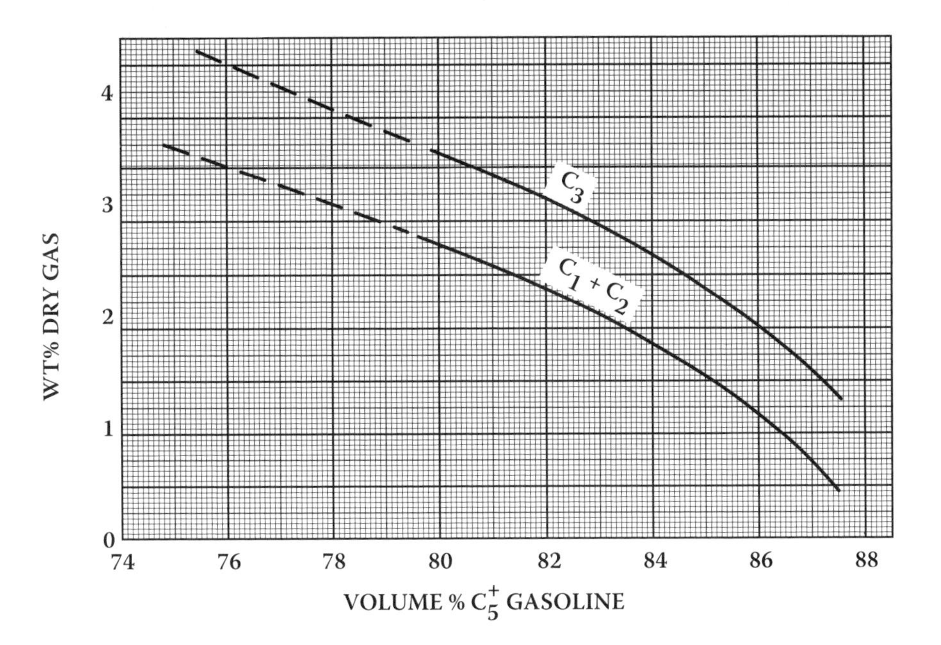

FIGURE 10.6 Catalytic reforming yield correlations.

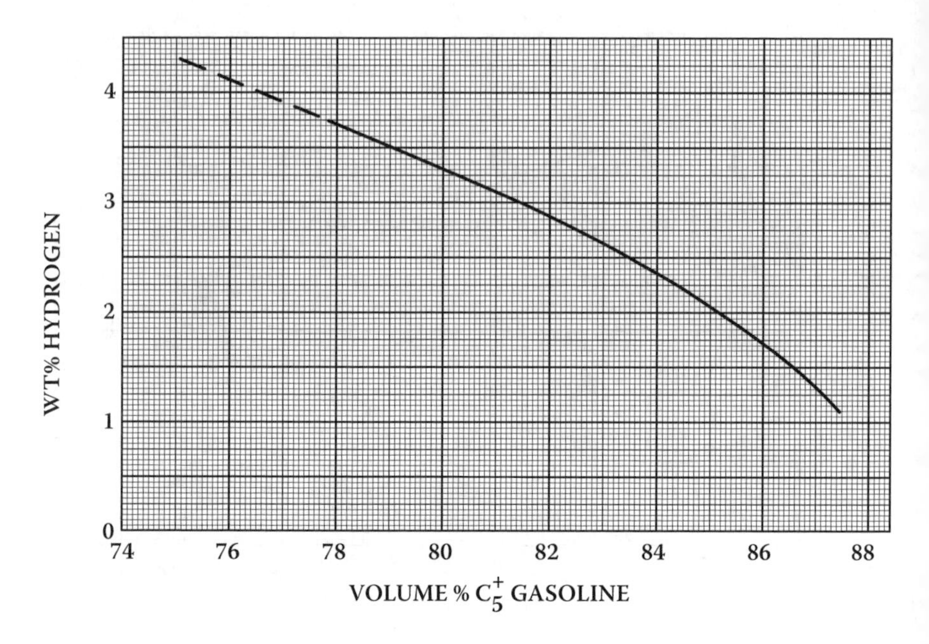

FIGURE 10.7 Catalytic reforming yield correlations.

Reaction temperatures of about 200 to 400°F (95 to 205°C) are preferred to higher temperatures because the equilibrium conversion to isomers is enhanced at the lower temperatures. At these relatively low temperatures, a very active catalyst is necessary to provide a reasonable reaction rate. The available catalysts used for isomerization contain platinum on various bases. Some types of catalysts (alumina support) require the continuous addition of very small amounts of organic chlorides to maintain high catalyst activities. This is converted to hydrogen chloride in the reactor, and consequently the feed to these units must be free of water and other oxygen sources in order to avoid catalyst deactivation and potential corrosion problems. A second type of catalyst uses a molecular sieve base and is reported to tolerate feeds saturated with water at ambient temperature [7]. A third type of catalyst contains platinum supported on a novel metal oxide base. This catalyst has a 150°F (83°C) higher operating temperature than conventional zeolitic isomerization catalysts and can be regenerated. Catalyst life is usually 3 years or more with all of these catalysts.

An atmosphere of hydrogen is used to minimize carbon deposits on the catalyst, but hydrogen consumption is negligible [1].

The composition of the reactor products can closely approach chemical equilibrium. The actual product distribution is dependent upon the type and age of the catalyst, the space velocity, and the reactor temperature. The pentane fraction of the reactor product is about 75 to 80 wt% isopentane, and the hexane fraction is about 86 to 90 wt% hexane isomers [9].

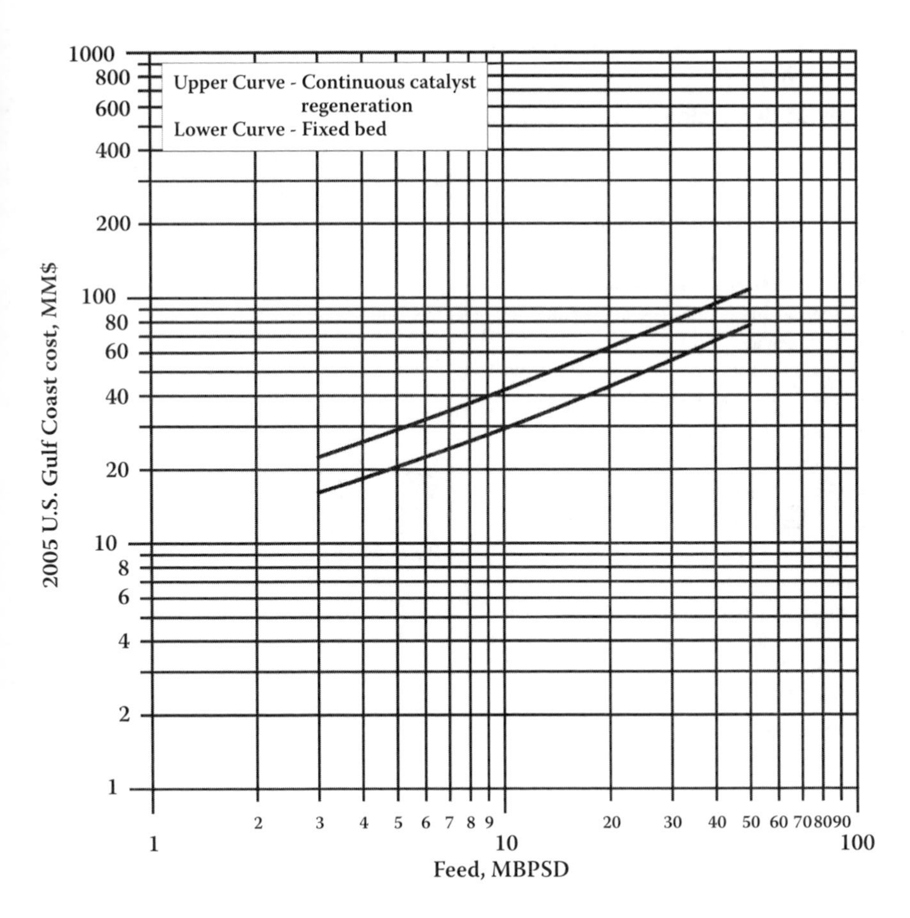

FIGURE 10.8 Catalytic reforming unit investment cost: 2005 U.S. Gulf Coast (see Table 10.2).

Following is a simplified conversion summary for a typical LSR cut. The values are on a relative weight basis and do not account for the weight loss resulting from hydrocracking to molecules lighter than pentane.

LSR component	Feed weight	Product weight	RONC (unleaded)
Isopentane	22	41	92
Normal pentane	33	12	62
2,2-Dimethylbutane	1	15	96
2,3-Dimethylbutane	2	5	84
2-Methylpentane	12	15	74
3-Methylpentane	10	7	74
Normal hexane	20	5	26
Total	100	100	

TABLE 10.2
Catalytic Reforming Unit Cost Data

Costs included	
1. All battery limit facilities required for producing 102 RON unleaded reformate from an HSR naphtha sulfur-free feed	
2. Product debutanizer	
3. All necessary control and instrumentation	
4. Preheat and product cooling facilities to accept feed and release products at ambient temperatures	
Costs not included	
1. Cooling water, steam, and power supply	
2. Initial catalyst charge	
3. Royalty	
4. Feed fractionation or desulfurization	
Catalyst charge	
Initial catalyst charge cost is approximately $280/BPD of feed.	
Royalties	
Running royalty is about $0.05 to $0.10/bbl.	
Paid-up royalty is about $50 to $100/BPD.	
Utility data (per bbl feed)	
Steam,[a] lb	30
Power, kWh	3
Cooling water, gal	400
Fuel gas (LHV), MMBtu	0.3
Catalyst replacement, $	0.16

[a] With some configurations, a net steam production is realized.
Note: See Figure 10.8.

If the normal pentane in the reactor product is separated and recycled, the product RON can be increased by about three numbers (83 to 86 RONC) [8]. If both normal pentane and normal hexane are recycled, the product clear RON can be improved to about 87 to 90. Separation of the normals from the isomers can be accomplished by fractionation or by vapor phase adsorption of the normals on a molecular sieve bed. The adsorption process is well developed in several large units [7].

Some hydrocracking occurs during the reactions, resulting in a loss of gasoline and the production of light gas. The amount of gas formed varies with the catalyst type and age and is sometimes a significant economic factor. The light gas produced is typically in the range of 1.0 to 4.0 wt% of the hydrocarbon feed to the reactor. For preliminary estimates, the composition of the gas produced can be assumed to be 95 wt% methane and 5 wt% ethane.

For refineries that do not have hydrocracking facilities to supply isobutane for alkylation unit feed, the necessary isobutane can be made from n-butane by isomerization. The process is very similar to that of LSR gasoline isomerization, but a feed deisobutanizer is used to concentrate the n-butane in the reactor charge. The reactor product is about 58 to 62 wt% isobutane.

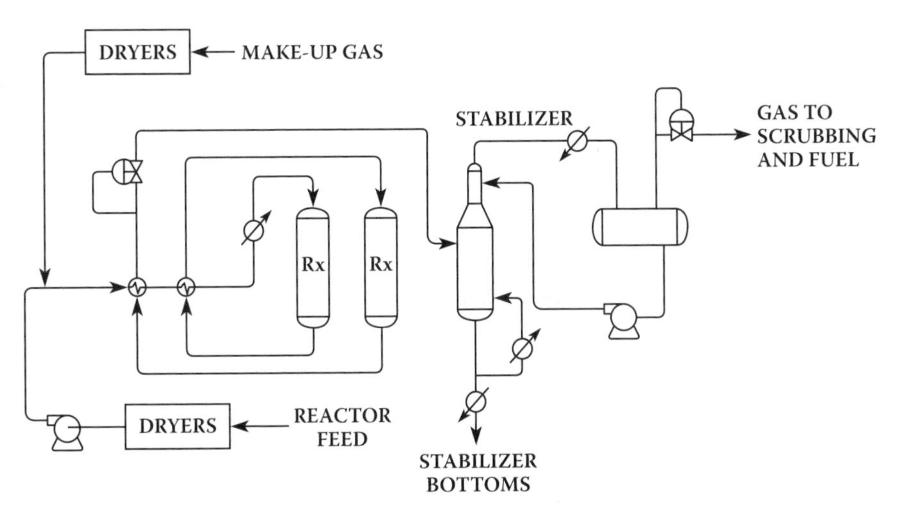

FIGURE 10.9 H-O-T Pernex™ isomerization unit. (Reprinted courtesy of UOP LLC.)

A representative flow scheme for an isomerization unit is shown in Figure 10.9. Typical operating conditions are [1,9]

Reactor temperature	200–400°F	95–205°C
Pressure	250–500 psig	1725–3450 kPa
Hydrogen/HC mole ratio	0.05/1	
Single-pass LHSV	1–2 v/hr/v	
Liquid product yield	> 98 wt%	

10.8 CAPITAL AND OPERATING COSTS

The capital construction and operating costs for an isomerization unit can be estimated from Figure 10.10 and Table 10.3.

10.9 ISOMERIZATION YIELDS

Yields vary with feedstock properties and operating severity. A typical product yield is given in Table 10.4 for 13-number improvement in both RON and MON clear for a 12 psi RVP C_3+ isomerate product.

10.10 CASE-STUDY PROBLEM: NAPHTHA HYDROTREATER, CATALYTIC REFORMER, AND ISOMERIZATION UNIT

In this case, the feed to the catalytic reformer consists of the heavy straight-run naphtha (190 to 380°F) from the crude unit and the coker naphtha. In practice, the coker naphtha would probably be separated into a C_5-180°F (C_5-82°C) fraction, which would not be reformed, and a 180 to 380°F (82 to 193°C) fraction, which would be used as reformer feed. As sufficient information is not available to estimate the

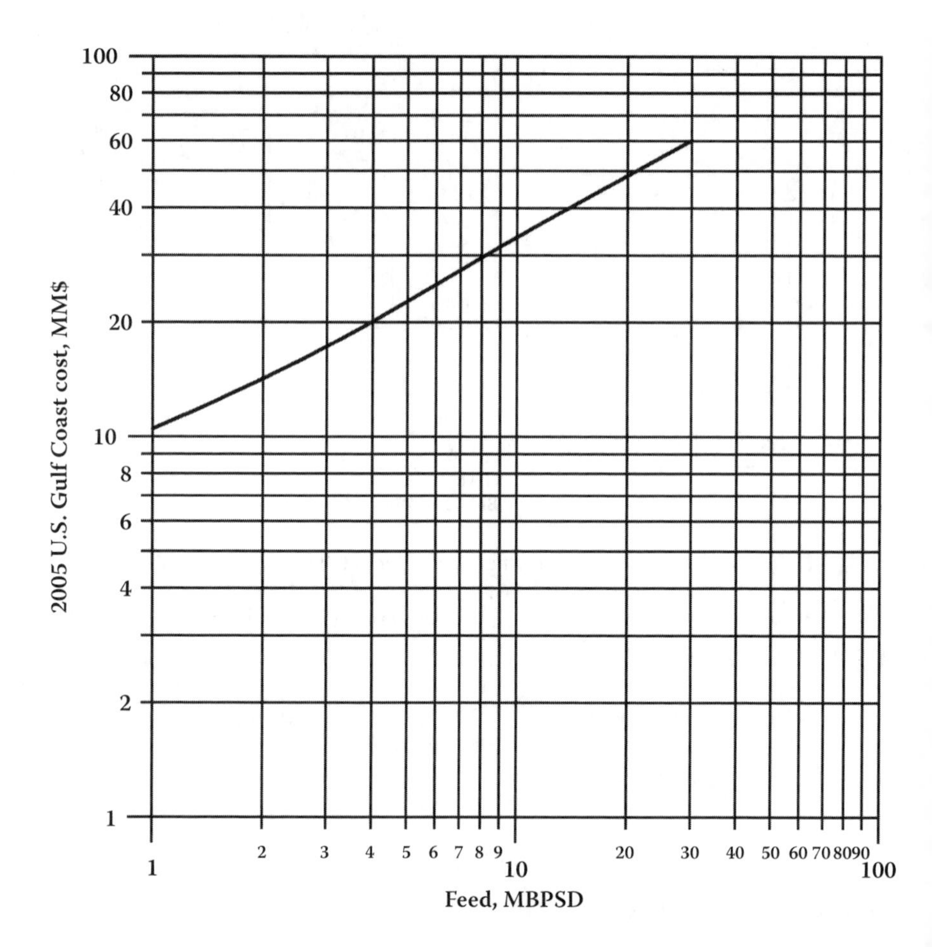

FIGURE 10.10 Paraffin isomerization units (platinum catalyst type) investment cost: 2005 U.S. Gulf Coast (see Table 10.3).

quantities and properties of these fractions, it is necessary for the purposes of this calculation to send all of the coker gasoline to the reformer. The C_5-180°F (C_5-82°C) fraction will undergo little octane improvement, but its gasoline quality will be helped by saturation of the olefins. The required severity of reforming is not known until after the gasoline blending calculations are made. Therefore, it is necessary to assume a value for the first time through. A severity of 94 RON clear is used for this calculation. The following procedure is used:

1. Calculate the characterization factor (K_W) of the feed.
2. Determine the C_5+ naphtha volume yield from Figure 10.3.
3. Determine the weight or volume yields of H_2, $C_1 + C_2$, C_3, iC_4, and nC_4 from Figure 10.4, Figure 10.5, and Figure 10.6.
4. Calculate the weight yield of all product streams except C_5+ naphtha.

TABLE 10.3
Paraffin Isomerization with Platinum Catalysts Cost Data

Costs Included	
1. Feed drying	
2. Drying of hydrogen makeup	
3. Complete preheat, reaction, and hydrogen circulation facilities	
4. Product stabilization	
5. Sufficient heat exchange to cool products to ambient temperatures	
6. Central controls	
7. Paid-up royalty	
Costs not included	
1. Hydrogen source	
2. Cooling water, steam, and power supply	
3. Feed desulfurization	
4. Initial catalyst charge, about \$150/BPD of reactor feed	
Royalties	
Paid-up royalty is about \$90 to \$160/BPD.	
Utility data (per bbl "fresh" feed)	
Power, kWh	1.0
Cooling water, gal (30°F)	600–1000
Fuel (LHV), MMBtu	0.20
Catalyst replacement, \$	0.08
Hydrogen makeup, scf	40

Note: See Figure 10.10.

TABLE 10.4
Isomerization Yields

Component	Vol% on feed
C_3	0.5
iC_4	1.5
nC_4	1.0
C_5–C_7	102.0

5. Determine the weight yield of C_5+ naphtha by difference.
6. Calculate the API gravity of C_5+ naphtha.
7. Make sulfur or nitrogen balance if needed to determine H_2S and NH_3 made and net hydrogen produced.
8. Estimate the utility requirements from Table 10.2.

The results are tabulated in Table 10.5 for the hydrotreater and catalytic reformer and Table 10.6 for the isomerization unit. Table 10.7 gives chemical and utility requirement data for the processes.

TABLE 10.5
Hydrotreater and Catalytic Reformer Material Balance: 100,000 BPCD Alaska North Slope Crude Oil Basis (Severity, 94 ROM clear; K_W = 11.7)

Component	vol%	BPD	°API	(lb/hr)/BPD	lb/hr	wt% S	lb/hr S
Feed							
190–380°F HSR	69.3	12,500	48.5	11.47	143,380	0.1	140
Coker naphtha	30.7	5,540	54.6	11.09	61,420	0.65	400
Total	100.0	18,040			204,800		540
Products							
H_2 wt%, total	1.7				3,480		
$C_1 + C_2$ wt%	1.0				2,050		
C_3, wt%	1.8	500		7.42	3,690		
iC_4	2.0	360		8.22	2,960		
nC_4	2.8	505		8.51	4,300		
C_5+ reformate	86.5	15,610	39.7	12.06	188,320		
Total		16,975			204,800		
Hydrogen[a]							
H_2S					574		540
H_2, net					3,446		

[a] H_2S = 32.06 = 16.84 lb-mol/hr; H_2 in H_2S = (16.84)(2) = 34 lb/hr

TABLE 10.6
Isomerization Unit Material Balance: 100,000 BPCD Alaska North Slope Crude Oil Basis (Severity, Once-Through)

Component	vol%	BPD	°API	(lb/hr)/BPD	lb/hr	wt% S	lb/hr S
Feed							
C_5-180°F	73.4	4,277	63.0	10.61	45,394	0.02	1
Coker light naphtha	26.6	1,551	65.0	10.51	6,303	0.41	64
Total	100.0	5,828			616,970		65
Products							
H_2S					69		65
C_3, wt%	0.5	29		7.39	215		
iC_4	0.8	47		8.22	382		
nC_4	2.2	128		8.51	1,092		
C_5+ reformate	98.4	5,735	65.6	10.46	60,007		
Total		16,975			204,800		65

TABLE 10.7
Hydrotreater, Catalytic Reformer, and Isomerization Unit Chemical and Utility Requirements

Utility	Reformer and HT		Isomerization	
	Per barrel feed	Per day	Per barrel feed	Per day
Steam, Mlb	0.036	631		
Power, MkWh	0.005	88	0.001	6
CW, Mgal	0.9	15,786	0.8	4662
Fuel, MMBtu	0.4	7,016	0.2	1166
Catalyst, $	0.12	2,105	0.05	291
H_2 makeup, Mscf			0.04	233

PROBLEMS

1. Reactor pressure is an important process variable in catalytic reforming. A common reaction in reforming is the conversion of methylcyclopentane to benzene. Calculate the barrels of benzene formed from one barrel of methylcyclopentane at the following reactor outlet conditions: (a) 900°F, 600 psig (482°C, 4140 kPa) and (b) 900°F, 300 psia (482°C, 207 kPa). The hydrogen feed rate to the reactor is 10,000 scf/bbl of methylcyclopentane.
   Assume the reaction is a single ideal gas-phase reaction and thermodynamic equilibrium is obtained. The National Bureau of Standards values for free energies of formation at 900°F are +66.09 kcal/g mol for methylcyclopentane and +50.78 kcal/g mol for benzene.
2. Determine the yield of n-butane and C_5+ gasoline when reforming 4500 BPD of HSR gasoline, K_W = 11.9, to a 92 clear RON.
3. A 180 to 380°F virgin naphtha stream with a mean average boiling point (MABP) of 275°F and 50.2°API is reformed to a 96 RON clear gasoline blending stock. Make an overall material balance around the reformer for a 10,000 BPD (1,590 m^3/d) feed rate.
4. Estimate the installed cost for a 6200 BPSD isomerization unit to increase the RON of an LSR naphtha by 13 numbers. Determine the utility requirements and estimate the direct operating costs per barrel of feed if two operators are required per shift at an hourly rate of $26.00 per operator.
5. Estimate the direct operating cost per barrel of feed for an 8400 BPSD catalytic reformer upgrading an HSR naphtha to a 96 RON clear product. The 50.2°API naphtha feed has an MABP of 275°F. Assume the reformer requires two operators per shift at an hourly rate of $26.00 per operator. Include royalty costs as a direct operating cost.
6. For problem 5, express the direct operating cost on a per barrel of C_5^+ reformate basis and compare with the cost of producing a barrel of 90 RON clear reformate from the same feed.

7. For the reforming units in problems 5 and 6, what is the single largest operating expense per barrel? What percentage of the cost is this?
8. Draw a flow diagram for an isomerization unit that increases by 20 research numbers the octane of the feed consisting of 45 vol% n-pentane and 55 vol% n-hexane.
9. Calculate the product gas composition of a reformer operated to maximize aromatics production with the following feedstock of saturated hydrocarbons:

	vol%
C_6H_{12}	10.1
C_6H_{14}	18.9
C_7H_{14}	12.8
C_7H_{16}	21.2
C_8H_{16}	17.7
C_8H_{18}	19.3
Total	100.0

10. Calculate the total operating costs per gallon of reformate for producing (a) a 92 clear RON product from 10,000 BPCD of the HSR naphtha from the assigned crude oil, and (b) a 98 clear RON product. Use the heating value of comparable products to obtain the values of gases produced.

NOTES

1. *Hydrocarbon Process. 49*(9), pp.195–197, 1970.
2. Thornton, D.P., *Petro/Chem. Eng. 41*(5), pp. 21–29, 1969.
3. *Hydrocarbon Process. 49*(9), p. 189, 1970.
4. Corlew, J.S., Personal communication, January 1993.
5. Weiszmann, J.A., *Handbook of Petroleum Refining Processes*, Myers, R.A. (Ed.), McGraw-Hill Book Co., New York, 1986, Sections 3.3–3.20.
6. Hennico, A., Mank, L., Mansuy, C., and Smith, D.H., *Oil Gas J. 90*(23) pp. 54–59, 1992.
7. Kouwenhoven, H.W. and Van Z. Langhout, W.C., *Chem. Eng. Prog. 67*(4), pp. 65–70, 1971.
8. Bour, G., Schwoerev, C.P., and Asselin, G.F., *Oil Gas J. 68*(43), pp. 57–61, 1970.
9. Greenough, P. and Rolfe, J.R.K., *Handbook of Petroleum Refining Processes*, Meyers, R.A. (Ed.), McGraw-Hill Book Co., New York, 1986, Sections 5.25–5.37.

ADDITIONAL READING

10. Fowle, M.J., Bent, R.D., Milner, B.E., and Masologites, G.P., *Petrol. Refiner. 31*(4), pp. 156–159, 1952.

11 Alkylation and Polymerization

The addition of an alkyl group to any compound is an alkylation reaction, but in petroleum terminology, the term *alkylation* is used for the reaction of low molecular-weight olefins with an isoparaffin to form higher molecular-weight isoparaffins [1]. Although this reaction is simply the reverse of cracking, the belief that paraffin hydrocarbons are chemically inert delayed its discovery until about 1935 [2]. The need for high-octane aviation fuels during World War II acted as a stimulus to the development of the alkylation process for production of isoparaffinic gasolines of high octane number.

Although alkylation can take place at high temperatures and pressures without catalysts, the only processes of commercial importance involve low-temperature alkylation conducted in the presence of either sulfuric or hydrofluoric acid. The reactions occurring in both processes are complex, and the product has a rather wide boiling range. By proper choice of operating conditions, most of the product can be made to fall within the gasoline boiling range, with motor octane numbers from 88 to 94 [2] and research octane numbers from 94 to 99.

11.1 ALKYLATION REACTIONS

In alkylation processes using hydrofluoric or sulfuric acids as catalysts, only isoparaffins with tertiary carbon atoms, such as isobutane or isopentane, react with the olefins. In practice, only isobutane is used because isopentane has a sufficiently high octane number and low vapor pressure to allow it to be effectively blended directly into finished gasolines.

The process using sulfuric acid as a catalyst is much more sensitive to temperature than the hydrofluoric acid process. With sulfuric acid, it is necessary to carry out the reactions at 40 to 70°F (5 to 21°C) or lower, to minimize oxidation reduction reactions that result in the formation of tars and the evolution of sulfur dioxide. When anhydrous hydrofluoric acid is the catalyst, the temperature is usually limited to 100°F (38°C) or below [2]. In both processes, the volume of acid employed is about equal to that of the liquid hydrocarbon charge, and sufficient pressure is maintained on the system to keep the hydrocarbons and acid in the liquid state. High isoparaffin/olefin ratios (4:1 to 15:1) are used to minimize polymerization and to increase product octane. Efficient agitation to promote contact between the acid and hydrocarbon phases is essential to high product quality and yields. Contact times from 10 to 40 min are in general use. The yield, volatility, and octane number of the product is regulated by adjusting the temperature, acid/hydrocarbon ratio, and isoparaffin/olefin ratio. At the same operating conditions, the products from the hydrofluoric and sulfuric acid alkylation processes are quite similar [3,4]. In practice, however, the plants are operated at different conditions, and the products are somewhat different. The effects of variables will be discussed for each process later, but for both processes the more important variables are

1. Reaction temperature
2. Acid strength
3. Isobutane concentration
4. Olefin space velocity

The principal reactions that occur in alkylation are the combinations of olefins with isoparaffins as follows:

$$\underset{\textbf{Isobutylene}}{CH_3-\overset{\overset{\displaystyle CH_3}{|}}{C}=CH_2} + \underset{\textbf{Isobutane}}{CH_3-\underset{\underset{\displaystyle H}{|}}{\overset{\overset{\displaystyle CH_3}{|}}{C}}-CH_3} \longrightarrow \underset{\textbf{2, 2, 4-trimethylpentane (isooctane)}}{CH_3-\underset{\underset{\displaystyle CH_3}{|}}{\overset{\overset{\displaystyle CH_3}{|}}{C}}-CH_2-\overset{\overset{\displaystyle CH_3}{|}}{CH}-CH_3}$$

$$\underset{\textbf{Propylene}}{CH_2=CH-CH_3} + \underset{\textbf{Isobutane}}{CH_3-\underset{\underset{\displaystyle H}{|}}{\overset{\overset{\displaystyle CH_3}{|}}{C}}-CH_3} \longrightarrow \underset{\textbf{2, 2-dimethylpentane (isoheptane)}}{CH_3-\underset{\underset{\displaystyle CH_3}{|}}{\overset{\overset{\displaystyle CH_3}{|}}{C}}-CH_2-CH_2-CH_3}$$

Another significant reaction in propylene alkylation is the combination of propylene with isobutane to form propane plus isobutylene. The isobutylene then reacts with more isobutane to form 2,2,4-trimethylpentane (isooctane). The first step involving the formation of propane is referred to as a hydrogen transfer reaction. Research on catalyst modifiers is being conducted to promote this step because it produces a higher octane alkylate than is obtained by formation of isoheptanes.

A number of theories have been advanced to explain the mechanisms of catalytic alkylation, and these are discussed in detail by Gruse and Stevens [2]. The one most widely accepted involves the formation of carbonium ions by transfer of protons from the acid catalyst to olefin molecules, followed by combination with isobutane to produce tertiary-butyl cations. The tertiary-butyl ion reacts with 2-butene to form C_8 carbonium ions capable of reacting with isobutane to form C_8 paraffins and tertiary-butyl ions. These tertiary-butyl ions then react with other 2-butene molecules to continue the chain. Figure 11.1 illustrates the above sequence using sulfuric acid, 2-butene, and isobutane as the example reaction. The alkylation reaction is highly

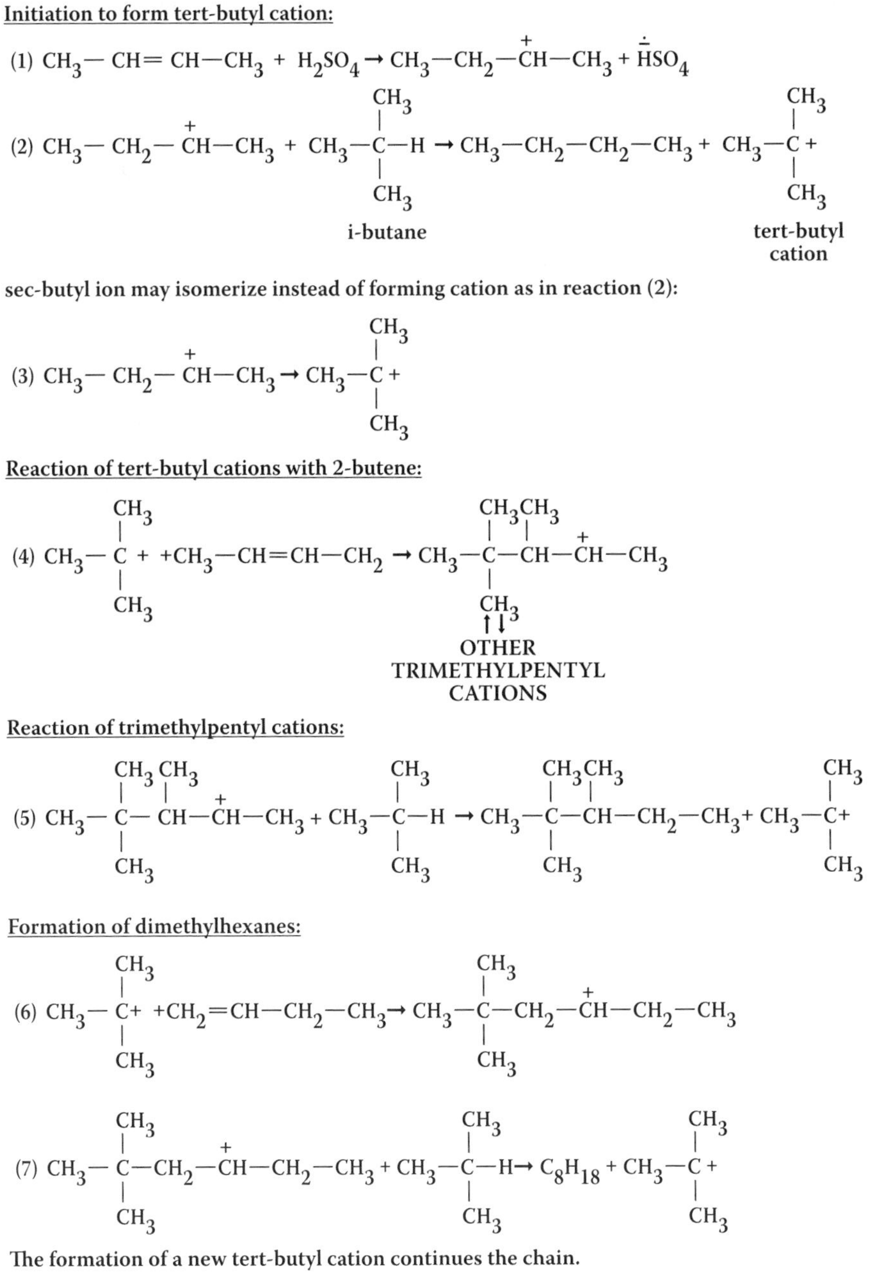

Initiation to form tert-butyl cation:

$$(1)\ CH_3-CH=CH-CH_3 + H_2SO_4 \rightarrow CH_3-CH_2-\overset{+}{C}H-CH_3 + \overset{-}{H}SO_4$$

$$(2)\ CH_3-CH_2-\overset{+}{C}H-CH_3 + \underset{\text{i-butane}}{CH_3-\overset{CH_3}{\underset{CH_3}{\overset{|}{\underset{|}{C}}}}-H} \rightarrow CH_3-CH_2-CH_2-CH_3 + \underset{\text{tert-butyl cation}}{CH_3-\overset{CH_3}{\underset{CH_3}{\overset{|}{\underset{|}{C}}}}+}$$

sec-butyl ion may isomerize instead of forming cation as in reaction (2):

$$(3)\ CH_3-CH_2-\overset{+}{C}H-CH_3 \rightarrow CH_3-\overset{CH_3}{\underset{CH_3}{\overset{|}{\underset{|}{C}}}}+$$

Reaction of tert-butyl cations with 2-butene:

$$(4)\ CH_3-\overset{CH_3}{\underset{CH_3}{\overset{|}{\underset{|}{C}}}} + {}^{+}CH_3-CH=CH-CH_2 \rightarrow CH_3-\overset{CH_3}{\underset{CH_3}{\overset{|}{\underset{|}{C}}}}-\overset{CH_3}{\overset{|}{C}H}-\overset{+}{C}H-CH_3$$

$\uparrow\downarrow$

OTHER TRIMETHYLPENTYL CATIONS

Reaction of trimethylpentyl cations:

$$(5)\ CH_3-\overset{CH_3}{\underset{CH_3}{\overset{|}{\underset{|}{C}}}}-\overset{CH_3}{\overset{|}{C}H}-\overset{+}{C}H-CH_3 + CH_3-\overset{CH_3}{\underset{CH_3}{\overset{|}{\underset{|}{C}}}}-H \rightarrow CH_3-\overset{CH_3}{\underset{CH_3}{\overset{|}{\underset{|}{C}}}}-\overset{CH_3}{\overset{|}{C}H}-CH_2-CH_3 + CH_3-\overset{CH_3}{\underset{CH_3}{\overset{|}{\underset{|}{C}}}}+$$

Formation of dimethylhexanes:

$$(6)\ CH_3-\overset{CH_3}{\underset{CH_3}{\overset{|}{\underset{|}{C}}}}+ + CH_2=CH-CH_2-CH_3 \rightarrow CH_3-\overset{CH_3}{\underset{CH_3}{\overset{|}{\underset{|}{C}}}}-CH_2-\overset{+}{C}H-CH_2-CH_3$$

$$(7)\ CH_3-\overset{CH_3}{\underset{CH_3}{\overset{|}{\underset{|}{C}}}}-CH_2-\overset{+}{C}H-CH_2-CH_3 + CH_3-\overset{CH_3}{\underset{CH_3}{\overset{|}{\underset{|}{C}}}}-H \rightarrow C_8H_{18} + CH_3-\overset{CH_3}{\underset{CH_3}{\overset{|}{\underset{|}{C}}}}+$$

The formation of a new tert-butyl cation continues the chain.

FIGURE 11.1 Alkylation chemistry.

exothermic, with the liberation of 124,000 to 140,000 Btu per barrel (929 MJ/m^3) of isobutane reacting [5].

11.2 PROCESS VARIABLES

The most important process variables are reaction temperature, acid strength, isobutene concentration, and olefin space velocity. Changes in these variables affect both product quality and yield.

Reaction temperature has a greater effect in sulfuric acid processes than in those using hydrofluoric acid. Low temperatures mean higher quality, and the effect of changing the sulfuric acid reactor temperature from 25 to 55°F (–4 to 13°C) is to decrease product octane from one to three numbers depending upon the efficiency of mixing in the reactor. In hydrofluoric acid alkylation, increasing the reactor temperature from 60 to 125°F (16 to 52°C), degrades the alkylate quality about three octane numbers [6].

In sulfuric acid alkylation, low temperatures cause the acid viscosity to become so great that good mixing of the reactants and subsequent separation of the emulsion is difficult. At temperatures above 70°F (21°C), polymerization of the olefins becomes significant and yields are decreased. For these reasons, the normal sulfuric acid reactor temperature is from 40 to 50°F (5 to 10°C) with a maximum of 70°F (21°C) and a minimum of 30°F (–1°C).

For hydrofluoric acid alkylation, temperature is less significant and reactor temperatures are usually in the range of 70 to 100°F (21 to 38°C).

Acid strength has varying effects on alkylate quality depending on the effectiveness of reactor mixing and the water content of the acid. In sulfuric acid alkylation, the best quality and highest yields are obtained with acid strengths of 93 to 95% by weight of acid, 1 to 2% water, and the remainder hydrocarbon diluents. The water concentration in the acid lowers its catalytic activity about three to five times as much as hydrocarbon diluents, thus an 88% acid containing 5% water is a much less effective catalyst than the same strength acid containing 2% water. The poorer the mixing in a reactor, the higher the acid strength necessary to keep acid dilution down [6]. Increasing acid strength from 89 to 93% by weight increases alkylate quality by one to two octane numbers.

In hydrofluoric acid alkylation, the highest octane number alkylate is attained in the 86 to 90% by weight acidity range. Commercial operations usually have acid concentrations between 83 and 92% hydrofluoric acid and contain less than 1% water.

Isobutane concentration is generally expressed in terms of isobutane/olefin ratio. High isobutane/olefin ratios increase octane number and yield, and reduce side reactions and acid consumption. In industrial practice, the isobutane/olefin ratio on reactor charge varies from 5:1 to 15:1. In reactors employing internal circulation to augment the reactor feed ratio, internal ratios from 100:1 to 1000:1 are realized.

Olefin space velocity is defined as the volume of olefin charged per hour divided by the volume of acid in the reactor. Lowering the olefin space velocity reduces the amount of high-boiling hydrocarbons produced, increases the product octane, and lowers acid consumption. Olefin space velocity is one way of expressing reaction time; another is by using contact time. Contact time is defined as the residence time of the fresh feed and externally recycled isobutane in the reactor. Contact time for hydrofluoric acid alkylation ranges from 5 to 25 min and for sulfuric acid alkylation from 5 to 40 min [7]. Although the relationship is only approximate, Mrstik, Smith, and Pinkerton [5] developed a correlating factor, F, which is useful in predicting trends in alkylate quality where operating variables are changed.

$$F = \frac{I_E (I/O)_F}{100(SV)_O}$$

where

I_E = isobutane in reactor effluent, liquid volume %
$(I/O)_F$ = volumetric isobutane/olefin ratio in feed
$(SV)_O$ = olefin space velocity, v/hr/v

The higher the value of F, the better the alkylate quality. Normal values of F range from 10 to 40.

11.3 ALKYLATION FEEDSTOCKS

Olefins and isobutane are used as alkylation unit feedstocks. The chief sources of olefins are catalytic cracking and coking operations. Butenes and propene are the most common olefins used, but pentenes (amylenes) are included in some cases. Some refineries include pentenes in alkylation unit feed to lower the FCC gasoline vapor pressure and reduce the bromine number in the final gasoline blend. The alkylation of pentenes is also considered as a way to reduce the C_5 olefin content of final gasoline blends and reduce its effects on ozone production and visual pollution in the atmosphere. Olefins can be produced by the dehydrogenation of paraffins, and isobutane is cracked commercially to provide alkylation unit feed. Hydrocrackers and catalytic crackers produce a great deal of the isobutane used in alkylation, but it is also obtained from catalytic reformers, crude distillation, and natural gas processing. In some cases, normal butane is isomerized to produce additional isobutane for alkylation unit feed.

11.4 ALKYLATION PRODUCTS

In addition to the alkylate stream, the products leaving the alkylation unit include the propane and normal butane that enter with the saturated and unsaturated feed streams, as well as a small quantity of tar produced by polymerization reactions.

The product streams leaving an alkylation unit are

1. LPG-grade propane liquid
2. Normal butane liquid
3. C_5+ alkylate
4. Tar

TABLE 11.1
Range of Operating Variables in Alkylation

	HF	H_2SO_4
Isobutane concentrations		
Vol% in reaction zone	30–80	40–80
External ratio in olefins	3–12	3–12
Internal ratio in olefins	—	50–1000
Olefin concentration		
Total HC contact time, min	8–20	20–30
Olefin space velocity, v/hr/v	—	0.1–0.6
Reactor temperature		
°F	60–115	35–60
°C	16–46	2–16
Reactor acid conc., wt%	80–95	88–95
Acid in emulsion, vol%	25–80	40–60

Source: Note 8.

TABLE 11.2
Theoretical Yields and Isobutane Requirements Based on Olefin Reacting

	Alkylate vol%	Isobutane vol%
Ethylene	188	139
Propene	181	128
Butene (mixed)	172	112
Pentenes (mixed)	165	96

Only about 0.1% by volume of olefin feed is converted into tar. This is not truly a tar but a thick, dark brown oil containing complex mixtures of conjugated cyclopentadienes with side chains [9].

Typical alkylation operating conditions are shown in Table 11.1, and theoretical yields of alkylates and isobutane requirements based on olefin reacted are given in Table 11.2.

11.5 CATALYSTS

Concentrated sulfuric and hydrofluoric acids are the only catalysts used commercially today for the production of high-octane alkylate gasoline, but other catalysts are used to produce ethylbenzene, cumene, and long-chain (C_{12} to C_{16}) alkylated benzenes [9].

As discussed in Section 11.1, the desirable reactions are the formation of C_8 carbonium ions and the subsequent formation of alkylate. The main undesirable reaction is polymerization of olefins. Only strong acids can catalyze the alkylation reaction, but weaker acids can cause polymerization to take place. Therefore, the acid strengths must be kept above 88% by weight H_2SO_4 or HF in order to prevent excessive polymerization. Sulfuric acid containing free SO_3 also causes undesired side reactions, and concentrations greater than 99.3% H_2SO_4 are not generally used [9].

Isobutane is soluble in the acid phase only to the extent of about 0.1% by weight in sulfuric acid and about 3% in hydrofluoric acid. Olefins are more soluble in the acid phase, and a slight amount of polymerization of the olefins is desirable, as the polymerization products dissolve in the acid and increase the solubility of isobutane in the acid phase.

If the concentration of the acid becomes less than 88%, some of the acid must be removed and replaced with stronger acid. In hydrofluoric acid units, the acid removed is redistilled and the polymerization products removed as a thick, dark acid-soluble oil (ASO). The concentrated HF is recycled in the unit, and the net consumption is about 0.3 lb per barrel of alkylate produced. Unit inventory of hydrofluoric acid is about 25 to 40 lb acid per BPD of feed [10].

The sulfuric acid removed usually is regenerated in a sulfuric acid plant, which is generally not a part of the alkylation unit. The acid consumption typically ranges from 13 to 30 lb per barrel of alkylate produced. Makeup acid is usually 98.5 to 99.3 wt% H_2SO_4.

11.6 HYDROFLUORIC ACID PROCESSES [8,11,15]

There are two commercial alkylation processes using hydrofluoric acid as the catalyst. They are designed and licensed by ConocoPhillips Petroleum Co. and UOP. Typical operating conditions are given in Table 11.3 and Table 11.4.

The basic flow scheme is the same for both the ConocoPhillips and the UOP processes (Figure 11.2; see also Photo 12, Appendix E).

Both the olefin and isobutane feeds are dehydrated by passing the feedstocks through a solid bed desiccant unit. Good dehydration is essential to minimize potential corrosion of process equipment, which results from addition of water to hydrofluoric acid.

After dehydration, the olefin and isobutane feeds are mixed with hydrofluoric acid at sufficient pressure to maintain all components in the liquid phase. The reaction mixture is allowed to settle into two liquid layers. The acid has a higher density than

TABLE 11.3
HF Alkylation Yields, Product Octanes, and Isobutane Requirements

	Vol/vol olefin		Clear octane no.	
	Isobutane	Alkylate	Research	Motor
Propylene	1.33	1.77	93	91
Butylenes	1.16	1.75	96	94

TABLE 11.4
HF Alkylate Properties

Gravity, °API	**71.4**	
RVP, psi (kPa)	**4.5**	**(31)**
ASTM distillation	**°F**	**°C**
IBP	110	43
5%	155	68
10%	172	78
20%	190	88
50%	217	103
70%	222	105
90%	245	119
EP	370	188

the hydrocarbon mixture and is withdrawn from the bottom of the settler and passed through a cooler to remove the heat gained from the exothermic reaction. The acid is then recycled and mixed with more fresh feed, thus completing the acid circuit.

A small slip stream of acid is withdrawn from the settler and fed to an acid rerun column to remove dissolved water and polymerized hydrocarbons. The acid rerun column contains about five trays and operates at 150 psig (1034 kPa) [3].

The overhead product from the rerun column is clear hydrofluoric acid, which is condensed and returned to the system.

The bottom product from the rerun column is a mixture of tar and an HF-water azeotrope. These components are separated in a tar settler (not shown on the flow diagram). The tar is used for fuel, and the HF-water mixture is neutralized with lime or caustic. This rerun operation is necessary to maintain the activity of the hydrofluoric acid catalyst.

The hydrocarbon layer removed from the top of the acid settler is a mixture of propane, isobutane, normal butane, and alkylate along with small amounts of hydrofluoric acid. These components are separated by fractionation, and the isobutane is recycled to the feed. Propane and normal butane products are passed through caustic treaters to remove trace quantities and hydrofluoric acid.

Although the flow sheet (Figure 11.2) shows the fractionation of propane, isobutane, normal butane, and alkylate to require three separate fractionators, many alkylation plants have a single tower where propane is taken off overhead, a partially purified isobutane recycle is withdrawn as a liquid several trays above the feed tray, a normal butane product is taken off as a vapor several trays below the feed tray, and the alkylate is removed from the bottom.

The design of the acid settler-cooler-reactor section is critical for good conversion in a hydrofluoric acid alkylation system. Different reactor system designs have been made over the years by both UOP and Phillips. Many of the reactor systems designed by UOP are similar to a horizontal shell and tube heat exchanger with cooling water flowing inside the tubes to maintain the reaction temperatures at the desired level. Good mixing is attained in the reactor by using a recirculating pump

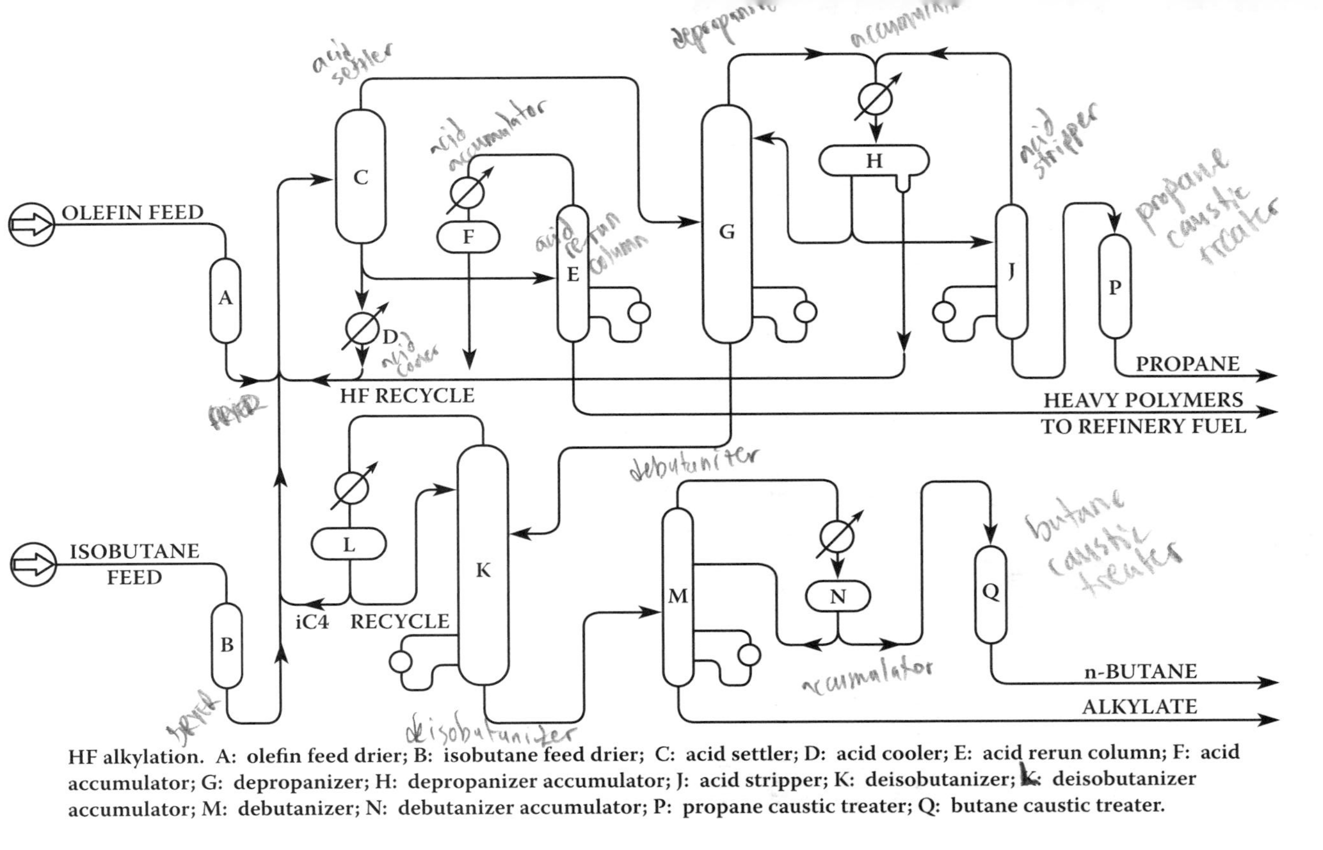

HF alkylation. **A: olefin feed drier; B: isobutane feed drier; C: acid settler; D: acid cooler; E: acid rerun column; F: acid accumulator; G: depropanizer; H: depropanizer accumulator; J: acid stripper; K: deisobutanizer; L: deisobutanizer accumulator; M: debutanizer; N: debutanizer accumulator; P: propane caustic treater; Q: butane caustic treater.**

FIGURE 11.2 Phillips hydrofluoric acid alkylation unit.

to force the mixture through the reactor at a rate about eight to ten times the mixed hydrocarbon feed rate to the reactor.

Reactor systems designed by ConocoPhillips usually have been similar to that illustrated in Figure 11.3. Acid circulation in this system is by gravity differential, and thus a relatively expensive acid circulation pump is not necessary.

In portions of the process system where it is possible to have HF-water mixtures, the process equipment is fabricated from Monel metal or Monel-clad steel. The other parts of the system are carbon steel.

Special precautions are taken to protect maintenance and operating personnel from injury by accidental contact with acid. These precautions include special seals on acid-containing equipment such as pumps and valve stems; rubber safety jackets, pants, gloves, and boots that must be worn by personnel entering an acid area; safety eyeglasses; caustic tubs for washing all hand tools; safety showers; special acid drain systems; and many others.

Careful attention to engineering design details and extensive operator training combined with the above precautions are necessary to provide safe operations for hydrofluoric acid alkylation units.

11.7 SULFURIC ACID ALKYLATION PROCESSES

The major alkylation processes using sulfuric acid as a catalyst are the autorefrigeration process, licensed by ExxonMobil Research and Engineering (similar to a process previously licensed by the M. W. Kellogg Co.), and the effluent refrigeration process, licensed by Stratford Engineering Corp. There are also some older units using time-tank reactors, but no new units of this type have been constructed recently.

The major differences between the autorefrigeration and effluent refrigeration processes are in the reactor designs and the point in the process at which propane and isobutane are evaporated to induce cooling and provide the process refrigeration required. The process flow diagram for the autorefrigeration process is shown in Figure 11.4.

The autorefrigeration process uses a multistage cascade reactor with mixers in each stage to emulsify the hydrocarbon-acid mixture. Olefin feed or a mixture of olefin feed and isobutane feed is introduced into the mixing compartments, and enough mixing energy is introduced to obtain sufficient contacting of the acid catalyst with the hydrocarbon reactants to obtain good reaction selectivity. The reaction is held at a pressure of approximately 10 psig (69 kPag) in order to maintain the temperature at about 40°F (5°C). In the Stratco, or similar type of reactor system, pressure is kept high enough [45 to 60 psig (310 to 420 kPag)] to prevent vaporization of the hydrocarbons [12]. In the ExxonMobil process, acid and isobutane enter the first stage of the reactor and pass in series through the remaining stages. The olefin hydrocarbon feed is split and injected into each of the stages. Exxon mixes the olefin feed with the recycle isobutane and introduces the mixture into the individual reactor sections to be contacted with the catalyst (Photo 13, Appendix E).

The gases vaporized to remove the heats of reaction and mixing energy are compressed and liquefied. A portion of this liquid is vaporized in an economizer to

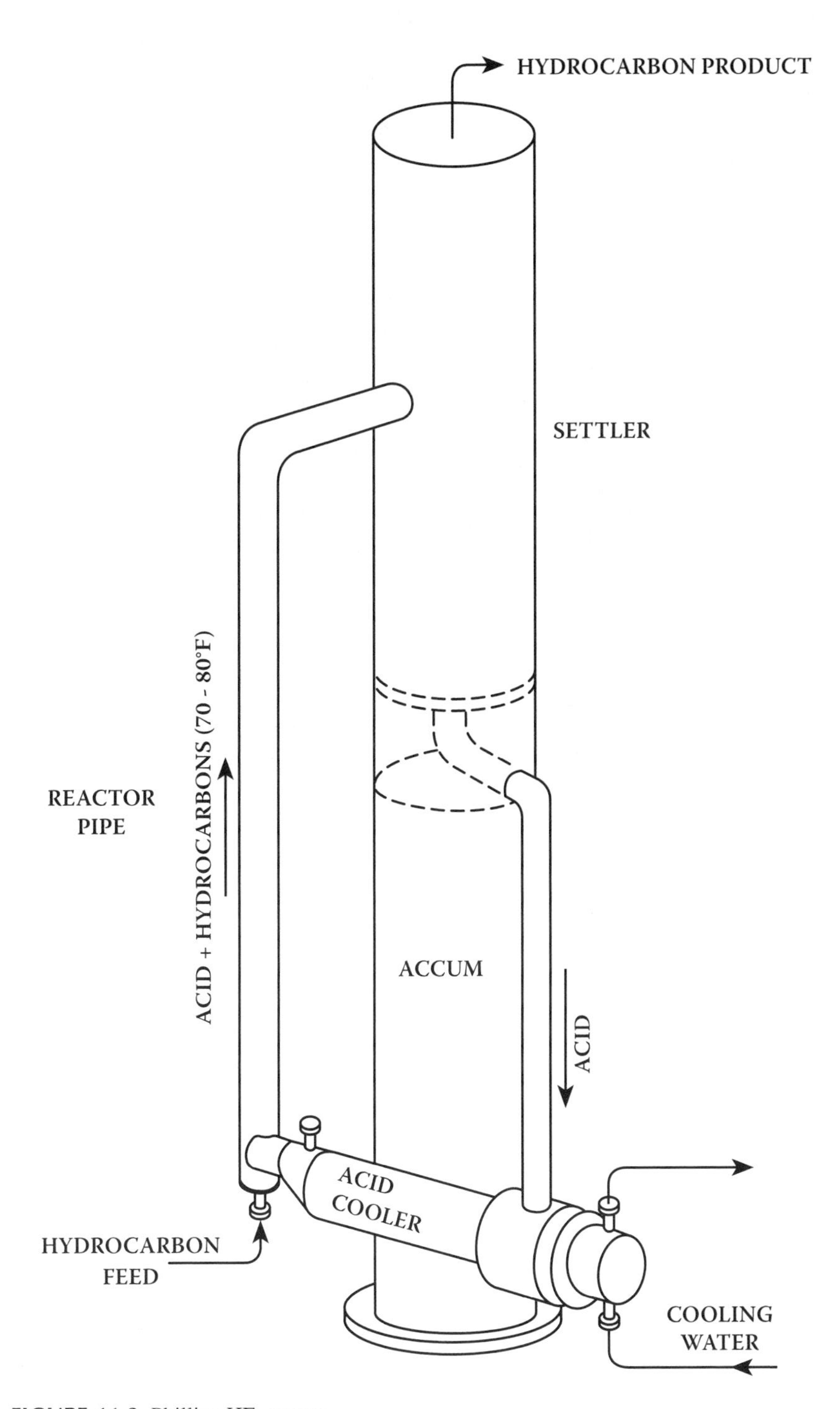

FIGURE 11.3 Phillips HF reactor.

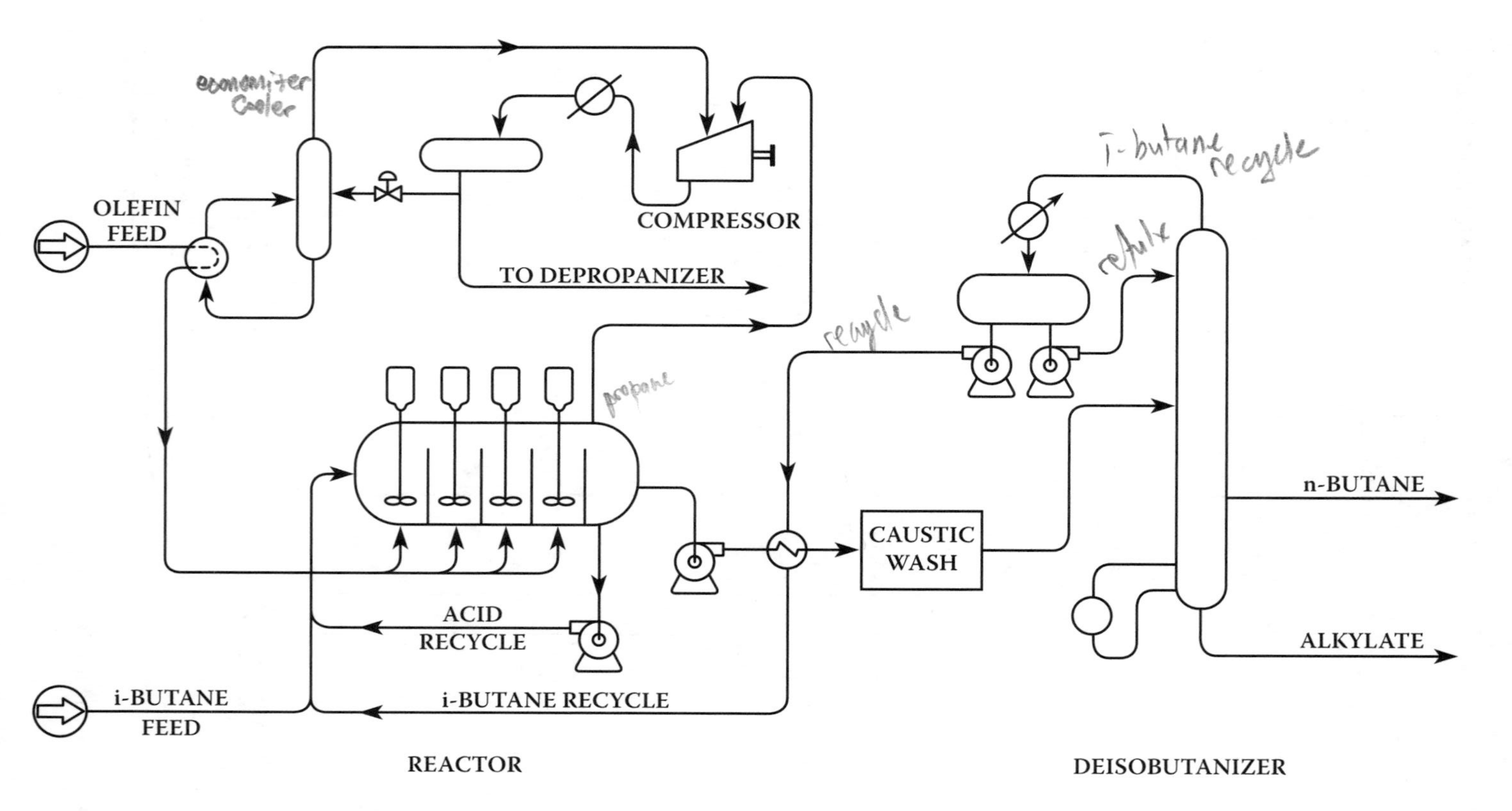

FIGURE 11.4 Autorefrigeration sulfuric acid alkylation unit.

cool the olefin hydrocarbon feed before it is sent to the reactor. The vapors are returned for recompression. The remainder of the liquefied hydrocarbon is sent to a depropanizer column for removal of the excess propane that accumulates in the system. The liquid isobutane from the bottom of the depropanizer is pumped to the first stage of the reactor.

The acid-hydrocarbon emulsion from the last reactor stage is separated into acid and hydrocarbon phases in a settler. The acid is removed from the system for reclamation, and the hydrocarbon phase is pumped through a caustic wash followed by a water wash (or a fresh acid wash followed by either caustic or alkaline water washes) to eliminate trace amounts of acid and then sent to a deisobutanizer [13]. The deisobutanizer separates the hydrocarbon feed stream into isobutane (which is returned to the reactor), n-butane, and alkylate product.

The effluent refrigeration process (Stratco) uses a single-stage reactor in which the temperature is maintained by cooling coils (Figure 11.5). The reactor contains an impeller that emulsifies the acid-hydrocarbon mixture and recirculates in the reactor. The average residence time in the reactor is on the order of 20 to 25 min.

Emulsion removed from the reactor is sent to a settler for phase separation. The acid is recirculated, and the pressure of the hydrocarbon phase is lowered to flash vaporize a portion of the stream and reduce the liquid temperature to about 30°F (−1°C) [8]. The cold liquid is used as a coolant in the reactor tube bundle.

The flashed gases are compressed and liquefied, then sent to the depropanizer, where LPG-grade propane and recycle isobutane are separated. The hydrocarbon liquid from the reactor tube bundle is separated into isobutane, n-butane, and alkylate streams in the deisobutanizer column. The isobutane is recycled, and n-butane and alkylate are product streams.

A separate distillation column can be used to separate the n-butane from the mixture, or it can be removed as a sidestream from the deisobutanizing column. The choice is a matter of economics because including a separate column to remove the n-butane increases the capital and operating costs. Separating n-butane as a sidestream from the deisobutanizing column can be restricted because the pentane content is usually too high to meet butane sales specifications. The sidestream n-butane can be used for gasoline blending [12].

Typical product yields and qualities are

Item	Propylene	Butylene	Pentylene
True alkylate, LV% on olefin	171–178	170–178	197–220
iC_4 consumed, LV% on olefin	119–132	110–117	96–133
Acid consumed (98.5 wt%), lb/bbl TA	34–42	13–25	25–42
RVP, psi	3.8	2.6	4.0
MONC	88–90	92–94	88–93
RONC	89–92	94–98	90–92

Source: Exxon Research and Engineering Co. and Stratco, Inc.

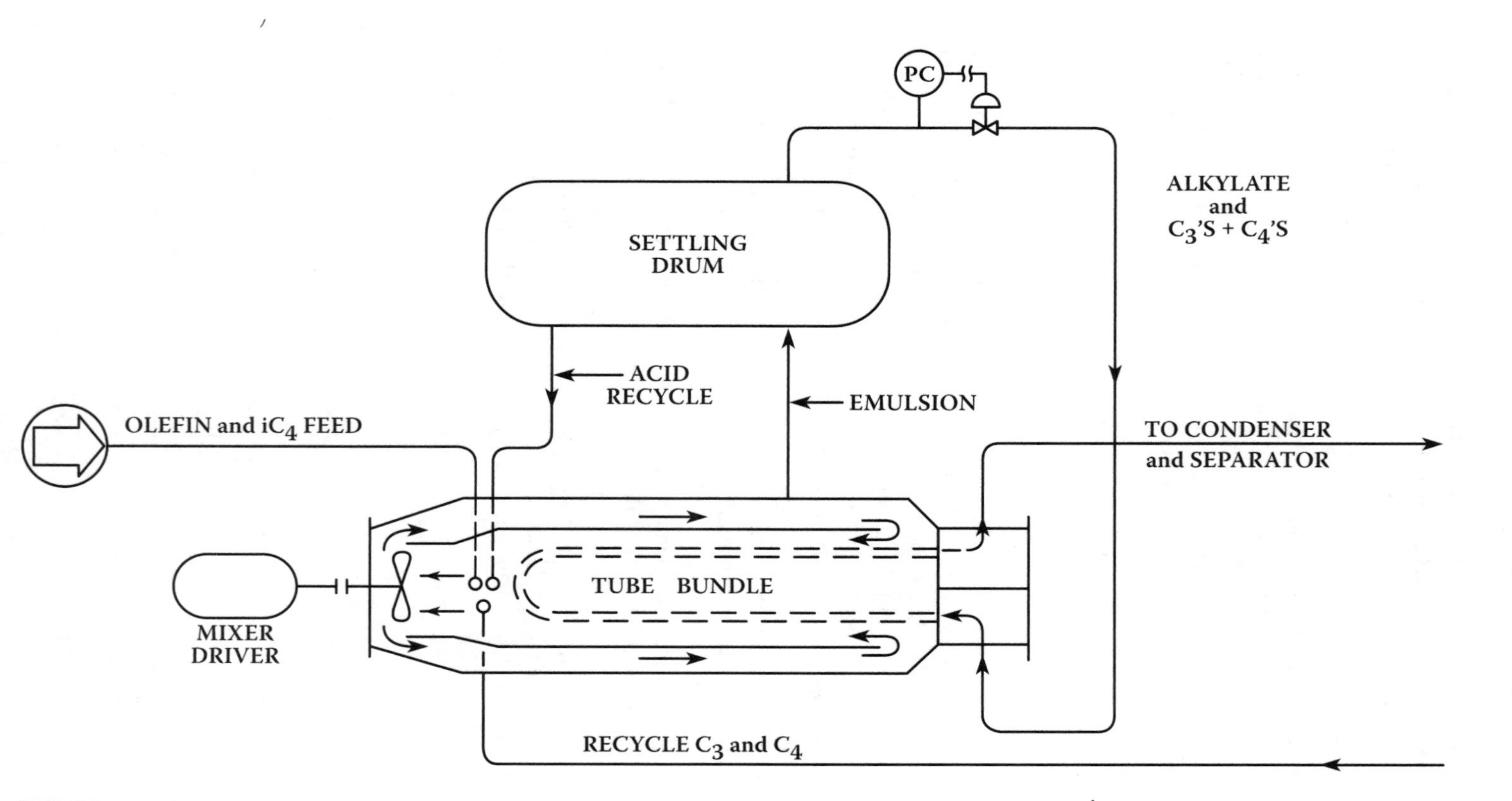

FIGURE 11.5 Stratco contactor.

11.8 COMPARISON OF PROCESSES

The most desirable alkylation process for a given refinery is governed by economics. In particular, the location of the refinery with respect to acid supply and disposal is very important. If the refinery is at a distance from either sulfuric acid suppliers or purchasers of spent sulfuric acid, the cost of transportation of fresh acid or the cost of disposing of the large quantities of spent acid can render the use of sulfuric acid economically unattractive. Only a small amount of makeup hydrofluoric acid is required for the HF process, as facilities are provided to regenerate the spent hydrofluoric acid. As a result, the cost of transporting hydrofluoric acid from a remote supplier is not a major cost. Albright, in a series of articles on alkylation processes, summarized the comparison of the processes [3,4,8,11,14]. The following discussion highlights the main issues.

The important question for a refinery is which alkylation process is best for the production of the desired product. Many factors are important, including total operating expenses; initial capital costs; alkylate quality; flexibility of operation; reactants available; yields and conversion of reactants; maintenance problems; safety; experience with a given process; and patents, licensing arrangements, and possible royalties.

Advocates of the hydrofluoric acid process argue that both capital and total operating costs are less than those of sulfuric acid processes for the following reasons:

1. Smaller and simpler reactor designs are feasible.
2. Cooling water can be used instead of refrigeration.
3. Smaller settling devices are needed for emulsions.
4. Essentially complete regeneration of the hydrofluoric acid catalyst occurs. Hence, hydrofluoric acid consumption and costs are very low. Disposal of spent acid is not necessary.
5. There is increased flexibility of operation relative to temperature, the external ratio of isobutane to olefin, and so on.
6. There is decreased need for turbulence or agitation when acid and hydrocarbon streams are combined.

Advocates of sulfuric acid processes counter the above arguments for hydrofluoric acid processes with the following:

1. Additional equipment is needed for the hydrofluoric acid process to recover or neutralize the hydrofluoric acid in various streams. Such equipment includes the hydrofluoric acid stripper tower, hydrofluoric acid regeneration tower, and neutralization facilities for the several product streams. With sulfuric acid, the entire effluent hydrocarbon stream is neutralized.
2. Equipment is required to dry the feed streams to a few ppm water in hydrofluoric acid processes. Drying is beneficial but not required in sulfuric acid processes. Normally, only feed coalescers are used to remove the free water that drops out of the chilled feed.
3. Additional equipment at increased cost is required for safety in a hydrofluoric acid unit. In some hydrofluoric acid plants, a closed cooling water

system is required as a safety measure in the event of hydrofluoric acid leakage into the system. Maintenance costs and the amount of safety equipment in hydrofluoric acid processes are greater.
4. Capital costs for hydrofluoric acid processes are slightly more than sulfuric acid processes when the cost of hydrofluoric acid mitigation equipment is included.
5. Isobutane in hydrofluoric acid processes is not fully used for production of alkylate, because self-alkylation occurs to a higher extent when hydrofluoric acid is used as a catalyst.
6. There are greater limitations on obtaining alkylates with high octane numbers with hydrofluoric acid processes. This is particularly true if isobutylene is removed from the feed by an upstream MTBE or ETBE unit.
7. Safety and environmental restrictions limit the use of hydrofluoric systems in highly populated areas [15].

11.9 ALKYLATION YIELDS AND COSTS

Typical alkylation yields based on the percentage of olefin in fresh feed are given in Table 11.5. These do not take into account effects of temperature and are based on isobutane olefin ratios of 10:1 for propylene, 6:1 for butylenes, and 10:1 for amylenes (pentalenes).

Cost curves for construction of alkylation units are shown in Figure 11.6. The costs are average costs and include the items given in Table 11.6. Power and chemical consumption are also given in Table 11.6.

11.10 SAFETY ISSUES

Hydrofluoric and sulfuric acids are hazardous materials to work with, and strict safety precautions must be observed. Hydrofluoric acid, in particular, has the problem that if liquid hydrofluoric acid is sprayed into the atmosphere, a vapor cloud can be formed that is difficult to disperse. There is concern that if equipment failure should cause this to occur, a light breeze could carry this vapor cloud into a populated area and be fatal to the people in that area. Recently, ConocoPhillips and UOP have developed methods that minimize the formation of a vapor cloud under these conditions; these methods include using additives for the hydrofluoric acid that prevents the formation of a vapor cloud and equipment design changes to significantly reduce the amount of hydrofluoric acid in the unit and to provide for emergency dumping of the acid in the system to underground storage tanks.

11.11 POLYMERIZATION

Propene and butanes can be polymerized to form a high-octane product boiling in the gasoline boiling range. The product is an olefin having unleaded octane numbers of 97 RON and 83 MON. The polymerization process was widely used in the 1930s and 1940s to convert low-boiling olefins into gasoline blending stocks but was

TABLE 11.5
Typical Alkylation Yields Based on Percent Olefin in Fresh Feed

	Propylene					Butylene					Amylene				
	MW	lb/gal	vol%	wt%	mol%	MW	lb/gal	vol%	wt%	mol%	MW	lb/gal	vol%	wt%	mol%
Feeds															
Olefin	42.0	4.38	100.00	100.00	100.00	56.0	5.00	100.00	100.00	100.00	70.1	5.46	100.00	100.00	100.00
Isobutane	58.1	4.69	160.00	171.32	124.07	58.1	4.69	120.00	112.56	108.76	58.1	4.69	140.00	120.25	145.06
Total			260.00	271.32	224.07			220.00	212.56	208.76			240.00	220.25	245.06
Products															
Propane	44.0	4.26	29.80	29.05	27.70	—	—	—	—	—	—	—	—	—	—
n-Butane	—	—	—	—	—	58.1	4.88	12.71	12.40	12.00	—	—	—	—	—
Pentane	72.1	5.21	7.17	8.53	4.98	72.1	5.21	6.80	7.08	5.50	72.1	5.25	45.20	43.48	42.24
Depent rerun alkylate	112.6	5.84	159.40	212.56	79.27	112.6	5.84	151.10	176.56	88.03	110.9	5.86	145.60	156.19	98.72
Alkyl bottoms	165.0	6.31	12.60	18.18	4.61	165	6.31	11.90	15.02	5.12	230	6.67	14.40	17.58	5.37
Tar, 20°API	360	7.78	1.70	3.00	0.35	360	7.78	0.96	1.50	0.23	360	7.78	2.11	3.00	0.58
Total			210.67	271.32	116.91			183.47	212.56	110.88			207.31	220.25	146.91
(All feeds)/(All products)			1.234	1.000	1.917			1.199	1.000	1.883			1.158	1.000	1.668

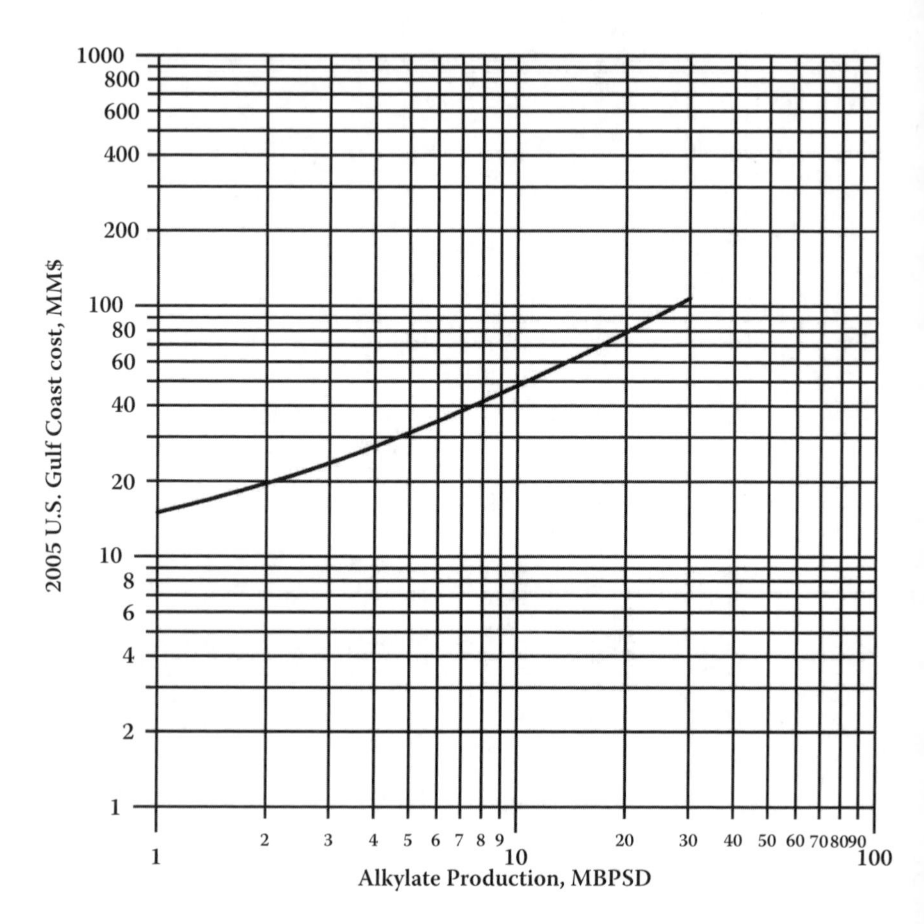

FIGURE 11.6 Alkylation unit investment cost: 2005 U.S. Gulf Coast (see Table 11.6).

supplanted by the alkylation process after World War II. The mandated reduction in the use of lead in gasoline and the increasing proportion of the market demand for unleaded gasolines created a need for low-cost processes to produce high-octane gasoline blending components. Polymerization produces about 0.7 barrels of polymer gasoline per barrel of olefin feed as compared with about 1.5 barrels of alkylate by alkylation, and the product has a high octane sensitivity, but capital and operating costs are much lower than for alkylation. As a result, polymerization processes are being added to some refineries.

Typical polymerization reactions are shown in Table 11.7, but although iC_4H_8 reacts to give primarily di-isobutylene, propene gives mostly trimers and dimers with only about 10% conversion to dimer [16].

The most widely used catalyst is phosphoric acid on an inert support. This can be in the form of phosphoric acid mixed with kieselguhr (a natural clay) or a film of liquid phosphoric acid on crushed quartz. Sulfur in the feed poisons the catalyst, and any basic materials neutralize the acid and increase catalyst consumption. Oxygen

TABLE 11.6
Alkylation Unit Cost Data

Costs included

1. All facilities required for producing alkylate from a feedstream of isobutane and C_3 to C_5 unsaturates in proper proportions.
2. All necessary controllers and instrumentation.
3. All BL process facilities.
4. Feed treating (molecular-sieve unit to remove moisture in feed).

Costs not included

1. Cooling water, steam, and power supply.
2. Feed and product storage.

	HF	H_2SO_4
Royalty		
$/bbl TA[a]	0.03	0–0.04
Utility data (per bbl TA[a])		
Steam, lb	200	200
Power, kWh	3.7	4.6
Cooling water, gal crclt.	3700	3300
Chemicals (per bbl TA[a])		
Acid, lb	0.3	30
Caustic, lb	0.2	0.2

[a] Total alkylate (C_5 + alkylate).

Note: See Figure 11.6.

TABLE 11.7
Polymerization Reactions

```
    C          C          C     C
    |          |          |     |
C=C—C  +  C=C—C  →  C—C—C=C—C
                          |
                          C

                                C
                                |
C=C—C  +  C=C—C  →  C—C—C=C—C
```

dissolved in the feed adversely affects the reactions and must be removed. Normal catalyst consumption rates are in the range of 1 lb of catalyst per 100 to 200 gal of polymer produced (830 to 1660 1/kg).

The feed, consisting of propane and butane as well as propene and butene, is contacted with an amine solution to remove hydrogen sulfide and caustic washed to remove mercaptans. It is then scrubbed with water to remove any caustic or amines and then dried by passing through a silica gel or molecular-sieve bed. Finally, a small amount of water (350 to 400 ppm) is added to promote ionization of the acid before the olefin feed stream is heated to about 400°F (204°C) and passed over the catalyst bed. Reactor pressures are about 500 psig (3450 kPa).

The polymerization reaction is highly exothermic, and temperature is controlled either by injecting a cold propane quench or by generating steam. The propane and butane in the feed act as diluents and a heat sink to help control the rate of reaction and the rate of heat release. Propane is also recycled to help control the temperature.

After leaving the reactor, the product is fractionated to separate the butane and lighter material from the polymer gasoline. Gasoline boiling-range polymer production is normally 90 to 97 wt% on olefin feed or about 0.7 bbl of polymer per barrel of olefin feed. A simplified process flow diagram for the UOP unit is shown in Figure 11.7 [18], and ranges of reaction conditions are given in Table 11.8 [16].

Insitut Francais du Petrole licenses a process to produce dimate (isohexene) from propene using a homogeneous aluminum alkyl catalyst that is not recovered. The process requires a feed stream that is better than 99% propane and propene because C_2s and C_4s poison the catalyst. Dienes and triple-bonded hydrocarbons can create problems, and in some cases, it is necessary to selectively hydrogenate the feed to eliminate these compounds. The major advantage of this process is the low capital cost because it operates at low pressures. A simplified process flow diagram is given in Figure 11.8 [6] and feed impurity limits in Table 11.9.

Product properties and yields are shown in Table 11.10 for alkylate, polymer, and dimer blending stocks. Table 11.11 shows polymerization unit costs.

11.12 CASE-STUDY PROBLEM: ALKYLATION AND POLYMERIZATION

Table 11.12 and Table 11.13 give the alkylation and polymerization units material balances, and Table 11.14 shows the alkylation and polymerization units chemical and utility requirements.

There is not sufficient isobutane available for alkylating both the butylenes and propylenes. Therefore, butylenes will be alkylated up to the amount of isobutane available, and the remaining butylenes plus the propylenes will be polymerized.

TABLE 11.8
Polymerization Operating Conditions

Temperature[a]	300–425°F (175–235°C)
Pressure	400–1,500 psi (2,760–10,340 kPa)
Space velocity	0.3 gal/lb (2.5 1/kg-hr)

[a] Usually 200–220°C.

Note: Higher pressures mean lower temperatures.

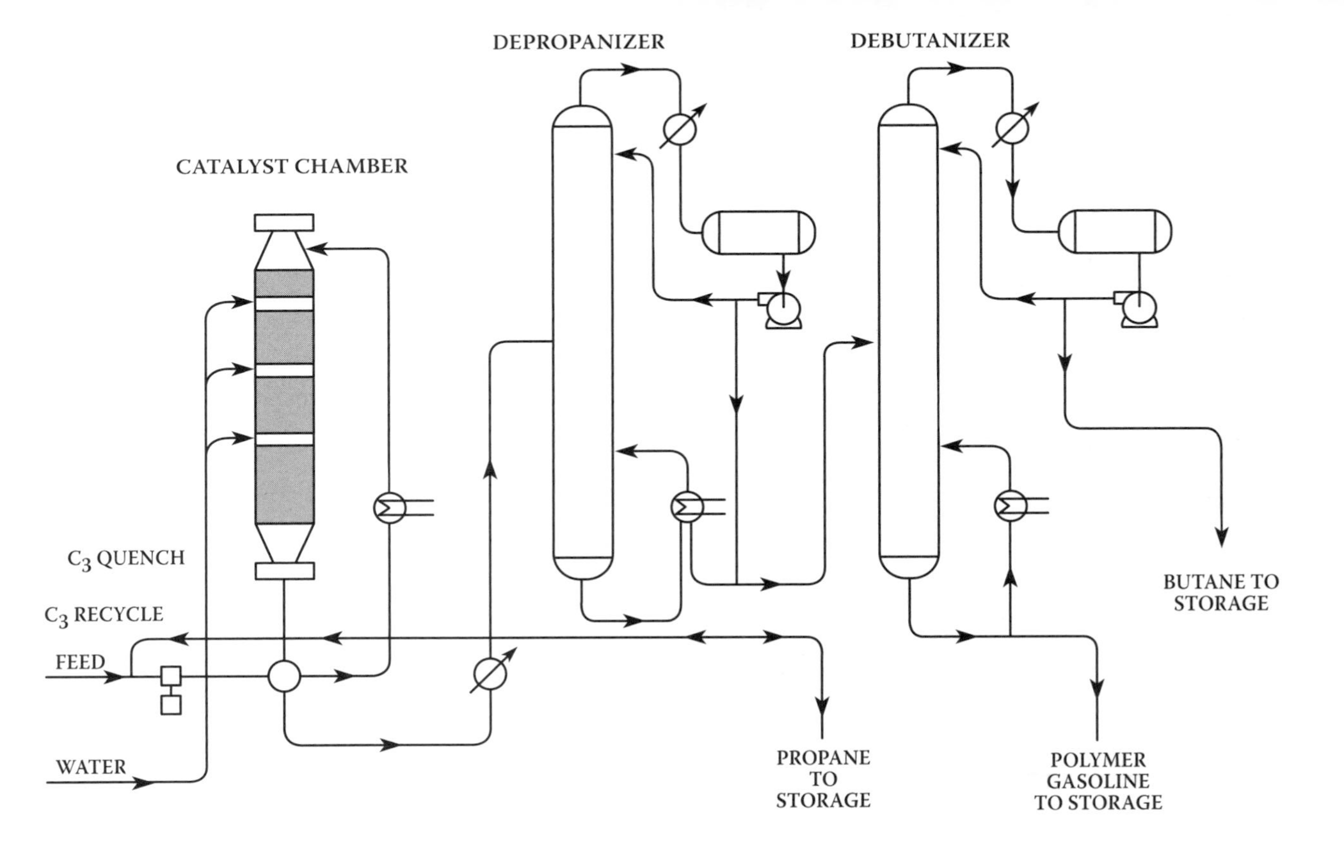

FIGURE 11.7 UOP solid phosphoric acid polymerization unit. (Courtesy of R.E. Payne, 37[9], 316–329 [1958].)

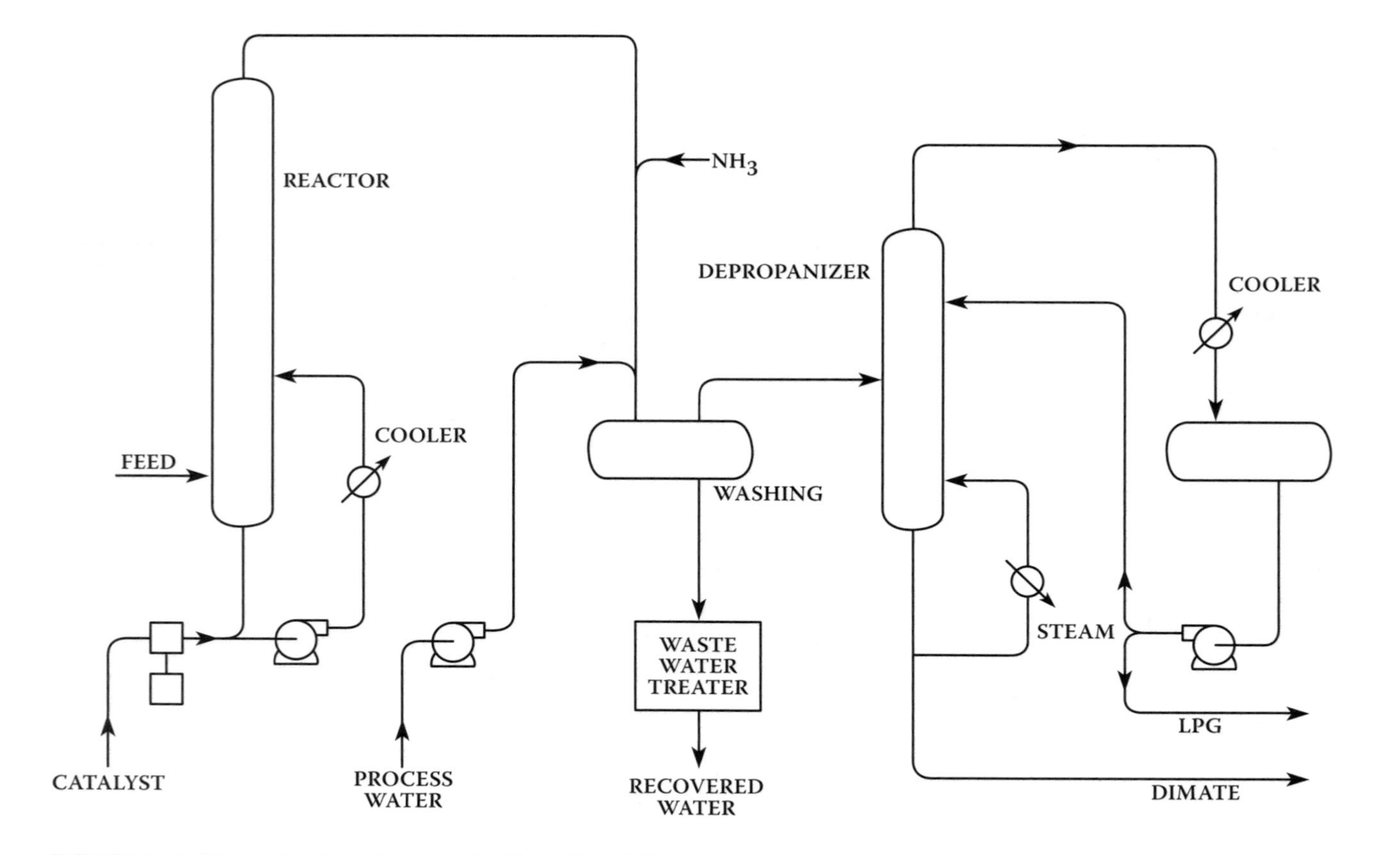

FIGURE 11.8 Dimersol polymerization unit. (From Note 17.)

TABLE 11.9
Dimersol Feed Impurities

Component	Ppm
H_2O	3
Acetylene	15
Sulfur	5
Propadiene	1
Butadiene	10

Source: Note 7.

TABLE 11.10
Volume Yields Per bbl Mixed Olefin Feed

	Alkylate	Polymer	Dimate
C_5+ gasoline	1.55	0.68	0.68
RON	94	97	97
MON	91	83	82

TABLE 11.11
Polymerization Unit Cost Data

Investment cost	
2005 U.S. Gulf Coast	
500 BPD	\$1,750,000
2000 BPD	\$5,000,000
Costs included	
Costs are for reactor section only but include preheat heat exchange and product cooling.	
Utility data (per bbl polymer)	
Fuel, MMBtu	0.036
MMkJ	0.038
Cooling water, gal crclt., 30°F Δt	290
m^3 crclt., 17°C Δt	1.1

TABLE 11.12
HF Alkylation Unit Material Balance

Component	BPD	°API	lb/h/BPD	lb/h
Feed				
$iC_4^=$	3129		8.22	25,716
C_4	3559		8.76	31,180
C_3	0			0
Total	6688			56,896
Products				
$C_4^=$	952		8.76	8,341
nC_4	331		8.51	2,820
C_5+ alkylate	4117	77.9	9.86	40,584
Alkylate bottoms	310			4,683
Tar	25			468
Total	5735			56,896

TABLE 11.13
Polymerization Unit Material Balance

Component	BPD	°API	lb/h/BPD	lb/h
Feed				
$C_3^=$	2094			15,475
$C_4^=$	952			8,341
Total	3046			23,816
C_5+ polymer	2071	47.9	11.50	23,816

TABLE 11.14
Alkylation and Polymerization Units Chemical and Utility Requirements

	Alkylation		Polymerization	
	Per/B TA	Per day	Per/B day	Per day
Steam, Mlb	0.011	45	0.036	75
Power, MkWh	0.0037	15		
Cooling water, Mgal	3.7	15,231	0.29	601
Fuel, MMBtu	1.04	4,281	0.036	75
HF acid, lb	0.3	1,235		
Caustic, lb	0.2	823		
Catalyst, $			1.34	1740

PROBLEMS

1. A refinery has available 2700 BPCD of butylenes and 2350 BPCD of isobutane for possible akylation unit feed. How many barrels of alkylate can be made from these feedstocks? What will be the production of other products?
2. An unsaturated feed stream consisting of 1750 BPCD of butylenes and 1550 BPCD of propylene is fed to an alkylation unit. How many BPCD of isobutane will be required for the unit? How much alkylate will be made?
3. Make an overall material balance for an alkylation unit with feed rates of 1710 BPCD of propylene, 3320 BPCD of butylenes, 1550 BPCD of amylene, and 9570 BPCD of isobutane.
4. Estimate the 2005 construction costs for building the alkylation unit of problem 3 and the chemical and utility requirements (a) if sulfuric acid catalyst is used and (b) if hydrofluoric acid catalyst is used.
5. Using the construction costs and chemical and utility requirements of problem 4, what are the costs per barrel of alkylate produced for direct operation, including royalties but not labor, and depreciation? Assume 16-year straight-line depreciation, with dismantling costs equal to salvage

value. Compare total costs per barrel of alkylate for hydrofluoric and sulfuric acid processes.

6. Compare the operating costs per barrel of alkylate produced from a 4000 BPSD sulfuric acid alkylation plant and a 4000 BPSD hydrofluoric acid alkylation plant. Obtain chemical costs from the latest issue of *Chemical Marketing Reporter.* Use the following utility costs: steam, \$3.15/Mlb; electric power, \$0.05/kWh; cooling water: makeup, \$0.45/M gal, circulating, see p. 356; fuel, \$2.25/MMBtu; labor costs, \$950/day.
7. A refinery has 5,560 BPSD of isobutane available for alkylation. There are also 2,110 BPSD of propylene, 2,200 BPSD of butylenes, and 14,090 BPSD of amylene available. What combination of feedstocks to the alkylation unit will maximize alkylate production?
8. Estimate the 2006 construction costs for building an alkylation unit to feed 2400 BPSD of propylene, 4000 BPSD of butylenes, and 8350 BPSD of isobutane.
9. Assuming 10-year straight-line depreciation and dismantling costs to equal salvage value, calculate the operating costs, including depreciation and running royalties, but not labor, per barrel of alkylate produced for hydrofluoric and sulfuric acid plants producing 10,000 BPSD of alkylate.
10. For problem 6, compare the daily operating costs for a 4000 BPSD feed rate hydrofluoric acid plant when producing alkylate from 98% propylene feedstock with that when using a 97.5% butylenes feedstock.
11. Make an overall volume and weight material balance for an alkylation unit with feed rates of 1550 BPCD of propylene, 3520 BPCD of butylenes, 1800 BPCD of amylene, and 9510 BPCD of isobutane.
12. For an alkylation plant with a 97.2% on-stream factor producing 9000 BPCD of alkylate gasoline having clear octane numbers of 95.9 MON and 97.3 RON, estimate its 2006 U.S. Gulf Coast construction cost and its annual chemical and utility requirements.
13. Make an overall weight and volume material balance for an alkylation unit with a feed rate of 2500 BPCD of butylenes and 2800 BPCD of isobutane.
14. A refinery has 8000 BPCD of isobutane available for alkylation. There are also 2120 BPSD of propylene, 3985 BPSD of butylenes, and 1880 BPSD of amylene available. What is the maximum yield of pentane-free alkylate that can be produced?

NOTES

1. Ipatieff, V.N. and Schmerling, L., *Advances in Catalysis,* Vol. I, Academic Press, New York, 1948, pp. 27–63.
2. Gruse, W.A. and Stevens, D.R., *Chemical Technology of Petroleum,* 3rd ed., McGraw-Hill Book Co., New York, 1960, pp. 153–163.
3. Albright, L.F., *Chem. Eng. 73*(19), pp. 205–210, 1966.
4. Albright, L.F., *Chem. Eng. 73*(21), pp. 209–215, 1966.
5. Mrstik, A.V., Smith, K.A., and Pinkerton, R.D., *Advan. Chem. Ser.* 5(97), 1951.
6. Payne, R.E., *Petrol. Refiner. 37*(9), 316–329, 1958.

7. Hengstebeck, R.J., *Petroleum Processing,* McGraw-Hill Book Co., New York, 1959, pp. 218–233.
8. *Petrol. Refiner. 31*(19), pp. 156–164, 1952.
9. Thomas, C.L., *Catalytic Processes and Proven Catalysts,* McGraw-Hill Book Co., New York, 1970, pp. 67–69.
10. *Oil Gas J., 90*(14), pp. 67–72, 1992.
11. *Petrol. Process. 12*(5), p. 146; *12*(8), p. 79, 1957.
12. Lerner, H. and Citarella, V.A., *Hydrocarbon Process. 70*(11), 89–94, 1991.
13. *Oil Gas J., 88*(18), pp. 62–66, 1990.
14. Albright, L.F., *Oil Gas J. 88*(46), pp. 79–92; *88*(48), pp. 70–77, 1990.
15. Van Zele, R.L. and Diener, R., *Hydrocarbon Process., 69*(6), pp. 92–98, 1990.
16. Thomas, C.L., *Catalytic Processes and Proven Catalysts,* McGraw-Hill Book Co., New York, 1970, pp. 87–96.
17. Andrews, J.W. and Bonnifay, P.L., *Hydrocarbon Process. 56*(4), 161–164 (1977).
18. *Petrol. Refiner. 39*(9), 232, 1960.

12 Product Blending

Almost all products from a fuels refinery are blended, even asphalt. Up until the late 1960s, products had to be blended in batches because computer capability was inadequate and on-line blending instrumentation was not available. Today, even small refineries use on-line blending because the equipment is relatively inexpensive and cost savings as compared to batch blending are significant.

Increased operating flexibility and profits result when refinery operations produce basic intermediate streams that can be blended to produce a variety of on-specification finished products. For example, naphthas can be blended into either gasoline or jet fuel, depending upon the product demand. Aside from lubricating oils, the major refinery products produced by blending are gasolines, jet fuels, heating oils, and diesel fuels. The objective of product blending is to allocate the available blending components in such a way as to meet product demands and specifications at the lowest cost and to produce incremental products that maximize overall profit. The volumes of products sold, even by a medium-sized refiner, are so large that savings of a fraction of a cent per gallon will produce a substantial increase in profit over the period of 1 year. For example, if a refiner sells about 1 billion gallons of gasoline per year (about 65,000 BPCD; several refiners sell more than that in the United States), a saving of one one-hundredth of a cent per gallon results in an additional profit of $100,000 per year.

Today, most refineries use computer-controlled in-line blending for blending gasolines and other high-volume products. Inventories of blending stocks, together with cost and physical property data, are maintained in the computer. When a certain volume of a given quality product is specified, the computer uses linear or geometric programming models to optimize the blending operations to select the blending components to produce the required volume of the specified product at the lowest cost.

To ensure that the blended streams meet the desired specifications, stream analyzers, measuring, for example, boiling point, specific gravity, Reid vapor pressure (RVP), and research and motor octane, are installed to provide feedback control of additives and blending streams.

Blending components to meet all critical specifications most economically is a trial-and-error procedure that is easy to handle with the use of a computer. The large number of variables makes it probable there will be a number of equivalent solutions that give the approximate equivalent total overall cost or profit. Optimization programs permit the computer to provide the optimum blend to minimize cost and maximize profit. Both linear and geometric programming techniques are used. Geometric programming is preferred if sufficient data are available to define the equations, because components blend nonlinearly, and values are functions of the quantities of the components and their characteristics.

The federal requirement that ethanol be used in motor gasoline has created difficulties for refineries because motor gasolines containing ethanol cannot be pipelined. The petroleum distribution and storage system contains water. Water picks up ethanol easily, and this can cause an ethanol-water phase to form, which is separate from the gasoline. This reduces the ethanol content of the gasoline. As a result, ethanol is delivered and stored separately until the ethanol is blended into the gasoline at the distribution truck loading dock [1]. This is called "splash blending."

A special blend of gasoline is produced for blending with ethanol at the distribution point. This blend is referred to as reformulated blend for oxygenate blending (RBOB). E85, blends of ethanol and gasoline containing from 70% to 85% ethanol, are designed to be burned in cars and trucks with engines designed to run on either reformulated gasoline (RFG) or E85. This reduces the U.S. need for imported crude oil by a significant amount, but when using E85, the mileage produced is only about 70% of that when operating on reformulated gasoline.

The same basic techniques are used for calculating the blending components for any of the blended refinery products. Gasoline is the largest volume refinery product and will be used as an example to help clarify the procedures.

For purposes of preliminary cost evaluation studies, calculations generally are not made on the percentage distilled specifications at intermediate percentages, even though these are important with respect to such operating characteristics as warm-up, acceleration, and economy. The allowable blending stocks are those with boiling ranges within the product specifications [e.g., C_4-380°F (C_4-193°C)] and the control criteria to meet Reid vapor pressure and octane requirements.

12.1 REID VAPOR PRESSURE

The desired RVP of a gasoline is obtained by blending n-butane with C_5-380°F (C_5-193°C) naphtha. The amount of n-butane required to give the needed RVP is calculated by

$$M_t(RVP) = \sum_{i=1}^{n} M_i(RVP)_i \tag{12.1}$$

where

M_t = total moles of blended product
$(RVP)_t$ = specification RVP for product, psi
M_i = moles of component i
$(RVP)_i$ = RVP of component i, psi or kPa

12.1.1 Example 1

Base Stock	BPD	lb/hr	MW	mol/hr	mol%	RVP	PVP
LSR gasoline	4,000	39,320	86	457	21.0	11.1	2.32
Reformate	6,000	69,900	115	617	28.4	2.8	0.80
Alkylate	3,000	30,690	104	595	13.4	4.6	0.62
FCC gasoline	8,000	87,520	108	810	37.2	4.4	1.64
Total	21,000			2179	100.0		

Blend for a 10 psi RVP (n-butane: MW = 58, RVP = 52 psi).

Butane requirement:

$$\begin{aligned}(2179)(5.38) + M(52.0) &= (2179 + M)(10)\\ 11{,}723 + 52.0M &= 21{,}790 + 10.0.M\\ 42.0M &= 10{,}067\\ M &= 240 \text{ moles } nC_4 \text{ required}\end{aligned}$$

	BPD	lb/hr	MW	mol/hr
n-Butane	1640	13,920	58	240

Total 10 psi RVP gasoline = 21,000 + 1,640 = 22,640 BPD

Blending property data for many refinery streams are given in Table 12.1.

The theoretical method for blending to the desired Reid vapor pressure requires that the average molecular weight of each of the streams be known. Although there are accepted ways of estimating the average molecular weight of a refinery stream from boiling point, gravity, and characterization factor, a more convenient way is to use the empirical method developed by Chevron Research Co. Vapor pressure blending indices (VPBIs) have been compiled as a function of the RVP of the blending streams and are given in Table 12.2. The Reid vapor pressure of the blend is closely approximated by the sum of all the products of the volume fraction (v) times the VPBI for each component. In equation form:

$$RVP_{blend} = \Sigma v_i(VPBI)_i \tag{12.2}$$

In the case where the volume of the butane to be blended for a given RVP is desired:

$$A(VPBI)_a + B(BPBI)_b + \cdots + W(VPBI)_w = (Y + W)(VPBI)_m \tag{12.3}$$

where

A = bbl of component a, and so on
W = bbl of n-butane (w)
Y = A + B + C + · · · (all components except n-butane)
$(VPBI)_m$ = VPBI corresponding to the desired RVP of the mixture
w = subscript indicating n-butane

12.1.2 Example 2

Component	BPCD	RVP	VPBI	vol X VPBI
n-Butane	W	51.6	138.0	138W
LSR gasoline	4,000	11.1	20.3	81,200
Reformate	6,000	2.8	3.62	21,720
Alkylate	3,000	4.6	6.73	20,190
FCC gasoline	8,000	4.4	6.37	50,960
Total	21,000 + W			174,070 + 138W

TABLE 12.1
Blending Component Values for Gasoline Blending Streams

No.	Component	RVP, psi	MON	RON	°API
1.	iC_4	71.0	92.0	93.0	
2.	nC_4	52.0	92.0	93.0	
3.	iC_5	19.4	90.8	93.2	
4.	nC_5	14.7	72.4	71.5	
5.	iC_6	6.4	78.4	79.2	
6.	LSR gasoline (C_5-180 × °F)	11.1	61.6	66.4	78.6
7.	LSR gasoline isomerized once-through	13.5	81.1	83.0	80.4
8.	HSR gasoline	1.0	58.7	62.3	48.2
9.	Light hydrocrackate	12.9	82.4	82.8	79.0
10.	Hydrocrackate, C_5-C_6	15.5	85.5	89.2	86.4
11.	Hydrocrackate, C_6-190°F	3.9	73.7	75.5	85.0
12.	Hydrocrackate, 190–250°F	1.7	75.6	79.0	55.5
13.	Heavy hydrocrackate	1.1	67.3	67.6	49.0
14.	Coker gasoline	3.6	60.2	67.2	57.2
15.	Light thermal gasoline	9.9	73.2	80.3	74.0
16.	C_6^+ light thermal gasoline	1.1	68.1	76.8	55.1
17.	FCC gasoline, 200–300°F	1.4	77.1	92.1	49.5
18.	Hydrog. light FCC gasoline, C_5^+	13.9	80.9	83.2	51.5
19.	Hydrog. C_5-200°F FCC gasoline	14.1	81.7	91.2	58.1
20.	Hydrog. light FCC gasoline, C_6^+	5.0	74.0	86.3	49.3
21.	Hydrog. C_5^+ FCC gasoline	13.1	80.7	91.0	54.8
22.	Hydrog. 300–400°F FCC gasoline	0.5	81.3	90.2	48.5
23.	Reformate, 94 RON	2.8	84.4	94.0	45.8
24.	Reformate, 98 RON	2.2	86.5	98.0	43.1
25.	Reformate, 100 RON	3.2	88.2	100.0	41.2
26.	Aromatic concentrate	1.1	94.0	107.0	
27.	Alkylate, $C_3^=$	5.7	87.3	90.8	
28.	Alkylate, $C_4^=$	4.6	95.9	97.3	70.3
29.	Alkylate, $C_3^=$, $C_4^=$	5.0	93.0	94.5	
30.	Alkylate, $C_5^=$	1.0	88.8	89.7	
31.	Polymer	8.7	84.0	96.9	59.5

$$\begin{aligned}
\text{For 10 psi RVP, } (VPBI)_m &= 17.8 \\
17.8(21{,}000 + W) &= 174{,}070 + 138W \\
(138 - 17.8)W &= 373{,}800 - 174{,}070 \\
120.2W &= 199{,}730 \\
W &= 1660 \text{ bbl n-butane required}
\end{aligned}$$

$$\text{Total 10 psi RVP gasoline} = 21{,}000 + 1{,}660 = 22{,}660 \text{ BPCD}$$

Although this differs slightly from the result obtained in Example 1, it agrees well within the limits required for normal refinery operation.

TABLE 12.2
Reid Vapor Blending Index Numbers for Gasoline and Turbine Fuels

Vapor Pressure, psi	0.0	0.1	0.2	0.3	0.4	0.5	0.6	0.7	0.8	0.9
0	0.00	0.05	0.13	0.22	0.31	0.42	0.52	0.64	0.75	0.87
1	1.00	1.12	1.25	1.38	1.52	1.66	1.79	1.94	2.08	2.23
2	2.37	2.52	2.67	2.83	2.98	3.14	3.30	3.46	3.62	3.78
3	3.94	4.11	4.28	4.44	4.61	4.78	4.95	5.13	5.30	5.48
4	5.65	5.83	6.01	6.19	6.37	6.55	6.73	6.92	7.10	7.29
5	7.47	7.66	7.85	8.04	8.23	8.42	8.61	8.80	9.00	9.19
6	9.39	9.58	9.78	9.98	10.2	10.4	10.6	10.8	11.0	11.2
7	11.4	11.6	11.8	12.0	12.2	12.4	12.6	12.8	13.0	13.2
8	13.4	13.7	13.9	14.1	14.3	14.5	14.7	14.9	15.2	15.4
9	15.6	15.8	16.0	16.2	16.4	16.7	16.9	17.1	17.3	17.6
10	17.8	18.0	18.2	18.4	18.7	18.9	19.1	19.4	19.6	19.8
11	20.0	20.3	20.5	20.7	20.9	21.2	21.4	21.6	21.9	22.1
12	22.3	22.6	22.8	23.0	23.3	23.5	23.7	24.0	24.2	24.4
13	24.7	24.9	25.2	25.4	25.6	25.9	26.1	26.4	26.6	26.8
14	27.1	27.3	27.6	27.8	28.0	28.3	28.5	28.8	29.0	29.3
15	29.5	29.8	30.0	30.2	30.5	30.8	31.0	31.2	31.5	31.8
16	32.0	32.2	32.5	32.8	33.0	33.2	33.5	33.8	34.0	34.3
17	34.5	34.8	35.0	35.3	35.5	35.8	36.0	36.3	36.6	36.8
18	37.1	37.3	37.6	37.8	38.1	38.4	38.6	38.9	39.1	39.4
19	39.7	39.9	40.2	40.4	40.7	41.0	41.2	41.5	41.8	42.0
20	42.3	42.6	42.8	43.1	43.4	43.6	43.9	44.2	44.4	44.7
21	45.0	45.2	45.5	45.8	46.0	46.3	46.6	46.8	47.1	47.4
22	47.6	47.9	48.2	48.4	48.7	49.0	49.3	49.5	49.8	50.1
23	50.4	50.6	50.9	51.2	51.5	51.7	52.0	52.3	52.6	52.8
24	53.1	53.4	53.7	54.0	54.2	54.5	54.8	55.1	55.3	55.6
25	55.9	56.2	56.5	56.7	57.0	57.3	57.5	57.9	58.1	58.4
26	58.7	59.0	59.3	59.6	59.8	60.1	60.4	60.7	61.0	61.3
27	61.5	61.8	62.1	62.4	62.7	63.0	63.3	63.5	63.8	64.1
28	64.4	64.7	65.0	65.3	65.6	65.8	66.1	66.4	66.7	67.0
29	67.3	67.6	67.9	68.2	68.4	68.8	69.0	69.3	69.6	69.9
30	70.2									
40	101									
(nC_4) 51.6	138									
(iC_4) 72.2	210									
(C_3) 190	705									

Equation:
$$VPBI = VP^{1.25}$$

Example:
Calculate the vapor pressure of a gasoline blend as follows:

Component	Volume Fraction	Vapor Pressure, psi	Vapor Pressure Blending Index No.	Volume Fraction x VPBI
n-Butane	0.050	51.6	138	6.90
Light Straight Run	0.450	6.75	10.9	4.90
Heavy Refined	0.500	1.00	1.00	0.50
Total	1.000	7.4	12.3	12.30

From the brochure, "31.0°API Iranian Heavy Crude Oil," by arrangement with Chevron Research Company.

Source: Note 3.

12.2 OCTANE BLENDING

Octane numbers are blended on a volumetric basis using the blending octane numbers of the components. True octane numbers do not blend linearly, and it is necessary to use blending octane numbers in making calculations. Blending octane numbers are based upon experience and are those numbers which, when added on a volumetric average basis, will give the true octane of the blend. True octane is defined as the octane number obtained using a CFR test engine.

The formula used for calculations is

$$B_t ON_t = \sum_{i=1}^{n} (B_i ON_i) \tag{12.4}$$

where

B_t = total gasoline blended, bbl
ON_t = desired octane of blend
B_i = bbl of component i
ON_i = blending octane number of component i

12.3 BLENDING FOR OTHER PROPERTIES

There are several methods of estimating the physical properties of a blend from the properties of the blending stocks. One of the most convenient methods of estimating properties that do not blend linearly is to substitute for the value of the inspection to be blended another value that has the property of blending approximately linearly. Such values are called blending factors or blending index numbers. The Chevron Research Co. has compiled factors or index numbers for vapor pressures, viscosities, flash points, and aniline points. These are tabulated in Table 12.2, Table 12.3, Table 12.4, and Table 12.5, respectively. This material is copyrighted and reproduced with permission of the Chevron Research Co. Table 12.6 shows the blending values of octane improvers.

Examples are given in each table of the use of blending index numbers. Because it is more complicated than the others, viscosity blending is more fully discussed below.

In blending some products, viscosity is one of the specifications that must be met. Viscosity is not an additive property, and it is necessary to use special techniques to estimate the viscosity of a mixture from the viscosities of its components. The method most commonly accepted is the use of special charts developed by and obtainable from ASTM.

Blending of viscosities may be calculated conveniently by using viscosity factors (VFs) from Table 12.3. It is usually true to a satisfactory approximation that the viscosity factor of the blend will be the sum of all the products of the volume fraction times the viscosity factor for each component. In equation form:

$$VF_{blend} = \Sigma\, (V_i \times VF_i) \tag{12.5}$$

Table 12.3 shows an example calculation.

Blending of kinematic viscosities (centistokes, or cSt) may be done at any temperature, but the viscosities of all components of the blend must be expressed at the same temperature. Blending of Saybolt Universal viscosities also may be done at any temperature and interchangeably with kinematic viscosities at the same temperature. Therefore, Table 12.3 may be used to convert viscosities expressed in centistokes to Saybolt Universal seconds (SUS) and vice versa.

TABLE 12.3
Viscosity Blending Index Numbers

FACTORS FOR VOLUME BLENDING OF VISCOSITIES AT CONSTANT TEMPERATURE CORRESPONDING TO VALUES OF KINEMATIC VISCOSITY

Centistokes	0.00	0.01	0.02	0.03	0.04	0.05	0.06	0.07	0.08	0.09
0.5	0.000	0.006	0.013	0.019	0.025	0.030	0.036	0.041	0.046	0.051
0.6	0.056	0.061	0.065	0.069	0.074	0.078	0.082	0.086	0.089	0.093
0.7	0.097	0.100	0.104	0.107	0.110	0.114	0.117	0.120	0.123	0.126
0.8	0.128	0.131	0.134	0.137	0.139	0.142	0.144	0.147	0.149	0.152
0.9	0.154	0.156	0.159	0.161	0.163	0.165	0.167	0.169	0.172	0.174
cSt	**0.0**	**0.1**	**0.2**	**0.3**	**0.4**	**0.5**	**0.6**	**0.7**	**0.8**	**0.9**
1	0.176	0.194	0.210	0.224	0.236	0.247	0.257	0.266	0.275	0.283
2	0.290	0.297	0.303	0.309	0.314	0.320	0.325	0.329	0.334	0.338
3	0.342	0.346	0.350	0.353	0.357	0.360	0.363	0.366	0.369	0.372
4	0.375	0.378	0.380	0.383	0.385	0.387	0.390	0.392	0.394	0.396
5	0.398	0.400	0.402	0.404	0.406	0.408	0.410	0.411	0.413	0.414
6	0.416	0.418	0.419	0.421	0.422	0.423	0.425	0.426	0.428	0.429
7	0.431	0.432	0.433	0.434	0.436	0.437	0.438	0.439	0.440	0.442
8	0.443	0.444	0.445	0.446	0.447	0.448	0.449	0.450	0.451	0.452
9	0.453	0.454	0.455	0.456	0.456	0.457	0.458	0.459	0.460	0.461
cSt	**0**	**1**	**2**	**3**	**4**	**5**	**6**	**7**	**8**	**9**
10	0.462	0.470	0.477	0.483	0.489	0.494	0.499	0.503	0.508	0.511
20	0.515	0.519	0.522	0.525	0.528	0.531	0.533	0.536	0.538	0.541
30	0.543	0.545	0.547	0.549	0.551	0.553	0.555	0.557	0.558	0.559
40	0.561	0.563	0.564	0.566	0.567	0.568	0.570	0.571	0.572	0.573
50	0.575	0.576	0.577	0.578	0.579	0.580	0.581	0.582	0.583	0.584
60	0.585	0.586	0.587	0.588	0.589	0.590	0.591	0.592	0.592	0.593
70	0.594	0.595	0.596	0.596	0.597	0.598	0.599	0.599	0.600	0.601
80	0.601	0.602	0.603	0.603	0.604	0.605	0.605	0.606	0.607	0.607
90	0.608	0.608	0.609	0.610	0.610	0.611	0.611	0.612	0.612	0.613
cSt	**0**	**10**	**20**	**30**	**40**	**50**	**60**	**70**	**80**	**90**
100	0.613	0.618	0.623	0.627	0.631	0.634	0.637	0.640	0.643	0.646
200	0.648	0.651	0.653	0.655	0.657	0.659	0.661	0.662	0.664	0.666
300	0.667	0.669	0.670	0.671	0.673	0.674	0.675	0.676	0.678	0.679
400	0.680	0.681	0.682	0.683	0.684	0.685	0.686	0.687	0.688	0.688
500	0.689	0.690	0.691	0.692	0.692	0.693	0.694	0.695	0.696	0.696
600	0.697	0.698	0.698	0.699	0.700	0.700	0.701	0.701	0.702	0.702
700	0.703	0.704	0.704	0.705	0.705	0.706	0.706	0.707	0.707	0.708
800	0.708	0.709	0.709	0.710	0.710	0.711	0.711	0.712	0.712	0.713
900	0.713	0.714	0.714	0.715	0.715	0.715	0.716	0.716	0.716	0.717
cSt	**0**	**100**	**200**	**300**	**400**	**500**	**600**	**700**	**800**	**900**
1,000	0.717	0.721	0.724	0.727	0.730	0.733	0.735	0.737	0.739	0.741
2,000	0.743	0.745	0.747	0.748	0.750	0.751	0.752	0.754	0.755	0.756
3,000	0.757	0.758	0.759	0.761	0.762	0.763	0.764	0.765	0.765	0.766
4,000	0.767	0.768	0.769	0.770	0.770	0.771	0.772	0.772	0.773	0.774
5,000	0.775	0.775	0.776	0.777	0.778	0.778	0.778	0.779	0.779	0.780
6,000	0.780	0.781	0.781	0.782	0.782	0.783	0.783	0.784	0.784	0.785
7,000	0.785	0.786	0.786	0.787	0.787	0.787	0.788	0.788	0.789	0.790
8,000	0.790	0.790	0.790	0.791	0.791	0.791	0.792	0.792	0.792	0.793
9,000	0.793	0.794	0.794	0.794	0.795	0.795	0.795	0.796	0.796	0.796
cSt	**0**	**1000**	**2000**	**3000**	**4000**	**5000**	**6000**	**7000**	**8000**	**9000**
10,000	0.796	0.799	0.802	0.804	0.806	0.808	0.810	0.812	0.814	0.815
20,000	0.817	0.818	0.820	0.821	0.822	0.823	0.824	0.825	0.826	0.827
30,000	0.828	0.829	0.830	0.831	0.832	0.833	0.833	0.834	0.835	0.836
40,000	0.836	0.837	0.838	0.838	0.839	0.839	0.840	0.841	0.841	0.842
50,000	0.842	0.843	0.843	0.844	0.844	0.845	0.845	0.846	0.846	0.847
60,000	0.847	0.848	0.848	0.848	0.849	0.849	0.850	0.850	0.850	0.851
70,000	0.851									
80,000	0.854									
90,000	0.858									

cSt	
100,000	0.860
200,000	0.877
300,000	0.887
400,000	0.894
500,000	0.899
600,000	0.903
700,000	0.906
800,000	0.909
900,000	0.912
cSt	
1,000,000	0.914
2,000,000	0.928
3,000,000	0.937
4,000,000	0.942
5,000,000	0.947
6,000,000	0.950
7,000,000	0.953
8,000,000	0.956
9,000,000	0.958
cSt	
10,000,000	0.960
20,000,000	0.973
30,000,000	0.980
40,000,000	0.985
50,000,000	0.989
60,000,000	0.992
70,000,000	0.995
80,000,000	0.997
90,000,000	0.999

Example: Calculate the viscosity of a three-component blend as follows:

Component	Vol. Fraction in Blend	Viscosity	Factor	Vol. Fraction x Factor
A	0.5	430 SFS at 122°F	0.700	0.350
B	0.3	82.5 SUS at 130°F	0.500	0.150
C	0.2	2.15 cSt at 130°F	0.300	0.060
Total	1.0	(39.5 cSt at 130°F) (or 183 SUS at 130°F) (or 26 SFS at 122°F)	(0.560)	0.560

FACTORS FOR VOLUME BLENDING OF VISCOSITIES AT CONSTANT TEMPERATURE CORRESPONDING TO VALUES OF SAYBOLT UNIVERSAL SECONDS

SUS	0.0	0.1	0.2	0.3	0.4	0.5	0.6	0.7	0.8	0.9
32	0.275	0.278	0.280	0.282	0.284	0.286	0.288	0.290	0.292	0.294
33	0.296	0.298	0.300	0.302	0.303	0.305	0.307	0.309	0.310	0.312
34	0.314	0.315	0.317	0.318	0.320	0.321	0.323	0.324	0.326	0.327
35	0.328	0.330	0.331	0.333	0.334	0.335	0.337	0.338	0.339	0.340
36	0.342	0.343	0.344	0.345	0.346	0.347	0.349	0.350	0.351	0.352
37	0.353	0.354	0.355	0.356	0.357	0.358	0.359	0.360	0.362	0.363
38	0.363	0.364	0.365	0.366	0.367	0.368	0.369	0.370	0.371	0.372
39	0.373	0.373	0.374	0.375	0.376	0.377	0.378	0.378	0.379	0.380
SUS	**0**	**1**	**2**	**3**	**4**	**5**	**6**	**7**	**8**	**9**
40	0.381	0.388	0.395	0.402	0.408	0.413	0.418	0.423	0.428	0.431
50	0.435	0.439	0.442	0.445	0.449	0.451	0.454	0.457	0.459	0.462
60	0.464	0.466	0.469	0.471	0.473	0.475	0.476	0.478	0.480	0.482
70	0.483	0.485	0.486	0.488	0.489	0.491	0.492	0.493	0.495	0.496
80	0.497	0.498	0.499	0.501	0.502	0.503	0.504	0.505	0.506	0.507
90	0.508	0.509	0.510	0.511	0.512	0.513	0.513	0.514	0.515	0.516
SUS	**0**	**10**	**20**	**30**	**40**	**50**	**60**	**70**	**80**	**90**
100	0.517	0.524	0.531	0.537	0.542	0.547	0.551	0.555	0.559	0.562
200	0.565	0.568	0.571	0.574	0.576	0.579	0.581	0.583	0.585	0.587
300	0.589	0.591	0.593	0.595	0.596	0.598	0.600	0.601	0.603	0.604
400	0.605	0.607	0.608	0.609	0.611	0.612	0.613	0.614	0.615	0.616
500	0.617	0.618	0.619	0.620	0.621	0.622	0.623	0.624	0.625	0.626
600	0.627	0.628	0.628	0.629	0.630	0.631	0.632	0.632	0.633	0.634
700	0.635	0.635	0.636	0.637	0.637	0.638	0.639	0.639	0.640	0.640
800	0.641	0.642	0.642	0.643	0.643	0.644	0.645	0.645	0.646	0.646
900	0.647	0.647	0.648	0.648	0.649	0.649	0.650	0.650	0.651	0.651
SUS	**0**	**100**	**200**	**300**	**400**	**500**	**600**	**700**	**800**	**900**
1,000	0.652	0.656	0.660	0.664	0.667	0.670	0.673	0.676	0.678	0.681
2,000	0.683	0.685	0.687	0.689	0.691	0.692	0.694	0.696	0.697	0.699
3,000	0.700	0.701	0.703	0.704	0.705	0.706	0.707	0.708	0.709	0.710
4,000	0.711	0.712	0.713	0.714	0.715	0.716	0.717	0.718	0.719	0.719
5,000	0.720	0.721	0.722	0.722	0.723	0.724	0.725	0.725	0.726	0.726
6,000	0.727	0.728	0.728	0.729	0.729	0.730	0.731	0.731	0.732	0.732
7,000	0.733	0.733	0.734	0.734	0.735	0.735	0.736	0.736	0.737	0.737
8,000	0.738	0.738	0.739	0.739	0.740	0.740	0.740	0.741	0.741	0.742
9,000	0.742	0.742	0.743	0.743	0.744	0.744	0.744	0.745	0.745	0.745
SUS	**0**	**1000**	**2000**	**3000**	**4000**	**5000**	**6000**	**7000**	**8000**	**9000**
10,000	0.746	0.749	0.752	0.755	0.758	0.760	0.762	0.764	0.766	0.768
20,000	0.770	0.771	0.773	0.774	0.776	0.777	0.778	0.779	0.781	0.782
30,000	0.783	0.784	0.785	0.786	0.787	0.788	0.789	0.790	0.790	0.791
40,000	0.792	0.793	0.793	0.794	0.795	0.795	0.796	0.797	0.797	0.798
50,000	0.799	0.799	0.800	0.800	0.801	0.802	0.802	0.803	0.803	0.804
60,000	0.804	0.805	0.805	0.806	0.806	0.807	0.807	0.807	0.808	0.808

SUS	
70,000	0.809
80,000	0.813
90,000	0.816
SUS	
100,000	0.819
200,000	0.838
300,000	0.849
400,000	0.856
500,000	0.862
600,000	0.867
700,000	0.870
800,000	0.874
900,000	0.877
SUS	
1,000,000	0.879
2,000,000	0.895
3,000,000	0.904
4,000,000	0.911
5,000,000	0.915
6,000,000	0.919
7,000,000	0.923
8,000,000	0.925
9,000,000	0.928
SUS	
10,000,000	0.930
20,000,000	0.944
30,000,000	0.952
40,000,000	0.957
50,000,000	0.961
60,000,000	0.965
70,000,000	0.968
80,000,000	0.970
90,000,000	0.972

FACTORS FOR VOLUME BLENDING OF VISCOSITIES AT 130°F CORRESPONDING TO VALUES OF SAYBOLT FUROL SECONDS AT 122°F

SFS at 122°F	0	1	2	3	4	5	6	7	8	9
20						0.558	0.561	0.563	0.566	0.568
30	0.570	0.572	0.574	0.576	0.578	0.580	0.582	0.584	0.585	0.587
40	0.588	0.590	0.591	0.593	0.594	0.595	0.597	0.598	0.599	0.600
50	0.601	0.602	0.604	0.605	0.606	0.607	0.608	0.609	0.610	0.610
60	0.611	0.612	0.613	0.614	0.615	0.616	0.616	0.617	0.618	0.619
70	0.619	0.620	0.621	0.622	0.622	0.623	0.624	0.624	0.625	0.626
80	0.626	0.627	0.627	0.628	0.629	0.629	0.630	0.630	0.631	0.632
90	0.632	0.633	0.633	0.634	0.634	0.635	0.635	0.636	0.636	0.637
SFS at 122°F	**0**	**10**	**20**	**30**	**40**	**50**	**60**	**70**	**80**	**90**
100	0.637	0.642	0.646	0.649	0.653	0.656	0.659	0.661	0.664	0.666
200	0.669	0.671	0.673	0.675	0.676	0.678	0.680	0.681	0.683	0.684
300	0.686	0.687	0.688	0.689	0.691	0.692	0.693	0.694	0.695	0.696
400	0.697	0.698	0.699	0.700	0.701	0.702	0.703	0.703	0.704	0.705
500	0.706	0.707	0.707	0.708	0.709	0.710	0.710	0.711	0.712	0.712
600	0.713	0.713	0.714	0.715	0.715	0.716	0.716	0.717	0.718	0.718
700	0.719	0.719	0.720	0.720	0.721	0.721	0.722	0.722	0.723	0.723
800	0.724	0.724	0.724	0.725	0.725	0.726	0.726	0.727	0.727	0.727
900	0.728	0.728	0.729	0.729	0.729	0.730	0.730	0.730	0.731	0.731
SFS at 122°F	**0**	**100**	**200**	**300**	**400**	**500**	**600**	**700**	**800**	**900**
1,000	0.732	0.735	0.738	0.741	0.743	0.746	0.748	0.750	0.752	0.754
2,000	0.755	0.757	0.759	0.760	0.761	0.763	0.764	0.764	0.766	0.767
3,000	0.769	0.770	0.771	0.772	0.773	0.773	0.774	0.775	0.776	0.777
4,000	0.778									
5,000	0.784									
6,000	0.790									
7,000	0.795									
8,000	0.798									
9,000	0.802									

Note: Values from this table are for 130°F, although the Saybolt Furol seconds are at 122°F. This table alone must not be used for any other temperatures. Values from this table may be used interchangeably with values for kinematic and Saybolt Universal viscosities if the latter are for 130°F.

For SFS at 210°F, assume SUS = 10 x SFS and use the Saybolt Universal table.

Source: Note 3.

TABLE 12.4
Flash Point Blending Index Numbers [2]

May be used to blend flash temperatures determined in any apparatus but, preferably, not to blend closed cup with open cup determinations.

Flash Point, °F	0	1	2	3	4	5	6	7	8	9
0	168,000	157,000	147,000	137,000	128,000	120,000	112,000	105,000	98,600	92,400
10	86,600	81,200	76,100	71,400	67,000	62,900	59,000	55,400	52,100	49,000
20	46,000	43,300	40,700	38,300	36,100	34,000	32,000	30,100	28,400	26,800
30	25,200	23,800	22,400	21,200	20,000	18,900	17,800	16,800	15,900	15,000
40	14,200	13,500	12,700	12,000	11,400	10,800	10,200	9,680	9,170	8,690
50	8,240	7,810	7,410	7,030	6,670	6,330	6,010	5,700	5,420	5,150
60	4,890	4,650	4,420	4,200	4,000	3,800	3,620	3,441	3,280	3,120
70	2,970	2,830	2,700	2,570	2,450	2,330	2,230	2,120	2,020	1,930
80	1,840	1,760	1,680	1,600	1,530	1,460	1,400	1,340	1,280	1,220
90	1,170	1,120	1,070	1,020	978	935	896	857	821	786
100	753	722	692	662	635	609	584	560	537	515
110	495	475	456	438	420	404	388	372	358	344
120	331	318	305	294	283	272	261	252	242	233
130	224	216	208	200	193	186	179	172	166	160
140	154	149	144	138	134	129	124	120	116	112
150	108	104	101	97.1	93.8	90.6	87.5	84.6	81.7	79.0
160	76.3	73.8	71.4	69.0	66.7	64.5	62.4	60.4	58.4	56.5
170	54.7	52.9	51.3	49.6	48.0	46.5	45.1	43.6	42.3	40.9
180	39.7	38.4	37.3	36.1	35.0	33.9	32.9	31.9	30.9	30.0
190	29.1	28.2	27.4	26.6	25.8	25.0	24.3	23.6	22.9	22.2
200	21.6	20.9	20.3	19.7	19.2	18.6	18.1	17.6	17.1	16.6
210	16.1	15.7	15.2	14.8	14.4	14.0	13.6	13.3	12.9	12.5
220	12.2	11.9	11.6	11.2	10.9	10.6	10.4	10.1	9.82	9.56
230	9.31	9.07	8.83	8.60	8.37	8.16	7.95	7.74	7.55	7.35
240	7.16	6.98	6.80	6.63	6.47	6.30	6.15	5.99	5.84	5.70
250	5.56	5.42	5.29	5.16	5.03	4.91	4.79	4.68	4.56	4.45
260	4.35	4.24	4.14	4.04	3.95	3.86	3.76	3.68	3.59	3.51
270	3.43	3.35	3.27	3.19	3.12	3.05	2.98	2.91	2.85	2.78
280	2.72	2.66	2.60	2.54	2.48	2.43	2.37	2.32	2.27	2.22
290	2.17	2.12	2.08	2.03	1.99	1.95	1.90	1.86	1.82	1.79

Flash Point, °F	0	10	20	30	40	50	60	70	80	90
300	1.75	1.41	1.15	0.943	0.777	0.643	0.535	0.448	0.376	0.317
400	0.269	0.229	0.196	0.168	0.145	0.125	0.108	0.094	0.082	0.072
500	0.063	0.056	0.049	0.044	0.039	0.035	0.031	0.028	0.025	0.022

Example:

Component	Volume	Flash Point, °F	Blending Index	Volume X Blending Index
A	0.30	100	753	226
B	0.10	90	1,170	117
C	0.60	130	224	134
Total	1.00	111	477	477

From the brochure, "31.0°API Iranian Heavy Crude Oil," by arrangement with Chevron Research Company. Copyright © 1971 by Chevron Oil Trading Company.

Source: Note 3.

Viscosity factors also are given in Table 12.3 for viscosities expressed in Saybolt Furol seconds (SFS). It is important that Saybolt Furol viscosities be blended only at 122°F (50°C). If SFS viscosities are at any other temperature, they must be converted to centistokes or SUS before blending.

Viscosity factors for SFS at 122°F (50°C) may be used interchangeably with viscosity factors for SUS at 130°F (54.4°C) and with centistokes at 130°F (54.4°C). Thus, Table 12.3 may be used also to convert viscosities in SFS at 122°F (50°C) to either kinematic or Saybolt Universal viscosities at 130°F (54.4°C).

A similar method has been developed by Reid and Allen of Chevron Research Co. for the estimation of "wax" pour points of distillate blends [3]. Pour point indices

TABLE 12.5
Aniline Point Blending Index Numbers

Aniline Point, °F	0	-1	-2	-3	-4	-5	-6	-7	-8	-9
-10	20.0	17.4	14.9	12.6	10.3	8.10	6.06	4.17	2.46	1.00
0	49.1	46.0	42.8	39.8	36.8	33.8	30.9	28.1	25.3	22.6

Aniline Point, °F	0	1	2	3	4	5	6	7	8	9
0	49.1	52.4	55.6	58.9	62.3	65.7	69.1	72.6	76.1	79.6
10	83.2	86.8	90.5	94.2	97.9	102	105	109	113	117
20	121	125	129	133	137	141	145	149	153	157
30	162	166	170	174	179	183	187	192	196	200
40	205	209	214	218	223	227	232	237	241	246
50	250	255	260	264	269	274	279	283	288	293
60	298	303	308	312	317	322	327	332	337	342
70	347	352	357	362	367	372	377	382	388	393
80	398	403	408	414	419	424	429	435	440	445
90	451	456	461	467	472	477	483	488	494	491
100	505	510	516	521	527	532	538	543	549	554
110	560	566	571	577	582	588	594	599	605	611
120	617	622	628	634	640	645	651	657	663	669
130	674	680	686	692	698	704	710	716	722	727
140	733	739	745	751	757	763	769	775	781	788
150	794	800	806	812	818	824	830	836	842	849
160	855	861	867	873	880	886	892	898	904	911
170	917	923	930	936	942	948	955	961	967	974
180	980	986	993	999	1,006	1,012	1,019	1,025	1,031	1,038
190	1,044	1,050	1,057	1,064	1,070	1,077	1,083	1,090	1,096	1,103
200	1,110	1,116	1,122	1,129	1,136	1,142	1,149	1,156	1,162	1,169
210	1,176	1,182	1,189	1,196	1,202	1,209	1,216	1,222	1,229	1.236
220	1.242	1.249	1,256	1,262	1,269	1,276	1,283	1.290	1,330	1,337
230	1.310	1,317	1.324	1,331	1,337	1,344	1,351	1,358	1,365	1,372
240	1,379	1.386	1,392	1,400	1,406	1,413	1,420	1,427	1,434	1,441

Mixed Aniline Point, °F	0	1	2	3	4	5	6	7	8	9
0	-736	-730	-723	-716	-709	-703	-696	-689	-682	-675
10	-668	-660	-653	-646	-639	-631	-623	-616	-608	-600
20	-593	-584	-577	-569	-561	-552	-544	-536	-528	-519
30	-511	-503	-494	-486	-477	-468	-460	-451	-442	-433
40	-425	-416	-407	-398	-389	-380	-371	-361	-352	-343
50	-334	-324	-315	-306	-296	-287	-277	-267	-258	-248
60	-239	-229	-219	-210	-200	-190	-180	-170	-160	-150
70	-140	-130	-120	-110	-100	-89.6	-79.4	-69.2	-58.9	-48.6
80	-38.3	-27.9	-17.5	-7.06	3.39	13.9	24.4	35.0	45.5	56.1
90	66.8	77.4	88.1	98.8	110	120	131	142	153	164
100	175	186	197	208	219	230	241	252	263	274
110	285	297	308	319	330	342	353	364	376	387
120	399	410	422	433	445	456	468	479	491	503
130	514	526	538	550	561	573	585	597	609	620
140	632	644	656	668	680	692	704	716	728	741

Example:	Component	Volume	Aniline Point, °F	Index	Volume X Index
	A	0.8	70	347	278
	B	0.2	40 (Mixed)	-425	-85
	Total	1.0	37 (Or 102 Mixed)	193	193

From the brochure, "31.0°API Iranian Heavy Crude Oil," by arrangement with Chevron Research Company. Copyright © 1971 by Chevron Oil Trading Company.

Source: Note 3.

TABLE 12.6
Blending Values of Octane Improvers [4,5]

Compound	RVP, psi	Blending octane RON	MON	(R + M)/2
Methanol	40	135	105	120
Ethanol	11	132	106	119
tert-Butanol (TBA)	6	106	89	98
MTBE	9	118	101	110
ETBE	4	118	102	110
TAME	1.5	111	98	105
TEL	—	10,000	13,000	

Source: Note 4 and Note 5.

for distillate stocks are given in Table 12.7. The pour point index of the blend is the sum of the products of the volume fraction times the pour point blending index (PPBI) for each component, or

$$PPBI_{blend} = \Sigma V_i PPBI_i \tag{12.6}$$

The viscosity of a blend can also be estimated by API Procedure 11A4.3, given on pp. 11–35 of the *API Technical Data Book–Petroleum Refining*.

12.4 CASE-STUDY PROBLEM: GASOLINE BLENDING

The requirements are to produce a 50/50 split of premium and regular gasoline lines having 91 and 87 posted octane numbers, respectively, and Reid vapor pressures of 10.2 psi (70.3 kPa).

For this split between regular and premium, the pool octane number needed is 89.0. The available basic blending stocks are then selected for blending. This is a trial-and-error process at this stage. After selecting the stocks, the quantity of n-butane required to give the desired vapor pressure is calculated first because the n-butane contributes significantly to the octane of the finished product.

The gasoline blending streams available from the various units are

Base stock	BPCD	MON	RON
Isomerate	5,735	81.1	83.0
Reformate	14,749	86.9	98.5
FCC C_5 gasoline	20,117	76.8	92.3
Light hydrocrackate	814	82.4	82.8
Alkylate	4,117	95.9	97.3
Polymer	2,071	84.0	96.9
Total:	47,603		

TABLE 12.7
Pour Point Blending Indices for Distillate Stocks

ASTM 50% Temp	300	350	375	400	425	450	475	500	525	550	575	600	625	650	675	700
Pour Point																
70	133	131	129	128	127	125	123	120	118	115	113	110	108	105	103	100
65	114	111	109	107	105	103	101	98	96	94	91	88	85	82	79	76
60	99	94	92	90	87	85	82	80	77	74	72	69	67	64	62	60
55	88	79	77	75	73	71	68	66	63	61	58	56	53	50	48	46
50	72	68	66	63	61	59	56	54	52	49	47	44	42	39	37	35
45	60	56	54	52	50	48	46	44	42	40	38	35	33	31	29	27
40	52	48	46	44	42	40	38	36	34	32	30	28	26	24	22	21
35	44	41	39	37	35	33	32	30	28	26	24	23	21	19	18	16
30	37	34	32	31	29	27	26	24	23	21	19	18	16	15	14	13
25	32	29	27	26	24	23	21	20	18	17	15	14	13	12	11	10
20	27	24	23	21	20	19	17	16	15	14	12	11	10	9.1	8.3	7.5
15	23	20	19	18	17	16	14	13	12	11	10	9.0	8.1	7.2	6.4	5.8
10	20	17	16	15	14	13	12	11	9.8	8.8	8.0	7.1	6.3	5.6	5.0	4.5
5	17	15	14	13	12	11	9.7	8.8	7.9	7.1	6.3	5.6	5.0	4.4	3.8	3.5
0	14	12	11	10	9.6	8.7	7.9	7.1	6.3	5.6	5.0	4.4	3.8	3.4	3.0	2.7
-5	12	10	9.5	8.7	8.0	7.2	6.5	5.8	5.1	4.5	3.9	3.4	3.0	2.7	2.4	2.1
-10	10	8.8	8.0	7.3	6.6	5.9	5.3	4.7	4.1	3.6	3.2	2.8	2.5	2.2	1.9	1.6
-15	8.8	7.4	6.8	6.1	5.5	4.9	4.4	3.9	3.4	3.0	2.6	2.2	1.9	1.7	1.4	1.2
-20	7.5	6.3	5.7	5.1	4.6	4.1	3.6	3.2	2.8	2.4	2.1	1.8	1.5	1.3	1.1	0.94
-25	6.4	5.3	4.7	4.2	3.7	3.3	2.9	2.5	2.2	1.9	1.7	1.4	1.2	1.0	0.90	0.72
-30	5.5	4.5	4.0	3.6	3.2	2.8	2.4	2.1	1.8	1.5	1.3	1.1	0.96	0.80	0.67	0.56
-35	4.6	3.7	3.3	2.9	2.6	2.3	2.0	1.7	1.4	1.2	1.0	0.90	0.75	0.62	0.51	0.43
-40	4.0	3.2	2.8	2.5	2.2	1.9	1.6	1.4	1.2	1.0	0.86	0.73	0.62	0.51	0.41	0.33
-45	3.3	2.7	2.4	2.1	1.8	1.5	1.3	1.1	0.98	0.82	0.68	0.58	0.48	0.38	0.31	0.25
-50	2.8	2.3	2.0	1.7	1.5	1.3	1.1	0.93	0.78	0.66	0.56	0.47	0.38	0.31	0.25	0.20
-55	2.5	1.9	1.7	1.4	1.2	1.1	0.90	0.77	0.65	0.55	0.46	0.37	0.30	0.24	0.19	0.15
-60	2.1	1.6	1.4	1.2	1.0	0.87	0.74	0.62	0.52	0.43	0.36	0.30	0.24	0.19	0.14	0.10
-65	1.8	1.4	1.2	1.0	0.85	0.72	0.60	0.50	0.41	0.34	0.28	0.23	0.18	0.14	0.10	0.07
-70	1.5	1.1	0.99	0.84	0.71	0.60	0.50	0.42	0.36	0.30	0.25	0.20	0.15	0.11	0.08	0.05

By arrangement with Chevron Research Company.

Source: Note 3.

For the first round of calculations, the blending stocks selected for the premium gasoline should total approximately the volume fraction of premium gasoline times the total blending stocks. The following stocks are used:

Component	Volume	RVP	VPBI	vol(VPBI)
n-Butane	W	51.6	138.0	138W
Isomerate	5,735	13.5	25.9	148,395
Reformate	14,749	2.2	2.7	39,517
FCC gasoline	20,117	4.4	6.4	128,199
Light hydrocrackate	814	12.9	24.4	19,895
Alkylate	4,117	4.6	6.7	27,732
Polymer	2,071	8.7	14.9	30,950
Total	47,603 + W			

$$
\begin{aligned}
18.2(47{,}603 + W) &= 394{,}688 + 138W \\
866{,}375 + 18.2W &= 394{,}688 + 138W \\
119.8W &= 471{,}687 \\
W &= 3{,}937 \text{ bbl C4}
\end{aligned}
$$

Total volume of 10.2 psi RVP premium gasoline = 51,540 BPCD.

12.4.1 Octane Calculations for Pool Gasoline

Component	BPCD	Vol frac.	MON	ΣMON	RON	ΣRON
n-Butane	3,937	0.077	92.0	7.05	93.0	7.12
Isomerate	5,735	0.111	81.1	9.02	83.0	9.23
Reformate	14,749	0.286	86.9	24.85	98.5	28.18
FCC C_5 + gasoline	20,117	0.390	76.8	29.97	92.3	36.02
Light hydrocrackate	814	0.016	82.4	1.30	82.8	1.31
Alkylate	4,117	0.080	95.9	7.66	97.3	7.77
Polymer	2,071	0.040	84.0	3.38	96.9	3.89
Total	51,540	1.000		83.23		95.53

Pool octane [(MON + RON) / 2)] = 88.38 PON

This is not acceptable, as the octane requirement for pool gasoline is 89 PON. There are several ways of correcting this. Among the possibilities are

1. Increase severity of reforming to produce a 98.8 or 100 RON clear reformate. (This does not appear to be attractive because the aromatics content of the gasoline would increase and the volume would decrease.)
2. Use an octane blending agent, such as MTBE, ETBE, or ethanol, to improve the pool octane.

Recalculating pool gasoline RVP and PON, after adding sufficient MTBE to increase the PON to 89.0, gives the following:

Component	Volume	RVP	VPBI	vol(VPBI)
n-Butane	W	51.6	138.0	138W
Isomerate	5,735	13.5	25.9	148,395
Reformate	14,749	2.2	2.7	39,517
FCC C_5 + naphtha	20,117	4.4	6.4	128,119
Light hydrocrackate	814	12.9	24.4	19,895
Alkylate	4,117	4.6	6.7	27,732
Polymer	2,071	8.7	14.9	30,950
MTBE	1,593	9.0	15.6	24,832
Total	49,195 + W			419,520 + 138W

$$419{,}520 + 138W = 18.2(49{,}195 + W)$$
$$119.8W = 895{,}349 - 419{,}520 = 475{,}829$$
$$W = 3{,}984 \text{ bbl}$$

Total pool 10.2 psi RVP, 89.0 PON gasoline = 53,179 BPCD.

12.5 CASE-STUDY PROBLEM: DIESEL AND JET FUEL BLENDING

In order to meet sulfur specifications for diesel fuel, the basic blending streams, atmospheric gas oil and light coker gas oil (LCGO), are hydrotreated to remove sulfur and to improve cetane number by saturation of olefinic compounds in the LCGO. The hydrotreater is operated to reduce the diesel fuel sulfur to < 0.0015 wt% (< 15 ppm). This hydrotreated product plus a small amount of alkylate bottoms is blended into diesel fuel. This diesel fuel blend also can be used for No. 2 home heating oil by putting heating oil additives into the base stock rather than using diesel fuel additives.

Laboratory tests are necessary to determine the aromatics content of diesel fuel oil, and these are not available to students. Therefore, an alternative specification of a cetane index ≥ 45.0 is used.

12.5.1 Diesel Fuel and Home Heating Oil Blend

Component	BPCD	°API	C.I.	lb/h/BPD	lb/h	wt% S	lb/h S
HT diesel	19,686	31.0	47.6	12.70	250,096	0.0015	4
Alkylate btms.	310	5.3	5.9	15.02	4,683	0.00	0
Total	19,996	30.5	46.9		254,780	0.0015	4

If additional diesel fuel or No. 2 fuel oil production is needed, FCC LGO can be diverted from hydrocracker feed, or the hydrocracker can be operated to produce diesel fuel rather than to maximize jet fuel production.

Jet fuel is blended from hydrotreated kerosine from the atmospheric distillation unit, the heavy naphtha fraction (350 to 400°F) from the atmospheric distillation unit, and the jet fuel fraction (400 to 525°F) from the hydrocracker.

Jet fuel sulfur specifications are assumed to be the same as those for highway diesel fuel oil.

12.5.2 Jet Fuel Blend

Component	BPCD	°API	lb/h/BPD	lb/h	wt% S	lb/h S
HC jet	8,441	32.2	12.61	106,439	0.00	0
Atm. kero.	11,956	38.5	12.14	145,194	0.0015	2
HSR naphtha	4,332	42.5	11.85	51,354	0.0030	2
Total	24,729			302,987		4

The relative quantities of diesel fuel and jet fuel can be varied over a reasonable range by changing the cut points and relative quantities of the side cut and bottoms streams from the middle distillate hydrotreater.

PROBLEMS

1. Using values from Table 12.1, calculate the number of barrels of n-butane that have to be added to a mixture of 1250 barrels of HSR gasoline, 750 barrels of LSR gasoline, and 620 barrels of C_5 FCC gasoline to produce a 9.0 psi Reid vapor pressure. What are the research and motor octane numbers of the blend?
2. For the blend of components in problem 1, what would be the posted octane number of the 9.0 psi RVP gasoline if 10 vol% ethanol was added to the gasoline mixture?
3. Calculate the amount of butane needed to produce a 12.5 psi RVP for a mixture of 2730 barrels of LSR gasoline, 2490 barrels of 94 RON reformate, 6100 barrels of heavy hydrocrackate, and 3600 barrels of C_5 + FCC gasoline. How much ETBE must be added to produce a 90 RON product?
4. What is the flash point of a mixture of 2500 barrels of oil with a flashpoint of 120°F, 3750 barrels with a flashpoint of 35°F, and 5000 barrels with a 150°F flashpoint?
5. Calculate the pour point of the following mixture:

Component	Barrels	ASTM 50% temp., °F	Pour point, °F
A	5200	575	10
B	3000	425	50
C	6500	500	65
D	3250	550	45

6. What is the viscosity of a blend of 2000 barrels of oil with a viscosity of 75.5 cSt at 130°F, 3000 barrels with 225 cSt at 130°F, and 5000 barrels with 6500 cSt at 130°F?
7. Calculate the octane numbers of the final blend and amount of butane needed for producing a 9.5 psi RVP gasoline from 5,100 BPSD of LSR gasoline, 3,000 BPSD light hydrocrackate, 4,250 BPSD alkylate, 10,280 BPSD heavy hydrocrackate, 14,500 BPSD FCC C_5 + gasoline, 14,200 BPSD of 96 RON reformate, and 2,500 BPSD of polymer gasoline.
8. Recommend the best method for increasing the clear posted octane number of the pool gasoline in problem 7 by three numbers. Estimate the cost involved. Assume any necessary processing units are available and have the necessary capacity.
9. Calculate the number of barrels of n-butane that have to be added to a mixture of 1000 barrels of light thermal gasoline, 1000 barrels of polymer gasoline, and 1000 barrels of C_4 = alkylate to produce a gasoline product having 10.0 psi Reid vapor pressure.
10. What is the posted octane number and Reid vapor pressure of the gasoline product of problem 3?
11. Calculate the clear octane numbers (RON and MON) and the amount of butane needed for a 12.0 psi RVP gasoline produced from the following:

	BPSD
LSR naphtha	4,200
Light hydrocrackate	1,800
C_5 + alkylate	4,500
Heavy hydrocrackate	9,150
Reformate (94 RON)	11,500
C_5 + FCC gasoline	15,600

12. Recommend the best method (lowest capital cost) for increasing the posted octane number of the pool gasoline in problem 11 by 5.5 octane numbers. Estimate the size of the unit and its 1994 construction cost.

NOTES

1. EIA, "Eliminating MTBE in Gasoline in 2006," February 22, 2006.
2. *Oil Gas J. 88*(25), pp. 42–50, 1990.
3. *31.0°API Iranian Heavy Crude Oil*, Chevron Oil Trading Co., 1971.
4. Hoffman, H.L., *Hydrocarbon Process. 59*(2), pp. 57–59, 1980.
5. Keller, J.L., *Hydrocarbon Process. 58*(5), pp. 127–138, 1979.

13 Supporting Processes

There are a number of processes that are not directly involved in the production of hydrocarbon fuels but serve in a supporting role. These include the hydrogen unit, to produce hydrogen for hydrocracking and hydrotreating; the gas processing unit, which separates the low-boiling hydrocarbons; the acid gas treating unit, which removes hydrogen sulfide and other acid gases from the hydrocarbon gas stream; the sulfur recovery unit; and effluent water-treating systems.

13.1 HYDROGEN PRODUCTION AND PURIFICATION

Many refineries produce sufficient quantities of hydrogen for hydrotreating from their naphtha-fed platinum catalyst reforming operations. Some of the more modern plants with extensive hydrotreating and hydrocracking operations, however, require more hydrogen than is produced by their catalytic reforming units. This supplemental hydrogen requirement can be provided by one of two processes: partial oxidation of heavy hydrocarbons such as fuel oil, or steam reforming of light ends such as methane (natural gas), ethane, or propane [1,2]. The steam reforming process employs catalysts, but it should not be confused with the catalytic reforming of naphtha for octane improvement. Relative hydrogen production costs by the two processes are primarily a function of feedstock cost. Steam reforming of methane usually produces hydrogen at a lower cost than partial oxidation of fuel oil. For this reason, steam reforming is more widely used in North America than partial oxidation for hydrogen production.

Steam reforming for hydrogen production is accomplished in four steps:

1. *Reforming*. This involves the catalytic reaction of methane with steam at temperatures in the range of 1400 to 1500°F (760 to 816°C), according to the following equation:

$$CH_4 + H_2O \rightarrow CO + 3H_2 \tag{13.1}$$

   This reaction is endothermic and is carried out by passing the gas through catalyst-filled tubes in a furnace. The catalyst is usually in the form of hollow cylindrical rings ranging up to 3/4 in. (1.9 cm) in diameter. It consists of 25 to 40% nickel oxide deposited on a low-silica refractory base.

2. *Shift conversion*. More steam is added to convert the CO from step 1 to an equivalent amount of hydrogen by the following reaction:

$$CO + H_2O \rightarrow CO_2 + H_2 \tag{13.2}$$

This is an exothermic reaction and is conducted in a fixed-bed catalytic reactor at about 650°F (343°C). Multiple catalyst beds in one reactor with external cooling between beds are commonly employed to prevent the temperature from getting too high, as this would adversely affect the equilibrium conversion. The catalyst used is a mixture of chromium and iron oxide.

3. *Gas purification.* The third step is removal of carbon dioxide by absorption in a circulating amine or hot potassium carbonate solution. Several other treating solutions are in use. The treating solution contacts the hydrogen and carbon dioxide gas in an absorber containing about 24 trays, or the equivalent amount of packing. Carbon dioxide is absorbed in the solution, which is then sent to a still for regeneration.
4. *Methanation.* In this step, the remaining small quantities of carbon monoxide and carbon dioxide are converted to methane by the following reactions:

$$CO + 3H_2 \rightarrow CH_4 + H_2O$$

$$CO_2 + 4H_2 \rightarrow CH_4 + 2H_2O \qquad (13.3)$$

This step is also conducted in a fixed-bed catalytic reactor at temperatures of about 700 to 800°F (370 to 427°C). Both reactions are exothermic, and if the feed concentration of CO and CO_2 is more than 3%, it is necessary to recycle some of the cooled exit gas to dissipate the heat of reaction. The catalyst contains 10 to 20% nickel on a refractory base.

The preceding description is somewhat idealized, because the actual reactions are more complicated than those shown. The actual process conditions of temperature, pressure, and steam/carbon ratios are compromises of several factors.

Figure 13.1 is a simplified flow diagram for steam reforming production of hydrogen. Investment and operating costs can be estimated from Figure 13.2 and Table 13.1.

Partial oxidation of fuel oils is accomplished by burning the fuel at high pressures (800 to 1300 psig, 5515 to 8962 kPa) with an amount of pure oxygen that is limited to that required to convert the fuel oil to carbon monoxide and hydrogen. Enough water (steam) is added to shift the carbon monoxide to hydrogen in a catalytic shift conversion step. The resulting carbon dioxide is removed by absorption in hot potassium carbonate or other solvents.

Ideally, the partial oxidation reactions are as follows:

$$2C_nH_m + nO_2 \rightarrow 2nCO + mH_2$$

$$2nCO + 2nH_2O \rightarrow 2nCO_2 + 2nH_2$$

A good summary of partial oxidation processes has been published [3]. In some refineries, a significant amount of hydrogen is present in gas streams vented from hydrocracking or hydrotreating operations. Recovery of this hydrogen should be

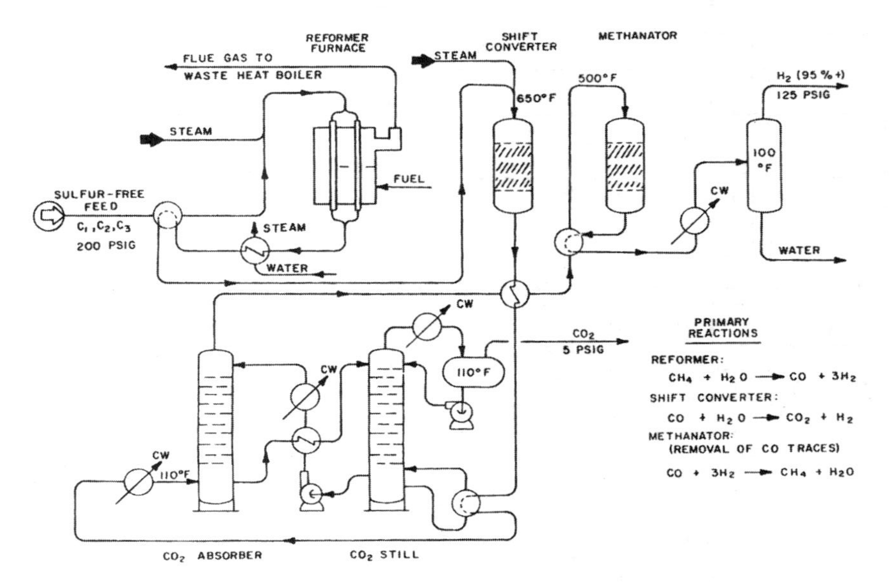

FIGURE 13.1 Hydrogen production by steam reforming.

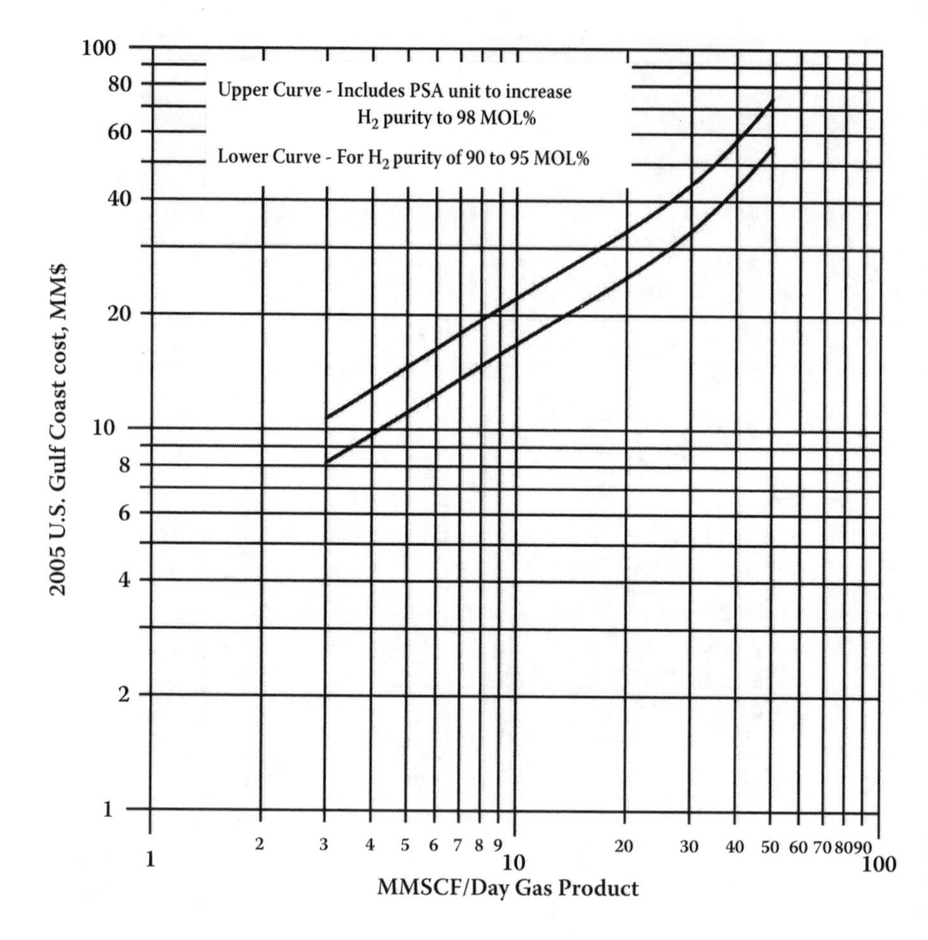

FIGURE 13.2 Hydrogen production by steam-methane reforming investment cost: 2005 U.S. Gulf Coast (see Table 13.1).

considered whenever it is necessary to supplement the catalytic reformer hydrogen to produce an overall refinery hydrogen balance. Three industrial processes are used for recovery of a concentrated hydrogen stream from a dilute gas, where the other major components are methane and other hydrocarbon vapors. The three processes are

1. Cryogenic phase separation
2. Adsorption
3. Diffusion

In the cryogenic phase separation method, the gas is cooled to about –200 to –250°F (–129 to –157°C) at a pressure ranging from 200 to 500 psig (1380 to 3450 kPa). The resulting vapor phase is typically 90 mol% hydrogen, and the liquid phase contains most of the methane and other hydrocarbons. The liquid phase is expanded to about 50 psig (345 kPa) and used to cool the feed gas. This revaporizes

TABLE 13.1
Hydrogen Production Unit Cost Data

Costs included	
1. Feed gas desulfurization	
2. Reformer, shift converter, methanator, waste heat boiler, amine (MEA) unit	
3. Hydrogen delivery to batter limits at 250 psig (1725 kPa), 100°F (38°C), 90 to 95% pure	
4. Initial catalyst charges	
Costs not included	
1. Boiler feed water treating	
2. Cooling water	
3. Dehydration of hydrogen product	
4. Power supply	
Utility data	
Power, kWh/lb H_2	0.15
kWh/kg H_2	0.33
Cooling water,[a] gal crclt./lb H_2	65.0
M^3/kgH_2	0.54
Fuel,[b] MBTU/lb H_2	45.0
MJ/kg H_2	105.0
Treated boiler feed water,[c] gal/lb H_2	1.0
M^3/kg H_2	0.01
Catalysts and chemicals, c/lb H_2	0.80
c/kg H_2	1.75
Feed gas,[d] mol/mol H_2	0.26

[a] 30°F (17°C) rise.
[b] LHV basis, heater efficiency included.
[c] For waste heat boiler.
[d] Fuel not included.
Note: See Figure 13.2.

the liquid phase. The cold vapor phase is also used to cool the feed gas. Carbon dioxide, hydrogen sulfide, and water vapor must be removed from the feed gas prior to chilling. The conditions required to achieve various hydrogen purities are described in the literature [4].

In the adsorption process, the hydrocarbons are adsorbed from the gas on a solid adsorbent (usually a molecular sieve), and the hydrogen leaves the adsorber at the desired purity. Several adsorbent vessels are used, and the feed gas flow is periodically switched from one vessel to another so that the adsorbent can be regenerated. The adsorbed methane and other impurities are released from the adsorbent by simple pressure reduction and purging [5]. Hence, this process is called pressure swing adsorption.

The diffusion process separates hydrogen from methane and other gases by allowing the hydrogen to permeate through a membrane composed of small synthetic hollow fibers [6]. The driving force for this process is the difference between the hydrogen partial pressures on each side of the membrane. Thus, a substantial pressure drop must be taken on the hydrogen product to achieve high recoveries.

The most economical method for hydrogen recovery from refinery gas streams depends on the volume of the gas to be processed, the desired hydrogen recovery and purity, and the type of components to be separated. For relatively small streams [less than 2 to 3 MMscf/d (57 to 85 MNm^3/d)] the diffusion process should be considered. For bulk removal of hydrocarbons from relatively large streams [(greater than 20 MMscf/d (570 MNm^3/d)], the cryogenic process should be considered. The adsorption process usually has advantages when the hydrogen must be recovered at purities over 95 mol%.

13.2 GAS PROCESSING UNIT

The main functions of refinery gas processing units are [6]

1. Recovery of valuable C_3, C_4, C_5, and C_6 components from the various gas streams generated by processing units such as crude distillation units, cokers, cat crackers, reformers, and hydrocrackers
2. Production of a desulfurized dry gas consisting mostly of methane and ethane that is suitable for use as a fuel gas or as feedstock for hydrogen production

In the typical gas processing unit shown in Figure 13.3, low-pressure [0 to 20 psig (0 to 138 kPa)] gases are collected and compressed to approximately 200 psig (1380 kPa) and fed to an absorber-deethanizer. This column usually contains 20 to 24 trays in the absorption section (top) and 16 to 20 trays in the stripping section (bottom). Lean absorption oil is fed to the top tray in sufficient quantity to absorb 85 to 90% of the C_3s and almost all of the C_4s and heavier components from the feed gas and from vapor rising from the stripping section. The lean absorption oil is usually a dehexanized naphtha with an end point (ASTM) of 350 to 380°F (177 to 193°C).

Due to the vapor–liquid equilibrium conditions on the top tray, a significant amount of the lighter hydrocarbons (such as C_7) are vaporized from the lean oil and leave the top of the column with the residue gas. This material is recovered in the sponge absorber. The sponge absorber usually contains 8 to 12 trays. A heavy molecular-weight, relatively nonvolatile material, such as kerosine or No. 2 fuel oil, is used as sponge oil. This sponge oil is derived as a side cut from the coker fractionator or cat cracker fractionator. The rich sponge oil is returned as a side feed to the column from which it was derived for stripping to the recovered lean oil light ends.

Sufficient reboil heat is added to the bottom of the stripping section of the absorber-deethanizer to eliminate any absorbed ethane and methane from the bottom liquid product. This deethanized rich oil then flows to a debutanizer column, where essentially all the recovered propane, propylene, butanes, and butylenes are fractionated and taken off as overhead product. This type of debutanizer usually operates in the range of 125 to 150 psig (862 to 1035 kPa) and contains 26 to 30 trays. The bottom product from the debutanizer contains pentanes and heavier hydrocarbons recovered from the gas feed to the absorber-deethanizer plus the lean oil. This material is fed to a naphtha splitter. Natural gasoline or straight-run naphtha are

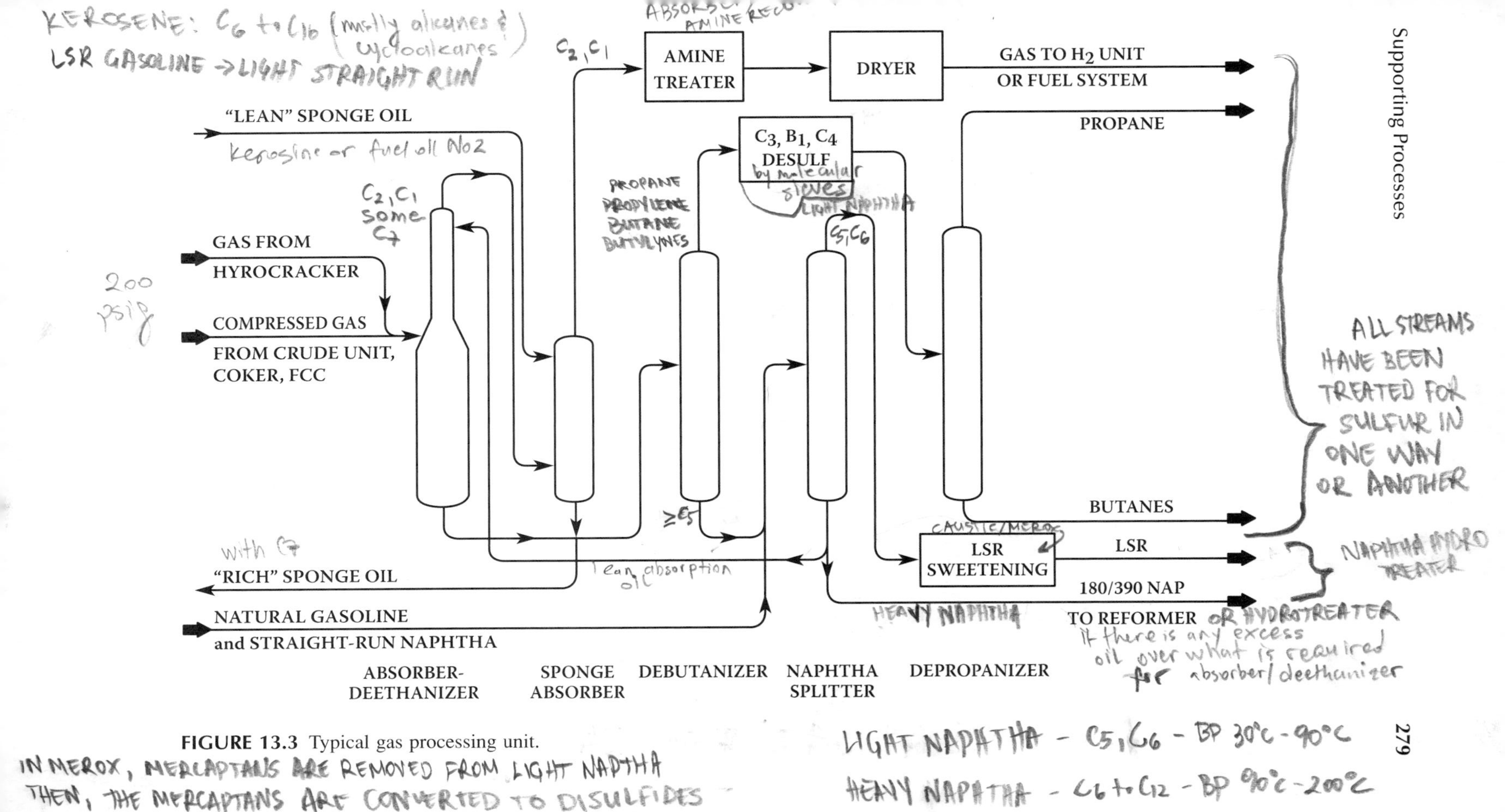

FIGURE 13.3 Typical gas processing unit.

sometimes fed to this same column. The naphtha splitter produces a C_5, C_6 light straight-run cut overhead and suitable lean absorption oil from the bottom. Bottoms product in excess of the lean oil requirement can be fed to a hydrotreater and reformer. The light straight-run product is desulfurized (or sweetened) and used directly as gasoline blend stock or else isomerized to improve octane and then used as gasoline blend stock.

The overhead C_3, C_4 product from the debutanizer is condensed, desulfurized, and fed to a depropanizer for separation into propane and butane. The desulfurization is usually accomplished by molecular sieve treating, which simultaneously dehydrates the stream.

Overhead gas from the sponge oil absorber is treated by contact with an aqueous solution of diethanolamine (DEA) or other solvents for removal of carbon dioxide and hydrogen sulfide. Extracted hydrogen sulfide is converted to elemental sulfur in a separate unit.

See Table 13.2 and Figure 13.4 for gas processing unit cost data.

13.3 ACID GAS REMOVAL

Gases from various operations in a refinery processing sour crudes contain hydrogen sulfide and occasionally carbonyl sulfide. Some hydrogen sulfide in refinery gases is formed as a result of conversion of sulfur compounds in processes such as hydrotreating, cracking, and coking. Until the period of about 1965 to 1970, it was common practice simply to burn this hydrogen sulfide along with other light gases as refinery fuel, because its removal from the gases and conversion to elemental sulfur was not economical. Today, however, air pollution regulations require that most of the hydrogen sulfide be removed from refinery fuel gas and converted to elemental sulfur.

In addition to hydrogen sulfide, many crudes contain some dissolved carbon dioxide that through distillation finds its way into the refinery fuel gas. These components—hydrogen sulfide and carbon dioxide—are generally termed acid gases. They are removed simultaneously from the fuel gas by a number of different processes, [7] some of which are the following:

13.3.1 Chemical Solvent Processes

1. Monoethanolamine (MEA)
2. Diethanolamine (DEA)
3. Methyl-diethanolamine (MDEA)
4. Diglycolamine (DGA)
5. Hot potassium carbonate

13.3.2 Physical Solvent Processes

1. Selexol
2. Propylene carbonate
3. Sulfinol
4. Rectisol

TABLE 13.2
Refinery Gas Processing Unit Cost Data

Costs Included

1. Compressors to raise gas pressure from 5 psig (35 kPa) to approximately 200 psig (1380 kPa)
2. Absorber-deethanizer utilizing naphtha as absorption oil
3. Sponge oil absorber
4. Debutanizer
5. Naphtha splitter[a]
6. C_3/C_4 molecular-sieve desulfurizer
7. Depropanizer
8. All related heat exchangers, pumps, scrubbers, accumulators, and so on
9. Initial charge of molecular sieves

Costs not included

1. Gas sweetening[b]
2. Butane splitter
3. Hot oil or steam supply for reboilers
4. Liquid product storage
5. Cooling water supply
6. Sponge oil distillation facilities (usually accomplished in cat cracker fractionator, coker fractionator, or crude tower)
7. LSR sweetening[c]

Utility data

Process fuel,[d]	MBTU/gal total liquid products	14.0
	MJ/M^3	3,900.0
Compressor,[e]	BHP/MMscfd gas	150.0
	$kWh/M^3/day$	0.1
Other power,[f]	kWh/gal total liquid products	0.06
	kWh/M^3	16.0
Cooling water,[g]	gal/gal (m^3/m^3) total liquid products	100.0

[a] When "outside" natural gasoline or straight-run naphtha is fed to the naphtha splitter, increase investment obtained from Figure 13.4 by $100 for each BPD of such material. This material is not to be included when determining the gal/Mscf parameter for Figure 13.4.

[b] See Figure 13.6 and Table 13.3.

[c] Add $40 for each BPD of LSR naphtha to be sweetened by caustic washing or "doctor treating."

[d] Process fuel requirement can frequently be obtained from heat exchange with hot oil from other refinery units. If sufficient detail is not available to ascertain that heat input can be obtained by exchange as stated above, assume that a fired heater must be provided.

[e] The gas compressor will generally be motor driven when gas volume is less than 5 MMscfd (142 MNm^3/d). In this case, it is necessary to add the compressor horsepower to the power requirement shown. For gas volumes in excess of 5 MMscfd (142 MN^3/d), the compressors may be driven by electric motors, gas turbines, reciprocating gas engines, or steam turbines. Appropriate adjustments should be made to the utility requirements to provide for operation of gas compressors.

[f] Power shown is for operation of process pumps only. All cooling is done with cooling water. No power is included for aerial coolers.

[g] 30°F (17°C) rise.

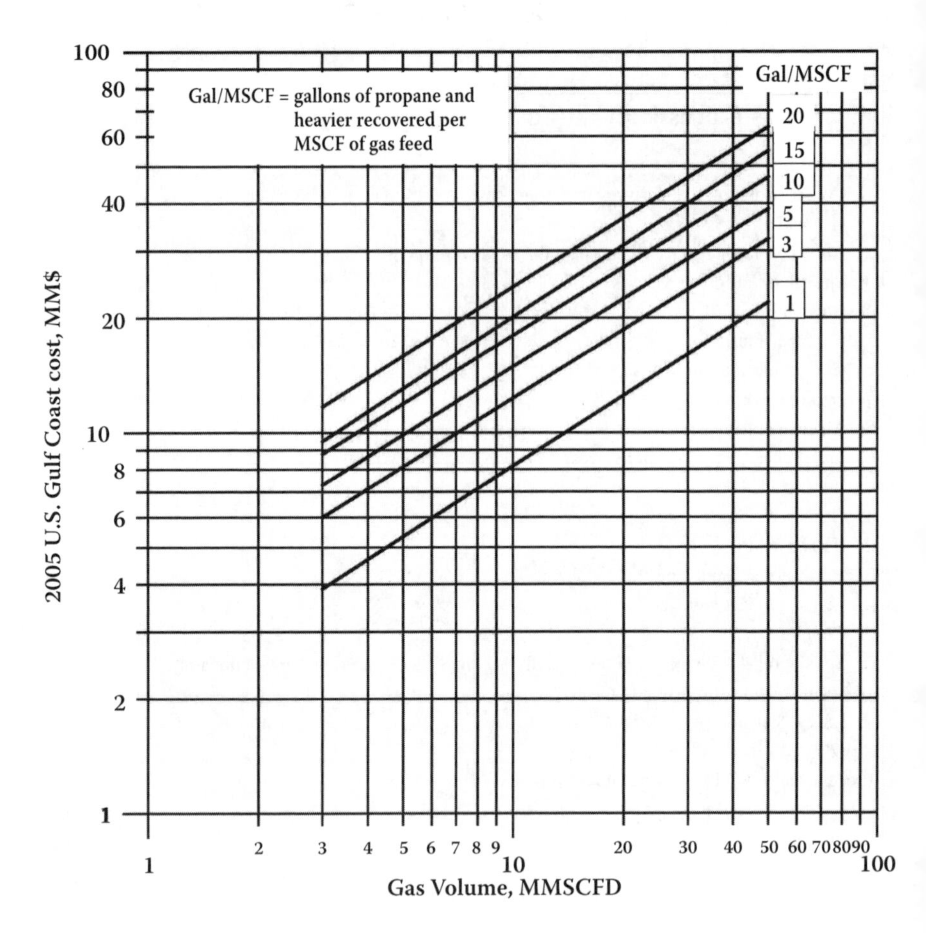

FIGURE 13.4 Refinery gas processing unit investment cost: 2005 U.S. Gulf Coast (see Table 13.2).

13.3.3 Dry Adsorbents Processes

1. Molecular sieve
2. Activated charcoal
3. Iron sponge
4. Zinc oxide

An excellent bibliography on this subject is given in the GPSA data book [8]. The selection of available processes for a given application involves many considerations. In general, the diethanolamine process has been the most widely used for refinery gas treating. This process uses an aqueous solution of diethanolamine with concentrations of the DEA in the range of 15 to 30 wt%. Methyl-diethanolamine can be used to replace diethanolamine to reduce the absorption of carbon dioxide and thereby produce an acid gas with a higher content of hydrogen sulfide. This provides some marginal improvement in Claus unit capacity or sulfur recovery efficiency. The solution is pumped to the top of an absorber containing about 20 to 24 trays or an equivalent amount of packing such as Pall rings. Hydrogen sulfide and carbon dioxide are removed from the gas by absorption into the solution. Rich solution from the absorber flows into a flash tank that is operated at a lower pressure than the absorber and permits any dissolved or entrained methane and ethane to be vented from the system.

The rich solution is then preheated and flows to a regenerator or still, where the acid gases are stripped from the solution by steam generated in a reboiler. The still also contains about 20 to 24 trays or an equivalent amount of packing. The acid gases are taken from the top of the still through a condenser, where most of the steam is condensed. This condensate is separated from the acid gases and returned to the top of the still as reflux. The acid gases are then sent to a sulfur recovery unit, where the hydrogen sulfide is converted to elemental sulfur. The lean solution is cooled and returned to the top of the absorber (Figure 13.5).

Operating conditions are usually such that the treated gas will contain less than 0.25 gr of sulfur per 100 scf (4 ppm H_2S) and less than 1.5 vol% carbon dioxide. This is controlled by the amount of solution circulated and the amount of steam generated in the reboiler. The solution rate is set at a value that is in the range of 0.15 to 0.40 gal/scf (20 to 53 l/Nm^3) of acid gases absorbed. The heat input to the reboiler is typically about 600 to 1200 Btu/gal (167 to 335 kJ/l) of circulating solution. It is generally considered advisable to have high solution rates to minimize corrosion. However, high solution rates result in increased investment and utility costs, and therefore a compromise is made in the design and operating conditions. Figure 13.6 and Table 13.3 give capital cost and operating cost information.

13.4 SULFUR RECOVERY PROCESSES

Until 1970, the major reason to recover sulfur from refinery gases was an economic one. The hydrogen sulfide was commonly used with other gases as a refinery fuel, and the sulfur dioxide concentrations in the flue gases were within acceptable limits. In those refineries with sulfur recovery units, the typical recovery was in the range of 90 to 93% of that contained in the hydrogen sulfide stream. Implementation of federal and state regulations now requires recovery of about 99% or more of the sulfur in refinery gas. This requires a two-stage process with a modified Claus unit for the first stage, followed by a second stage such as the Shell Claus off-gas treatment (SCOT) process [9,10].

13.4.1 MODIFIED CLAUS PROCESS

The most practical method for converting hydrogen sulfide to elemental sulfur is the modified Claus process. The original process was reported by Chance and Claus in 1885 [11,12]. Today, various modifications of the process are used, depending upon the hydrogen sulfide concentrations in the gas. The modified Claus process best suited for acid gas containing 50% or more hydrogen sulfide is the partial combustion (once-through) process. Figure 13.7 is a simplified flow diagram of this process, and Figure 13.8 and Table 13.4 give unit cost data.

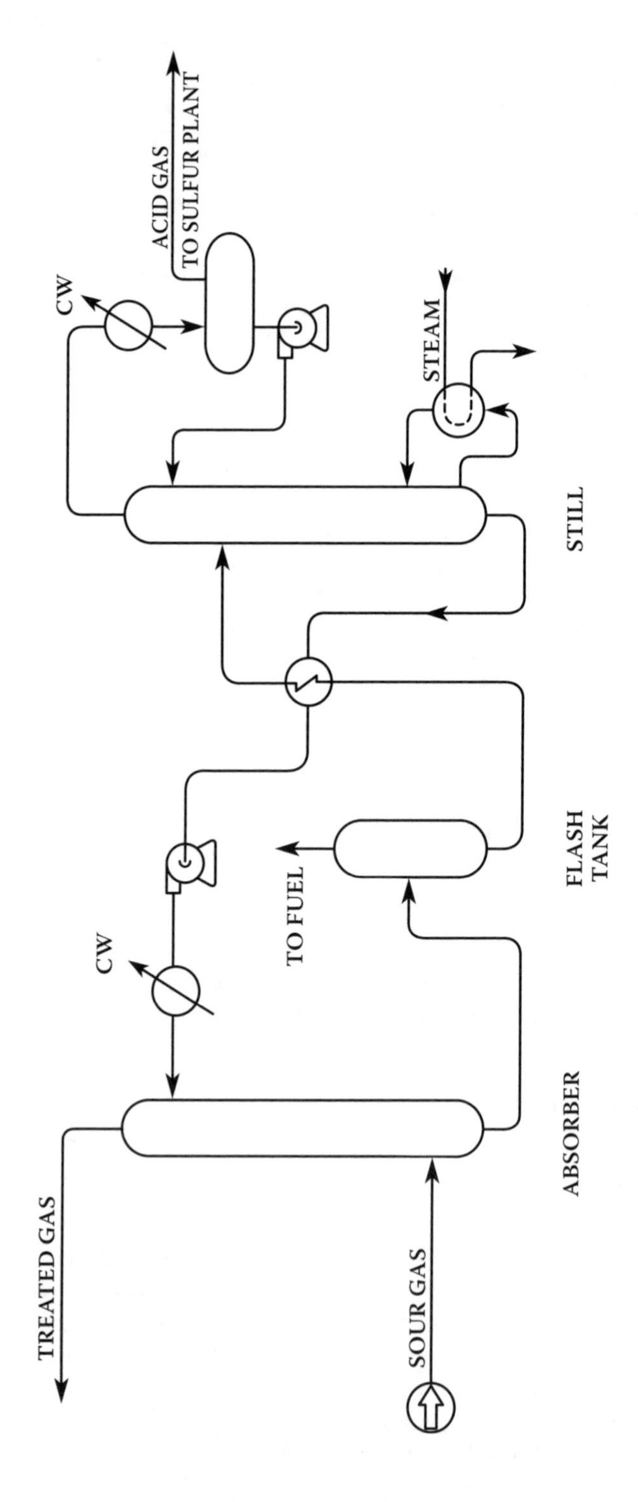

FIGURE 13.5 Amine treating unit.

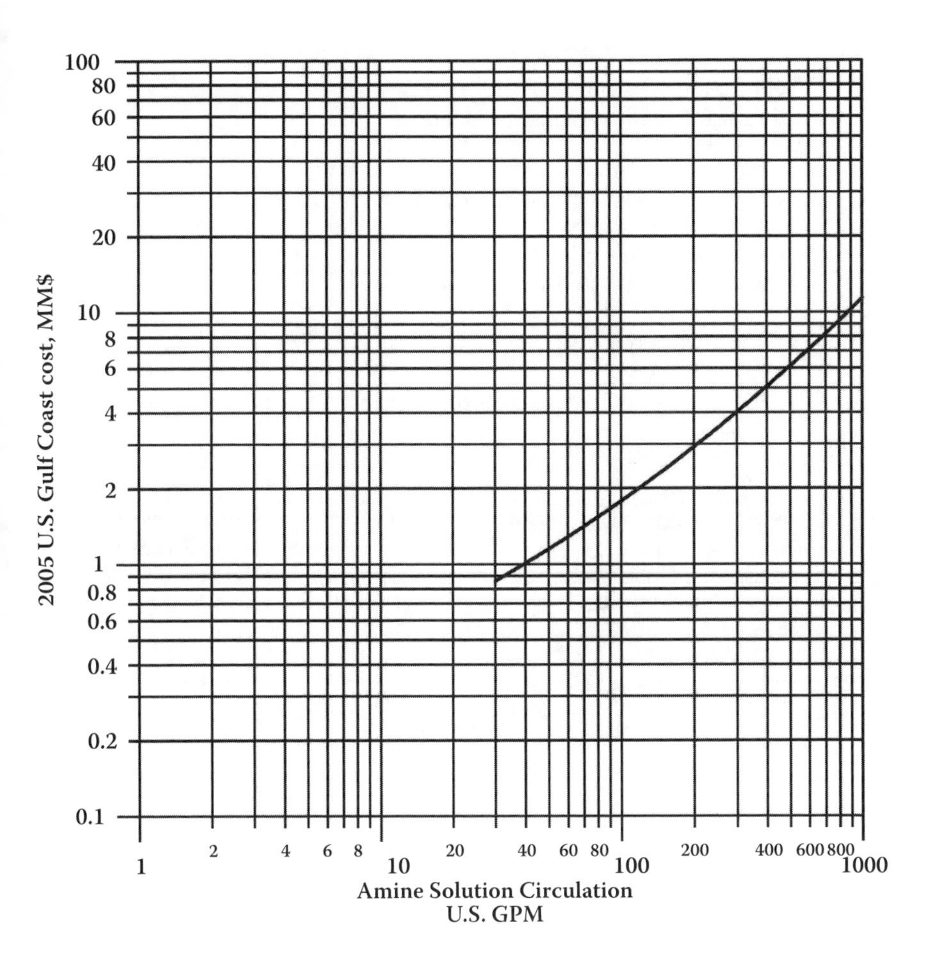

FIGURE 13.6 Amine gas treating unit investment cost: 2005 U.S. Gulf Coast (see Figure 13.5).

Present refinery practice generally provides for removal of hydrogen sulfide from refinery gas streams by solvent absorption, as discussed in the previous section. The acid gas stream recovered from these treating processes contains some carbon dioxide and minor amounts of hydrocarbons, but in most cases, the hydrogen sulfide content is over 50%. Therefore, the once-through Claus process is used in most sour crude refineries to convert the hydrogen sulfide to elemental sulfur.

In the partial combustion (once-through) process, the hydrogen sulfide-rich gas stream is burned with one third the stoichiometric quantity of air, and the hot gases are passed over an alumina catalyst to react sulfur dioxide with unburned hydrogen sulfide to free sulfur. The reactions are

$$\text{Burner:}\quad 2H_2S + 3O_2 \rightarrow 2H_2O + 2SO_2$$

$$\text{Reactor:}\quad 2H_2S + SO_2 \rightarrow 2H_2O + 3S \tag{13.4}$$

TABLE 13.3
Amine Gas Treating Unit Cost Data

Costs included	
1. Conventional, single flow, MEA, or DEA treating system	
2. Electric motor-driven pumps	
3. Steam-heated reboiler	
4. Water-cooled reflux condenser and solution cooler	
Costs not included	
1. Acid gas disposal	
2. Cooling water supply	
3. Steam (or hot oil) supply for regenerator reboiler	
Utility data	
Power,[a] kWh/gal solution circulated	0.01
kWh/M^3	2.64
Fuel,[b] BTU/gal solution circulated	1,000.00
MJ/M^3	280.00
Cooling water,[c] gal/gal (m^3/m^3) solution circulated	4.40
Amine makeup,[d] lb/MMscf inlet gas	2.50
kg/MMNm^3	0.04

[a] Assumes amine pumps driven by electric motors and cooling done with water.
[b] Reboiler heat usually supplied as 60 psig steam.
[c] 30°F (17°C) rise.
[d] In actual practice, amine solution circulation varies in the range of 0.15 to 0.40 gal per scf of acid gas (H_2S plus CO_2) removed. For preliminary estimates, a value of 0.30 gal of solution circulation per scf of acid gas can be assumed.
Note: See Figure 13.5.

The burner is located in a reaction chamber, which may be either a separate vessel or a part of the waste heat boiler. The purpose of the reaction chamber is to allow sufficient time for the combustion reaction to be completed before the gas temperature is reduced in the waste heat boiler.

Ammonia frequently is present in the Claus unit feed streams and must be completely destroyed in the reaction furnace to avoid plugging of equipment with ammonium salts. Specially designed burners and combustion zones have been developed for this purpose [13].

The waste heat boiler removes most of the exothermic reaction heat from gases by steam generation. Many types of waste heat boilers are in use. Usually they are arranged so that the gas flows through several tube passes in series with chambers, or "channels," where a portion of the gases many be withdrawn at elevated temperatures to use for reheating the main gas flow stream prior to the catalytic converters. Some elemental sulfur is often condensed and removed from the gas in the waste heat boiler. In some plants, a separate condenser is used after the waste heat boiler. The gas temperature entering the first catalytic converter is controlled at about 425 to 475°F (218 to 246°C), which is necessary to maintain the catalyst bed above sulfur dewpoint in order to avoid saturating the catalyst with sulfur and thereby deactivating

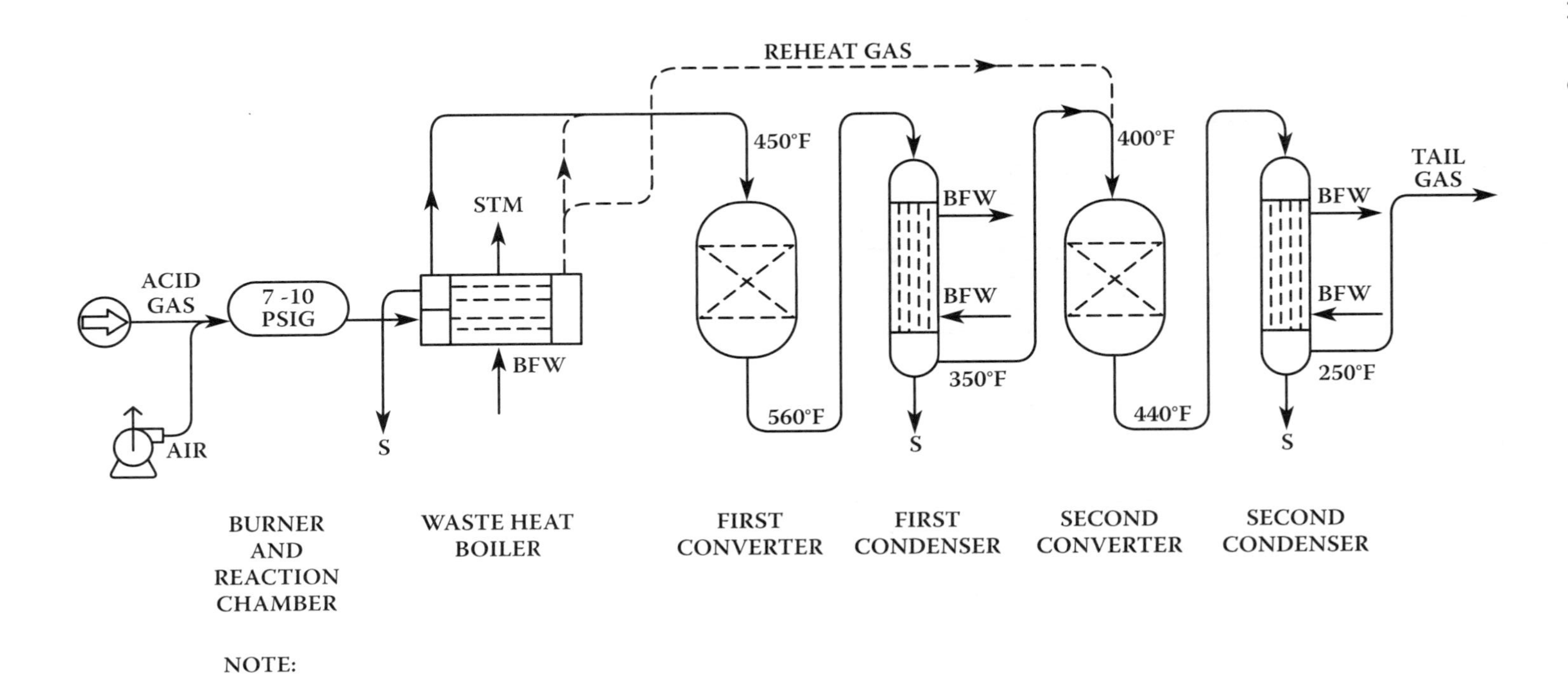

FIGURE 13.7 Once-through Claus sulfur process.

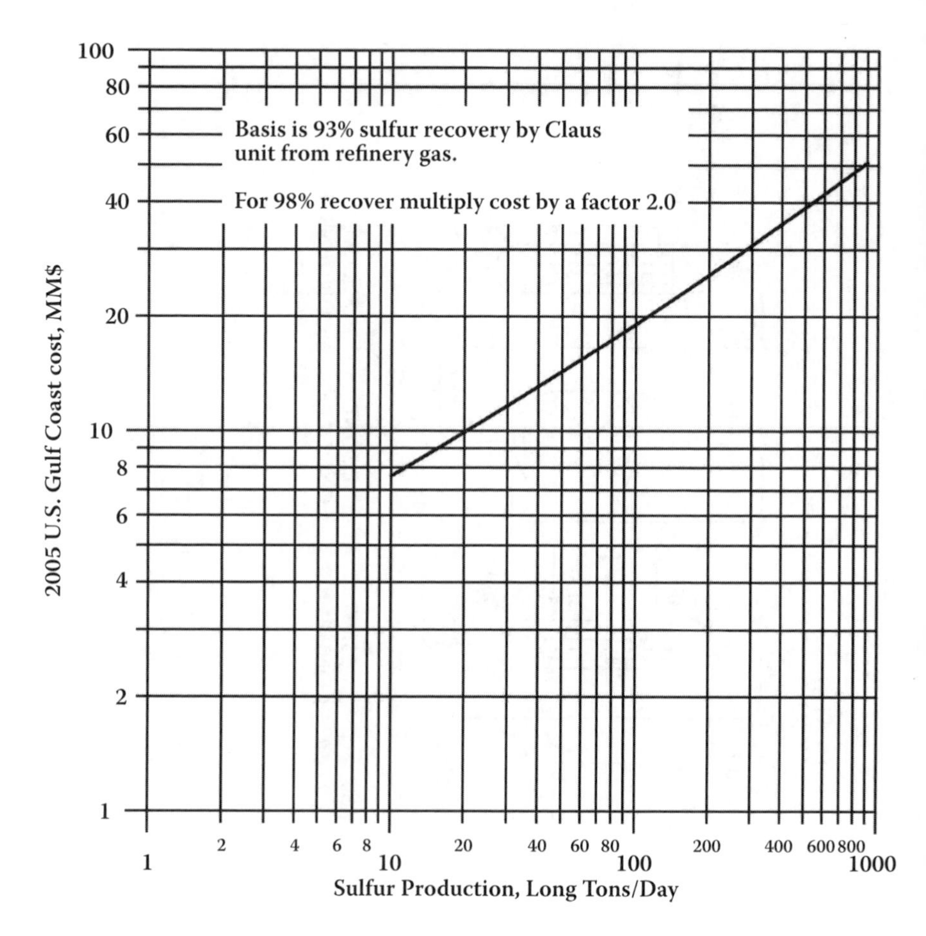

FIGURE 13.8 Claus sulfur plant investment cost: 2005 U.S. Gulf Coast (see Table 13.4).

the catalyst. The reaction between hydrogen sulfide and sulfur dioxide in the converter is also exothermic. Gases from the converter are cooled in the following condenser for removal of most of the elemental sulfur as liquid.

The condenser outlet temperatures must be maintained above about 275°F (135°C) to avoid solidifying the sulfur. Two converters and condensers in series are indicated in Figure 13.7, but most plants have three converters.

Modifications of the once-through process include various reheat methods for the converter feed temperature control, such as heat exchange with converter outlet gases, in-line burners, and fired reheaters. Overall recovery is usually not over 96% and is limited by thermodynamic considerations, as described in the literature [11].

When refineries process crude oils with sulfur contents higher than design, it is necessary to recover more sulfur in the Claus unit. An economical method for increasing the capacity of Claus units is the substitution of oxygen for a portion of the combustion air needed in the reaction furnace. This modification can increase the capacity for sulfur production by 50% or more at a relatively small capital cost [14].

TABLE 13.4
Claus Sulfur Recovery Unit Cost Data

Costs included	
1. Once-through modified Claus unit designed for 94 to 96% recovery	
2. Three converters (reactors) with initial charge of catalyst	
3. Incinerator and 150-ft (46-m)-tall stack	
4. Sulfur receiving tank and loading pump	
5. Waste heat boiler	
Costs not included	
1. Boiler feed water treating	
2. Boiler blowdown disposal	
3. Solid sulfur storage or reclaiming	
4. Sulfur loading facilities (except for loading pump)	
5. Supply of power and water	
6. Tail gas clean-up process unit	
Utility data	
Power, kWh/long ton of sulfur	100
kWh/MT	98
Boiler feed water, gal/long ton of sulfur	820
M^3/MT	3
Waste heat steam production at 250 psia (1740 kPa)	
lb/long ton sulfur	6500
kg/MT	2900
Fuel	None
Cooling water	None

Note: See Figure 13.8

13.4.2 Carbon-Sulfur Compounds

Carbonyl sulfide (COS) and carbon disulfide (CS_2) have presented problems in many Claus plant operations due to the fact that they cannot be converted completely to elemental sulfur and carbon dioxide. These compounds may be formed in the combustion step by reaction of hydrocarbons and carbon dioxide, as shown below:

$$CH_4 + SO_2 \rightarrow COS + H_2O + H_2$$

$$CO_2 + H_2 \rightarrow COS + H_2O$$

$$CH_4 + 2S_2 \rightarrow CS_2 + 2H_2S \quad (13.5)$$

Many more complex reactions are possible. These compounds, if unconverted, represent a loss of recoverable sulfur and an increase in sulfur emission to the atmosphere. Special alumina catalyst is significantly more effective in converting both COS and CS_2 to elemental sulfur than the conventional bromide catalyst [6].

The conversion of COS and CS_2 in the Claus unit may not be sufficient to meet allowable emission limits in some locations. Modifications to the tail gas unit have been designed to reduce these components to less than 20 ppmv in the outlet. Catalyst containing tungsten is used in the tail gas unit, which is a SCOT process, as described in the following paragraph [15].

13.4.3 SCOT Process

As discussed above, Claus unit tail gas contains small amounts of carbonyl sulfide and carbon disulfide in addition to hydrogen sulfide and sulfur dioxide. In Shell's SCOT process (Figure 13.9), the Claus tail gas is combined with a relatively small quantity of hydrogen or a mixture of carbon monoxide plus hydrogen and heated to about 480 to 570°F (250 to 300°C). This hot gas then flows through a fixed catalyst bed, where the various sulfur compounds are converted to hydrogen sulfide by reaction with hydrogen. The reactor effluent is cooled to ambient temperature, and the hydrogen sulfide is selectively absorbed from the gas with an aqueous amine solvent. The hydrogen sulfide is regenerated from the solvent in a conventional amine still and recycled to the Claus unit feed. The hydrogen sulfide content of the gas exiting the amine absorber is typically in the range of 50 to 400 ppm mol basis [16]. This gas is incinerated to convert the hydrogen sulfide to sulfur dioxide before venting.

Recoveries in excess of 99% can be obtained by addition of tail gas processing units such as the SCOT process. The addition of a SCOT unit increases the cost of the sulfur recovery facilities by a factor of about 2.0.

13.5 ECOLOGICAL CONSIDERATIONS IN PETROLEUM REFINING

Since the end of World War II, petroleum refineries have made special efforts to minimize discharge of wastes into the surrounding environment. This voluntary control of emissions was done on the basis of safety, fuel economy, and the economic advantage of good maintenance. Government regulations now require refineries to add facilities for stringent control of emissions or losses of substances deemed to be undesirable or potentially hazardous. Related extra capital costs vary widely, but for a new refinery, the costs are frequently considered to be in the range of 15 to 20% of the total investment. The total additional cost for installation, operation, and maintenance of these facilities is estimated to be about $0.15 to $0.30 (U.S., 2005) per gallon of refined products.

The potentially harmful substances that must be carefully controlled include discharge of liquid hydrocarbons into streams, rivers, lakes, and oceans and relief of hydrocarbon vapors into the atmosphere. Wastewater must be essentially free of toxic or carcinogenic chemicals. Benzene content of water must be reduced to less than 10 ppm by weight. Stack gases from boilers, process furnaces, FCC regenerators, and internal combustion engines must meet strict limits relative to potentially harmful compounds and particulates. Hydrocarbons, oxides of sulfur (SOX), and oxides of nitrogen (NOX) are examples of stack gas components that are limited to very low amounts. Noise levels must also be controlled. Some of the more common control methods are discussed in the following sections.

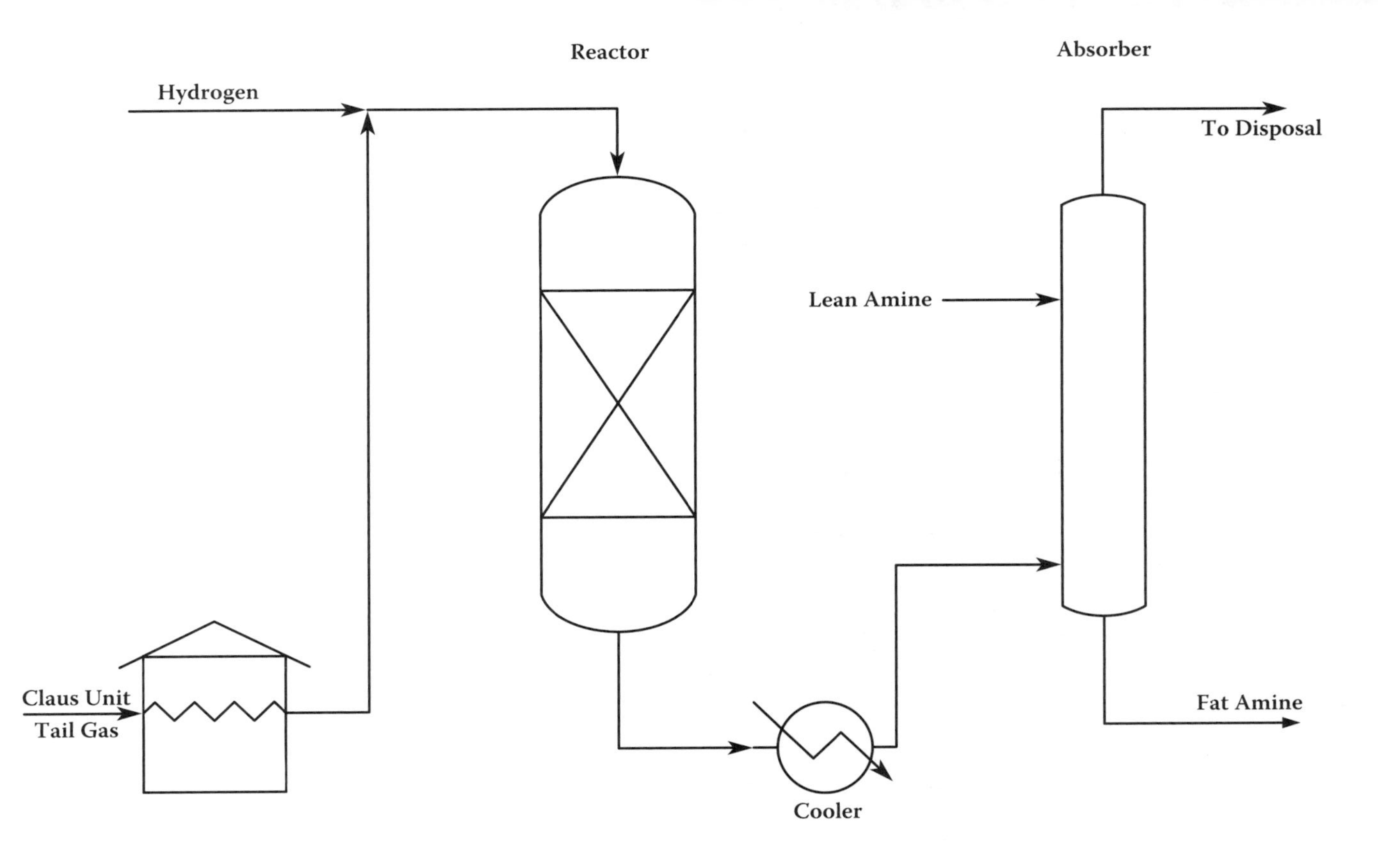

FIGURE 13.9 Shell Claus Off-gas Treatment (SCOT) process.

13.6 WASTEWATER TREATMENT

Typical sources of wastewater in refineries are

1. Runoff surface drainage from leaks, open drains, and spills carried away by rain
2. Crude and product storage tank water drains
3. Desalter water
4. Water drains from atmospheric still reflux drums
5. Water drains from barometric sumps or accumulators on vacuum tower ejectors
6. Water from hydraulic decoking of coke drums
7. Condensed steam from coke drum purging operations
8. Product fractionator reflux drums on units such as catalytic crackers, hydrotreaters, alkylation units, light ends recovery, and others
9. Cooling tower and boiler water blowdown

Surface water is collected in open trenches and sewer systems, and water from process vessels is collected in pipe drain systems. Practically all vessels, tanks, pumps, and low spots in piping are connected to a closed drain system. Any water that may be contaminated with oil is skimmed in large concrete sumps called API separators. The skimmed oil is pumped to slop tanks and then reprocessed. Some water from the API separators is used in the desalters, and the balance is purified by coagulation of impurities in flotation tanks. In this step, a mixture of ferric hydroxide and aluminum hydroxide is used to cause the impurities to form a froth or slurry, which floats to the top of the water. The froth is withdrawn and thickened or settled. The resulting sludge is then incinerated.

Water from the flotation tanks is oxygenated under pressure and then fed to digestion tanks, which may also receive sanitary sewage from the refinery. A controlled flock of bacteria is maintained in the digestion tanks to consume any remaining oils or phenolic compounds. A certain amount of the bacteria is continuously withdrawn from the digestion tanks and incinerated. Water from the digestion tanks may be given a final "polish" in sand filters and reused in the refinery or aerated to increase the oxygen content and subsequently discharged to natural drainage.

Oil-free water blowdown from cooling towers and boilers is neutralized and either evaporated in solar ponds, injected into disposal walls, or diluted with other treated wastewater to lower the dissolved solids content and then aerated and discharged to natural drainage.

Acid sludge from sources such as alkylation units is collected in a separate system and neutralized before going to the API separators.

Sour water drained from vessels such as the atmospheric still reflux drums is stripped in a bubble tower with either gas or steam to eliminate dissolved hydrogen sulfide and other organic sulfur compounds before feeding this water to the API separators. The stripped vapors are processed for sulfur recovery, as described in Section 13.3 and Section 13.4. As a precaution against spills, all storage tanks are normally surrounded by earthen dikes of sufficient size to retain the entire volume of oil that the tank can hold.

13.7 CONTROL OF ATMOSPHERIC POLLUTION

The major sources of potential atmospheric pollution include combustion gases exhausted from boilers, process furnaces, the FCC regenerator, and hydrocarbon vapors vented from process equipment and storage tanks. The sulfur dioxide content of combustion gases is controlled to local regulations by limiting the sulfur content of the fuel. Tail gases from Claus sulfur recovery units are further processed in a tail gas cleaning unit (TGCU), converting low concentrations of hydrogen sulfide and sulfur dioxide into elemental sulfur and thus achieving over 99% recovery of the inlet sulfur. The final tail gas is then incinerated and vented through tall stacks, often 200 ft or more in height, and at sufficient velocity so that resulting ground level sulfur dioxide concentrations are well within safe values.

Hydrocarbon vapors from process equipment and storage tanks are collected in closed piping systems and used for refinery fuel, or in the event of high venting rates during a process upset, the vapors are burned in a flare or burn pit, with special provisions to prevent visible smoke and to ensure complete combustion.

Fluid catalytic crackers are provided with two- to three-stage cyclones to minimize loss of catalyst dust to the atmosphere. In some cases, electrostatic precipitators are employed along with waste heat boilers to eliminate essentially all visible dust from catalytic cracker regenerator flue gases.

Crushing and screening of coke from delayed coking units is generally done in the wet condition to prevent dust losses to the air. The final coke product is often stored in buildings to prevent wind from carrying fine particles into the atmosphere.

13.8 NOISE LEVEL CONTROL

Noise in refineries originates from rotating machinery such as cooler fans, turbines, compressors, engines, and motors. High-velocity flow of fluids through valves, nozzles, and piping also contributes to the general noise level. To control these noises, equipment causing the noise is enclosed or insulated. Proper intake and exhaust silencers are provided on blowers, combustion engines, and turbines. In newer refineries, the land area used is sufficient so that, combined with the above noise control measures, essentially no noise is heard outside the refinery boundaries.

13.9 CASE-STUDY PROBLEM: SATURATED GAS RECOVERY, AMINE, AND SULFUR RECOVERY UNITS

In order to select the size of the auxiliary units, it is necessary to summarize the outputs of all the major process units to find the quantities of the light components, sulfur, and hydrogen available. The hydrogen requirements of the various hydrotreaters and the hydrocracker must be summarized to determine the hydrogen consumed. See Table 13.5 through Table 13.10.

The unsaturated gas plant is included in the cost of the FCC unit, and therefore the saturated gas plant is the one to be sized for capital and operating costs.

The amine gas treating unit removes the acid gases (hydrogen sulfide and carbon dioxide) from the gaseous streams in the saturated and unsaturated gas plants. No

TABLE 13.5
Light Ends Summary for Saturated Gas Plant

	H_2 (lb/h)	C_2 and lighter (lb/h)	C_3 (lb/h)	iC_4 (BPCD)	nC_4 (BPCD)	S in H_2 S (lb/h)
Crude units		372	2,341	159	629	
Coker		16,988	6,218	89	418	1,873
Hydrocracker		340	1,263	307	151	1,391
Reformer and HT	3480	2,050	3,690	360	505	574
Mid dist. HT		1,385	2,500	634	315	1,895
FCC HT		1,193	1,476	374	186	5,037
Isomerization			215	47	128	69
Alkylation					331	
Total	3480	22,328	17,703	1970	2663	10,839

TABLE 13.6
Saturated Gas Processing Unit Material Balance

Feed to unit	BPCD	lb/h	MW	mol/h	MMscfd	gal liq/Mscf gas feed
H_2S		9,377	34	276	2.51	
C_2 and lighter		6,691	23	291	2.65	
C_3	1472	11,105	44	252	2.30	4.7
iC_4	1559	15,873	58	274	2.49	5.0
nC_4	2008	18,674	58	322	2.93	6.5
Total	5039	62,065			13.05	16.2

TABLE 13.7
Saturated Gas Processing Unit and Amine Unit Utility Requirements

	Gas processing unit		Amine unit	
	Per unit	Per day	Per unit	Per day
Fuel, MMBtu	0.014	9,787	0.001	939
Comp. bhp	150	6,657		
Power, MkWh	0.00006	13	0.00001	9
Cooling water circ., Mgal	0.01	21,164	0.0044	2348
Amine, lb			2.5	116

TABLE 13.8
Claus and SCOT Units Material Balance

	scfm	lb/h	S, lb/h	LT/d
Feed				
H_2S	2158	11,615	10,933	
N_2		19,742		
O_2		5,944		
Total		37,757		
Products				
N_2		19,742		
H_2O		6,687		
Sulfur		10,933	10,933	117.1
Total		37,757		

TABLE 13.9
Claus and SCOT Units Utility Requirements

	Claus unit		SCOT unit	
	Per unit	Per day	Per unit	Per day
Fuel, MMBtu			0.001	94
Power, MkWh	0.100	12	0.00001	1
Cooling water circ., Mgal			0.0044	235
Boiler water, Mgal	0.82	2.4		
Steam prod., Mlb	6.5	773.5		
Amine, lb			2.5	11.6

TABLE 13.10
Hydrogen Unit Utility Requirements

	Per lb H_2	Total per day	Total per min
Power, MkWh	0.00015	31.7	
Cooling water circ., Mgal	0.065	13,720	5.66
Fuel, MMBtu	0.045	9,499	
BFW, Mgal	0.001	211	
Feed gas, Mscf	0.26/Mscf	10,920	
Cat. and chem., $	0.008	1,773	

Note: Unit capacity: 42,000,000 scf/d H_2 = 221,636 lb/d H_2.

information is available concerning the concentration of carbon dioxide, and it will not be considered in this problem. Typically, the carbon dioxide content of the gas to the unsaturated gas plant is about 1% by volume, but here the operation of this unit is included in the FCC operating utilities. The gases from the saturated gas plant are treated in this unit.

SCOT unit utilities depend on the amount of CO_2 in the Claus tail gas and the purity of the hydrogen being fed to the reactor. The type of amine solution used also varies the amount of utilities. The utilities for the SCOT unit can be estimated using the quantities given for amine units in Table 13.3.

A hydrogen balance around the hydrogen consuming the producing units (hydrotreaters, hydrocrackers, isomerization, and catalytic reformers) shows a hydrogen consumption of 55,691 Mscf/d, and hydrogen production from the reformer of 20,493 Mscf/d, with a net hydrogen deficiency of 35,197 Mscf/d. Environmental restrictions that limit the aromatic hydrogen content of gasolines may require reformer severity reduction in the future with a corresponding reduction in hydrogen production from this source. Therefore, a hydrogen unit will be built with a capacity of 42,000 Mscf/d of hydrogen, an approximately 20% safety factor. A steam reforming unit producing 95% pure hydrogen and using natural gas as feed will be constructed for the additional hydrogen needed.

NOTES

1. *Hydrocarbon Process. 71*(4), p. 140, 1992.
2. Johansen, T. et al., *Hydrocarbon Process. 61*(8), pp. 119–127, 1992.
3. Strelzoff, S., *Hydrocarbon Process. 53*(12), pp. 79–87, 1974.
4. Lehman, L. and Van Baush, E., *Chem. Eng. Prog. 72*(1), pp. 44–49, 1976.
5. Heck, J., *Oil Gas J. 78*(6), pp. 122–130, 1980.
6. Valdes, A.R., *Hydrocarbon Process. 43*(3), pp. 104–108, 1964.
7. Kohl, A. and Riesenfeld, F., *Gas Purification,* 5th ed., Gulf Publishing Co., 1993.
8. Gas Proc. Suppliers Assoc., *Eng. Data Book*, 10th ed., Tulsa, OK, 1987, Section 21.
9. Goar, B.G., *Oil Gas J. 73*(33), pp. 109–112, 1975.
10. Goar, B.G., *Oil Gas J. 73*(34), pp. 96–103, 1975.
11. Gamson, B.W. and Elkins, R.H., *Chem. Eng. Progr. 49*(4), pp. 203–215, 1953.
12. Thomas, C.L., *Catalytic Processes and Proven Catalysts,* Academic Press, New York, 1970, p. 184.
13. Bourdon, J.C., *Hydrocarbon Process. 76*(4), pp. 57–62, 1997.
14. Goar et al., *Oil Gas J. 83*(39), pp. 39–42, 1985.
15. Johnson, J.E. et al., GCC, February 1999, University of Oklahoma.
16. *Hydrocarbon Process. 71*(4), p. 126, 1992.

14 Refinery Economics and Planning

David Geddes

Modern refinery operations are very complex because of feedstock sources and qualities, multiple sources of feedstock, sophisticated processing technology, and increasingly stringent product specifications. Not only are they complex, but also, the various plants and various blended products are interrelated. This makes many refinery economic decisions difficult, because individual process units or product blends often cannot be evaluated in isolation, but require a means to determine the interrelated effects on the entire refinery. For example, consider the case of a refiner wanting to install new crude oil distillation capacity. If additional crude oil is processed, the refinery may not be able to pay off the investment by selling additional products because downstream bottlenecks limit the amount of crude oil being processed. A sitewide model of the refinery is, therefore, usually required in order to properly determine refinery economics.

Most refining companies have assigned economic decision analysis to refinery planning departments. Typically, these groups are located at the individual refineries, often with support from a planning group located at corporate headquarters.

These planners typically use a mathematical model representation of the entire refinery. This representation includes crude oil assays, process units, product blending, and utilities. To perform this function, refiners have typically used linear programming (LP) models. These models are used instead of detailed nonlinear simulation models because they can solve very quickly, are relatively easy to maintain, and provide sufficient accuracy for economic decision making.

14.1 LINEAR PROGRAMMING OVERVIEW

Linear programming is a mathematical technique that provides a means for finding an optimal solution to a set of linear equations (see Figure 14.1). In the case of the application of this technique to refining, these equations represent the technology and economics of the oil refinery. The linear programming approach is to develop a model, which consists of a set of linear equations, and also an objective function that represents the economics of the problem. The set of linear equations, or constraint equations, defines a feasible region that has an infinite number of solutions. The objective function is used to assign a relative value to each solution, and the LP solution technique is used to find the best, or optimal, solution.

The LP technique has been applied to a wide variety of applications, not only in refining, but also petrochemicals, airlines, and a broad range of other industries:

- Sitewide optimization
- Resource allocation

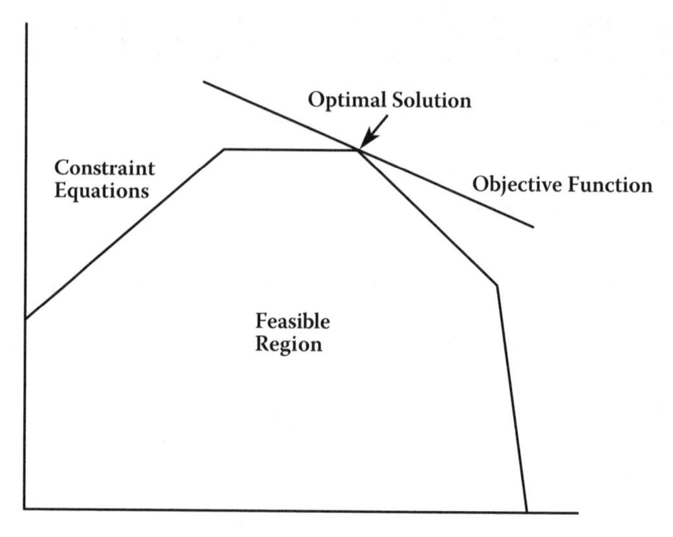

FIGURE 14.1 LP model illustration.

- Blending
- Scheduling
- Process optimization
- Supply and distribution

All of these applications can lead to significant profitability improvements for these industries.

As a simple illustration of this technique, consider an oil refinery feeding two different crude oils. Crude Oil A costs $50.00/bbl and Crude Oil B costs $40.00/bbl. The refinery produces gasoline, diesel oil, and fuel oil. Gasoline sells for $70.00/bbl, diesel oil sells for $65.00/bbl, and fuel oil sells for $60.00/bbl. The yields for each crude oil are shown in the following table:

	Crude Oil A	Crude Oil B
Gasoline	0.1	0.2
Diesel oil	0.3	0.2
Fuel oil	0.6	0.6

The maximum crude oil processing capacity is 100,000 B/d. This example can be formulated as a linear program as shown below:

Gasoline	$0.1\ X_1$	+	$0.2\ X_2$	=	X_3
Diesel oil	$0.3\ X_1$	+	$0.2\ X_2$	=	X_4
Fuel oil	$0.6\ X_1$	+	$0.6\ X_2$	=	X_5
Crude capacity	$1\ X_1$	+	$1\ X_2$	$\leq$	100

In the table X_1 represent the flow rate for Crude Oil A, X_2 represents the flow rate for Crude Oil B, X_3 is the product rate for gasoline, X_4 the product rate for diesel oil, and X_5 the product rate for fuel oil. Note there are more unknowns than equations in this example. Therefore, there are an infinite number of solutions to the problem. To determine the best of the many solutions, an objective function row is required:

$$\text{Objective function} = -50\ X_1 + -40\ X_2 + 70\ X_3 + 65\ X_4 + 60\ X_5$$

The objective function will determine the best or optimum solution within the feasible region defined by the other equations.

14.2 REFINERY LINEAR PROGRAMMING MODELS

Refinery linear programming models consist of a series of many linear equations. Typical models contain in the range of 1000 to 5000 equations. These equations are used to describe the process yields, utility requirements, and blending operations. Because the primary purpose of the refinery LP is to determine refinerywide economics, the LP requires only a level of detail to make proper economic decisions.

There are two different modeling structures typically used to represent the process model. The first of these is simple vector models. The second is delta-base models, which can be used to represent more complex process. Delta-base models are used to model process unit yield changes as a function of feedstock quality or process severity.

An example of a vector yield structure is shown below:

	Yield Vector
Feedstock	
Butylene	–1.0000
Isobutane	–1.2000
Product	
n-Butane	0.1271
Pentane	0.0680
Alkylate	1.5110
Alky bottoms	0.1190
Tar	0.0096
Utilities	
Steam, 1 lb	7.28
Power, kWh	2.45
Cooling water, m gal	2.48
Fuel, MMBtu	0.69

Although many process units can be represented with these simplified yield structures, other process units are more complicated, and the yields for these units are a function of feedstock quality or operating severity. For these cases, the refining industry typically uses delta-base yield models. These models include a base yield vector similar to that shown earlier. The base yield vector represents yields for a defined feedstock quality or operating severity. The models also include one or more delta vectors to adjust the yield for changes in feedstock quality or severity. The following example, using hydrocracker yields for the C5–180 and 180–400 products, illustrates this technique:

(Activity)	Feed (10)	Base Yield (10)	Delta Kw (10)	Delta API (5.0)
Feed	1.0	−1.0		
Hydrogen		−1500		
C5-180		8.1	1.0	3.6
180–400		28.0	−5.5	11.0
Kw	−12.1	10.9	1.2	
API	−22.0	20.0		4.0

In the above example, a base yield vector is defined using a feedstock with a characterization factor, Kw, of 10.9, and an API gravity of 20.0. The actual feed, however, has a characterization factor of 12.1 and an API gravity of 22.0. The first delta vector will adjust the base vector yields as a function of characterization factor, and the second vector adjusts for API. Structure is introduced into the model to perform the following:

1. Assuming the LP optimizer calculating the feed rate is 10, then the feed column and base yield columns will also be multiplied by 10. This is done because the Feed row must sum to zero.
2. The Kw row must also sum to zero; therefore, the Delta Kw column will also be multiplied by 10.
3. Because the API row must sum to zero, the Delta API column will be multiplied by 5.

The volume percentage yields of C5-180 and 180–400 are therefore calculated as follows:

$$\text{C5-180} = (10 \times 8.1 + 10 \times 1.0 + 5 \times 3.6)10 = 10.9$$

$$180\text{–}400 = (10 \times 28.0 - 10 \times 5.5 + 5 \times 11.0)10 = 28.0$$

One advantage of delta-base yield models is that these models can be easily recalibrated to current plant conditions, by periodic updating of the base vector using plant data.

In addition to process units, blending operations are also modeled in detail. An example of this type of LP structure is shown below:

	n-butane	Lt hycrk	Reformate	FCC gaso	Alkylate		Spec
N-butane	−1						
Lt hydrocrackate		−1					
Reformate			−1				
FCC gasoline				−1			
Alkylate					−1		
Premium gasoline blend	1	1	1	1	1		
Research octane number	93.0	82.8	94.0	92.3	97.3	≤	95
Motor octane number	92.0	82.4	84.4	76.8	95.9	≤	90
Reid vapor pressure	138.0	20.3	3.62	1.12	6.37	≤	17.8

The LP structure for blending requires that the calculated property, for example, research octane number, must be less than or equal to 95. The LP will determine the optimum quantity of blend components and will also produce a blend that meets the specifications for research octane, motor octane, and Reid vapor pressure.

The objective function for refinery LP models is often defined as the product revenue, less feedstock costs, less utility costs. In some cases, capital charges and other factors are also incorporated into the objective function.

Seemingly, linear equations would be inadequate to model complex refinery processes. However, accurate linear models can usually be constructed, often by building models that apply to the normal range of operations, but are not intended to cover the full range of possible operations. Modern linear programming software systems also include the capability to model certain nonlinearities that are important for refining applications, such as nonlinear blending, process characterization, and tank mixing.

Model building is normally done using multiple Excel spreadsheets to describe the technology and economics of a refinery. Commercial LP modeling systems then read these spreadsheets, validate the input data, generate an LP matrix, optimize the matrix, and prepare the results in final format. These commercial systems allow the user to quickly and easily run multiple cases. This capability is important because planners often must quickly run a series of cases to fully explore the important parameters of a given situation. For example, in working on crude evaluations, it is not unusual to evaluate 50 or more crude oils by running a series of cases.

14.3 ECONOMICS AND PLANNING APPLICATIONS

Refinery economics and planning applications generally fall into the categories shown below:

1. Crude oil evaluation
2. Production planning
3. Day-to-day operations optimization
4. Product blending and pricing
5. Shutdown planning
6. Multirefinery supply and distribution
7. Yearly budgeting
8. Investment studies
9. Environmental studies
10. Technology evaluation

In a typical refinery, somewhere in the range of 100 to 200 LP cases are run each week. Most of these cases relate to crude oil evaluation, production planning, and day-to-day operations planning. A description of each of these applications is presented in the following paragraphs.

14.3.1 Crude Oil Evaluation

Crude oil selection is considered one of the most, if not the most, important business processes for an operating refinery with the opportunity to purchase a variety of crude oils. The cost of crude oil is typically 70 to 80% of the total operating costs for a refiner; therefore, it is vital that the most profitable crude oils for that refinery are purchased. Although the cost of crude oil is typically the same for various refineries in a particular location, the value of crude oil varies for each refinery. The value of crude oil is a function of many factors. These factors include the quality of the crude oil as described by its crude oil assay, the amount of the crude oil to be processed, refinery product demand and price, the refinery configuration, and the other crude oils being processed.

Because many refineries are located in coastal locations with access to crude oil tankers, refiners often have the ability to purchase many different crude oils. It is not unusual for refiners to evaluate 50 to 100 crude oils and select the most profitable ones for their refinery.

There are several methods used to evaluate crude oils. One common technique is to determine the difference in value of increments of crude oils compared to a reference crude oil. This is referred to as a breakeven analysis. These incremental crude oil volumes are evaluated in addition to a fixed crude slate that often consists of a group of crude oils that are normally fed to the refinery. Perhaps this fixed crude slate represents crude oils purchased on a long-term contract, whereas the crude oils being evaluated are available on the spot market.

Regardless of the method used, refiners use the LP to determine the break-even value of crude oils. The break-even value is the value of the crude that produces the same refinery profitability as for a reference crude. An example of this is shown in the table below:

	Break-even Calculation		
Amount	**CASE 1**	**CASE 2**	**CASE 3**
Crude A	40,000B/D	40,000B/D	40,000B/D
Crude B	40,000	40,000	40,000
Crude C	20,000	20,000	20,000
Reference crude ($40/B)	10,000		
New crude 1 ($40/B)		10,000	
New crude 2 ($40/B)			10,000
Gross margin ($1,000/D)	200	195	190
Break-even value	Base	40.00 + (195 – 200) / 10	40 + (190 – 200) / 10
$/B	= $40.00	= 39.5	= 39.00

This break-even analysis indicates we should pay $0.50/B less for the New Crude 1 than for the reference crude in order to produce the same level of profitability. (The gross margin shown above is normally calculated by the LP as product revenue minus feedstock cost minus utility costs.) Break-even values can be communicated to the crude oil supply department in the following format:

Crude Oil Break-even Values

Reference crude	$40.00 (Base)
New crude 1	$39.50
New crude 2	$39.00

The oil supply department can then prepare a table as follows:

	Market Price	Break even	Margin Gain
Reference crude	$38.00	$40.00	$2.00
New crude 1	39.00	39.50	0.50
New crude 2	40.00	39.00	−1.00

We know the market prices for crude oil will vary on a daily basis, but using the break-even values, the oil trader will know the relative crude oil values and can therefore make purchase decisions on this basis. Other economic factors include price basis risk, crude oil forward prices, feedstock quality variability, logistical reliability, working capital cost, and expected ship demurrage costs.

Other crude oil evaluation methods are also used, such as crude oil replacement economics, crude oil absolute economics, and cargo replacement economics, and are valid for a given situation. Because the crude oil value can vary more than $1–2/bbl, depending on the evaluation method, it is important to use an evaluation technique that closely resembles the method by which crude oil will be processed. Many planners also calculate break-even values for each crude at several crude oil volumes. Because the break-even value of crude oil decreases with increasing amount of crude processed, this is a very useful practice in order to gain a better understanding of the crude oil value and provide guidance in decisions regarding the amounts of incremental crude oils to process.

14.3.2 Monthly Production Planning

Production planning is a routine business process commonly performed on a biweekly or monthly basis. This activity is a coordination activity between the crude oil acquisition, product sales, and refinery operations departments. The refinery planner will determine the crude oils that are scheduled for delivery during the specified period. The sales group will estimate sales of each product for the same period. Finally, the refinery operations group will determine the operating rates for the various refinery units, and determine, for example, if any of the process units are scheduled for shutdown during the period.

The refinery planner will input all of this information into the refinery LP and determine if all of the products can be produced with the crude oil scheduled for delivery. The LP will also provide refinery operations with guidance for how the various refinery units should operate to maximize profitability. This includes process unit operating rates, crude distillation cut points, and reactor operating severities, and so on.

14.3.3 Day-to-Day Operations Optimization

This business process provides economic guidance to refinery operations on a day-to-day basis. Although the production plan is used as the starting point, unexpected operating changes will create the need to modify the plan. For example, if there is an unexpected plant shutdown, the refinery operating department needs guidance on how to make adjustments in the most profitable manner.

14.3.4 Product Blending and Pricing

Although the production plan includes product blending, many refiners use separate blending models, for example, gasoline blending models used for day-to-day product blending, as well as blend stock purchases and sales. This is necessary, in part, because product blending is a batch operation, which is different from the continuous operations of the refinery process units. In addition, some refiners purchase blend components on a spot basis. Therefore, for many refiners, product blending is considered a separate business process.

In general, product pricing used for refinery economics is the expected market price. However, there are special situations where refiners need to use the LP to determine the break-even price for a particular product. For example, if a refinery can sell an increment of jet fuel at a price $5.00 below the market price, should this be done? The break-even example for this case is shown below:

	Product Break-even Analysis	
	BASE CASE	CASE 1
Regular gasoline ($60/bbl)	50,000	50,000
Premium gasoline ($65/bbl)	50,000	50,000
Jet fuel ($50/bbl)	20,000	20,000
Alternate jet fuel sale ($45/bbl)	0	5,000
Gross margin ($1,000/D)	100.00	98.00
Break-even value	45.00	45 + (100 – 98) / 5 = 45.40

This analysis has determined the break-even value is $45.40/bbl. The refinery requires at least this price to make a profitable sale.

The blending LP is also used to aid in decisions related to the purchase of blend components, for initial product blend recipes to be supplied to the on-line blend systems, to determine reformer severities, and for blend component inventory projections. The blending LP can be a single blend LP or a multiperiod LP. Multiperiod blending is useful in planning blend stock inventory strategies over a 1- or 2-week time horizon.

14.3.5 Shutdown Planning

Shutdown planning can be complicated because of the need to take advantage of inventory in order to minimize the impact of the shutdown on the overall refinery production rates. For example, consider the case of a shutdown in a fluid catalytic cracking unit. In planning for this event, the planner needs to utilize tankage to store FCC feedstock during the shutdown. Multiperiod models are also used to evaluate gasoline blend stocks that need to be inventoried prior to the shutdown in order to minimize the dropoff in gasoline production during the shutdown period. Shutdown planning, therefore, requires the ability to model key inventories in the refinery in addition to the process units. The system also includes multiple time periods in order to model periods with different process flow rates. Process unit flow rates will vary for the periods before, during, and after the shutdown. This is illustrated in Figure 14.2.

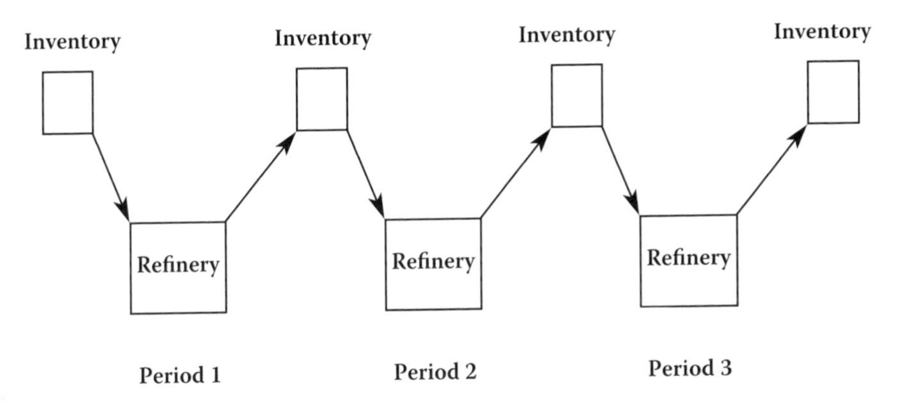

FIGURE 14.2 Multiperiod model.

For complex refinery shutdowns, a multiperiod refinery model is invaluable in analyzing the various possible options:

- What crude oils to process, and when
- How to use the available tankage
- How to sequence unit shutdowns
- How product supplies will be affected

14.3.6 Multirefinery Supply and Distribution

Many cases, refining companies own more than one refinery. For many refineries that do not share limited feedstock sources, supply the same markets, or transfer materials between refineries, independently planning for each refinery is sufficient. However, if there is interaction between refineries, then these interactions need to be considered. Examples of refinery interactions are listed below:

- A limited supply of crude oil needs to be allocated between refineries.
- Intermediate materials or blend stocks can be transferred between refineries, for example, to more fully utilize process unit capacity.
- Markets can economically be supplied by more than one refinery, and the most profitable distribution of products needs to be determined.
- Inventory trade-offs can be made between refineries and wholesale distribution terminals.

These options are illustrated in Figure 14.3.

In Figure 14.3, crude oil AA is produced locally and is only available to Refinery A. Crude oil BB is a potential feedstock to all three refineries. Crude oil CC is only available to Refineries B and C because of transportation limitations. A multirefinery LP can be used to determine the proper allocation of the three crude oils to the refineries.

Similarly, a multirefinery LP can be used to determine which refinery should supply the two markets shown in Figure 14.3. Intermediate material from Refinery C can be

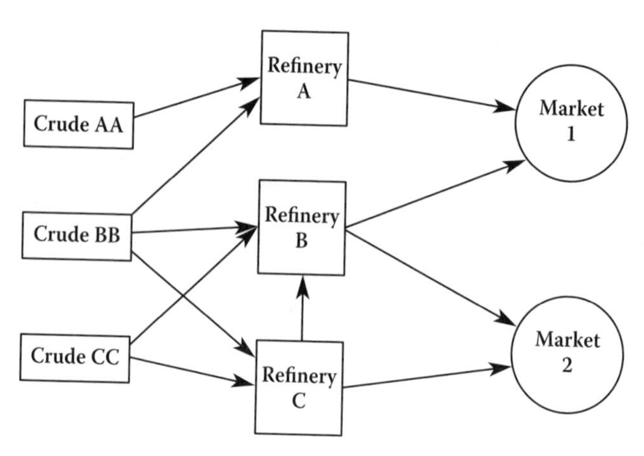

FIGURE 14.3 Multirefinery network.

transported to Refinery B to take advantage of spare process unit capacity at Refinery B. The multirefinery LP can be used to determine when this option should be used. This decision is likely related to the final product demand for each refinery, which is driven by the distribution of products to the two markets.

Another example of multiplant optimization is a refinery olefins plant complex. In this situation, the refinery supplies most of the naphtha feedstock for the olefins complex. Aromatic by-products from the olefins plant are valuable blend components for the refinery. Capturing these synergies can greatly increase the overall profitability of the two facilities compared to stand-alone operations. This is the reason many companies locate olefins plants next to a refinery. Naphtha-based olefins plants often perform feedstock evaluation using the LP to evaluate the additional naphtha feedstock purchases.

14.3.7 Yearly Budgeting

Oil refiners typically develop yearly budgets in order to forecast economic performance for the upcoming fiscal year. The refinery LP is typically used to prepare economic projections for the year. These projections are used for financial forecast. The forecast can be used to determine working capital requirements and to determine capital expenditure financing requirements.

14.3.8 Investment Studies

When investments in new refineries or refinery expansions are considered, the first step in the project is to conduct a feasibility study to determine the economics of the project. The feasibility study will typically include the following steps:

- Market study to forecast product demand
- Configuration study to evaluate various processing options, for example, resid processing options such as delayed coking and visbreaking; or vacuum gas oil processing options, such as fluid catalytic cracking versus hydrocracking

- Capital cost estimate for selected configuration
- Detailed economic forecast

LP models are used for the configuration study. For refinery expansion studies, the refiner often has determined a certain process unit needs to be expanded. The LP model is used to determine the impact of the expanded new unit on the overall refinery operations.

14.3.9 Environmental Studies

The refinery LP is also used for environmental studies. Because the refinery LP is a sitewide model that includes process and blending interactions, it is often the preferred tool for these kinds of studies. There are a number of studies that can be done, for example:

- EPA-mandated reformulated gasoline
- Processing requirements to reduce product sulfur content, for example, for gasoline and diesel oil
- Processing adjustments to reduce CO_2 emissions related to global warming

14.3.10 Technology Evaluation

Because of the complicated nature of refinery operations, the refinery LP is often required for technology evaluations. One example is the replacement of catalyst in fluid cracking units, hydrotreaters, or catalytic reformers. Because the new catalysts will likely change the reactor yield slate, the impact of this change is best determined with the LP in order to account for the interactions between plants and between blending operations.

PROBLEMS

1. An oil refinery feeds two different crude oils. Crude Oil 1 costs \$45.00/bbl and Crude Oil 2 costs \$41.00/bbl. The refinery produces gasoline, diesel oil, and fuel oil. Gasoline sells for \$65.00/bbl, diesel oil sells for \$60.00/bbl, and fuel oil sells for \$50.00/bbl. The yields for each crude oil are shown in the following table:

	Crude Oil 1	Crude Oil 2
Gasoline	0.15	0.25
Diesel oil	0.25	0.25
Fuel oil	0.6	0.50

The maximum sales volume for gasoline is 30,000BPD, and the maximum sales volume for fuel oil is 100,000BPD. Formulate this as a linear program application.

2. Develop a vector-based alkylation model feeding propylene feedstock using the yield data shown in the alkylation and polymerization chapter.
3. Develop a delta-base process model for a delayed coking unit using the data in coking and thermal processing chapter. It is only necessary to describe the coke and gasoline yields.
4. Calculate the crude oil break-even value for Crude Oil D using the information shown below:

	CASE 1 Amount	CASE 2
Crude A	50,000BPD	40,000BPD
Crude B	40,000	40,000
Crude C	20,000	20,000
Reference crude ($50/bbl)	10,000	
Crude oil D ($40/bbl)		10,000
Gross margin ($1,000/D)	165	150

5. The sales department has an opportunity to sell an additional 10,000 BPD gasoline at $55.00/bbl. Using the data from the refinery LP shown below, determine if this sell is profitable.

	BASE CASE	CASE 1
Regular gasoline ($60/bbl)	50,000	50,000
Premium gasoline ($65/bbl)	50,000	50,000
Jet fuel ($50/bbl)	20,000	20,000
Alternate regular gasoline ($55/bbl)	0	10,000
Gross margin ($1,000/D)	100.00	102.00

NOTES

1. Luenberger, D.G., *Linear and Nonlinear Programming,* 2nd ed., 2005.
2. Tucker, M.A., *LP Modeling – Past, Present and Future,* 2001 NPRA Computer Conference.
3. Personal communications, Rangnow & Associates, LLC, 2006.
4. Kutz, T., *Early Planning to Evaluate Alternatives for a Complex Refinery Shutdown,* 2004 NPRA Annual Meeting.

ADDITIONAL READING

5. Trierwiler, L.D., *Advances in Crude Oil LP Modeling,* 2001 NPRA Computer Conference.
6. Rangnow, D. and Davis, J., *Integrating Market Opportunities with Refining Capabilities to Maximize Asset Value,* 2001 NPRA Computer Conference.
7. Gallagher, C.P. and Geddes, D.D., Modern Planning and Scheduling in the Refining and Petrochemical Industries, *Proceedings of the International Conference on Petroleum Refining and Petrochemical Processing,* 1991, Vol. 1, pp. 28–34.

15 Lubricating Oil Blending Stocks

The large number of natural lubricating and specialty oils sold today are produced by blending a small number of lubricating oil base stocks and additives. The lube oil base stocks are prepared from selected crude oils by distillation and special processing to meet the desired qualifications. The additives are chemicals used to give the base stocks desirable characteristics that they lack or to enhance and improve existing properties. The properties considered important are

1. Viscosity
2. Viscosity change with temperature (viscosity index, or VI)
3. Pour point
4. Oxidation resistance
5. Flash point
6. Boiling temperature
7. Acidity (neutralization number)

Viscosity. The viscosity of a fluid is a measure of its internal resistance to flow. The higher the viscosity, the thicker the oil and the thicker the film of the oil that clings to a surface. Depending upon the service for which it is used, the oil needs to be very thin and free-flowing or thick with a high resistance to flow. From a given crude oil, the higher the boiling-point range of a fraction, the greater the viscosity of the fraction. Therefore, the viscosity of a blending stock can be selected by the distillation boiling range of the cut.

Viscosity index. The rate of change of viscosity with temperature is expressed by the viscosity index of the oil. The higher the VI, the smaller its change in viscosity for a given change in temperature. The VIs of natural oils range from negative values for oils from naphthenic crudes to about 100 for paraffinic crudes. Specially processed oils and chemical additives can have VIs of 130 and higher. Additives, such as polyisobutylenes and polymethacrylic acid ester, are frequently mixed with lube blending stocks to improve the viscosity–temperature properties of the finished oils. Motor oils must be thin enough at low temperature to permit easy starting and viscous enough at engine operating temperatures [180 to 250°F (80 to 120°C)] to reduce friction and wear by providing a continuous liquid film between metal surfaces.

Pour point. The lowest temperature at which an oil will flow under standardized test conditions is reported in 5°F (3°C) increments as the pour point of the oil. For motor oils, a low pour point is very important to obtain ease of starting and proper start-up lubrication on cold days.

There are two types of pour points: a viscosity pour point and a wax pour point. The viscosity pour point is approached gradually as the temperature is lowered and the viscosity of the oil increases until it will not flow under the standardized test conditions. The wax pour point occurs abruptly as the paraffin wax crystals precipitate from solution and the oil solidifies. Additives that affect wax crystal properties can be used to lower the pour point of a paraffin base oil.

A related test is the cloud point, which reports the temperature at which wax or other solid materials begin to separate from solution. For paraffinic oils, this is the starting temperature of crystallization of paraffin waxes.

Oxidation resistance. The high temperatures encountered in internal combustion engine operation promote the rapid oxidation of motor oils. This is especially true for the oil coming in contact with the piston heads, where temperatures can range from 500 to 750°F (260 to 400°C). Oxidation causes the formation of coke and varnishlike asphaltic materials from paraffin base oils and sludge from naphthenic base oils [1]. Antioxidant additives, such as phenolic compounds and zinc dithiophosphates, are added to the oil blends to suppress oxidation and its effects.

Flash point. The flash point of an oil has little significance with respect to engine performance and serves mainly to give an indication of hydrocarbon emissions or of the source of the oils in the blend; for example, whether it is a blend of high- and low-viscosity oils to give an intermediate viscosity or is comprised of a blend of center cut oils. Low flash points indicate greater hydrocarbon emissions during use.

Boiling temperature. The higher the boiling temperature range of a fraction, the higher the molecular weights of the components and, for a given crude oil, the greater viscosity. The boiling ranges and viscosities of the fractions are the major factors in selecting the cut points for the lube oil blending stocks on the vacuum distillation unit.

Acidity. The corrosion of bearing metals is largely due to acid attack on the oxides of the bearing metals [2]. These organic acids are formed by the oxidation of lube oil hydrocarbons under engine operating conditions and by acids produced as by-products of the combustion process that are introduced into the crankcase by piston blow-by. Motor oils contain buffering materials to neutralize these corrosive acids. Usually, the dispersant and detergent additives are formulated to include alkaline materials, which serve to neutralize the acid contaminants. Lube oil blending stocks from paraffinic crude oils have excellent thermal and oxidation stability and exhibit lower acidities than do oils from naphthenic crude oils. The neutralization number is used as the measure of the organic acidity of an oil; the higher the number, the greater the acidity.

15.1 LUBE OIL PROCESSING

The first step in the processing of lubricating oils is the separation on the crude oil distillation units of the individual fractions according to viscosity and boiling-range specifications. The heavier lube oil raw stocks are included in the vacuum fractionating tower bottoms with the asphaltenes, resins, and other undesirable materials.

The raw lube oil fractions from most crude oils contain components that have undesirable characteristics for finished lubricating oils. These must be removed or reconstituted by processes such as liquid–liquid extraction, crystallization, selective hydrocracking, or hydrogenation. The undesirable characteristics include high pour points, large viscosity changes with temperature (low VI), poor oxygen stability, poor color, high cloud points, high organic acidity, and high carbon- and sludge-forming tendencies.

The processes used to change these characteristics are

1. Solvent deasphalting to reduce carbon- and sludge-forming tendencies
2. Solvent extraction and hydrocracking to improve viscosity index
3. Solvent dewaxing and selective hydrocracking to lower cloud and pour points
4. Hydrotreating and clay treating to improve color and oxygen stability
5. Hydrotreating and clay treating to lower organic acidity

Although the main effects of the processes are as described above, there are also secondary effects. For example, although the main result of solvent dewaxing is the lowering of the cloud and pour points of the oil, solvent dewaxing also slightly reduces the VI of the oil.

For economic reasons, as well as process ones, the process sequence is usually in the order of deasphalting, solvent extraction, dewaxing, and finishing. However, dewaxing and finishing processes are frequently reversed [3]. In general, the processes increase in cost and complexity in this same order.

15.2 PROPANE DEASPHALTING

The lighter distillate feedstocks for producing lubricating oil base stocks can be sent directly to the solvent extraction units, but the atmospheric and vacuum still bottoms require deasphalting to remove the asphaltenes and resins before undergoing solvent extraction. In some cases, the highest boiling distillate stream may also contain sufficient asphaltenes and resins to justify deasphalting.

Propane usually is used as the solvent in deasphalting but it may also be used with ethane or butane in order to obtain the desired solvent properties. Propane has unusual solvent properties in that from 100 to 140°F (40 to 60°C), paraffins are very soluble in propane, but the solubility decreases with an increase in temperature until at the critical temperature of propane [206°F (96.8°C)], all hydrocarbons become insoluble. In the range of 100 to 206°F (40 to 96.8°C), the high molecular-weight asphaltenes and resins are largely insoluble in propane. Separation by distillation is generally by molecular weight of the components and solvent extraction is by type of molecule. Propane deasphalting falls in between these categories because separation is a function of both molecular weight and type of molecular structure.

The feedstock is contacted with four to eight volumes of liquid propane at the desired operating temperature. The extract phase contains from 15 to 20% by weight of oil with the remainder solvent. The heavier the feedstock, the higher the ratio of propane to oil required.

The raffinate phase contains from 30 to 50% propane by volume and is not a true solution but an emulsion of precipitated asphaltic material in propane [4].

As in most other refinery processes, the basic extraction section of the process is relatively simple, consisting of a cylindrical tower with angle iron baffles arranged in staggered horizontal rows [5] or containing perforated baffles [4] using a countercurrent flow of oil and solvent. Some units use the rotating disc contactor (RDC) for this purpose.

A typical propane deasphalting unit (Figure 15.1) injects propane into the bottom of the treater tower, and the vacuum tower bottoms feed enters near the top of the tower. As the propane rises through the tower, it dissolves the oil from the residuum and carries it out of the top of the tower. Between the residuum feed point and the top of the tower, heating coils increase the temperature of the propane–oil extract phase, thus reducing the solubility of the oil in the propane. This causes some of the oil to be expelled from the extract phase, creating a reflux stream. The reflux flows down the tower and increases the sharpness of separation between the oil portion of the residuum and the asphaltene and resin portion. The asphaltene and resin phase leaving the bottom of the tower is the raffinate, and the propane–oil mixture leaving the top is the extract.

The solvent recovery system of the propane deasphalting process, as in all solvent extraction processes, is much more complicated and costly to operate than the treating section. Two-stage flash systems or supercritical techniques are used to recover the propane from the raffinate and extract phases. In the flash system, the first stages are operated at pressures high enough to condense the propane vapors with cooling water as the heat exchange medium. In the high-pressure raffinate flash tower, foaming and asphalt entrainment can be a major problem. To minimize this, the flash tower is operated at about 550°F (290°C) to keep the asphalt viscosity at a reasonably low level.

The second stage strips the remaining propane from the raffinate and extract at near-atmospheric pressure. This propane is compressed and condensed before being returned to the propane accumulator drum.

The propane deasphalting tower is operated at a pressure sufficiently high to maintain the solvent in the liquid phase. This is usually about 500 psig (3448 kPa).

The asphalt recovered from the raffinate can be blended with other asphalts into heavy fuels or used as a feed to the coking unit.

The heavy oil product from vacuum residuum is called bright stock. It is a high-viscosity blending stock that, after further processing, is used in the formulation of heavy-duty lubricants for truck, automobile, and aircraft services.

15.3 VISCOSITY INDEX IMPROVEMENT AND SOLVENT EXTRACTION

There are three solvents used for the extraction of aromatics from lube oil feedstocks, and the solvent recovery portions of the systems are different for each. The solvents are furfural, phenol, and N-methyl-2-pyrrolidone (NMP). The purpose of solvent extraction is to improve the viscosity index, oxidation resistance, and color of the lube oil base stock and to reduce the carbon- and sludge-forming tendencies of the lubricants by separating the aromatic portion from the naphthenic and paraffinic portions of the feedstock.

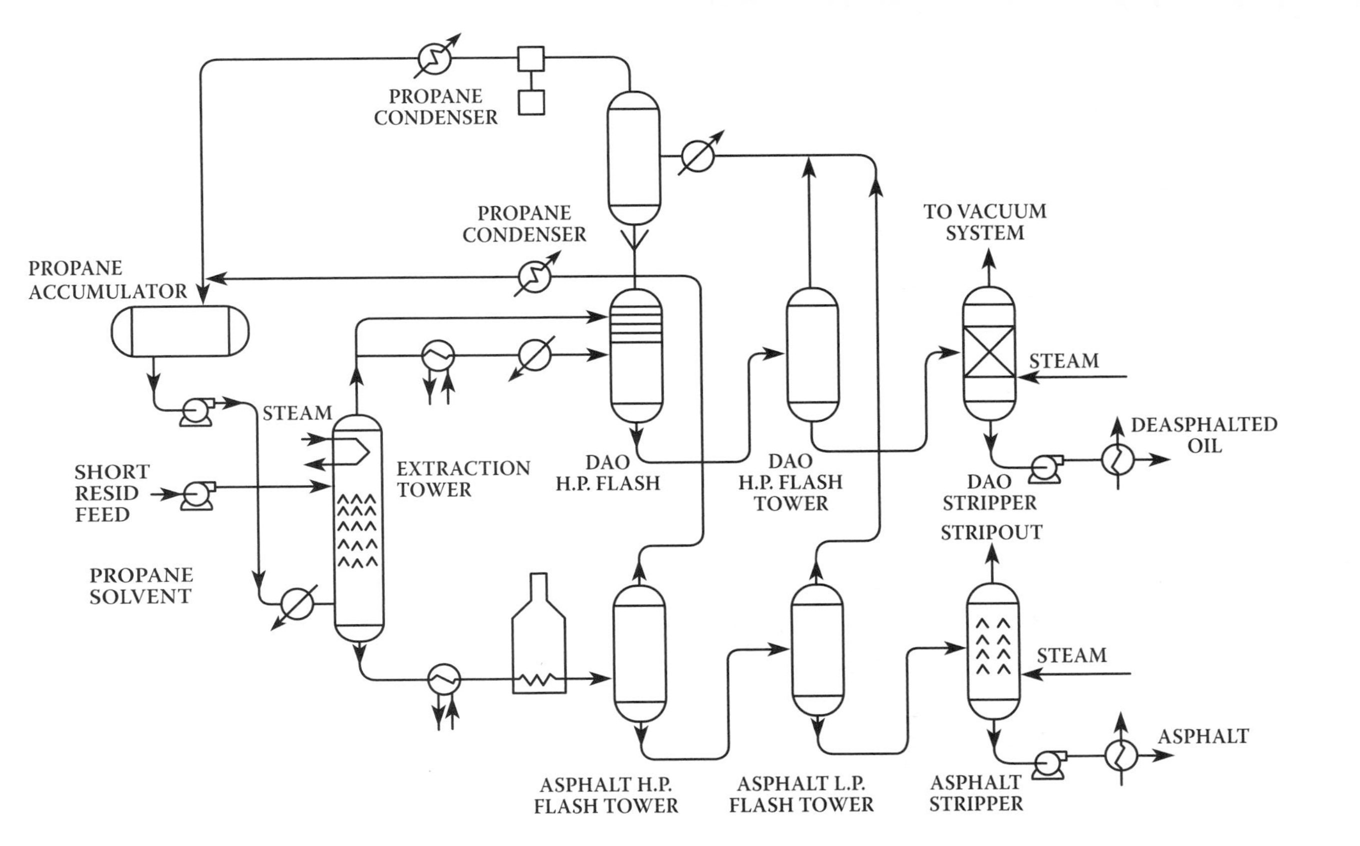

FIGURE 15.1 Typical propane deasphalter.

15.3.1 Furfural Extraction

The process flow through the furfural extraction unit is similar to that of the propane deasphalting unit except for the solvent recovery section, which is more complex. The oil feedstock is introduced into a continuous countercurrent extractor at a temperature that is a function of the viscosity of the feed; the greater the viscosity, the higher the temperature used. The extraction unit is usually a raschig ring-packed tower or a rotating disc contactor with a temperature gradient from top to bottom of 60 to 90°F (30 to 50°C). The temperature at the top of the tower is a function of the miscibility temperature of the furfural and oil. It is usually in the range of 220 to 300°F (105 to 150°C).

The oil phase is the continuous phase, and the furfural-dispersed phase passes downward through the oil. Extract is recycled at a ratio of 0.5:1 to improve the extraction efficiency.

Furfural-to-oil ratios range from 2:1 for light stocks to 4.5:1 for heavy stocks. Solvent losses are normally less than 0.02 wt% of raffinate and extract flow rates. Furfural is easily oxidized, and inert gas blankets are maintained on the system to reduce oxidation and polymerization. Sometimes, deaeration towers are used to remove dissolved oxygen from the feed. Furfural is subject to thermal decomposition, and skin temperatures of heat exchange equipment used to transfer heat to furfural-containing streams must be carefully controlled to prevent polymerization of the furfural and fouling of the heat exchange surfaces.

The furfural is removed from the raffinate and extract streams by flashing and steam-stripping. Furfural forms an azeotrope with water and this results in a unique furfural recovery system. Furfural is purified in the furfural tower by distilling overhead the water–furfural azeotrope vapor, which upon condensing separates into water-rich and furfural-rich layers. The furfural-rich layer is recycled to the furfural tower as reflux, and the furfural in the water-rich layer is separated from the water by steam-stripping. The overhead vapors, consisting of the azeotrope, are condensed and returned to the furfural–water separator. The bottoms product from the furfural tower is the pure furfural stream, which is sent to the furfural solvent drum.

The most important operating variables for the furfural extraction unit are the furfural-to-oil (F/O) ratio, extraction temperature, and extract recycle ratio. The F/O ratio has the greatest effect on the quality and yield of the raffinate, whereas the temperature is selected as a function of the viscosity of the oil and the miscibility temperature. The extract recycle ratio determines to some extent the rejection point for the oil and the sharpness of separation between the aromatics and naphthenes and paraffins.

15.3.2 Phenol Extraction

The process flow for the phenol extraction unit is somewhat similar to that of the furfural extraction unit but differs markedly in the solvent recovery section, because phenol is easier to recover than is furfural.

The distillate or deasphalted oil feed is introduced near the bottom of the extraction tower and phenol enters near the top. The oil-rich phase rises through the tower, and the phenol-rich phase descends the tower. Trays or packing are used to provide intimate contact between the two phases. Some of the newer phenol extraction units use either rotating disc contactors or centrifugal extractors to contact the two phases. Both the RDC and the centrifugal extractors offer the advantage that much smaller volumes are needed for the separations. As all lube oil refineries operate on a blocked operation basis (that is, charging one feedstock at a time), the lower inventories make it possible to change from one feedstock to another with a minimum loss of time and small loss of off-specification product.

The extraction tower and RDC are operated with a temperature gradient that improves separation by creating an internal reflux. The phenol is introduced into the tower at a higher temperature than the oil. The temperature of the phenol-rich phase decreases as it proceeds down the column and the solubility of the oil in this phase decreases. The oil coming out of the phenol-rich phase reverses direction and rises to the top as reflux. The tower top temperature is kept below the miscible temperature of the mixture, and the tower bottom temperature is usually maintained at about 20°F (10°C) lower than the top.

Phenol will dissolve some of the paraffins and naphthenes as well as the aromatics. Water acts as an antisolvent to increase the selectivity of the phenol, and typically from 3 to 8 vol% water is added to the phenol. A decrease in reaction temperature has a similar effect. Raffinate yield is increased by increasing water content or decreasing temperature.

The important extraction tower operating variables are

Phenol-to-oil ratio (treat rate)
Extraction temperature
Percentage of water in phenol

Treat rates vary from 1:1 to 2.5:1, depending upon the quality and viscosity of the feed and the quality of the product desired. Increasing the treat rate for a given stock improves the VI of the product and decreases the yield. Phenol is recovered from the extract and raffinate streams by distillation and gas or steam stripping. Phenol losses average from 0.02 to 0.04% of circulation rate.

15.3.3 NMP Extraction

The NMP extraction process uses N-methyl-pyrrolidone as the solvent to remove the condensed ring aromatics and polar components from the lubricating oil distillates and bright stocks. This process was developed as a replacement for phenol extraction because of the safety, health, and environmental problems associated with the use of phenol. Several differences between the characteristics of NMP and phenol make it necessary to modify the phenol plant design. These differences include a 40°F (22°C) higher boiling point for NMP, a 115°F (64°C) lower melting point, complete miscibility of NMP with water, no azeotrope formation of NMP with water, and a 69% lower viscosity than phenol at 122°F (50°C). A simplified process flow diagram in which the solvent is recovered by flashing and stripping with steam at pressures above atmospheric is shown in Figure 15.2 [1].

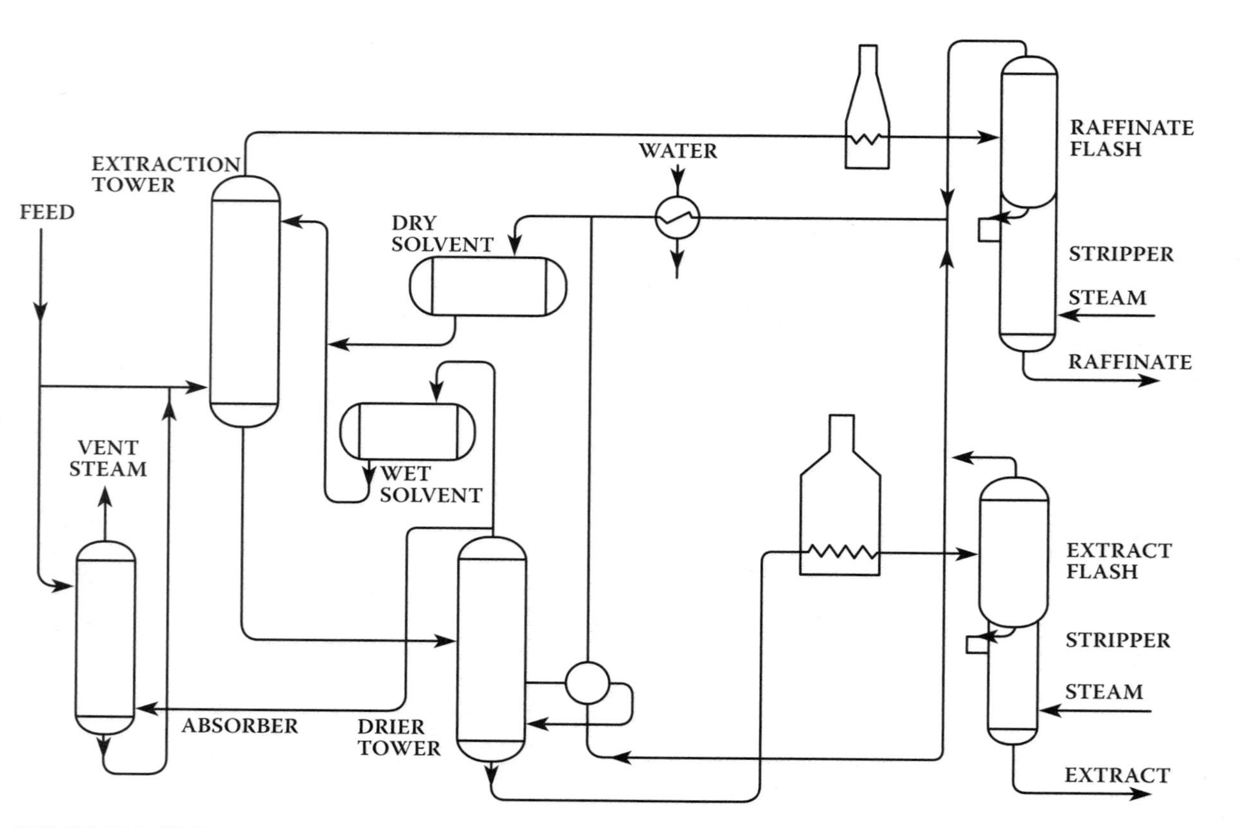

FIGURE 15.2 NMP extraction unit using steam-stripping for solvent recovery.

TABLE 15.1
Viscosity Index of Hydrocarbons

Type of hydrocarbon	Viscosity index
n-Paraffins	175
i-Paraffins	155
Mononaphthenes	142
Dinaphthenes+	70
Aromatics	50

Source: Note 7.

A portion of the distillate or deasphalted oil feed is used as the lean oil in an absorption tower to remove the NMP from the exiting stripping steam. The rich oil from the absorption tower is combined with the remainder of the feed, which is heated to the desired temperature before being introduced near the bottom of the treater tower. The hot solvent enters near the top of the tower. Specially designed cascade weir trays are used to mix and remix the NMP-rich and oil-rich phases as they pass through the tower.

The solvent is stripped from the raffinate and extract by distillation and steam-stripping. Recovery of NMP is better than that for phenol, and NMP losses are only 25 to 50% those of phenol.

The lower viscosity of NMP gives greater throughput for a given size tower. This results in lower construction costs for a grassroots plant and up to a 25% increase in throughput for converted phenol plants.

Solvent-to-oil ratios for a given feedstock and quality product are the same for NMP and phenol extraction, but raffinate oil yields average 3 to 5% higher for the NMP extraction.

15.4 VISCOSITY INDEX IMPROVEMENT AND HYDROCRACKING

Components of lubricating oil fractions that have high viscosity indices are the mononaphthalenes and isoparaffins (see Table 15.1) [7,11]. Hydrocracking of vacuum gas oils increases the paraffin concentration and the viscosity index of a lube oil feedstock [8] and produces increasing quantities of the mononaphthalenes and isoparaffins as hydrocracking severity increases. Hydrogenation of polyaromatic aromatic compounds to polynaphthenic ring compounds, breaking of polynaphthenic rings, and isomerization of n-paraffins are promoted by high conversion, low space velocity, and low reaction temperature [11].

15.5 DEWAXING

All lube stocks, except those from a relatively few highly naphthenic crude oils, must be dewaxed or they will not flow properly at ambient temperatures. Dewaxing is one of the most important and most difficult processes in lubricating oil manufacturing. There are two types of processes in use today. One uses refrigeration to crystallize the wax and solvent to dilute the oil portion sufficiently to permit rapid filtration to separate the wax from the oil. The other uses a selective hydrocracking process to crack the wax molecules to light hydrocarbons.

15.5.1 Solvent Dewaxing

There are two principal solvents used in the United States in solvent dewaxing processes: propane and ketones. Dichloroethane-methylene is also used in some other countries. The ketone processes use either a mixture of methyl ethyl ketone (MEK) with methyl isobutyl ketone (MIBK) or MEK with toluene. The solvents act as a diluent for the high molecular-weight oil fractions to reduce the viscosity of the mixture and provide sufficient liquid volume to permit pumping and filtering. The process operations for both solvent processes are similar but differ in the equipment used in the chilling and solvent recovery portions of the process. About 85% of the dewaxing installations use ketones as the solvent, and the other 15% use propane. The comparative advantages and disadvantages of the processes are as follows [4]:

15.5.2 Propane

1. Readily available, less expensive, and easier to recover.
2. Direct chilling can be accomplished by vaporization of the solvent, thus reducing the capital and maintenance costs of scraped-surface chillers.
3. High filtration rates can be obtained because of its low viscosity at very low temperatures.
4. Rejects asphaltenes and resins in the feed.
5. Large differences between filtration temperatures and pour point of finished oils [25 to 45°F (5 to 25°C)].
6. Requires use of a dewaxing aid.

15.5.3 Ketones

1. Small difference between filtration temperature and pour point of dewaxed oil [9 to 18°F (5 to 10°C)].
   a. Lower pour point capability
   b. Greater recovery of heat by heat exchange
   c. Lower refrigeration requirements
2. Fast chilling rate. Shock chilling can be used to improve process operations.
3. Good filtration rates but lower than for propane.

15.5.4 Ketone Dewaxing

The most widely used processes use mixtures of MEK–toluene and MEK–MIBK for the solvent. MEK–benzene was used originally, but health hazards associated with the handling of benzene as well as its cost resulted in the use of MEK–toluene in its place. MEK–MIBK mixtures have fewer health problems than the earlier used compositions. A simplified process flow diagram is shown in Figure 15.3.

Solvent is added to the solvent-extracted oil feed, and the mixture is chilled in a series of scraper-surface exchangers and chillers. Additional solvent is added to the feed to maintain sufficient liquid for easy handling as the temperature is decreased and the wax crystallizes from solution. Cold filtrate from the rotary filters is used as the heat exchange medium in the exchangers and a refrigerant, usually propane, in the chillers. The cold slurry is fed to the rotary filters for removal of the crystalline wax. The filter cake is washed with cold solvent, and an inert gas blanket, at a slight positive pressure, is maintained on the filters and also used to "blow back" the wax cake from the filter cloth and release it before it reaches the filter drum scraper knife.

The solvent is recovered from the filtrate and the wax by heating and two-stage flashing, followed by steam-stripping.

The water introduced into the system by the steam-stripping is removed continuously by a series of phase separations and distillations. The overheads from the steam-strippers are combined and condensed. The condensate separates into two phases in the decanter drum with the aqueous phase containing about 3.5 mol% ketones and the ketone-rich phase about 3 mol% water. The ketone phase is fed into the top of the solvent dehydration tower as reflux, and the water-rich phase is fed into the ketone stripper. The ketone is stripped from the water-rich phase with steam, and water is discharged from the bottom of the tower into the sewer.

The products from the dewaxing unit are a dewaxed oil and a slack wax. The dewaxed oil next must go through a finishing step to improve its color and color stability. The slack wax either is used for catalytic cracker feed or undergoes a deoiling operation before being sold as industrial wax.

15.5.5 Dilchill Dewaxing

The Dilchill dewaxing process was developed by ExxonMobil. It is a modification of the ketone dewaxing process that uses shock chilling of the waxy oil feed by direct injection with very cold solvent in a highly agitated mixer. The wax crystals formed are larger and more dense than the crystals formed in scraped-surface chillers. Higher filtration rates and better oil removal are achieved, resulting in lower capital and operating costs as well as higher dewaxed oil yields. The scraped-surface coolers are eliminated, but the chillers are still necessary.

15.5.6 Propane Dewaxing

Propane dewaxing differs from ketone dewaxing in that the propane is used both as a diluent and a refrigerant for direct chilling. As a diluent, it dilutes the oil phase so it is fluid and acts as the carrying medium after the wax crystallizes. As a refrigerant, it chills the oil by being evaporated batchwise from the propane–waxy oil solution. The cooling rate usually averages about 3°F (1.5°C) per min, but ranges from almost 7°F (4°C) per min at the beginning of the chilling cycle to about 2°F (1°C) at the end. In propane dewaxing, the chilling rate is limited to avoid shock chilling because it results in poor wax crystal formation and growth. The amount of

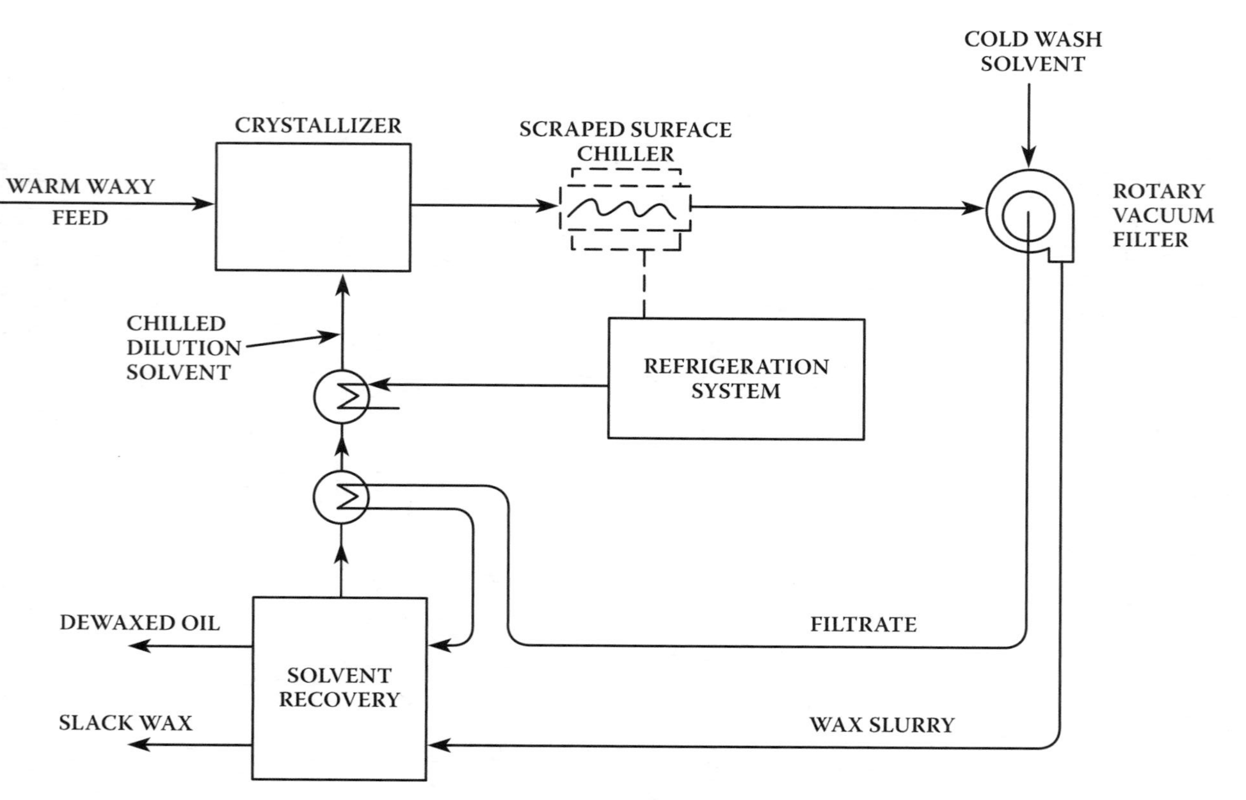

FIGURE 15.3 ExxonMobil Dilchill dewaxing unit. (From Note 1.)

dilution required is a function of the viscosity of the oil and ranges from a 1.5:1 ratio of propane to fresh feed for light stocks to 3:1 for heavy stocks [5].

Waxy oil feed is mixed with propane, cooled to about 80°F (27°C), and charged into the warm solution drum under sufficient pressure to prevent vaporization of the propane. From the solution drum, it is charged into one of the two batch chillers, where it is cooled at a controlled rate by evaporation of the propane. It usually takes about 30 min to chill the mixture to the desired temperature. The slurry is then discharged into the filter feed drum. The cycle is arranged so that while one chiller is being used to chill a batch, the other chiller is first discharged into the filter feed drum, then refilled from the warm solution drum.

The wax is separated from the oil on rotary filters. The wax cake is washed with cold propane while on the filter to remove and recover as much oil as practical. Cold propane vapors are used to release the filter cake.

Most of the propane in the dewaxed oil and wax stream is recovered by heating each of the streams to about 320°F (160°C) and flashing at a pressure sufficiently high that the propane can be condensed with refinery cooling water. The remaining propane is recovered by low-pressure flashing and a final steam-stripping of each of the streams. The propane from the low-pressure flash is compressed, condensed, and returned to the propane storage drum.

A dewaxing aid such as Paraflow™ (a methacrylate polymer) is frequently added to the waxy oil feed at a level of 0.05 to 0.2 liquid vol% on feed to modify the wax crystal structure. This results in higher filtration rates and denser wax cakes, from which it is easier to wash the oil.

15.5.7 Selective Hydrocracking

There are two types of selective hydrocracking processes for dewaxing oil; one uses a single catalyst for pour point reduction only and the other uses two catalysts to reduce the pour point and improve the oxygen stability of the product [9,10].

For the pour point reduction operation, both processes use synthetic shape-selective zeolite catalysts that selectively crack the n-paraffins and slightly branched paraffins. Zeolites with openings about 6 Å in diameter provide rapid cracking rates for n-paraffins, with the rate decreasing rapidly as the amount of branching increases.

The ExxonMobil Lube Dewaxing Process uses a fixed-bed reactor packed with two catalysts, and the process flow is similar to that of hydrotreating unit. Severity of operation is controlled by the furnace outlet temperature (reactor temperature). Essentially no methane or ethane is formed in the reaction. A similar British Petroleum (BP) process produces propane, butane, and pentane in the ratio 2:4:3 by weight. Reaction conditions for the ExxonMobil and BP processes appear to be similar with the following considered typical ranges:

Reactor temp., °F (°C)	560–700 (290–370)
Reactor pres., psig (kPa)	300–2,000 (2,070–13,800)
H_2 partial pres., psig (kPa)	250–1,500 (1,725–10,340)
Gas rate, scf/bbl (Nm^3/t)	500–5,000 (100–1,000)
H_2 consumption, scf/bbl (Nm^3/t)	20–40 (100–200)

Yields of dewaxed oil of similar pour point from the same feedstocks vary from 0 to 15% greater than from solvent dewaxing (SDW), with the increase reflecting the difficulty of separating the oil from the wax in the SDW processes.

The feed to the selective hydrocracking unit is solvent-extracted oil from the aromatic extraction units. The advantages claimed over conventional solvent dewaxing units include

1. Production of very low pour and cloud point oils from paraffinic stocks.
2. Lower capital investment.
3. Improved lube oil base stock yields.
4. A separate hydrofinishing operation is not necessary.

15.6 HYDROFINISHING

Hydrotreating of dewaxed lube oil stocks is needed to remove chemically active compounds that affect the color and color stability of lube oils. Most hydrotreating operations use cobalt-molybdate catalysts and are operated at a severity set by the color improvement needed. Organic nitrogen compounds seriously affect the color and color stability of oils, and their removal is a major requirement of the operation [4].

The process flow is the same as that for a typical hydrotreating unit. Representative operating conditions are

Reactor temp, °F(°C)	400–650 (200–340)
Reactor pres., psig (kPa)Y	500–800 (3450–5525)
LHSV, v/h/v	0.5–2.0
H_2 as rate, scf/bbl (Nm^3/t)	500 (100)

Usually, finished oil yields are approximately 98% of dewaxed oil feed.

15.7 FINISHING BY CLAY CONTACTING

Many older lube oil processing plants use contacting of the dewaxed oil with activated clays at elevated temperatures to improve the stability of the finished oils in engine service. Polar compounds (aromatic and sulfur and nitrogen containing molecules) are adsorbed on the clay and removed by filtration. Spent clay disposal and operating restrictions have generally caused the clay treating to be replaced by hydrofinishing.

15.8 ENVIRONMENTAL IMPACTS

Stricter environmental requirement and mileage standards for new cars have created a growing demand for high-quality multigrade motor oils that have lower volatility and oil consumption characteristics and reduced thickening of the oil by oxidation during service, which increases fuel consumption. Solvent-refined oils have difficulty in meeting the new standards and are increasingly being replaced with hydrocracked and poly-alpha-olefin-based oils, especially in the lower viscosity grades such as 5W-30 and 10W-30 multigrades [11].

The low-viscosity multigrade oils are typically blended from low-viscosity mineral-based oils, which have high volatilities and tendency to high oil consumption and rapid thickening by oxidation during service. Hydrocracking of base stocks, followed by solvent extraction to remove partially hydrocracked aromatic compounds [7], offers a more cost-effective route than production of poly-alpha-olefins [12].

NOTES

1. Bushnell, J.D. and Fiocco, R.J., *Hydrocarbon Process.* 59(5), pp. 119–123, 1980.
2. Asseff, P.A., *Petroleum Additives and Their Functions,* April 1966, Lubrizol Corp., Cleveland.
3. Sequeria, A., Preprints Div. Petro. Chem., Amer. Chem. Soc. Meeting, Washington, D.C., August 23–28, 1992, pp. 1286–1292.
4. Soudel, M., *Hydrocarbon Process. 53*(12), pp. 59–66, 1974.
5. Nelson, W.L., *Petroleum Refining Engineering,* 4th ed., McGraw-Hill Book Co., New York, 1958, pp. 62–70.
6. Rhodes, A.K., *Oil Gas J.* 91(1), pp. 45–51, 1993.
7. Ushio, M. et al., Preprints Div. Petro. Chem., Amer. Chem. Soc. Meeting, Washington, D.C., August 23–28, 1992, pp. 1293–1302.
8. Rossi, A., Preprints Div. Petro. Chem., Amer. Chem. Soc. Meeting, Washington, D.C., August 23–28, 1992, pp. 1322–1336.
9. Hargrove, J.D., Elkes, G.J., and Richardson, A.H., *Oil Gas J. 77*(2), pp. 103–105, 1979.
10. Smith, K.W., Starr, W.C., and Chen, N.Y., *Oil Gas J. 78*(21), pp. 75–84, 1980.
11. Yates, N.C., Kiovsky, T.E., and Bales, J.R., Preprints Div. Petro. Chem., Amer. Chem. Soc. Meeting, Washington, D.C., August 23–28, 1992, pp. 1303–1312.
12. Bushnell, J.D. and Eagen, J.F., Commercial Experience with Dilchill Dewaxing, Presented at NPRA Meeting, September 11–12, 1975.

16 Petrochemical Feedstocks

The preparation of hydrocarbon feedstocks for petrochemical manufacture can be a significant operation in today's petroleum refineries. There are three major classes of these feedstocks according to use and method of preparation. These are aromatics, unsaturates (olefins and diolefins), and saturates (paraffins and cycloparaffins). Aromatics are produced using the same catalytic reforming units used to upgrade the octanes of heavy straight-run naphtha gasoline blending stocks. For petrochemical use, the aromatic fraction is concentrated by solvent extraction techniques. Some of the unsaturates are produced by the fluid catalytic cracking unit, but most have to be produced from refinery feedstocks by steam cracking or low molecular polymerization of low molecular-weight components.

To provide better continuity from regular refinery operations, the petrochemical feedstock preparation will be covered in this order: (1) aromatics, (2) unsaturates, and (3) saturates.

16.1 AROMATICS PRODUCTION

For aromatics production for petrochemical use, the catalytic reformer can be operated at higher severity levels than for motor gasoline production. Highly naphthenic feedstocks also aid in high aromatics yields because the dehydration of naphthenes is the most efficient reaction taking place and provides the highest yield of aromatics. Table 16.1 illustrates the increase in aromatics yields as the reformer operating severity is increased [the clear research octane number (RON) is the measure of operating severity].

The C_6 to C_8 aromatics (benzene, toluene, xylene, and ethylbenzene) are the large-volume aromatics used by the petrochemical industry, with benzene having the greatest demand. The product from the catalytic reformer contains all of these aromatics, and it is separated into its pure components by a combination of solvent extraction, distillation, and crystallization. In addition, because of the much greater demand for benzene, the excess of the toluene and xylenes over market needs can be converted to benzene by hydrodealkylation.

16.1.1 Solvent Extraction of Aromatics

Present separation methods for recovery of aromatics from hydrocarbon streams use liquid–liquid solvent extraction to separate the aromatic fraction from the other hydrocarbons. Most of the processes used by U.S. refineries use either polyglycols or sulfolane as the extracting solvent. The polyglycol processes are the Udex process

TABLE 16.1
Aromatics Yield as a Function of Severity

Product RON	Percent aromatics
90	54
95	60
100	67
103	74

developed by Dow Chemical Co. and licensed by UOP and the Tetra process licensed by the Linde Division of Union Carbide. The solvents used are tetraethylene glycol for the Tetra process and usually diethylene glycol for the Udex process, although dipropylene glycol and triethylene glycol can also be used. The Sulfolane process was originally developed by the Royal Dutch/Shell group and is licensed worldwide by UOP. It uses sulfolane (tetrahydrothiophene 1-1 dioxide) as the solvent, and a simplified process flow diagram is shown in Figure 16.1a.

The important requirements for a solvent are [1]

1. High selectivity for aromatics versus nonaromatics
2. High capacity (solubility of aromatics)
3. Capability to form two phases at reasonable temperatures
4. Capability of rapid phase separation
5. Good thermal stability
6. Noncorrosivity and nonreactivity

The Udex process, using polyglycols, was the first of the solvent extraction processes to find widespread usage. Many of these units are still in operation in U.S. refineries but, since 1963, most new units have been constructed using the Sulfolane or Tetra processes (Figure 16.1b). At a given selectivity, the solubility of aromatics in Sulfolane is about double that of triethylene glycol. The higher solubility permits lower solvent circulation rates and lower operating costs.

Both the polyglycol and Sulfolane solvent processes use a combination of liquid–liquid extraction and extractive stripping to separate the aromatics from the other hydrocarbons [1,2,3] because of the characteristics of polar solvents. As the concentration of aromatics in the solvent increases, the solubility of the nonaromatic hydrocarbons in the extract phase also increases. This results in a decrease in the selectivity of the solvent and a carryover of some of the nonaromatic hydrocarbons with the extract phase to an extractive stripper. In the extractive stripper, an extractive distillation occurs, and the nonaromatic hydrocarbons are stripped from the aromatic–solvent mixture and returned as reflux to the extraction column. The solvent is then recovered from the nonaromatic hydrocarbon-free extract stream, leaving an extract that contains less than 1000 ppm (< 0.1%) aliphatics (see Figure 16.2).

Aromatics recoveries are typically equal to or better than 99.9, 99.0, and 97.0% for benzene, toluene, and xylenes, respectively.

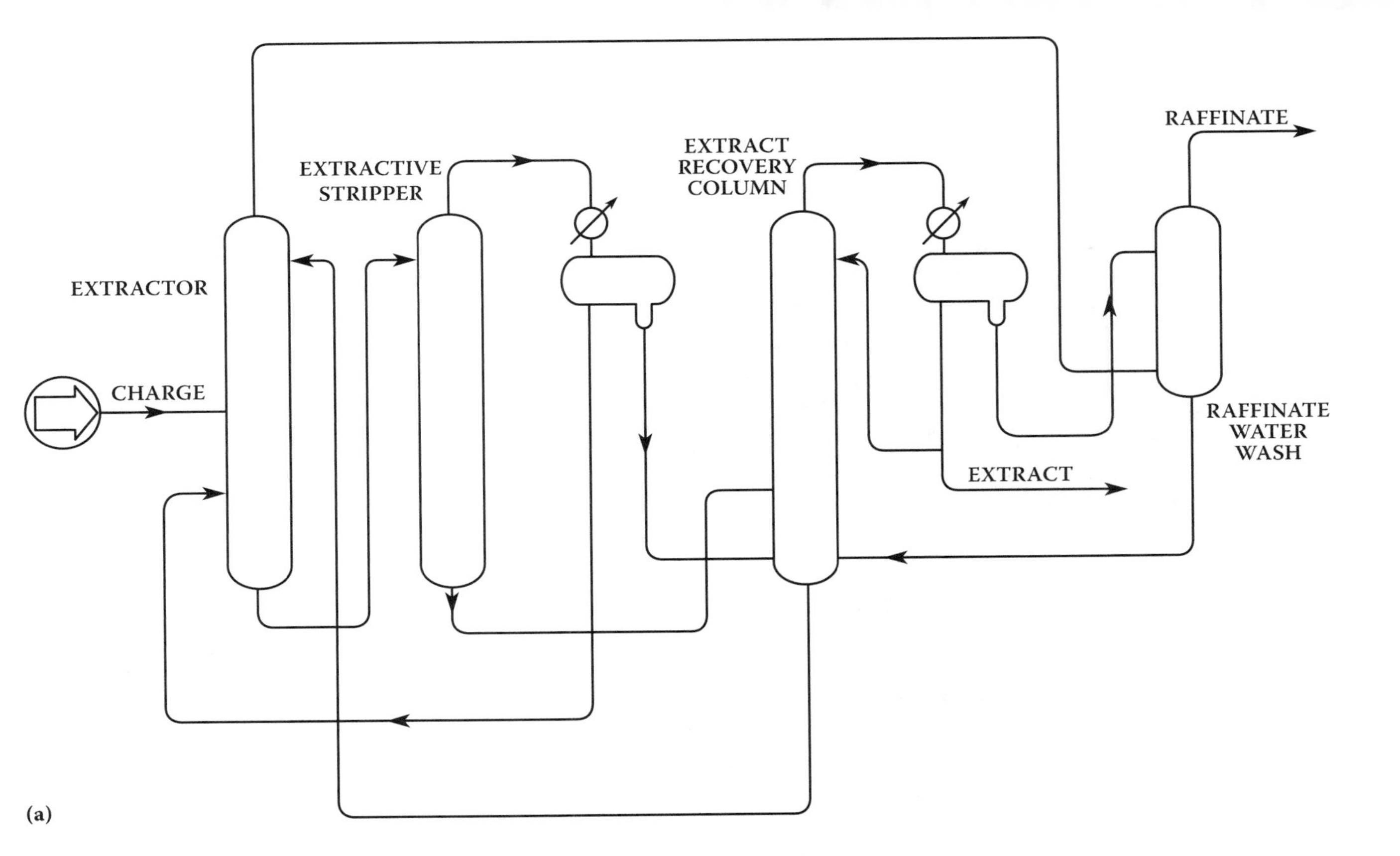

FIGURE 16.1A Sulfolane aromatics extraction unit.

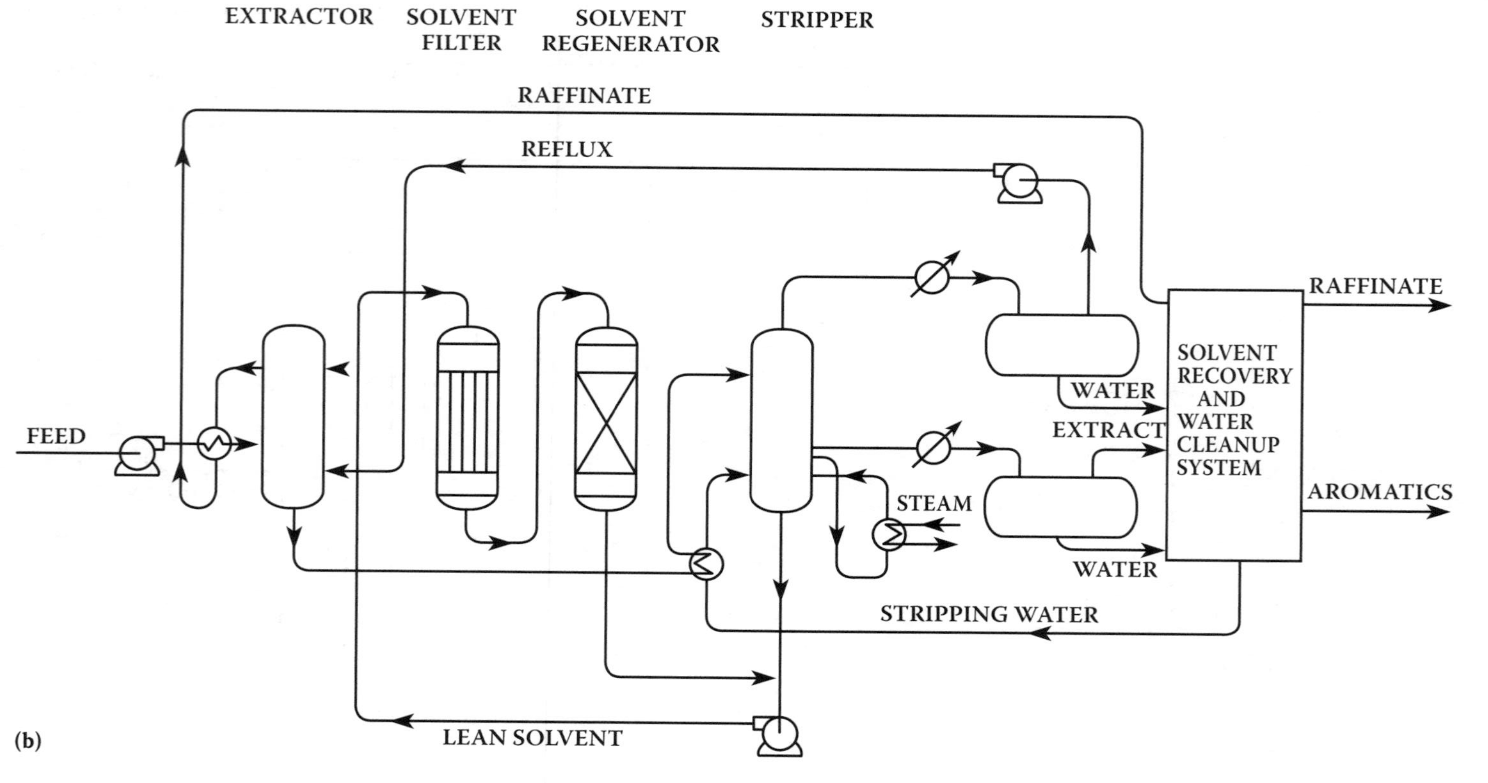

FIGURE 16.1B Tetra aromatics extraction unit.

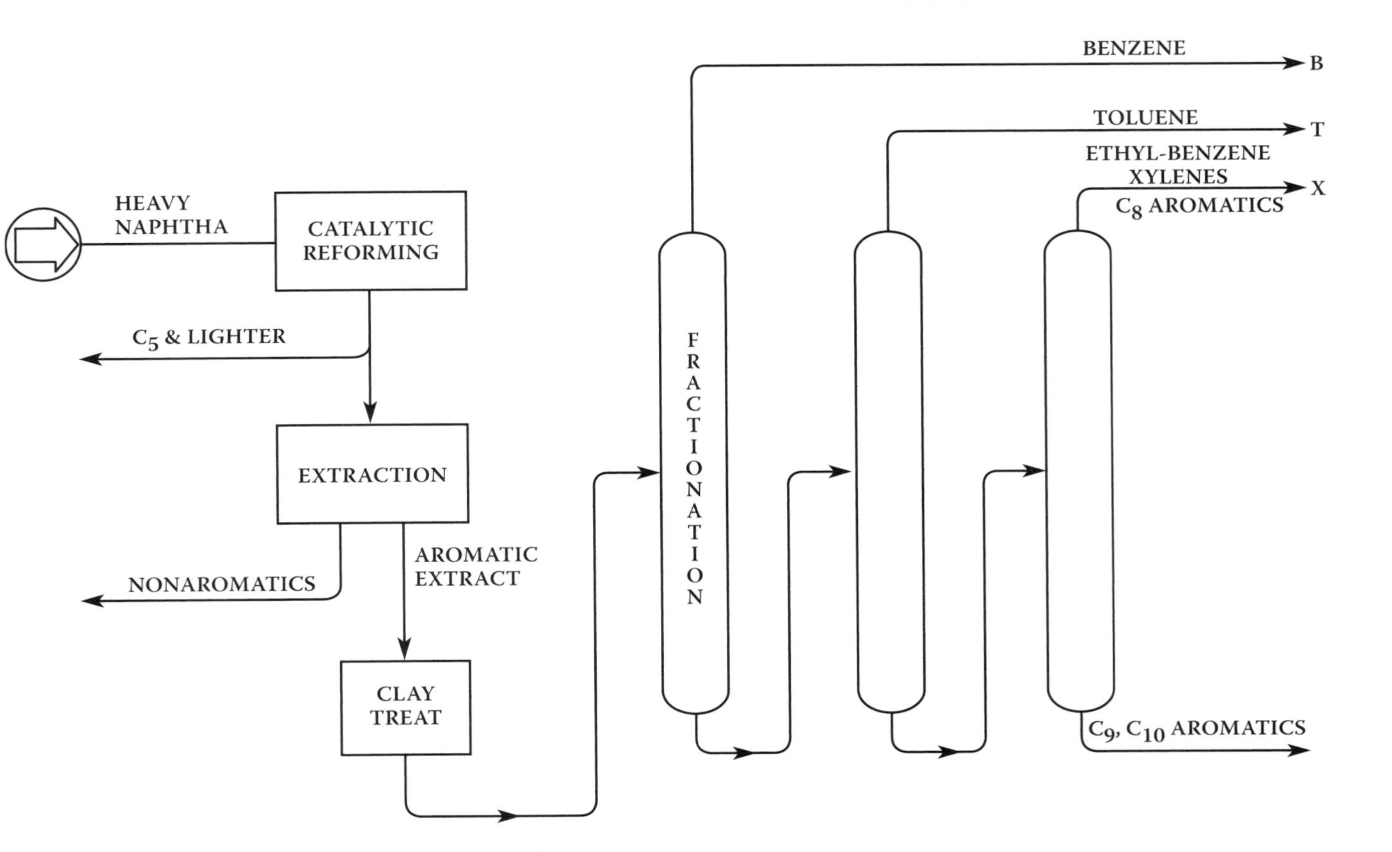

FIGURE 16.2 Sequence of aromatics recovery operations.

Because of the high affinity of the polar solvents for water and the low solubilities in the nonaromatic hydrocarbon raffinate phase, the solvent is recovered from the raffinate phase by washing with water. The water is returned to either the extractive stripper (Tetra) or extract recovery column (Sulfolane) for recovery of the solvent.

The water content of the solvents is carefully controlled and used to increase the selectivity of the solvents. The water content of the polyglycols is kept in the range of 2 to 10 wt% and that of sulfolane is maintained at about 1.5 wt%.

The solvent quality is maintained in the Sulfolane process by withdrawing a slipstream from the main solvent recirculation stream and processing it in a solvent reclaiming tower to remove any high-boiling contaminants that accumulated. The Tetra process solvent quality is maintained by using a combination of filtration and adsorption to remove accumulated impurities.

Sulfolane and the polyglycols have thermal stability problems, and the skin temperatures of heat exchange equipment must be limited. For sulfolane, skin temperatures must be kept below 450°F (232°C).

16.1.2 Aromatics Separation

Benzene and toluene can be recovered from the extract product stream of the extraction unit by distillation. The boiling points of the C_8 aromatics are so close together (Table 16.2) that separation by distillation becomes more difficult and a combination of distillation and crystallization or adsorption is used. The ethylbenzene is first separated from the mixed xylenes in a three-unit fractionation tower with 120 trays per unit for a total of 360 trays. Each unit is about 200 ft in height, and the units are connected so they operate as a single fractionation tower of 360 trays. High reflux ratios are needed to provide the desired separation efficiency, because the difference between the boiling points of ethylbenzene and p-xylene is only about 3.9°F (2°C). This is a very energy-intensive operation, and with today's energy costs, it is usually less costly to make ethylbenzene by alkylating benzene with ethylene.

A typical processing sequence for the separation of C_8 aromatics is shown in Figure 16.3. After removal of the ethylbenzene by super-fractionation, the o-xylene along with the higher-boiling C_9^+ aromatics are separated from the p- and m-xylenes by fractionation. The boiling point of o-xylene is more than 9.6°F (5°C) greater

TABLE 16.2
Boiling and Melting Points of Aromatics

	Boiling points		Melting points	
	°F	°C	°F	°C
Benzene	176.2	80.1	42.0	5.5
Toluene	231.2	110.6	−139.0	−95.0
Ethylbenzene	277.1	136.2	−139.0	−95.0
p-Xylene	281.0	138.4	55.9	13.3
m-Xylene	282.4	139.1	−54.2	−47.9
o-Xylene	291.9	144.4	−13.3	−25.2

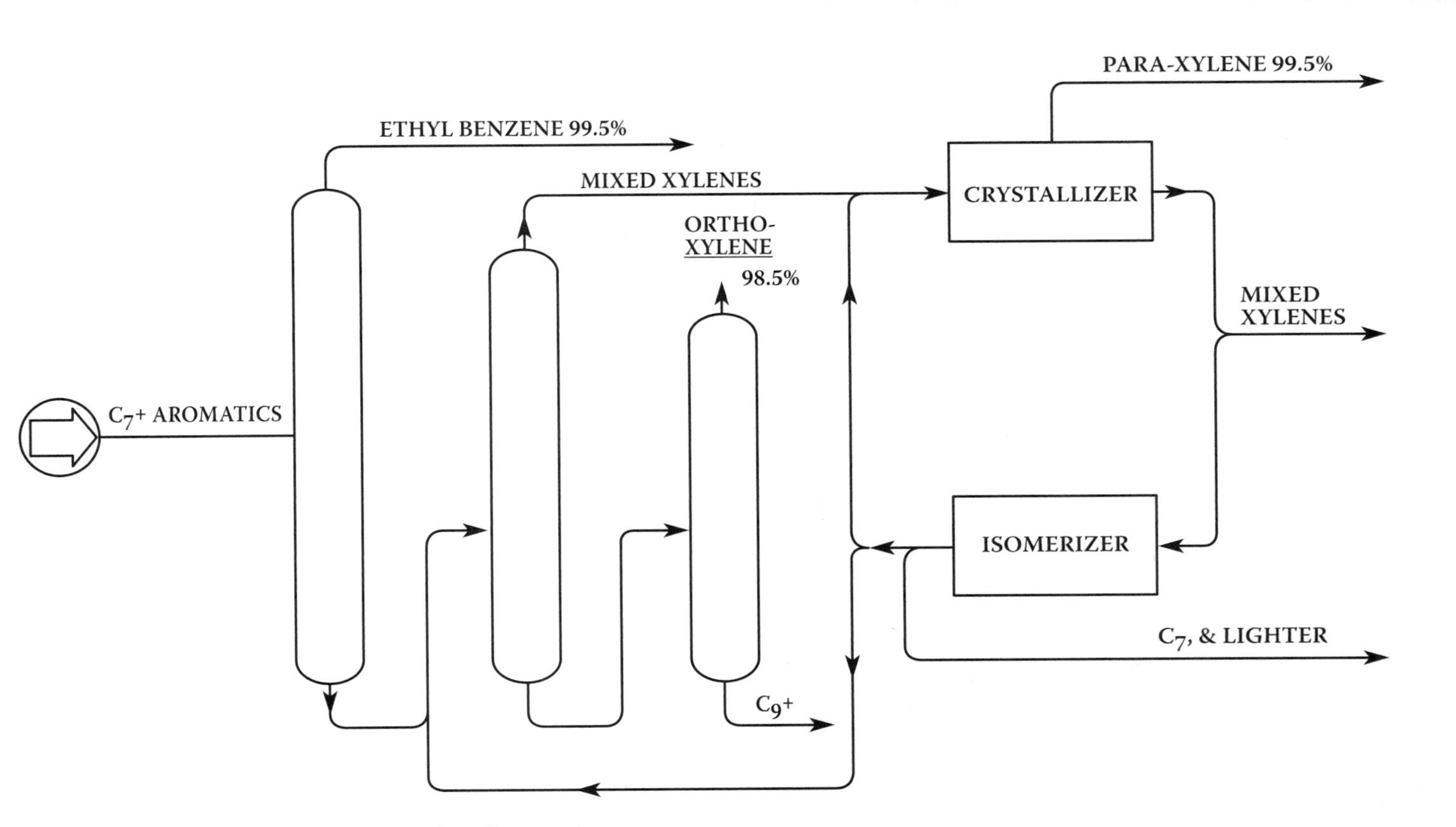

FIGURE 16.3 Processing sequence to produce C_8 aromatics.

than that of its closest isomer, m-xylene, and it can be separated economically in a 160-plate distillation column. The overhead stream from this column is a mixture of m- and p-xylenes, and the bottoms stream contains o-xylene and the C_9^+ aromatics. The bottoms stream is processed further in a 50-tray return column to separate 99+% purity o-xylene from the C_9^+ aromatics.

The mixed m- and p-xylene overhead stream from the fractionator is sent to either the Parex unit or the crystallization unit to separate the m- and p-xylenes. The UOP Parex process uses adsorption of p-xylene on a molecular-sieve adsorbent to separate in m- and p-xylenes. The adsorbent is selected such that the p-xylene molecules will be adsorbed and the m-xylene molecules will pass through the adsorbent bed. Recoveries of 96% per pass of pure p-xylene can be achieved [4] as compared with 60 to 65% recovery of p-xylenes by the freeze-crystallization process.

The p-xylene usually is stripped from the adsorbent with p-diethylbenzene or a mixture of diethylbenzene isomers. The UOP process uses a simulated moving adsorbant-bed design.

If fractional crystallization is used for separation, the mixed m- and p-xylene overhead stream from the fractionator is fed to the crystallization unit to separate the m- and p-xylenes. The solidification point of p-xylene is 55.9°F (13.3°C), whereas that of m-xylene is –54.2°F (–47.9°C). A simplified flow diagram of the solvent extraction process is shown in Figure 16.4 [1].

The xylenes feed is dried, chilled to about –40°F (–40°C), and charged to staged crystallizers. Ethylene is used as the refrigerant in indirect chilled surface-scraped heat exchangers, and the mixture is chilled to –60°F (–50°C) in the first crystallizer stage and to –90°F (–68°C) in the second stage. The second-stage crystallizer effluent is charged to a combination solid and screen bowl centrifuge to separate the p-xylene crystals from the mother liquor. The operation of this centrifuge is critical to the cost-effectiveness of this process, because it controls the subsequent purification costs and the overall refrigeration requirements of this separation [1].

The p-xylene crystals from the centrifuge are purified in a series of partial remelting, recrystallization, and centrifugation stages. The temperatures are increased for each remelting and recrystallization stage, and the filtrate from each stage is recycled to the preceding stage. This countercurrent operation gives a high-purity p-xylene product from the last stage. The m-xylene-rich filtrate stream is used as feedstock for a xylene isomerization unit or sent to gasoline blending.

The m-xylene-rich stream from the crystallizer and the o-xylene stream from the rerun column can be converted to p-xylene by a xylene isomerization process if demand for p-xylene warrants it. The process flow diagram for this process is similar to the isomerization process in Chapter 10. The catalyst used is a zeolite–non-noble metal catalyst developed by Mobil.

The aromatic-rich feedstock is mixed with hydrogen, heated to 700 to 800°F (375 to 425°C), and charged to a fixed-bed reactor containing a non-noble metal catalyst at a pressure of about 150 psig (1034 kPa). The mixture is isomerized to a near-equilibrium mixture of o-, m-, and p-xylenes [1,5]. After cooling, the xylenes are separated from the hydrogen-rich gas in a flash separator and from other hydrocarbons in distillation columns. The xylene-rich stream is returned to the p-xylene crystallization unit or the Parex unit as feed.

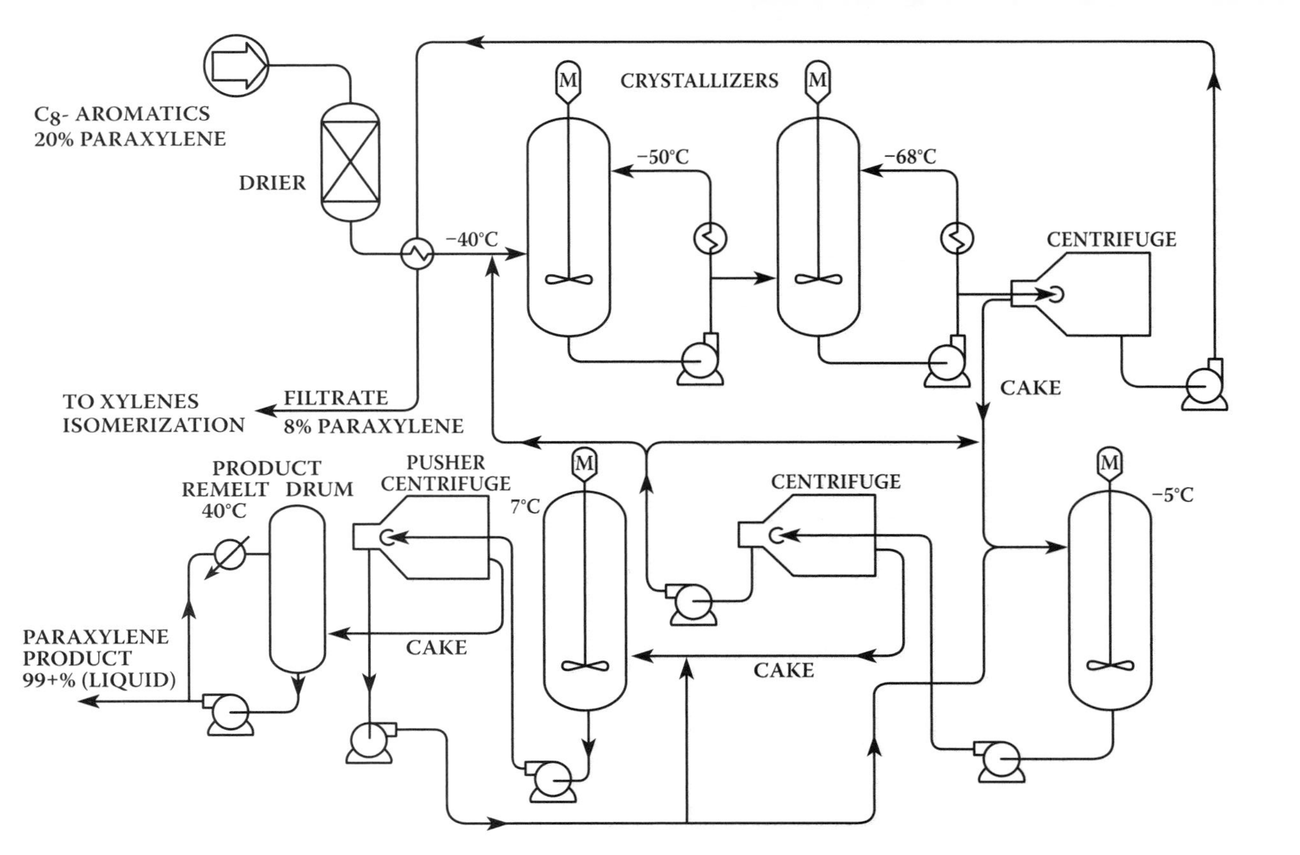

FIGURE 16.4 p-Xylene by crystallization.

Other isomerization processes are used to maximize p-xylene production by both isomerizing o- and m-xylenes and by converting ethylbenzene into xylenes [1]. A two-stage process is generally used for conversion of aromatic hydrocarbons with two or more carbons in the side chains, with the first stage for partial hydrogenation and the second for dehydrogenation. A single-stage process can also be used. The catalysts are either silica-alumina, containing a small amount of platinum, or molecular-sieve silica-alumina catalysts containing about 1% platinum. The equilibrium concentration of p-xylene obtained in the effluent is about 24%.

16.1.3 Benzene

The primary source of benzene is from the refinery catalytic reforming unit, but substantial amounts of benzene are also produced by the hydrodealkylation of toluene. The process flow for a hydrodealkylation unit (HDA) is also similar to that of an isomerization unit. The feed can be toluene or a mixture of toluene and xylenes. It is heated to 1175°F (630°C) at 600 psig (4140 kPa) and charged to an open noncatalytic reaction ($L/D > 20$), where thermal dealkylation of toluene and xylenes takes place during the residence time of 25 to 30 sec. The hydrogenation step in the dealkylation reaction is highly exothermic [22,000 Btu/lb-mole (12,220 cal/g-mole) of H_2 consumed], and temperature is controlled by injection of quench hydrogen at several points along the reactor. The hydrodealkylation reaction results in the conversion of about 90% of the aromatics in the feed and selectively favors about 95% of the conversion to benzene. A small amount of polymer is also formed, primarily diphenyl. A small amount of hydrogen sulfide or carbon disulfide is added to the feed to prevent catastrophic corrosion of the furnace tubes, and a small amount of the polymer is recycled to minimize polymer formation [1].

16.2 UNSATURATE PRODUCTION

Although a substantial amount of C_3 and C_4 olefins are produced by the fluid catalytic cracking unit in the refinery and some C_2 and C_3 olefins by the cat cracker and coker, the steam cracking of gas oils and naphthas is the most important process for producing a wide range of unsaturated hydrocarbons for petrochemical use. Ethylene is also produced by thermal cracking ethane and propane.

Steam cracking is the thermal cracking and reforming of hydrocarbons with steam at low pressure, high temperature, and very short residence times (generally < 1 sec).

The hydrocarbon is mixed with steam in the steam/hydrocarbon weight ratio of 0.2:0.8 and fed into a steam-cracking furnace. Residence times in the cracking zone range from 0.3 to 0.8 sec with a coil outlet temperature between 1400 and 1500°F (760 and 840°C) and a coil outlet pressure in the range from 10 to 20 psig (69 to 138 kPa). The coil outlet stream is quickly quenched to about 600°F (350°C) to stop the cracking and polymerization reactions. Recent designs of steam crackers utilize heat transfer in a steam generator or transfer-line heat exchange with a low-pressure drop to lower the coil outlet temperature quickly with a maximum amount of heat recovery. The quenched furnace outlet stream is sent to the primary fractionator, where it is separated into a gas stream and liquid product streams according to boiling-point range. The gases are separated into individual components as desired by compression

TABLE 16.3
Steam Cracking Conditions

	Naphtha		Wax	
Coil outlet temp., °F (°C)	1100–1650	(730–900)	1100–1200	(595–650)
Pres., psig (kPa)	45	(310)	15–30	(103–207)
Steam/HC, wt%	0.3–0.9		0.12–0.15	
C_3 conversion, wt%	60–80		8–10	

and high-pressure fractionation. Streams involved in low-temperature rectification must be desiccant dried to remove water and prevent ice and hydrate formation.

Paraffins obtained from the dewaxing of lubricating oil base stocks are frequently steam cracked to produce a wide range of linear olefins, usually with from 11 to 16 carbon atoms per molecule. Mild steam cracking conditions are employed, because the object is to minimize the formation of high molecular-weight olefins. As with catalytic cracking, the reactions favor the breaking of the paraffinic chain near its center with the formation of one paraffinic molecule and one olefinic molecule. Typical operating conditions are given in Table 16.3 [6].

The purity of the feedstock has a major effect on the product quality. The petrolatum (high molecular-weight microcrystalline waxes) obtained from the dewaxing of lubricating oil base stocks contains from 85 to 90% straight- and branched-chain paraffins, with the remainder naphthenes and aromatics. Aromatics are especially undesirable, and the wax stream must be deoiled before being used as a steam-cracker feedstock. By choosing a feedstock with an initial boiling point greater than the end point of the desired product, it not only increases the yield of the desired olefins but also makes it possible to separate the product olefins from the feed paraffins by distillation. A typical product stream contains above 80% monoolefins, with only a small percentage less than 11 carbons or more than 16 carbons.

Linear monoolefins are also produced by several catalytic processes. The overall process consists of a vapor-phase catalytic dehydrogenation unit, followed by an extraction unit to extract the linear olefins from the paraffin hydrocarbons by adsorption on a bed of solid adsorbent material. The olefins are desorbed with a hydrocarbon boiling at a lower temperature than the olefin product to make it easy to separate the olefin product from the desorbant. The desorbant hydrocarbon is recycled internally in the extraction section. The olefin product contains about 96 wt% linear olefins, of which about 98% are monoolefins. Approximately 93 wt% of the linear paraffins in the feed are converted to linear olefins [1,7].

16.3 SATURATED PARAFFINS

16.3.1 NORMAL PARAFFINS

Normal paraffins are recovered from petroleum fractions by vapor-phase adsorption on molecular sieves having an average pore diameter of 5 Å. A simplified process flow diagram for the ExxonMobil Ensorb process is shown in Figure 16.5. The

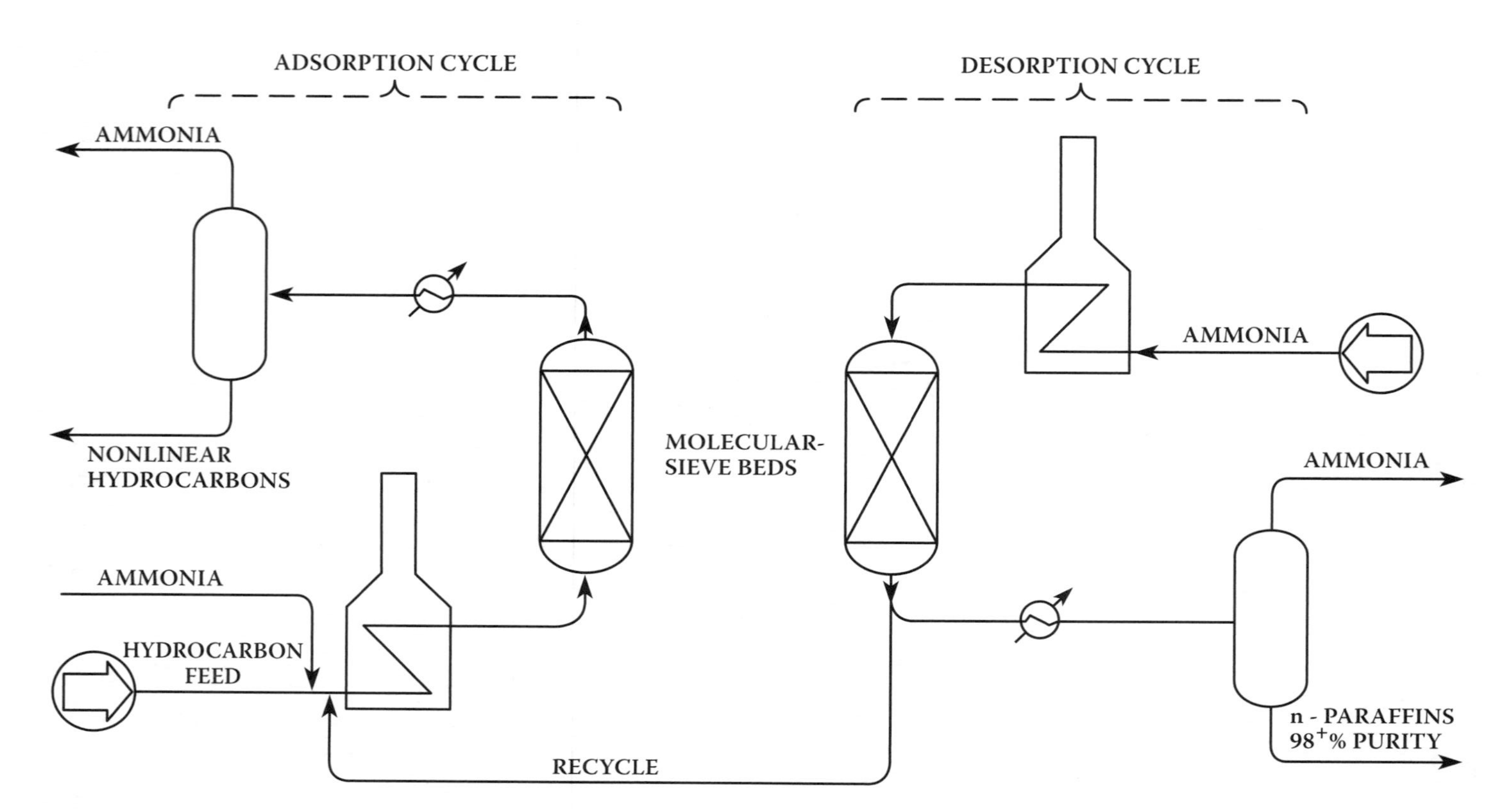

FIGURE 16.5 n-Paraffin recovery by adsorption.

adsorption takes place in the vapor phase at pressures of 5 to 10 psig (35 to 69 kPa) and temperatures from 575 to 650°F (300 to 350°C). Ammonia is used to desorb the normal paraffins. A semibatch operation using two beds of 5 Å molecular sieves permits continuous operation, with one bed being desorbed while the other bed is on-stream. The sieves gradually lose capacity because of contamination with polymerization products and are regenerated by controlled burn-off of the high molecular-weight hydrocarbons. Run times of about 12 months between regenerations are common.

The greatest demand for n-paraffins are for use in the manufacture of biodegradable detergents using straight-chain paraffins with from 11 to 14 carbons per molecule.

16.3.2 Cycloparaffins

Cycloparaffins are usually prepared by the hydrogenation of the corresponding aromatic compound. For example, cyclohexane is prepared by the hydrogenation of benzene. There are a number of processes licensed by the hydrogenation of benzene [1].

The hydrogenation of benzene is carried out over platinum of Raney nickel supported on alumina or silica-alumina. The hydrogenation reaction is highly exothermic, with the release of about 1150 Btu/lb (640 kcal/kg) of benzene converted to cyclohexane. Reactor temperature is controlled by recycling and injecting a portion of the cyclohexane produced into the reactors [1].

Tetralin and decalin are prepared by the hydrogenation of naphthalene using the same hydrogenation catalysts as used for producing cyclohexane. Here, benzothiophene is a common impurity in naphthalene and causes the rapid deactivation of the nickel and platinum catalysts. It is necessary under these conditions to use a cobalt-molybdenum catalyst that has been presulfided to activate it [1].

NOTES

1. Thomas, C.L., *Catalytic Processes and Proven Catalysts*, Academic Press, New York, 1970.
2. Broughton, D.B. and Asselin, G.F., *Proceedings of the Seventh World Petroleum Congress*, Vol. 4, pp. 65–73, 1967.
3. *Hydrocarbon Process.* 59(9), pp. 203–204, 1980.
4. Mowry, J.R., *Handbook of Petroleum Refining Processes*, Meyers, R.A. (Ed.), McGraw-Hill Book Co., New York, 1986, pp. 10–12.
5. Prescott, J.H., *Chem. Eng.*, June, pp. 138–140, 1969.
6. Schutt, H.C. and Zdonik, S.B., *Oil Gas J.* 54(7), p. 98, 1956.
7. *Hydrocarbon Process.* 59(11), p. 185, 1980.

17 Additives Production from Refinery Feedstocks

17.1 USE OF ALCOHOLS AND ETHERS

In the United States during the 1970s, various high-octane additives began to be blended into motor gasoline to maintain octane levels as the use of TEL was reduced. Methanol and ethanol were used because of their availability and previous use in special situations. These alcohols were gradually replaced or supplemented with various ethers. Methyl tertiary butyl ether (MTBE), ethyl tertiary butyl ether (ETBE), and tertiary amyl methyl ether (TAME) evolved as having more desirable blending and combustion characteristics than the alcohols.

One of the main advantages of ethers relative to alcohols is the lower blending vapor pressure of the ethers. These are summarized in Table 17.1. The reason for the high blending vapor pressure of the alcohols is that they form minimum-boiling azeotropes with some of the hydrocarbons in the gasoline. Ethers form significantly fewer azeotropes, and those that are formed reduce vapor pressure rather than increasing it. Maintaining gasoline vapor pressures below about 8 psi RVP (55 kPa) is considered an important factor in minimizing hydrocarbon vapor losses to the atmosphere. Atmospheric hydrocarbons in the presence of nitrogen oxides and sunlight are believed to form ozone through a sequence of complex chemical reactions.

The potential benefits of adding oxygenated hydrocarbons to gasoline can be summarized as follows:

1. Less carbon monoxide emissions are possible.
2. Unburned hydrocarbon emissions are reduced.
3. Ozone content in the lower atmosphere of highly populated areas may be reduced part of the time during certain weather conditions.
4. Octane requirements for good engine performance can be met without use of TEL.
5. Most gasoline produced from FCUs (catalytic cracking) contains relatively volatile olefins such as isobutylene and isoamylene. These components are some of the more reactive ozone formers when vaporized. Reacting these olefins with alcohol to produce ethers decreases gasoline volatility and increases octane.

The disadvantages are as follows:

1. The combined capital and operating costs for producing methanol from methane and MTBE or others from existing sources of isobutylene are

TABLE 17.1
Typical Properties of Some Fuel Ethers and Alcohols

	MTBE	ETBE	TAME	MeOH	EtOH
Blending octane, (R + M)/2	109	110	105	118	114
Blending RVP, psig	8	4	3	60+	19
RVP psig	8	4	1.5	4.6	2.4
Atm. boiling point, °F	131	161	187	148	173
Molecular weight	88.15	102.18	102.18	32.04	46.07
wt% O_2	18.2	15.7	15.7	50.0	33.7
Density, 60/60	0.74	0.77	0.77	0.79	0.79

Note: Blending values for octane and RVP vary to a small extent depending on concentration in gasoline and gasoline composition. Table values are typical from several published sources [3,4,5].

more than the costs for converting the isobutylene to alkylate. This incremental cost difference is increased very significantly if the isobutylene is obtained by increasing the severity of cat cracker operation or by converting butanes to isobutylene.

2. Compared to using butane and butylenes for alkylate production or blending, carbon dioxide emissions to the atmosphere are significantly increased when the fuel required to produce the MTBE from methane and butane is added to the vehicular emissions. This increase in carbon dioxide emission and consumption of hydrocarbons is on the order of 25 to 50% per gallon of fuel product. The actual value depends on whether isobutylene, isobutane, or normal butane is available as feedstock.

Beginning about 1990, regional and federal environmental regulatory establishments in the United States required gasoline to contain specified minimum amounts of chemically combined oxygen. The published objective was to reduce the amount of carbon monoxide released to the atmosphere. A secondary potential benefit was a desired reduction of atmospheric ozone due to assumed lower reactivity of resulting volatile emissions compared to the replaced hydrocarbons.

These requirements have resulted in most refiners blending MTBE or TAME in the gasoline. Small refineries usually purchase MTBE from merchant plants. In larger refineries, MTBE and TAME are made by reacting isobutylene or isoamylene in the light naphtha from the FCU with purchased methanol.

Other sources of isobutylene include

1. By-product from steam cracking of naphtha or light hydrocarbons for production of ethylene and propylene
2. Catalytic dehydrogenation of isobutane
3. Conversion of tertiary butyl alcohol recovered as a by-product in the manufacture of propylene oxides [1]
4. Light naphtha from refinery coking units

The direct blending of ethanol is used extensively in Brazil and to a lesser extent in the United States and Canada. The ethanol used is produced by fermentation of agricultural products (corn, sugar cane, and so on).

The U.S. Environmental Protection Agency (EPA) has approved several different blends of oxygenates in gasoline. Some of these are listed below.

Oxygenate component	Nominal vol% in gasoline	Maximum wt% O_2 in gasoline
Ethanol	10	3.5
MTBE	15	2.7
Methanol + TBA	4.75 + 4.75	3.5

The upper limits of oxygen content were based on the theory that NOX emissions would increase if these limits are exceeded. The data regarding this concern are not conclusive, and therefore, the limits may not be necessary and may be removed [2].

17.2 ETHER PRODUCTION REACTIONS

Alcohols and isoolefins will react to form ethers according to the following equation:

$$R_1-\overset{\displaystyle R_2}{\overset{|}{C}}=CH_2+R_3OH \rightarrow R_1-\underset{\displaystyle CH_3}{\underset{|}{\overset{\displaystyle R_2}{\overset{|}{C}}}}-OR_3 \tag{17.1}$$

In Equation 17.1, R_1, R_2, and R_3 represent different alkyl groups. These reactions are reversible. Production of ethers is favored by low temperatures and is exothermic. Therefore, cooling of the reactors is required to assume optimum conversion. Reactor temperatures are controlled in the range of 140 to 200°F (60 to 93°C). Pressures are about 200 psig (1380 kPa), just high enough to maintain liquid phase conditions. Acidic resin ion exchange catalysts are used to promote the reactions [6,7].

The general reaction of methanol with an isoolefin is the addition of a methoxy group to the unsaturated tertiary carbon atom. The reactions for producing MTBE and TAME are shown below.

17.2.1 Methyl Tertiary Butyl Ether

$$H_3C-\overset{\displaystyle CH_3}{\overset{|}{C}}=CH_2+CH_3CH \rightarrow H_3C-\underset{\displaystyle CH_3}{\underset{|}{\overset{\displaystyle CH_3}{\overset{|}{C}}}}-OCH_3 \tag{17.2}$$

17.2.2 Tertiary Amyl Methyl Ether

$$\begin{array}{ccccccccccc} & & C & & CH_3 & & & & & CH_3 & \\ & & | & & | & & & & & | & \\ H_3C & - & C & - & C = CH_2 & + & CH_3CH & \rightarrow & H_3C - & C & - OCH_3 \\ & & | & & & & & & & | & \\ & & C & & & & & & & C_2H_5 & \end{array} \tag{17.3}$$

Ethanol reacts in the same way with isobutylene to form ethyl tertiary butyl ether:

$$\begin{array}{ccccccccc} & & CH_3 & & & & & CH_3 & \\ & & | & & & & & | & \\ H_3C & - & C = CH_2 & + & C_2H_5OH & \rightarrow & H_3C - & C & - OC_2H_5 \\ & & & & & & & | & \\ & & & & & & & CH_3 & \end{array} \tag{17.4}$$

17.3 ETHER PRODUCTION PROCESSES

At lease six variations of commercial processes are described in the literature [7,8,9,10]. All of the processes can be modified to react isobutylene or isoamylene with methanol or ethanol to produce the corresponding ether. All use an acidic ion exchange resin catalyst under controlled temperature and pressure conditions. Temperature control of the exothermic reaction is important to maximize conversion and minimize undesirable side reactions and catalyst deactivation. The reaction is usually carried out in two stages with a small excess of alcohol to achieve isoolefin conversions of over 99%.

The basic difference between the various processes is in reactor design and the method of temperature control. The licensed processes include

1. Fixed-bed adiabatic reactors with downflow liquid phase and a cooled external recirculation loop to control temperature
2. Fixed-bed adiabatic reactors, as above, but with a limited amount of vaporization at the outlet to control the temperature
3. Catalyst in tubes, with external cooling
4. Upflow liquid phase "expanded bed" with an external cooling loop

In addition to the above variations, some of the processes use a structured packing containing catalyst in a distillation tower as a second-stage reactor. One such process carrying the trademark Ethermax is licensed by UOP LLC [3,8]. The process is based on technology developments of Huels AG of Germany, Koch Engineering Co., and UOP. Figure 17.1 is a simplified flow diagram of the process. The following process description is based on feeding isobutylene and methanol to produce MTBE.

Isobutylene and a small excess stoichiometric amount of methanol join a controlled quantity of recycle from the reactor effluent and are cooled prior to entering the top of the primary reactor.

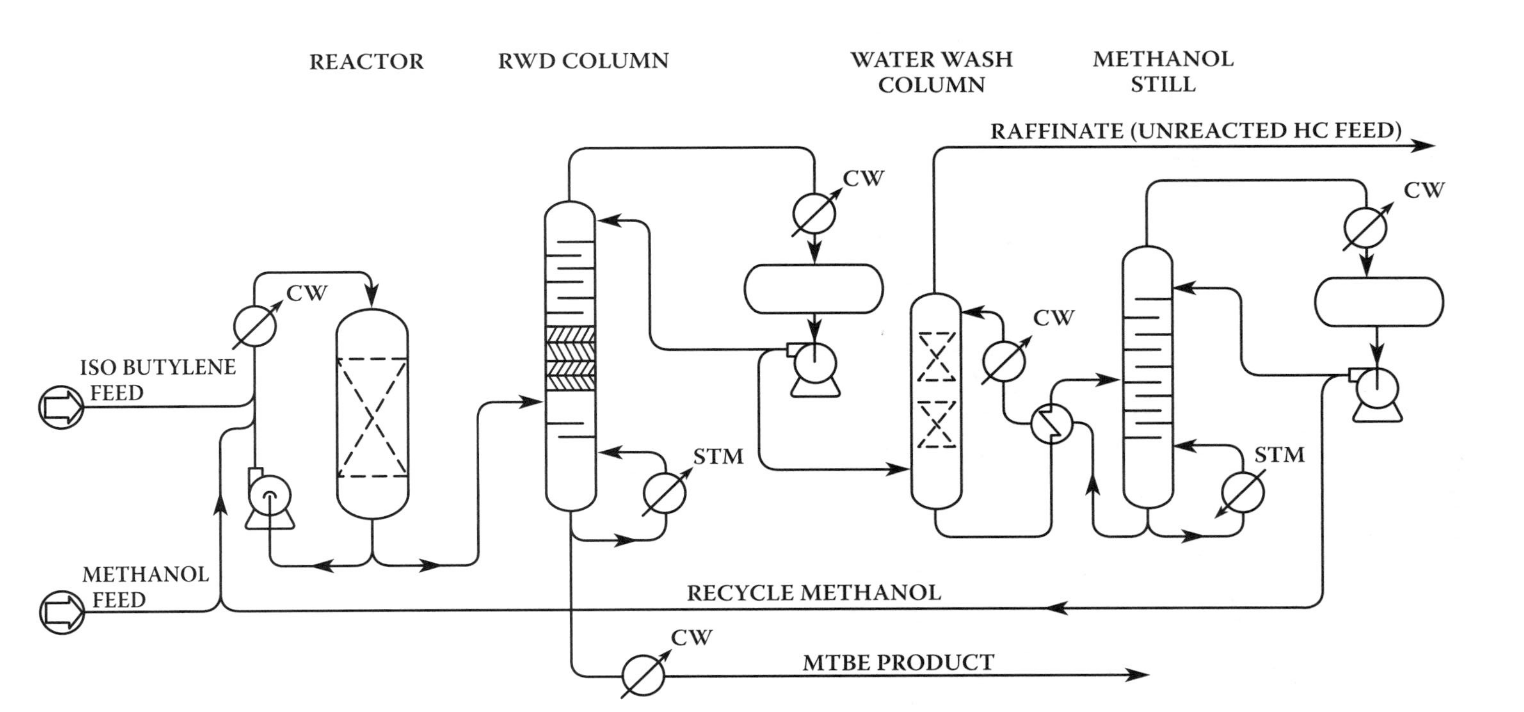

FIGURE 17.1 UOP LLC Ethermax process.

The combined feeds and recycle are all liquids. The resin catalyst in the primary reactor is a fixed bed of small beads. The reactants flow down through the catalyst bed and exit the bottom of the reactor. Effluent from the primary reactor contains MTBE, methanol, and unreacted C_4 olefins and usually some C_4 paraffins, which were in the feed.

A significant amount of the effluent is cooled and recycled to control the reactor temperature. The net effluent feeds a fractionator with a section containing catalyst. This is Koch's proprietary reaction with distillation (RWD) technology; therefore, the fractionator is called the RWD column. The catalyst section, located above the feed entry, is simply structured packing with conventional MTBE resin catalyst between the corrugated mesh plates. MTBE is withdrawn as the bottom product, and unreacted methanol vapor and isobutylene vapor flow up into the catalyst reaction to be converted to MTBE.

The advantages of an RWD column include essentially complete isoolefin conversions. A very detailed discussion of this technology was published by DeGarmo and coworkers [11]. The excess methanol and unreacted hydrocarbons are withdrawn from the RWD column reflux accumulator and fed to a methanol recovery tower. In this tower, the excess methanol is extracted by contact with water. The resultant methanol-water mixture is distilled to recover the methanol, which is then recycled to the primary reaction.

17.4 YIELDS

The overall conversion of isobutylene to MTBE is over 99%, and the methanol consumption is essentially stoichiometric. The volume yields are given below [12]:

$$\begin{aligned}
&1.00\text{ bbl } iC_4^{=} + 0.43\text{ bbl MeOH} \rightarrow 1.26\text{ bbl MTBE}\\
&1.00\text{ bbl } iC_4^{=} + 0.62\text{ bbl EtOH} \rightarrow 1.44\text{ bbl ETBE}\\
&1.00\text{ bbl } iC_5^{=} + 0.38\text{ bbl MeOH} \rightarrow 1.25\text{ bbl TAME}\\
&1.00\text{ bbl } iC_5^{=} + 0.55\text{ bbl EtOH} \rightarrow 1.44\text{ bbl TAEE}
\end{aligned} \tag{17.5}$$

17.5 COSTS FOR ETHER PRODUCTION

Capital costs can be approximated from Figure 17.2 and Table 17.2. The capital and operating costs have been found to correlate best with total hydrocarbon feed (excluding alcohol) as opposed to isoolefin feed or ether production. Therefore, the isoolefin content of the feedstock must be known to estimate the cost for producing a given amount of ether. For example, the C_4 cut of FCC gasoline typically contains 15 to 18 vol% isobutylene. The C_5 to 430°F (220°C) FCC gasoline typically contains about 8 to 12 vol% isoamylene.

17.6 PRODUCTION OF ISOBUTYLENE

In a refinery having both cat cracking and coking operations, the total available isobutylene and isoamylene is only enough to produce oxygenates (MTBE, ETBE,

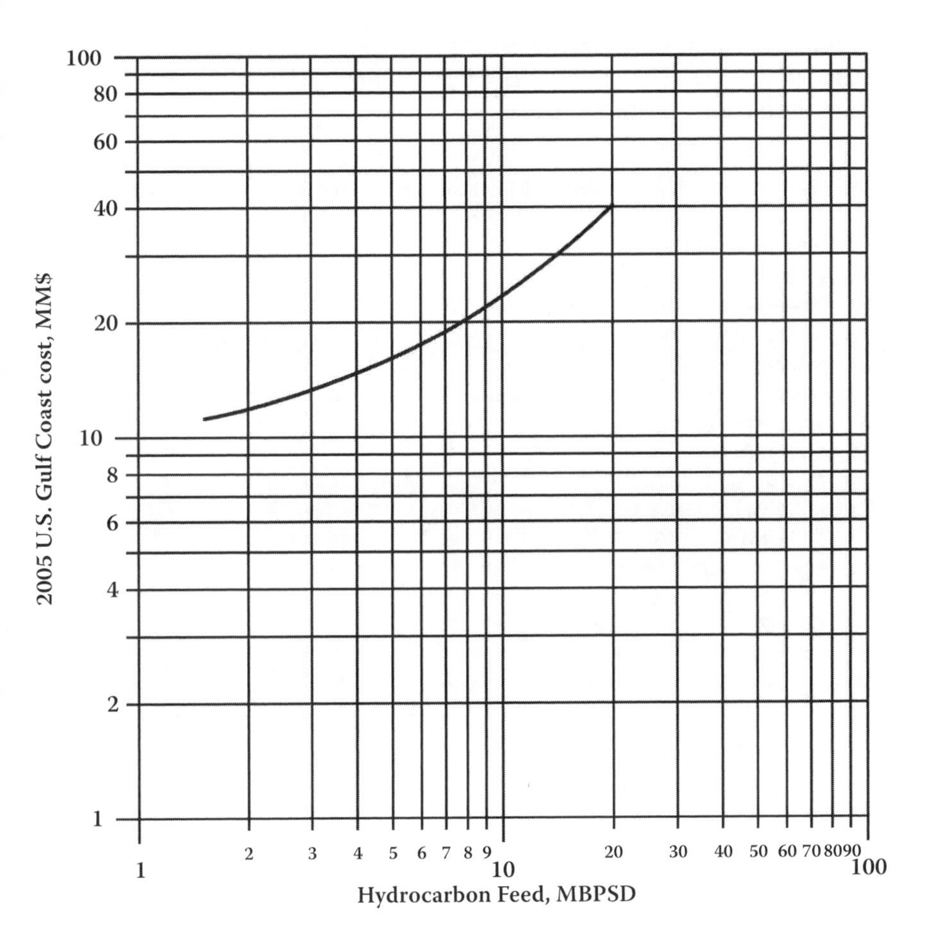

FIGURE 17.2 Ether production facility cost (see Table 17.2).

TAME) equivalent to an oxygen content of 0.3 to 0.6 wt% of the gasoline pool. In order to meet a 2.0 wt% oxygen requirement, additional isobutylene is produced by dehydrogenation of isobutane according to the following reaction:

$$H_3C-\underset{\displaystyle H}{\overset{\displaystyle CH_3}{\underset{|}{\overset{|}{C}}}}-CH_3 \rightarrow H_2C=\overset{\displaystyle CH_3}{\overset{|}{C}}-CH_3+H_2 \tag{17.6}$$

The resulting isobutylene is separated from the hydrogen and then combined with either methanol or ethanol to produce MTBE or ETBE as previously described, or used to make isooctene and isooctane.

The reaction is endothermic (about 53,000 Btu/lb mol at 1000°F). Relatively high temperatures and low pressures are required to obtain reasonable conversion. Figure 17.3 illustrates the conversions theoretically possible, assuming no minor

TABLE 17.2
Ether Production Unit Cost Data

Cost included

1. Two-stage reaction system with MTBE column and methanol recovery system similar to Figure 17.1
2. Initial catalyst fill and paid-up royalty

Costs not included

1. Cooling water, power, and steam supply
2. Feed prefractionation
3. Feed or product storage
4. Raffinate treating for saturation of olefins or trace oxygenates removal

Utility data (per barrel of hydrocarbon feed excluding alcohol)	
Power, kWh/bbl feed	1.0
kWh/MT feed	10.0
Steam, lb/bbl feed, 150 psig	350
kg/MT feed, 1035 kPa	1590
Cooling water, gal/bbl feed (30°F rise)	500
kg/MT feed (17°C rise)	1900
Catalyst replacement, $/bbl feed	0.70
$/MT feed	7.00
Fuel	None

Note: See Figure 17.2.

side reactions occur. The pressures shown are the sum of the equilibrium partial pressures of the isobutane plus isobutylene plus hydrogen. This is the reactor outlet total pressure only if there are no other molecular species present. With current catalysts, it is important to limit the temperatures to about 1100 to 1200°F (593 to 649°C) to minimize formation of undesirable species, which add to the cost of downstream separation facilities. These unwanted products can amount to about 5 to 10 wt% of the feed. They may include primarily methane, ethane, ethene, propane, propene, propadiene, some butane (from the feed), and butene 1 and 2.

A potential alternative to dehydrogenation of isoparaffins for isoolefin production is the isomerization of normal olefins. At present, commercial processes for this alternative are in the development stages.

17.7 COMMERCIAL DEHYDROGENATION PROCESSES

Although thermal, noncatalytic process configurations can be used to produce isobutylene from isobutane, catalytic processes have been favored because of their greater selectivity. Three catalytic processes are used in major installations in North America. These are Houdry's CATOFIN, ConocoPhillips Petroleum's STAR, and UOP's OLEFLEX. A fourth catalytic process licensed by Snamprogetti Yarsintez was developed in Russia.

These processes have single-pass conversions in the range of about 50 to 60%, with selectivities of about 88 to 90%. This means that about 88 to 90 mol% of the

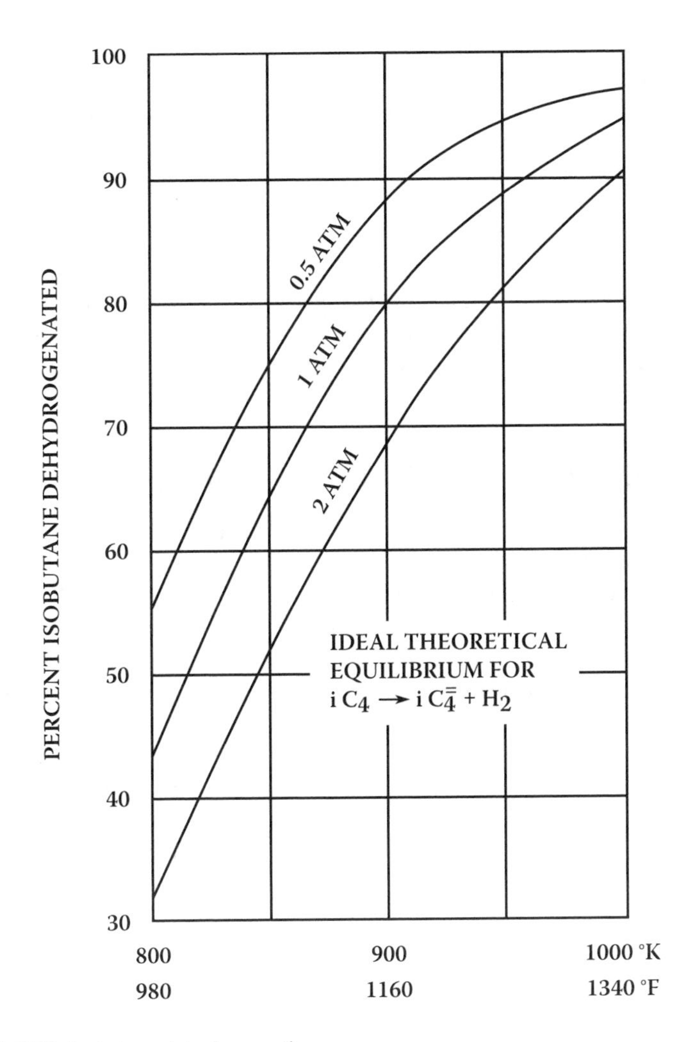

FIGURE 17.3 Isobutane dehydrogenation.

converted isobutane becomes isobutylene. Each one of these processes utilizes very different technology for providing the endothermic heat of reaction and for regenerating the catalyst by burning off the small amount of coke that accumulates on the catalyst. Coke yields are generally less than 0.5 wt% of the feed. A very good table comparing these processes has been made by Maples [13]. The main features of each process are described in the following sections.

17.8 HOUDRY'S CATOFIN [1]

This process, originally developed by Houdry, is now owned by United Catalysts Inc. and licensed by ABB Lummus Crest. The process utilizes three or more horizontal reactors with fixed beds of catalyst. The pelleted catalyst is chromic oxide on alumina. The reactors operate in parallel on offset time cycles in the following sequence:

1. On-stream dehydrogenation of isobutane
2. Steam purge
3. Reheat catalyst with preheated air and simultaneous combustion of coke formed during Step 1
4. Evacuation of residual air
5. Return to on-stream operation

The switching times on individual reactors are offset so that the overall process is continuous. The overall cycle time depends on the size of the reactors and the required feed rate. The total cycle may be repeated on each reactor about two or three times per hour. The sequencing and switching of valves is fully automated with appropriate safety interlocks.

17.9 CONOCOPHILLIPS PETROLEUM'S STAR [5]

This process utilizes a pelleted catalyst containing a noble metal on an alumina base. The catalyst is contained as a fixed bed in tubes that are externally fired to provide the heat of reaction. This allows the reaction to be conducted at essentially isothermal conditions, thereby reducing undesirable thermal cracking reactions. Steam is fed through the catalyst with the hydrocarbon. This lowers the partial pressure of the reactants, consequently allowing higher conversions at a given temperature, about as indicated on Figure 17.2. Steam addition also reduces coke formation and helps to hold the reaction temperature more uniformly. Use of steam as a partial pressure diluent instead of other gases, such as nitrogen, reduces the cost of the downstream separation and recovery facilities, because the steam can be condensed from the reactor effluent. The catalyst does deactivate due to coke deposition, and therefore periodic regeneration is required.

Multiple passes of tubes are provided in the reactor/furnace so that only a small portion of the reactor is in the regeneration step at any one time, thereby allowing the process to be operated continuously. The regeneration is done approximately as follows:

1. Stop hydrocarbon feed while continuing steam flow long enough to purge the reactants
2. Add air to the steam flow
3. Stop air and continue steam flow to purge air
4. Restart hydrocarbon feed

The process equipment may be economically sized so that a given number of catalyst tubes ("bank") can be regenerated in about 1 hr, and regeneration of each such bank occurs only two or three times in a 24-hr period.

17.10 UOP LLC OLEFLEX [3,15]

This process, developed and licensed by UOP LLC, is an extension of two successful UOP LLC technologies:

1. UOP LLC trademarked Pacol process for producing monoolefins from kerosine range paraffins
2. UOP LLC trademarked CCR Platforming process

The catalyst contains platinum on a unique spherical alumina support. The reaction section has three reactors in series with preheat before the first reactor and reheat before the second and third reactors to provide the heat of reaction. The reactants plus hydrogen recycle gas flow radially through the very slowly downward-moving beds of catalyst. Sufficient hydrogen partial pressure is maintained so that coke formation on the catalyst is less than 0.02 wt% of the hydrocarbon feed. This reduces catalyst regeneration requirements to the order of once in a period of about 3 to 10 days. Consequently, catalyst regeneration can be completely discontinued for several days for maintenance, if necessary, without interrupting olefin production. However, normal operation involves the continuous transfer of catalyst from reactor 1 to reactor 2 to reactor 3 to continuous regeneration and then return to reactor 1.

17.11 SNAMPROGETTI YARSINTEZ PROCESS [16]

This process was developed in the former Soviet Union and is essentially similar to a fluid-bed cat cracker. The catalyst is a promoted chromium oxide powder. Isobutane feed vapor and superheated steam fluidize the catalyst in the reactor where the dehydrogeneration occurs. Spent catalyst is withdrawn continuously from the reactor and transferred to the regenerator. Coke is burned off the catalyst in the regenerator, and additional fuel is used to increase the catalyst temperature before returning it to the reactor.

17.12 COSTS TO PRODUCE ISOBUTYLENE FROM ISOBUTANE

Capital costs can vary by 100% for any given installation because of the substantial differences between the various process technologies. As an example, facilities for converting 12,000 BPD of isobutane into about 9,700 BPD of isobutylene may cost somewhere in the range of $112 million to $225 million (basis U.S. Gulf Coast, 2005). This yield implies a conversion of 90% at 90% selectivity. The cost would include dehydrogenation, separation and recycle of unconverted isobutane, initial catalyst, and paid-up royalties. The costs would not include utility supplies, feed or product storage, or butane isomerization.

Utility requirements also vary substantially. Typical 2005 U.S. Gulf Coast costs for labor, utilities, catalysts, and chemicals are in the range of $5 to $10 per barrel of isobutylene produced. These costs do not include the cost of the isobutane feed.

17.13 INTERNATIONAL UNION OF PURE AND APPLIED CHEMISTS NOMENCLATURE

The use of common nomenclature for hydrocarbons and other organic compounds is the general practice in the refining industry and has been followed in this chapter.

TABLE 17.3
Nomenclature for Some Common Ethers

Acronym	Common name	IUPAC name
MTBE	Methyl tertiary butyl ether	2 methyl 2 methoxy propane
ETBE	Ethyl tertiary butyl ether	2 ethyl 2 methoxy propane
TAME	Tertiary amyl methyl ether	2 methyl 2 methoxy butane
TAEE	Tertiary amyl ethyl ether	2 ethyl 2 methoxy butane

Table 17.3 shows a comparison of the common designation and IUPAC nomenclature for some of the ethers that can be blended in gasoline.

17.14 ISOOCTENE AND ISOOCTANE PRODUCTION

In the United States during the late 1990s and early 2000s, states and the U.S. Congress became very concerned about the appearance of MTBE in groundwaters from which cities and individuals obtained their drinking water. Even though the MTBE concentrations were less than those permitted by EPA, there was enough MTBE in the water to give it an unpleasant taste and odor. A number of states outlawed the use of MTBE, and court cases were filed against oil companies. The fear of lawsuits and possible damages have caused the general disuse of ethers to supply octane and oxygen requirements for gasoline blending. The only acceptable oxygenate blending stock is ethanol.

MTBE production equipment in refineries can be modified to use the isobutylene, normally reacted with methanol to produce MTBE, to be isomerized into isooctene and then hydrogenated to make isooctane. This isooctane blending stream has a RON of 99 and a MON of 96, with a blending octane number (BON) of about 97.5.

The same acidic ion exchange resin catalyst used to make MTBE can be used to produce isooctene, but usually a similar catalyst more selective for isooctene production is used. A hydrotreating reactor can be added to the isooctene for conversion to isooctane, or the isooctene can be sent to an existing hydrotreater for conversion.

A simplified process flow diagram is shown in Figure 17.4 [17].

NOTES

1. Clark, R.G. et al., *Energy Prog. 7*(3), pp. 164–169, 1987.
2. *Oil Gas J. 91*(3), p. 46, 1993).
3. Gregor, J.H., Bakas, S.T., and Vlowetz, M.A., *Proceedings of 71st GPA Annual Convention,* pp. 230–235, 1992.
4. Prezelj, M., *Hydrocarbon Process. 66*(9), pp. 68–70, 1987.
5. Unzelman, G.H., *Oil Gas J.* April 10, pp. 23–37, 1989.
6. Ancillotti, F., *J. Cat. 46*, pp. 49–57, 1977.
7. Des Couriers, J., Etherification of C_4/C_5 Cuts: Which Route to Choose, NPRA Annual Meeting, March 25–27, 1990, San Antonio.
8. *Hydrocarbon Process. 72*(3), pp. 193–195, 1993.

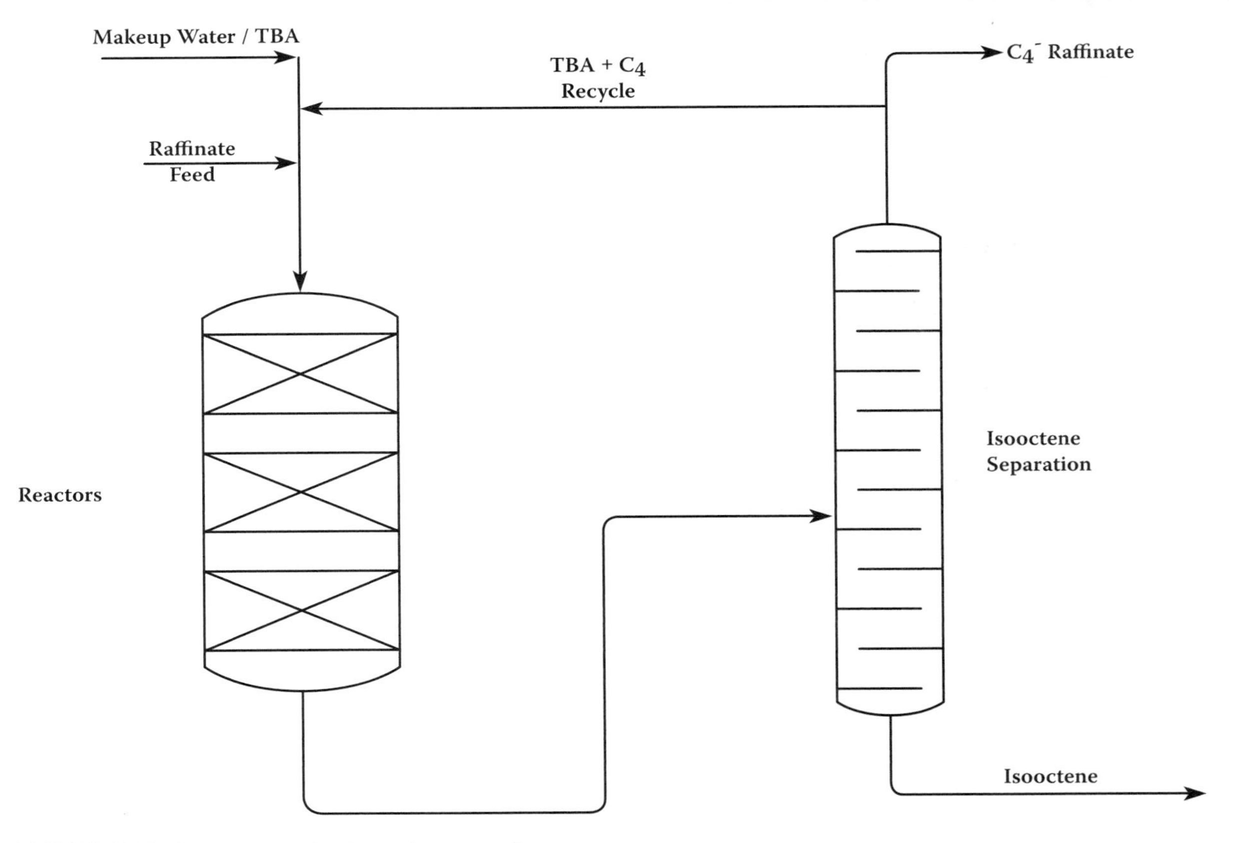

FIGURE 17.4 Isooctene production unit process flow.

9. Pescarollo, E., Trotta, R., and Sarathy, P.R., *Hydrocarbon Process. 72*(2), pp. 53–59, 1993.
10. Rock, K.L., Dunn, R.O. and Makovec, D.J., *Proceedings of 70th GPA Annual Convention*, pp. 222–241, 1991.
11. DeGarmo, J.L., Parulekar, V.N., and Pinjola, V., *Chem. Eng. Prog. 3*, pp. 43–50, 1992.
12. Piel, W.J., *Oil Gas J.* 87, December 4, pp. 40–43, 1989.
13. Maples, R.E., *Petroleum Refinery Process Economics,* PennWell Publishing Co., Tulsa, OK, 1993.
14. Dunn, R.O., Brinkmeyer, F.M., and Schuette, G.F., The Phillips Steam Active Reforming (STAR) Process, NPRA Annual Meeting, March 22–24, 1992, New Orleans.
15. *Hydrocarbon Process. 72*(3), p. 171, 1993.
16. Sarathy, P.R. and Suffridge, G.S., *Hydrocarbon Process. 72*(2), pp. 43–51, 1993.
17. Sloan, H.D. et al., Iso-octane Production from C_4's as an Alternative to MTBE, NPRA Annual Meeting, March 26–28, 2000.

18 Cost Estimation

Although detailed discussions of various capital cost estimating methods are not part of the intended scope of this work, some comments are pertinent. All capital cost estimates of industrial process plants can be classified as one of four types:

1. Rule-of-thumb estimates
2. Cost-curve estimates
3. Major equipment factor estimates
4. Definitive estimates

The capital cost data presented in this work are of the second type—cost-curve estimates.

18.1 RULE-OF-THUMB ESTIMATES

The rule-of-thumb estimates are, in most cases, only an approximation of the order of magnitude of cost. These estimates are simply a fixed cost per unit of feed or product. Some examples are

Complete coal-fired electric power plant: $3,250/kW
Complete synthetic ammonia plant: $260,000/TPD
Complete petroleum refinery: $33,000/BPD

These rule-of-thumb factors are useful for quick ballpark costs. Many assumptions are implicit in these values, and the average deviation from actual practice can often be more than 50%.

18.2 COST-CURVE ESTIMATES

The cost-curve method of estimating corrects for the major deficiency illustrated above by reflecting the significant effect of size or capacity on cost. These curves indicate that costs of similar process units or plants are related to capacity by an equation of the following form:

$$\frac{\text{Plant A cost}}{\text{Plant B cost}} = \frac{(\text{Plant A capacity})}{(\text{Plant B capacity})^{X}} \tag{18.1}$$

The relationship was reported by Lang [1], who suggested an average value of 0.6 for the exponent (X). Other authors have further described this function [2,3]. Cost curves of this type have been presented for petroleum refinery costs in the past [4,5].

The curves presented herein have been adjusted to eliminate certain costs such as utilities, storage, offsite facilities, and location cost differentials. Separate data are given for estimating the costs of these items. The facilities included have been defined in an attempt to improve accuracy.

It is important to note that most of the cost plots have an exponent that differs somewhat from the 0.6 value. Some of the plots actually show a curvature in the log–log slope that indicates that the cost exponent for these process units varies with capacity. Variations in the log–log slope (cost exponent) range from about 0.5 for small-capacity units up to almost 1.0 for very large units. This curvature, which is not indicated in the previously published cost curves, is due to paralleling equipment in large units and to disproportionately higher costs of very large equipment such as vessels, valves, and pumps. The curvature in the log–log slope of cost plots has been described by Chase [6].

The cost-curve method of estimating, if carefully used and properly adjusted for local construction conditions, can predict actual costs within 25%. Except in unusual circumstances, errors will probably not exceed 50%.

18.3 MAJOR EQUIPMENT FACTOR ESTIMATES

Major equipment factor estimates are made by applying multipliers to the costs of all major equipment required for the plant or process facility. Different factors are applicable to different types of equipment, such as pumps, heat exchangers, and pressure vessels [7]. Equipment size also has an effect on the factors.

It is obvious that prices of major equipment must first be developed to use this method. This requires that heat and material balances be completed in order to develop the size and basic specifications for the major equipment.

This method of estimating, if carefully followed, can predict actual costs within 10 to 20%.

A shortcut modification of this method uses a single factor for all equipment. A commonly used factor for petroleum refining facilities is 4.5. The accuracy of using this shortcut is, of course, lower than when using individual factors.

18.4 DEFINITIVE ESTIMATES

Definitive cost estimates are the most time-consuming and difficult to prepare but are also the most accurate. These estimates require preparation of plot plans, detailed flow sheets, and preliminary construction drawings. Scale models are sometimes used. All material and equipment are listed and priced. The number of worker-hours for each construction activity is estimated. Indirect field costs, such as crane rentals, costs of tools, and supervision, are also estimated.

This type of estimate usually results in an accuracy of ±5%.

18.5 SUMMARY FORM FOR COST ESTIMATES

The items to be considered when estimating investment from cost curves are

Process units
Storage facilities
Steam systems
Cooling water systems
  Subtotal A
Offsites
  Subtotal B
Special costs
  Subtotal C
Location factor
  Subtotal D
Contingency
  Total

18.6 STORAGE FACILITIES

Storage facilities represent a significant item of investment costs in most refineries. Storage capacity for crude oil and products varies widely at different refineries. In order to properly determine storage requirements, the following must be considered: the number and type of products, method of marketing, source of crude oil, and location and size of the refinery.

Installed costs for "tank farms" vary from $50 to $130 per barrel of storage capacity. This includes tanks, piping, transfer pumps, dikes, fire protection equipment, and tank-car or truck loading facilities. The value is applicable to low vapor-pressure products such as gasoline and heavier liquids. Installed costs for butane storage ranges from $100 to $130 per barrel, depending on size. Costs for propane storage range from $120 to $160 per barrel.

18.7 LAND AND STORAGE REQUIREMENTS

Each refinery has its own land and storage requirements, depending on location with respect to markets and crude supply, methods of transportation of the crude and products, and number and size of processing units. Availability of storage tanks for short-term leasing is also a factor, as the maximum amount of storage required is usually based on shutdown of processing units for turnaround at 18- to 24-month intervals rather than on day-to-day processing requirements. If sufficient rental tankage is available to cover turnaround periods, the total storage and land requirements can be materially reduced, as the land area required for storage tanks is a major portion of refinery land requirements.

Three types of tankage are required: crude, intermediate, and product. For a typical refinery that receives the majority of its crude by pipeline and distributes its products in the same manner, about 13 days of crude storage and 25 days of product storage should be provided. The 25 days of product storage is based on a 3-week shutdown of a major process unit. This generally occurs only every 18 months to 2 years, but sufficient storage is necessary to provide products to customers over this extended period. A rule-of-thumb figure for total tankage, including intermediate storage, is approximately 50 barrels of storage per BPD crude oil processed.

The trend to producing many types of gasolines and oxygenated components can increase blending and product storage requirements by about 50%.

Nelson indicates that in 1973, the average refinery had 69 days of total storage capacity [8]. In 2005, the range was about 25 to 90 days of total storage.

The land requirements are frequently dictated by considerations other than process or storage because of the desire to provide increased security, to be isolated from neighboring buildings, and so on. If operational matters are the prime consideration, the land necessary for operational and storage facilities is about 4 acres per 1000 BPCD crude capacity.

Nelson has summarized land used for 32 refineries from 1948 to 1971 [9]. The land in use when the refineries were built ranged as a function of refinery complexity from 0.8 acre to 5.7 acres per 1000 BPD crude capacity. Land actually purchased, though, was much more than this, and varied from 8 to 30 acres per 1000 BPD. This additional land provided a buffer zone between the refinery and adjacent property and allowed for expanding the capacity and complexity of the refinery. In a summary article on land and storage costs [10], Nelson suggests that 5 acres of land per 1000 BPCD crude capacity be used for planning purposes.

Land provided for growth and expansion to other processes, such as petrochemicals, should not be included in the investment costs, against which the return on investment is calculated. Land purchased for future use should be charged against the operation for which it is intended against overall company operations.

18.8 STEAM SYSTEMS

An investment cost of $105 per lb/hr of total steam generation capacity is used for preliminary estimates. This represents the total installed costs for gas- or oil-fired, forced-draft boilers, operating at 250 to 300 psig, and all appurtenant items such as water treating, deaerating, feed pumps, yard piping for steam, condensate, and stack gas cleanup.

Total fuel requirements for steam generation can be assumed to be 1200 Btu (lower heating value, or LHV) per pound of steam.

A contingency of 25% should be applied to preliminary estimates of steam requirements.

Water makeup to the boilers is usually 5 to 10% of the steam produced.

18.9 COOLING WATER SYSTEMS

An investment cost of $130 per gpm of total water circulation is recommended for preliminary estimates. This represents the total installed costs for a conventional induced-draft cooling tower, water pumps, water treating equipment, and water piping. Special costs for water supply and blowdown disposal are not included.

The daily power requirements (kWh/day) for cooling water pumps and fans is estimated by multiplying the circulation rate in gpm by 0.6. This power requirement is usually a significant item in total plant power load and should not be ignored.

The cooling tower makeup water is about 5% of the circulation. This is also a significant item and should not be overlooked.

An "omission factor," or contingency, of 15% should be applied to the cooling water circulation requirements.

18.10 OTHER UTILITY SYSTEMS

Other utility systems required in a refinery are electric power distribution, instrument air, drinking water, fire water, sewers, waste collection, and others. Because these are difficult to estimate without detailed drawings, the cost is normally included in the offsite facilities.

18.10.1 OFFSITES

Offsites are the facilities required in a refinery that are not included in the costs of major facilities. A typical list of offsites is shown below:

Electric power distribution
Fuel oil and fuel gas facilities
Water supply, treatment, and disposal
Plant air systems
Fire protection systems
Flare, drain, and waste containment systems
Plant communication systems
Roads and walks
Railroads
Fences
Buildings
Vehicles
Product and additives blending facilities
Product loading facilities

Obviously, the offsite requirements vary widely between different refineries.

The values shown below can be considered as typical for grassroots refineries when estimated as outlined in this text.

Crude oil feed (BPSD)	Offsite costs (percent of total major facilities costs[a])
Less than 30,000	50
30,000–100,000	30
More than 100,000	20

[a] Major facilities as defined herein include process units, storage facilities, cooling water systems, and steam systems.

Offsite costs for the addition of individual process units in an existing refinery can be assumed to be about 20 to 25% of the process unit costs.

18.10.2 Special Costs

Special costs include the following: land, spare parts, inspection, project management, chemicals, miscellaneous supplies, and office and laboratory furniture. For preliminary estimates, these costs can be estimated as 4% of the cost of the process units, storage, steam systems, cooling water systems, and offsites. Engineering costs and contractor fees are included in the various individual cost items.

18.10.3 Contingencies

Most professional cost estimators recommend that a contingency of at least 15% be applied to the final total cost determined by cost-curve estimates of the type presented herein.

The term *contingencies* covers many loopholes in cost estimates of process plants. The major loopholes include cost data inaccuracies when applied to specific cases and lack of complete definition of facilities required.

18.10.4 Escalation

All cost data presented in this book are based on U.S. Gulf Coast construction averages for the year 2005. This statement applies to the process unit cost curves, as well as values given for items such as cooling water systems, steam plant system, storage facilities, and catalyst costs. Therefore, in any attempt to use the data for current estimates, some form of escalation or inflation factor must be applied. Many cost index numbers are available from the federal government and from other published sources. Of these, the Chemical Engineering Plant Cost Index and the Nelson-Farrar Refinery (Inflation) Index are the most readily available and probably the most commonly used by estimators and engineers in the U.S. refining industry.

The use of these indices is subject to errors inherent in any generalized estimating procedure, but some such factor must obviously be incorporated in projecting costs from a previous time basis to a current period. It should be noted that the contingencies discussed in the previous section are not intended to cover escalation.

Escalation or inflation of refinery investment costs is influenced by items that tend to increase costs as well as by items that tend to decrease costs. Items that increase costs include obvious major factors, such as

1. Increased cost of steel, concrete, and other basic materials on a per-ton basis
2. Increased cost of construction labor and engineering on a per-hour basis
3. Increased costs of excessive safety standards and exaggerated pollution control regulations
4. Increase in the number of reports and amount of superfluous data necessary to obtain local, state, and federal construction permits

Items that tend to decrease costs are basically all related to technological improvements. These include

1. Process improvements developed by the engineers in research, design, and operation
2. More efficient use of engineering and construction personnel

Examples of such process improvements include improvement of fractionator tray capacities, improved catalysts that allow smaller reactors, and improved instrumentation, allowing for consistently higher plant feed rates.

18.10.5 Plant Location

Plant location has a significant influence on plant costs. The main factors contributing to these variations are climate and its effect on design requirements and construction conditions; local rules, regulations, codes, and taxes; and availability and productivity of construction labor.

Relative hydrocarbon process plant costs on a 2005 basis at various locations are given below:

Location	Relative cost
U.S. Gulf Coast	1.0
Los Angeles	1.4
Portland, Seattle	1.2
Chicago	1.3
St. Louis	1.5
Detroit	1.3
New York	1.7
Philadelphia	1.5
Alaska, North Slope	3.0
Alaska, Anchorage	2.0

18.11 APPLICATION OF COST ESTIMATION TECHNIQUES

Although economic evaluation will not be discussed until the next chapter, an example problem illustrating the methods to estimate capital and operating costs and return on investment is included here to aid in clarifying the principles discussed in this chapter. The illustrative problem is relatively simple, but the same techniques and procedures can be applied to the most complex refinery economic evaluation.

18.11.1 Statement of Problem

For the following example of a simplified refinery, calculate

1. The products available for sale
2. Investment

3. Operating costs
4. Simple rate of return on investment
5. True rate of return on investment

Also prepare a basic block flow diagram (Figure 18.1). The following data are available:

1. Crude charge rate: 30,000 BPSD
2. Crude oil sulfur content: 1.0 wt%
3. Full-range naphtha in crude: 4,000 BPSD
   240° MBP
   56°API gravity
   11.8 K_W
4. Light gas oil in crude: 4,000 BPSD
5. Heavy gas oil in crude: 4,000 BPSD
6. Vacuum gas oil in crude: 6,000 BPSD
7. Vacuum residual in crude: 12,000 BPSD
8. On-stream factor: 93.15%
9. Cost of makeup water: $0.32/1000 gal
10. Cost of power: $0.12 kWh
11. LHV of heavy gas oil: 5.5 MMBtu/bbl
12. Replacement cost for desulfurizer catalyst is $1.60/lb
13. Replacement cost for reformer catalyst is $8/lb
14. Insurance annual cost is 0.5% of plant investment
15. Local taxes annual cost is 1.0% of plant investment
16. Maintenance annual cost is 5.5% of plant investment
17. Misc. supplies annual cost is 0.15% of plant investment
18. Average annual salary plus payroll burden for plant staff and operators is $72,000
19. Value of crude oil and products at refinery is

	$/bbl
Crude	60.00
Gasoline	83.00
Light gas oil	79.00
Heavy gas oil	75.00
Vacuum gas oil	73.00
Vacuum residual	55.00

20. Depreciation allowance: 15 year, straight-line
21. Corporate income tax: 50% of taxable income
22. Location: St. Louis 2007
23. Construction period: 2007
24. Escalation rate (applicable to construction costs only) is 3% per year

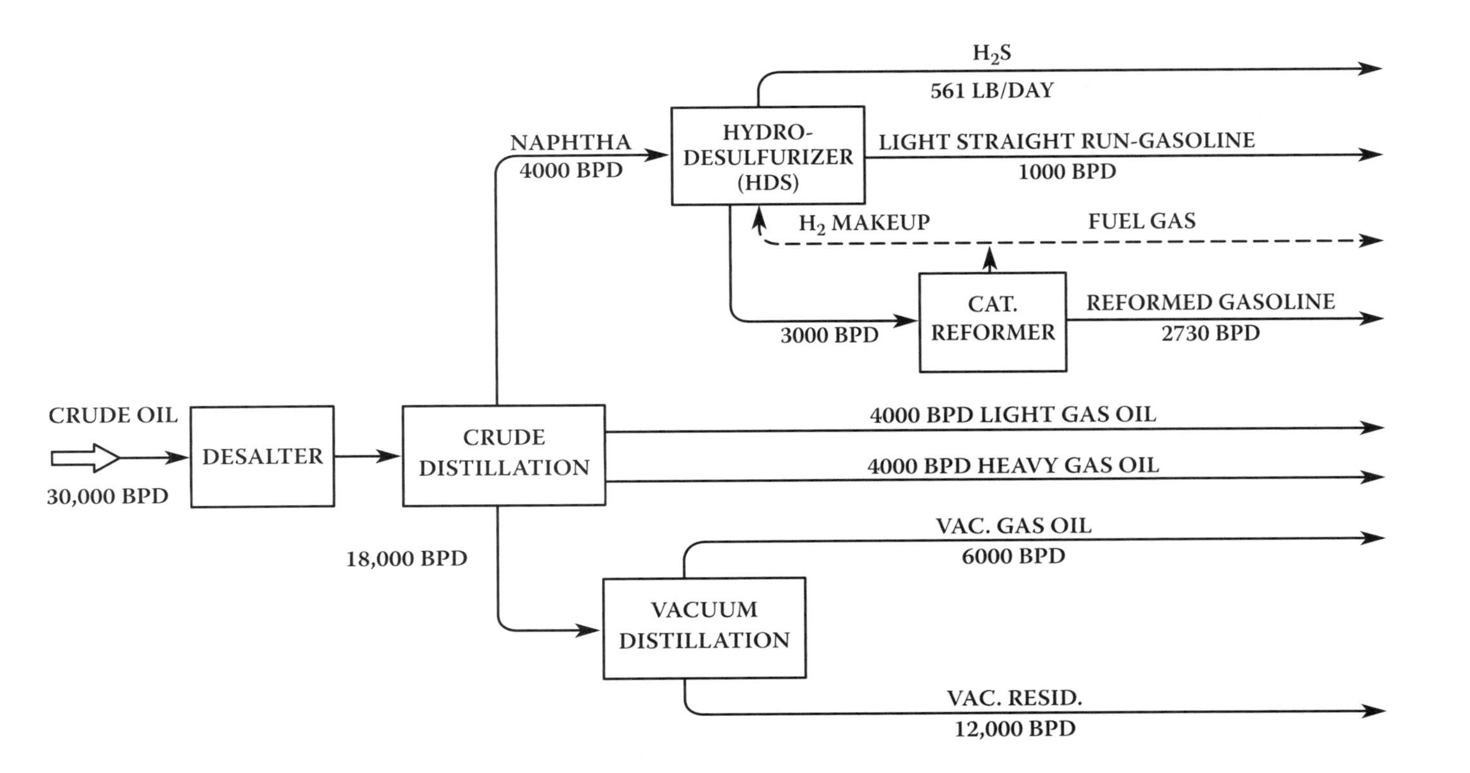

FIGURE 18.1 Block flow diagram for example problem.

18.11.2 Process Requirements

The crude oil is to be desalted and fractionated to produce full-range naphtha, light gas oil, heavy gas oil, and "atmospheric bottom." The latter cut is fed to a vacuum unit for fractionation into vacuum gas oil and vacuum residuals. The full-range naphtha is to be hydrodesulfurized. After desulfurization, the light straight-run (LSR) portion (i.e., the material boiling below 180°F) of the full-range naphtha is separated for blending into the gasoline product. The balance of the naphtha is fed to a catalytic reformer, which is operated to produce a reformate having a research octane number (clear) of 93. The reformate plus the LSR are mixed to make the final gasoline product. Propane and lighter hydrocarbons, including the hydrogen, which are produced in the catalytic reformer, are consumed as fuel. The necessary hydrogen makeup for the hydrodesulfurizer is, of course, taken from these gases before they are burned as fuel. The balance of the fuel requirement is derived from light gas oil. The hydrogen sulfide produced in the hydrodesulfurizer is a relatively small amount and is burned in an incinerator. No other product treating is required. It can be assumed that sufficient tankage for approximately 12 days' storage of all products is required. The total storage requirement will thus be approximately 360,000 barrels.

The preceding information, in addition to that contained in this book, is sufficient for solution of the problem.

18.11.3 Catalytic Reformer

Calculate the properties of feed to the reformer, given total naphtha stream properties as follows:

Midboiling point: 240°F
API gravity: 56°API
K_W: 11.8

The material boiling below 180°F (LSR) is not fed to the reformer. After desulfurizing the total naphtha, the LSR is fractionated out. The problem is to estimate the volume and weight of the LSR, assuming a distillation curve is not available. This LSR is then deducted from the total naphtha to find the net reformer feed. This is done as shown in the following steps:

1. Assume that butane and lighter hydrocarbons in the naphtha are negligible. Thus, the lightest material would be isopentane (iC_5) with a boiling point of 82°F. Hence, the midboiling point of the LSR would be approximately (82 + 180)(0.5) = 131°F.
2. Assume that the LSR has the same K_W as the total naphtha (i.e., 11.8).
3. From general charts (see Reference 5, Chapter 2) relating K_W, mean average boiling point, and gravity, find the gravity of the LSR. This is 76.5°API.
4. The gravity of the naphtha fraction boiling above 180°F is next determined by a similar procedure. The midboiling point for the total naphtha is given above as 240°F, and the initial boiling point was estimated in Step 1 as 82°F. Therefore, the naphtha end point can be estimated as shown:

$$240 + (240 - 82) = 398°F$$

Now the approximate midboiling point of the reformer feed is estimated as

$$(180 + 398)(0.5) = 289°F$$

Using a K_W of 11.8, the reformer feed gravity is found (as in Step 3) to be 52.5°API.

5. With the above estimates of gravity of both the LSR and reformer feed, it is now possible to estimate the relative amounts of each cut that will exist in the total naphtha stream. This is done by weight and volume balances as shown below. Use simplified nomenclature:

V_{LSR} = gallons LSR
V_{RF} = gallons reformer bed
V_N = gallons total naphtha
$W_{LS}R$ = pounds LSR
W_{RF} = pounds reformer feed
W_N = pounds total naphtha
lb/gal LSR = 5.93(67.5°API)
lb/gal RF = 6.41 (52.5°API)
lb/gal V_N = 6.29 (56°API)

Volume balance (hourly basis):

$$V_N = (4000)(42 / 24) = 7000 \text{ gal/hr}$$
$$V_{LSR} + V_{RF} = 7000 \text{ gal/hr}$$

Weight balance (hourly basis):

$$V_N (6.29) = (7000)(6.29) = 44{,}030 \text{ lb/hr}$$
$$V_{LSR}(5.93) + V_{RF} (6.41) = 44{,}030 \text{ lb/hr}$$

Simultaneous solution of the two equations for V_{LSR} and V_{RF} gives:

$$V_{LSR} = 1750 \text{ gal/hr} = 1000 \text{ BPD}$$
$$V_{RF} = 5250 \text{ gal/hr} = 3000 \text{ BPD}$$

6. The above information should then be tabulated as shown below to ascertain that all items balance.

Stream	°API	lb/gal	gal/hr	lb/hr	BPD
LSR	67.5	5.93	1750	10,378	1000
Reformer feed	52.5	6.41	5250	33,652	3000
Total naphtha			7000	44,030	4000

7. Before proceeding, it should be emphasized that the above methods for approximating the naphtha split into LSR and reformer feed are satisfactory for preliminary cost and yield computations, such as this example, but not for final design calculations.
8. The reformer feed properties can now be used with the yield curves in the text. The following yields are based on production of 93 RON reformate. From the yield curves:

   vol% C_5^+: 86.0
   vol% C_4s: 5.0 ($iC_4 / nC_4 = 41.5 / 58.5$)
   wt% C_1 and C_2: 1.1
   wt% C_3: 1.92
   wt% H_2: 1.75

   With the above data, complete the following table:

Component	gal/hr	lb/gal	lb/hr	BPD	Mscf/day
H_2			589		2682
C_1 and C_2			370		145[a]
C_3	153	4.23	646		133
iC_4	109	4.69	511	62	
nC_4	154	4.86	748	88	
C_5^+	4515	6.82	30,788[b]	2580	
Total			33,652	2730	2960
Feed	5250	6.29	33,652	3000	—

[a] Assume lb C_1/lb C_2 = 0.5 (i.e., C_1, C_2 = 23.3 mol wt).
[b] lb/hr C_5^+ obtained by difference from total feed less other products.

18.11.4 NAPHTHA DESULFURIZER

Assume crude oil = 1.0% sulfur. Then, from the curve for miscellaneous crudes, naphtha contains 0.05% sulfur (240°F MBP). Calculate the amount of sulfur produced. Assume K_W for desulfurized feed is 11.8. This, combined with 240°F MBP, gives a naphtha API gravity of 56°API (see reformer calculations).

56°API = 6.29 lb/gal
wt% S in naphtha = S_N = (4000)(42)(6.29)(0.0005) = 528 lb/day
Maximum H_2S formed = 32(528) = 561 lb/day
Theoretical H_2 required = 561 − 528
= 33 lb/day
= 16.5 mol/day
= 6.26 Mscf/day

The makeup H_2 required (see Reference 9) is about 100 to 150 scf/bbl, or

(4000)(0.15) = 600 Mscf/day

TABLE 18.1
Summary of Investment and Utilities Costs

	BPSD	\$ (×10³) 2005 Gulf Coast	Cooling water (gpm)	lb/hr stm.	kW	Fuel (MMBtu/hr)
Desalter	30,000	2,800	63		20	
Crude unit	30,000	50,000	3,125	12,500	1125	63
Vacuum unit	18,000	22,000	1,875	7,500	225	23
Naphtha desulfurizer	4,000	10,000	833	1,000	333	17
Reformer	3,000	17,000	833	3,750	375	38
Initial catalyst (desulfurizer)		Included				
Initial catalyst (reformer)		960				
Subtotal		102,760	6,729	24,750	2078	141
Cooling water system, 7,740 gpm[a]		1,071			194	
Steam system, 30,900 lb/hr[b]		3,214				37
Subtotal		107,045	6,729	24,750	2272	178
Storage[c]	12 days[d]	23,400				
Subtotal		130,045				
Offsites[c]	(30%)	39,134				
Subtotal		169,579				
Location factor	1.5					
Spec cost factor	1.04[e]					
Contingency	1.15[e]					
Escalation	(1.03)[f]					
Total		313,047				

[a] Add 15% excess capacity to calculated cooling water circulation.
[b] Add 25% excess capacity to calculated steam supply.
[c] Individual values for utilities in the storage and offsite categories are accounted for by notes a and b.
[d] 360,000 barrels at an average cost of \$50.00/bbl.
[e] These factors are compounded.
[f] This is the projected cost at the location in St. Louis in 2007. No paid-up royalties are included.

Hydrogen from catalytic reformer is 2682 Mscf/day, which is more than adequate. Table 18.1 gives a summary of investment and utilities costs.

18.11.5 Calculation of Direct Annual Operating Costs

After completing the investment and yield calculations, the annual operating costs of the refinery can be determined. Operating costs can be considered to include three major categories:

1. Costs that vary as a function of plant throughput and on-stream time. These include water makeup to the boilers and cooling tower, electric power, fuel, running royalties, and catalyst consumption.

2. Costs that are a function of the plant investment. These include insurance, local taxes, maintenance (both material and labor), and miscellaneous supplies.
3. Costs that are determined by the size and complexity of the refinery. These include operating, clerical, technical, and supervisory personnel.

The following sections illustrate development of the above costs.

18.11.5.1 On-Stream Time

Refineries generally have an on-stream (full capacity) factor of about 92 to 96%. For this example, a factor of 93.15% (340 days per year) is used.

18.11.5.2 Water Makeup

1. To cooling tower (30°FΔt):

   1% evaporation for 10°FΔt
   1/2% windage loss
   1% blowdown to control solids concentration
   Cooling tower, makeup = (3) (1%) + 1/2% + 1% = 4 1/2%
   Makeup = 0.045 × 7740 gpm = 348 gpm

2. To boiler:

   Average boiler blowdown to control solids concentration can be assumed to be 5%.
   Boiler makeup = (0.05)(30,900) = 1545 lb/hr = 3.1 gpm.
   Total makeup water = 351 gpm.
   Average cost to provide makeup water is approximately $0.32/1000 gal.
   Therefore, annual water makeup cost is $(351)(1440)(340)(0.32)(10^{-3})$ = $25,068

18.11.5.3 Power

Industrial power costs range from $0.10/kWh (in locations where there is hydro-electric power) to $0.24/kWh. For this example, use $0.24/kWh.

Power cost = (2212)(24)(340)(0.12)
= $2,040,653 per year

18.11.5.4 Fuel

In this example, no separate charge will be made for fuel, because it is assumed that the refinery will use some of the heavy gas oil products for fuel. The amount of gas oil consumed must be calculated, so that this quantity can be deducted from the products available for sale.

From the summary tabulation of utilities, we require 178 MMBtu/hr for full-load operation. This fuel is supplied by combustion of reformer off-gas supplemented with heavy gas oil. Some of the reformer off-gas is consumed in the hydrodesulfurizer, and this quantity (hydrogen portion only) must be deducted from available fuel. A fuel balance is made as shown below to determine the amount of heavy gas oil consumed as fuel.

Step 1: From reformer calculations the available fuel gas is

Component	Total (lb/hr)	HDS[a] usage (lb/hr)	Available for fuel (lb/hr)
H_2	589	132	457
C_1	123		123
C_2	246		246
C_3	646		646
Total	1604	132	1472

[a] From desulfurizer calculations, hydrogen makeup was 600 Mscfd.

In petroleum work, "standard conditions" are 60°F and 14.696 psig. At these conditions, 1 lb mole = 379.5 scf. Thus, hydrogen consumed in the HDS unit is

$$\frac{60,000}{(24)(379.5)}(2) = 132 \text{ lb/hr}$$

Step 2: The calculated lower heating value (LHV) of available fuel gas is

Component	Total (lb/hr)	LHV[a] (Btu/lb)	LHV (MMBtu/hr)
H_2	457	51,600	23.6
C_1	123	21,500	2.6
C_2	246	20,420	5.0
C_3	646	19,930	12.9
Total	1472		44.1

[a] From Note 11 and Note 12 or other convenient source.

Step 3: Heavy gas oil required for fuel. Assume 5.5 MMBtu/bbl LHV, then:

$$\frac{178 - 44.1}{5.5}(24) = 584 \text{ BPSD}$$

Step 4: Heavy gas oil remaining for sale:

$$4000 - 584 = 3416 \text{ BPSD}$$

18.11.5.5 Royalties

The reformer is a proprietary process, and therefore, royalties must be paid. On a running basis, these range from \$0.08 to \$0.15 per barrel of feed. For this example, use a value of \$0.10:

$$\text{Annual cost} = (0.10)(3000)(340) = \$102{,}000$$

18.11.5.6 Catalyst Consumption

Catalyst consumption costs are as follows:

Desulfurizer: 0.002 lb/bbl; \$1.60/lb:
Annual cost = (4000)(340)(0.002)(1) = \$4,320
Reformer: 0.004 lb/bbl; \$7 per lb:
Annual cost = (3000)(340)(0.004)(7) = \$28,560
Total catalyst cost: \$4,320 + \$28,560 = \$32,880/yr

18.11.5.7 Insurance

This cost usually is 0.5% of the plant investment per year.

$$(\$313{,}047{,}000)(0.005) = \$1{,}565{,}235$$

18.11.5.8 Local Taxes

Usually, local taxes account for 1% of the plant investment per year:

$$(\$313{,}047{,}000)(0.01) = \$3{,}130{,}470/\text{yr}$$

18.11.5.9 Maintenance

This cost varies between 3 and 8% of plant investment per year. For this example, use an average value of 5.5% (includes material and labor):

$$(\$313{,}047{,}000)(0.055) = \$17{,}217{,}585/\text{yr}$$

18.11.5.10 Miscellaneous Supplies

This item includes miscellaneous chemicals used for corrosion control, drinking water, office supplies, and so on. An average value is 0.15% of the plant investment per year.

$$(\$313{,}047{,}000)(0.0015) = \$469{,}705/\text{yr}$$

18.11.5.11 Plant Staff and Operators

The number of staff personnel and operators depends on plant complexity and location. For this example, the following staff could be considered typical of a modern refinery:

	No. per shift	Total payroll
Refinery manager		1
Operations manager		1
Maintenance manager		1
Engineers		3
Operators	4	18
Lab personnel		2
Technicians		2
Clerical personnel		4
Total		32

Assume the average annual salary plus payroll burden is $72,000 per person. Total annual cost for staff and operators is thus:

$$(\$72,000)(32) = \$2,304,000/\text{yr}$$

Note that maintenance personnel are not listed above because this cost was included with the maintenance item. Also note that it takes about 4 1/2 workers on the payroll for each shift job to cover vacations, holidays, illness, and fishing time.

Table 18.2 gives a summary of direct annual operating costs.

18.11.5.12 Calculation of Income before Income Tax

Sales are summarized in the following table:

Product	BPD	MBPY	$/bbl	$/yr (×10³)
Gasoline				
LSR	1,000			
Reformate	2,730			
Total	3,730	1,268	83.00	105,244
Light gas oil	4,000	1,360	79.00	107,440
Heavy gas oil	3,416	1,161	75.00	87,075
Vacuum gas oil	6,000	2,040	73.00	148,920
Vacuum residual	12,000	4,080	55.00	224,400
Total				673,079
Crude cost	30,000	10,200	60.00	612,000
Direct operating costs				29,577
Income before income tax				31,502

TABLE 18.2
Summary of Direct Annual Operating Costs

	$/yr (×10³)
Makeup water	25
Power	2,041
Fuel[a]	—
Royalties	102
Catalyst	33
Insurance	1,565
Local taxes	3,130
Maintenance	17,218
Miscellaneous supplies	470
Plant staff and operators	2,304
Subtotal	26,888
Contingency (10%)	2,689
Total[b]	29,579

[a] Fuel quantity is deducted from available heavy gas oil for sale.
[b] Additional items such as corporate overhead, research and development, and sales expense are omitted from this example.

18.11.5.13 Calculation of Return on Investment

Investment = $313,047,000

	$/yr (×10³)
Income before tax	31,502
Less depreciation allowance[a]	20,870
Taxable income	10,632
Income tax at 50%	5,316
Income after tax	5,316
Plus depreciation allowance	20,870
Cash flow	26,186
Return on investment (% per year)	8.00
Payout period (years)	11.95

[a] 15 year, straight-line.
Note: True rate of return (discounted cash flow) basis: 20-year life, no salvage value (see Appendix D). Interest on capital during construction period and average feedstock and product inventories are not considered in above product. These items would result in an increase in investment and a decrease in the rate of return.

A project that has a simple payout time of 12 years and a useful operating life of only 20 years has too low a rate of return for the risk involved. Therefore, the calculation of a true rate of return or discounted cash flow DCFROR is not necessary.

This demonstrates that, under the stated typical 2005 conditions, a simple 30,000 BPD refinery is not economically justified.

PROBLEMS

1. Calculate the chemical and utility requirements for a crude unit consisting of both atmospheric and vacuum fractionation towers charging 100,000 BPCD of the assigned crude oil and producing typical side, overhead, and bottom streams. Using assigned utility and chemical costs, calculate the cost per barrel of charge and total annual costs for chemicals and utilities.
2. Calculate the DCFROR on a 100,000 BPCD crude unit charging the assigned crude oil and producing typical side, overhead, and bottom streams. Use product values from *Oil and Gas Journal,* and for those streams not listed, base on equivalent heating value of closest comparable listed product. Construction costs should be brought to current value by use of *Nelson Indices*. Include total operating and materials costs. Use a 30-year useful life with salvage value equal to cost of removal. Construction began 2 years before unit start-up. Land costs are $400,000, and the labor requirement is three persons per shift.

NOTES

1. Lang, H.L., *Chem. Eng.* *55*(6), p. 112, 1948.
2. Williams, W., Jr., *Chem. Eng.* *54*(12), pp. 124–125, 1947.
3. Woodler, A.B. and Woodcock, J.W., *Eur. Chem. News,* September 10, pp. 8–9, 1965.
4. Guthrie, K.M., *Chem.* *77*(13), pp. 140–156, 1970.
5. Nelson, W.L., *Oil Gas J.* *63*(20), pp. 1939–1940, 1965.
6. Chase, J.D., *Chem. Eng.* *77*(7), pp. 113–118, 1970.
7. Guthrie, K.M., *Chem. Eng.* *76*(6), pp. 114–142, 1969.
8. Nelson, W.L., *Oil Gas J.* *71*(17), pp. 88–89, 1973.
9. Ibid., *70*(49), pp. 56–57, 1972.
10. Ibid., *72*(30), pp. 160–162, 1974.
11. Maxwell, J.B., *Data Book on Hydrocarbons,* Van Nostrand, New York, 1950.
12. Perry, R.H., Chilton, C.H., and Kirkpatrick, S.D., *Chemical Engineers' Handbook*, 6th ed., McGraw-Hill Book Co., New York, 1983.

ADDITIONAL READING

13. Stermole, F.J. and Stermole, J.M., *Economic Evaluation and Investment Decision Methods*, 11th ed., Investment Evaluation Corp., Golden, CO, 2006.

19 Economic Evaluation

Economic evaluations are generally carried out to determine if a proposed investment meets the profitability criteria of the company or to evaluate alternatives. There are a number of methods of evaluation, and a good summary of the advantages and disadvantages of each is given in Perry's *Chemical Engineers' Handbook* [1]. Most companies do not rely upon one method alone but utilize several to obtain a more objective viewpoint.

As this chapter is primarily concerned with cost estimation procedures, there will be no attempt to go into the theory of economics, but equations will be presented that are used for the economic evaluation calculations. There is a certain amount of basic information needed to undertake the calculations for an economic evaluation of a project [2,3,4,5].

19.1 DEFINITIONS

19.1.1 Depreciation

Depreciation arises from two causes: deterioration and obsolescence. These two causes do not necessarily operate at the same rate, and the one having the faster rate determines the economic life of the project. Depreciation is an expense, and there are several permissible ways of allocating it. For engineering purposes, depreciation is usually calculated by the straight-line method for the economic life of the project. Frequently, economic lives of 10 years or less are assumed for projects of less than $250,000.

19.1.2 Working Capital

The working capital (WC) consists of feed and product inventories, cash for wages and materials, accounts receivable, and spare parts. A reasonable figure is the sum of the above items for a 30-day period.

19.1.3 Annual Cash Flow

The annual cash flow (ACF) is the sum of the earnings after taxes and the depreciation for a 1-year period.

19.1.4 Sensitivity Analysis

Uncertainties in the cost of equipment, labor, operation, and raw materials as well as in future prices received for products can have a major effect on the evaluation of investments. It is important in appraising the risks involved to know how the outcome would be affected by errors in estimation, and a sensitivity analysis is made to show the changes in the rate of return due to errors of estimation of investment costs and raw material and product prices. These will be affected by the type of cost analysis performed (rough estimate or detailed analysis), stability of the raw material and product markets, and the economic life of the project. Each company will have its own bases for sensitivity analyses, but when investment costs are derived from the installed-cost figures in this book, the following values are reasonable:

	Decrease by (%)	Increase by (%)
Investment cost	15	20
Raw materials costs	3	5
Product prices	5	5
Operating volumes	5	2

19.1.5 Product and Raw Material Cost Estimation

It is very important that price estimation and projections for raw materials and products be as realistic as possible. Current posted prices may not be representative of future conditions, or even of the present value to be received from an addition to the quantities available on the market. A more realistic method is to use the average of the published low over the past several months.

19.2 RETURN ON ORIGINAL INVESTMENT

This method is also known as the engineer's method, du Pont method, or the capitalized earning rate. It does not take into account the time value of money but, because of this, offers a more realistic comparison of returns during the latter years of the investment. The return on original investment is defined as

$$\text{ROI} = \frac{\text{Average yearly profit} \times 100}{\text{Original fixed investment} + \text{working capital}} \tag{19.1}$$

The return on investment should be reported to two significant figures.

19.3 PAYOUT TIME

The payout time is also referred to as the cash recovery period or years to pay out. It is calculated by the following formula and is expressed to the nearest one tenth year:

$$\text{Payout time} = \frac{\text{Original depreciable fixed investment}}{\text{Annual cash flow}} \tag{19.2}$$

If the annual cash flow varies, the payout time can be calculated by adding the cash income after income taxes for consecutive years until the sum is equal to the total investment. The results can be reported to a fractional year by indicating at what point during the year the cash flow will completely offset the depreciable investment.

19.4 DISCOUNTED CASH FLOW RATE OF RETURN

The discounted cash flow rate of return (DCFROR) method is also called the investors' return on investment, internal rate of return, profitability index, and interest rate of return. A trial-and-error solution is necessary to calculate the average rate of interest earned on the company's outstanding investment in the project. It can also be considered as the maximum interest rate at which funds could be borrowed for investment in the project with the project breaking even at the end of its expected life.

The discounted cash flow is basically the ratio of the average annual profit during construction and earning life to the sum of the average fixed investment, working capital, and interest charged on the fixed and working capital that reflects the time value of money. This ratio is expressed as a percentage rather than a fraction. Discounted cash flow is discussed in detail, with an example of its use, in Appendix E.

In order to compare investments having different lives or with variations in return during their operating lives, it is necessary to convert rates of return to a common time basis for comparison. Although any time may be taken for this comparison, the plant start-up time is usually taken as the most satisfactory. Expenditures prior to start-up and income and expenditures after start-up are converted to their worth at start-up. The discussion to follow is based upon the predicted start-up time being the basis of calculation.

19.4.1 EXPENDITURES PRIOR TO START-UP

The expenditures prior to start-up can be placed in two categories: those that occur uniformly over the period of time before start-up and lump-sum payments that occur in an instant at some point before the start-up time. Construction costs are generally assumed to be disbursed uniformly between the start of construction and the start-up time, although equivalent results can be obtained if they are considered to be a lump-sum disbursement taking place halfway between the start of construction and start-up. The present worth of construction costs that are assumed to occur uniformly over a period of years, T, prior to start-up, can be calculated using either continuous interest compounding or discrete (annual) interest compounding.

Continuous interest compounding:

$$P_0 = \left(\frac{CC}{T}\right)\left(\frac{e^{iT} - 1}{i}\right)$$

Discrete (annual) interest compounding:

$$P_0 = \left(\frac{CC}{T}\right)\left(\frac{\ln[1/(1+i)]}{[1/1+ i)]^{T} - [1/(1+ i)]^{(t-1)}}\right)$$

where
P = worth at start-up time
CC = total construction cost
T = length of construction period in years before start-up
i = annual interest rate

The cost of the land is a lump-sum disbursement. In the equation given, it is assumed that the land payment coincides with the start of construction. If the disbursement is made at some other time, then the proper value should be substituted for T.

Continuous interest compounding:

$$P_0 = (LC)e^{iT}$$

Discrete (annual) interest compounding:

$$P_0 = (LC(1 + i)^T$$

where
LC = land cost
T = years before start-up time that payment was made
i = annual interest rate

19.4.2 Expenditures at Start-Up

Any costs that occur at start-up time do not have to be factored, but have a present worth equal to their cost. The major investment at this time is the working capital, but there also may be some costs involved with the start-up of the plant that would be invested at this time.

19.4.3 Income after Start-Up

The business income is normally spread throughout the year, and a realistic interpretation is that 1/365 of the annual earnings is being received at the end of each day. The present-worth factors for this type of incremental income are essentially equal to the continuous-income present-worth factors [6]. Even though the present worth of the income should be computed on a continuous-income basis, it is a matter of individual policy as to whether continuous or discrete compounding of interest is used [6,7]. The income for each year can be converted to the reference point by the appropriate equation.

Continuous-income, continuous-interest:

$$P_0 = (ACF)\left(\frac{e^i - 1}{i}\right)e^{-in}$$

Continuous-income, with interest compounded annually:

$$P_0 = (ACF)\left(\frac{[1/(1+\ i)^n - [1/(1+\ i)]^{n\text{-}1)}}{\ln[1/(1+\ i)]}\right)$$

where

ACF = annual cash flow for year N
n = years after start-up
i = annual interest rate

For the special case where the income occurs uniformly over the life of the project after start-up, the calculations can be simplified.

Uniform continuous-income, continuous-interest:

$$P_0 = (ACF)\left(\frac{e^i - 1}{i}\right)\left(\frac{e^{in} - 1}{ie^{in}}\right)$$

Uniform continuous-income, with interest compounded annually:

$$P_0 = (ACF)\frac{[1/(1+1)]^n - [1/1+\ i)]^{n\text{-}1}}{\ln[1/(1+i)]} \tag{19.3}$$

There are certain costs that are assumed not to depreciate and which are recoverable at the end of the normal service life of the project. Among these are the cost of land, working capital, and salvage value of equipment. These recoverable values must be corrected to their present worth at the start-up time.

Continuous interest:

$$P_0 = (SV + LC + WC)e^{-in} \tag{19.4}$$

Interest compounded annually:

$$P_0 = (SV +\ LC +\ WC)\left(\frac{1}{(1+\ i)}\right)^n \tag{19.5}$$

where

SV = salvage value, \$
LC = land cost, \$
WC = working capital, \$
i = annual interest rate, decimal/yr
n = economic life, yr

For many studies, the salvage value is assumed equal to dismantling costs and is not included in the recoverable value of the project.

It is necessary to use a trial-and-error solution to calculate the discounted cash flow rate of return because the interest rate must be determined that makes the present value at the start-up time of all earnings equal to that of all investments. An example of a typical balance for continuous-income and continuous-interest with uniform annual net income is

$$
\begin{aligned}
&(LC)(e^{iT}) + (CC)\left(\frac{e^{iT}-1}{i}\right)\left(\frac{1}{T}\right) + WC \\
&\quad - (ACF)\left(\frac{e^{i}-1}{i}\right)\left(\frac{e^{in}-1}{ie^{in}}\right) - (SV + LC + WC)\left(\frac{1}{e^{in}}\right) = 0
\end{aligned}
\tag{19.6}
$$

All of the values are known except i, the effective interest rate. An interest rate is assumed, and if the results give a positive value, the trial rate of return is too high and the calculations should be repeated using a lower value for i. If the calculated value is negative, the trial rate is too low and a higher rate of return should be tested. Continue the trial calculations until the rate of return is found that gives a value close to zero.

The return on investment should be reported only to two significant figures.

19.5 CASE-STUDY PROBLEM: ECONOMIC EVALUATION

The estimated 2005 construction costs of the refinery process units and their utility requirements are listed in Table 19.1.

The cooling water and steam systems and water makeup requirements were calculated according to the guidelines set up in Chapter 18.

The estimated refinery construction start is expected to be August 2006, and the process start-up date is anticipated to be in August 2008. Inflation rates of 3% per year are used to bring the costs to their values in 2008.

The working capital is assumed to be equal to 10% of the construction costs. A review of the refinery personnel requirements indicates that approximately 139 people will be required to operate the refinery, exclusive of maintenance personnel. The maintenance personnel are included in the 4.5% annual maintenance costs. An average annual salary of $80,000, including fringe benefits, is used.

A 20-year life will be assumed for the refinery with a dismantling cost equal to salvage value. Straight-line depreciation will be used.

The federal tax rate is 38%, and the state rate is 7%.

Investment costs and utility requirements are summarized in Table 19.1, operating costs in Table 19.2.

A summary of overall costs and realizations is given in Table 19.3, annual costs and revenues in Table 19.4, and total investment costs in Table 19.5, together with payout time and rates of return on investment.

TABLE 19.1
Investment Costs and Utilities

	BPCD	Percent on-stream[a]	BPSD	$ ($\times 10^3$), 2005	Power, (MkWh/d)	Cooling water, gpm	Process steam (Mlb/h)	Fuel (Mlb/h)	MMBtu/h
Desalter	100,000	96.9	103,200	5,500	10	277	87		
Atmospheric crude still	100,000	96.9	103,200	80,000	50	417		25.0	292
Vacuum pipe still	54,147	96.9	55,879	49,400	11	2,336		21.0	92
Coker	20,366	96.1	21,193	149,000	25	990		24.00	119
Middle distillate HT[b]	32,053	96.8	33,113	71,800	192	11,130		13.4	267
FCC HT[b]	36,144	96.8	37,339	78,000	217	12,550		15.0	301
FCC unit[b]	35,421	95.7	37,012	134,200	213	12,300		(44.3)	
HC unit[b]	10,621	97.1	10,928	122,900	138	3,319		33.2	89
Naphtha HT[b]	18,040	96.8	18,636	27,800	[c]	[c]	[c]	[c]	[c]
Reformer[b]	18,040	96.8	18,636	71,500	90	7,520		27.1	300
Isomerization[b]	5,828	96.8	6,021	29,000	6	3,237		49	
Alkylation[b]	5,735	97.2	5,900	32,000	15	10,577		1.91	178
Polymerization	3,170	96.1	3,170	3,000		417		3.1	3
H_2 unit, MMscfd	42	97.1	43.3	53,300	32	5,660	147		396
Satd, gas plant, MMscfd	13	96.9	13.4	27,300	13	14,697			408
Amine treater, gpm	652	96.9	673	9,100	9	1,630			39
Claus sulfur, LT/d	117	96.9	121	13,500			6	(31.7)	
SCOT unit, LT/d		96.9		13,500					
Subtotal				970,800	1,021	87,057	249	136.7	2,484
Cooling water system (+15%)				19,500	63	13,135	5,016		
Steam system (+25%)				33,300			34	34.2	205
Subtotal				990,300	1,084	100,679	5,299	170.9	2,689

continued

TABLE 19.1 (continued)
Investment Costs and Utilities

	BPCD	Percent on-stream[a]	BPSD	$ (×10³), 2005	Power, (MkWh/d)	Cooling water, gpm	Process steam	Fuel (Mlb/h)	MMBtu/h
Storage				460,000					
Subtotal				1,450,300					
Offsites (15%)				217,545					
Subtotal				1,667,845					
Location (1.2)									
Special costs (1.04)	(0.4352)(1,667,845) =			725,850					
Contingency (1.15)									
Total				2,393,695					

[a] Nelson, W.L., *Oil Gas J.* *69*(11), p. 86, 1971.

[b] Catalyst cost for initial charge and paid-up royalty included in unit construction cost; $1,459,000 for middle distillate HT, $1,644,000 for FCC HT, $3,940,000 for FCC unit, $3,763,000 for HC, $812,000 for naphtha HT, $812,000 for naphtha HT, $5,170,000 for reformer, $1,602,000 for isomerization, and $229,000 for alkylation unit royalty.

[c] Included with reformer utilities.

TABLE 19.2
Summary of Operating Costs

	$/yr (×10³)
Chemicals and catalysts	10,205
Water makeup	922
Power	31,653
Fuel	100,127
Insurance	13,078
Local taxes	26,157
Maintenance	117,704
Miscellaneous supplies	3,923
Plant staff and operators	11,120
Total[a]:	314,889

[a] Additional items, such as corporate overhead and research and development vary among refineries and companies. These items are omitted here.

TABLE 19.3
Refinery Annual Summary (2005 Prices)

	BPCD	($/bbl)	$/yr (×10³)
Raw materials (costs)			
North Slope crude	100,000	63.50	2,317,750
n-Butane	796	53.80	15,631
Methane, Mscf/day	10,920	3.60[a]	39,312
MTBE	1,595	37.80	22,006
Total			2,394,699
Products (realizations)			
Fuel gas, MMBtu/day	25,415	3.60	91,494
LPG	3,115	53.10	60,373
Gasoline	53,182	92.50	1,795,557
Jet fuel	24,729	96.30	869,212
Distillate fuel (No. 2)	19,996	83.35	605,000
Coke, ton/day	825	20.00[b]	6,023
Sulfur, LT/day	117	18.00[c]	770
Total			3,428,429

TABLE 19.4
Costs and Revenues

	$/yr (×10³)
Gross income	3,428,429
Production costs	
Raw materials	2,394,699
Operating costs	314,889
Depreciation	130,783
Total	2,840,371
Income before tax	588,058
Less federal and state income taxes	294,029
Net income	294,029
Cash flow	424,812

TABLE 19.5
Total Investment

Construction costs	2,615,655,000
Land cost	5,000,000
Working capital	261,565,500
Total	2,882,220,500
Return on original investment	14.7%
Payout time	6.8 years
Discounted cash flow rate of return	8.0%

19.6 CASE-STUDY PROBLEM: ECONOMIC SOLUTION

19.6.1 Storage Costs

Based on 50 bbl of storage per BPCD crude oil processed, assume 21 days storage provided for n-butane (n-butane: 5,260 BPD × 21 days = 110,460 bbl). Total storage (5,000,000 bbl) less spheroid storages (110,460 bbl) yields 4,889,540 bbl general storage.

	Cost, $ (×10³)
General storage at $90/bbl	
(4,889,540)($90) =	440,059
Spheroid storage at $105/bbl	
(110,460) ($125) =	13,808
Total storage =	453,867

19.6.2 Investment Cost

2005 investment cost = \$2,393,695,000 (Table 19.3)
Inflation rate estimated at 3% per year
Completion data scheduled for August 2007
Estimated completed cost = (\$2,393,695,000) $(1.03)^3$ = \$2,615,655,000

19.6.3 Water Makeup Cost

1. Cooling tower (makeup = 5%)
   Makeup = (0.05)(104,434 gpm) = 5,222 gpm
2. To boiler (boiler makeup = 5%; total steam produced = 212,700 lb/hr)
   Makeup = (212,700)(0.05) = 10,635 lb/hr = 21 gpm
3. Process water = 240 gpm
   Total makeup water, then, is 5483 gpm.
   Annual water makeup cost:
   (5483 gpm/1000) (1440 min/day) (365 days/yr) (\$0.32/kgal) = \$922,197

19.6.4 Chemical and Catalysts Costs

	\$/CD
Desalter	
Demulsifier: 500 lb/day × \$2.40/lb	1,200
Caustic: 200 lb/day × \$1.05/lb	210
Mid. dist. HT: 32,053 BPD × 0.05	1,603
FCC HT: 36,144 × 0.05	1,807
FCC: 35,421 BPD × 0.24	8,501
Naphtha HT: 18,040 BPD × 0.05	902
Catalytic reformer: 18,040 BPD × 0.16	2,886
Isomerization: 5,828 BPD × 0.08	466
Alkylation	
2,240 lb HF/day × \$1.10/lb	2.464
1,490 lb NaOH/day × \$0.30/lb	447
Hydrocracker: 16,930 BPD × 0.32	5,418
Hydrogen unit (Table 13.10)	1,773
Amine treater: 116 lb/day × \$2.30/lb	267
SCOT unit: 6 lb/day × \$2.30/lb	14
Total	27,958

(27,958) (365) = \$10,204,670/yr

19.6.5 Power Costs

Power usage = 1,084,000 kWh/day at \$0.08 per kWh
Annual cost = (1,084,000) (365) (\$0.08) = \$31,652,800

19.6.6 Fuel

Fuel requirements = 2,689 MMBtu/hr
Fuel gas purchased = (2,689 MMBtu/hr) ($4.25/MMBtu) = $11,430/hr
Annual cost = ($11,430) (24) (365) = $100,126,800/yr

19.6.7 Insurance Costs

Average of 0.5% of plant investment per year:

($2,615,655,000) (0.005) = $13,078,275

19.6.8 Local Taxes

Average of 1% of plant investment per year:

($2,615,655,000) (0.01) = $26,156,550

19.6.9 Maintenance Costs

Average of 4.5% of plant investment per year, including material and labor:

($2,615,655,000) (0.045) = $117,704,475

19.6.10 Miscellaneous Supplies Costs

Average of 0.15% of plant investment per year:

($2,615,655,000) (0.0015) = $3,923,483

From the above analysis, it is apparent that the estimated rate of return on the investment does not warrant the risk involved. There will have to be a greater spread between the product prices and the raw materials costs to make the building of a 100,000 BPCD refinery an investment worth making.

PROBLEMS

1. Determine the discounted cash flow rate of return, return on investment, and payout time for a refinery with the following specifications:

   Construction cost = $210,000,000
   Working capital = $21,000,000
   Land cost = $4,200,000
   Annual raw materials cost = $283,000,000
   Annual gross income = $433,000,000
   Annual operating costs = $31,300,000

Assume a 3-year construction time, a 20-year useful life with straight-line depreciation, and federal and state income taxes of 48% and 8%, respectively. Dismantling costs will equal salvage value.

2. For the conditions listed in problem 1, what would be the effect of the DCFRR if the construction took (a) 4 years? (b) 2 years?
3. What would be the DCFRR, payout time, and return on investment for the refinery listed in problem 1 if (a) the refinery were operated at 95% capacity? (b) operating costs were decreased by 1%?
4. Using the specifications of problem 1, make a sensitivity analysis as described in Section 14.1. What factors will be of greatest concern with respect to accurate forecasting?
5. If the declining balance method of depreciation is used instead of the straight-line method for the refinery in problem 1, what would the rates of return be during the first 10 years of operation?

NOTES

1. Perry, R.H., Chilton, C.H., and Kirkpatrick, S.D., *Chemical Engineers' Handbook*, 7th ed., McGraw-Hill Book Co., New York, 1997, pp. 63–78.
2. Dickinson, W.H., *AIEE Trans.*, paper 60–55, pp. 110–124, April 1960.
3. Happel, J., *Chem. Eng. Progr. 51*(12), pp. 533–539, 1995.
4. Stermole, F.J. and Stermole, J.M., *Economic Evaluation and Investment Decision Methods*, 11th ed., Investment Evaluation Corp., Golden, CO, 2006.
5. Weaver, J.B. and Reilly, R.J., *Chem. Eng. Progr. 52*(10), pp. 405–412, 1956.
6. Congelliere, R.H., *Chem. Eng. 77*(25), pp. 109–112, 1970.
7. Helen, F.C. and Black, J.H., *Cost and Optimization Engineering*, 2nd ed., McGraw-Hill Book Co., New York, 1983.

Appendix A

Definitions of Refining Terms

Acid heat A test that is indicative of unsaturated components in petroleum distillates. The test measures the amount of reaction of unsaturated hydrocarbons with sulfuric acid (H_2SO_4).

Acid number An ASTM test that determines the organic acidity of a refinery stream.

AGO Atmospheric gas oil. A diesel fuel and No. 2 heating oil blending stock obtained from the crude oil as a sidestream from the atmospheric distillation tower.

Alkylate The product of an alkylation process. A high-quality gasoline blending stock with a high octane and low sensitivity to driving conditions.

Alkylate bottoms A thick, dark brown oil containing high molecular-weight polymerization products of alkylation reactions.

Alkylation A polymerization process uniting olefins and isoparaffins; particularly, the reacting of butylenes and isobutane using concentrated sulfuric or hydrofluoric acid as a catalyst to produce a high-octane, low-sensitivity blend agent for gasoline.

Aluminum chloride treating A quality improvement process for steam cracked naphthas using aluminum chloride ($AlCl_3$) as a catalyst. The process improves the color and odor of the naphtha by the polymerization of undesirable olefins into resins. The process is also used when production of resins is desirable.

Aniline point The minimum temperature for complete miscibility of equal volumes of aniline and the test sample. The test is considered an indication of the paraffinicity of the sample. The aniline point is also used as a classification of the ignition quality of diesel fuels.

API gravity An arbitrary gravity scale defined as

$$°API = \frac{141.5}{\text{Specific gravity @60/60°F}} - 131.5$$

This scale allows representation of the gravity of oils, which on the specific gravity 60/60°F scale varies only over a range of 0.776 by a scale that ranges from less than 0 (heavy residual oil) to 340 (methane).

ARC Atmospheric reduced crude. The bottoms stream from the atmospheric distillation tower.

Aromatic A type of hydrocarbon molecular compound containing at least one benzene ring.

ASO Acid soluble oil. Polymers produced from side reactions in the alkylation process.

ASTM distillation A standardized laboratory batch distillation for naphthas and middle distillates carried out at atmospheric pressure without fractionation.

ASTM distillation range Several distillation tests are commonly referred to as "ASTM distillations." These are usually used in product specifications. These ASTM distillations give results in terms of percentage distilled versus temperature for a simple laboratory distillation with no fractionation. The values do not correspond to those of refinery process distillations, where fractionation is significant.

Barrel A volumetric measure of refinery feedstocks and products equal to 42 U.S. gal.

Barrels per calendar day (BPCD or B/CD) Average flow rates based on operating 365 days per year.

Barrels per stream day (BPSD or B/SD) Flow rates based on actual on-stream time of a unit or group of units. This notation equals barrels per calendar day divided by the service factor.

Battery limits (BL) The periphery of the area surrounding any process unit, which includes the equipment for that particular process.

B-B Butane–butylene fraction.

Bbl Abbreviation for a quantity of 42 U.S. gal.

BFOE Barrels fuel oil equivalent, based on net heating value (LHV) of 6,050,000 Btu per BFOE.

Bitumen That portion of petroleum, asphalt, and tar products that will dissolve completely in carbon disulfide (CS_2). This property permits a complete separation from foreign products not soluble in carbon disulfide.

Blending One of the final operations in refining, in which two or more different components are mixed together to obtain the desired range of properties in the final product.

Blending octane number When blended into gasoline in relatively small quantities, high-octane materials behave as though they had an octane number higher than shown by laboratory tests on the pure material. The effective octane number of the material in the blend is known as the blending octane number.

Blocked operation Operation of a unit, e.g., a pipe still, under periodic change of feed (one charge stock is processed at a time rather than mixing charge stocks) or internal conditions in order to obtain a required range of raw products. Blocked operation is demanded by critical specifications of various finished products. This frequently results in a more efficient operation because each charge stock can be processed at its optimum operation conditions.

Bottoms In general, the higher-boiling residue that is removed from the bottom of a fractionating tower.

Bright stock Heavy lube oils (frequently the vacuum still bottoms) from which asphaltic compounds, aromatics, and waxy paraffins have been removed. Bright stock is one of the feeds to a lube oil blending plant.

Bromine index Measure of the amount of bromine reactive material in a sample; ASTM D-2710.

Bromine number A test that indicates the degree of unsaturation in the sample (olefins and diolefins); ASTM D-1159.

C Clear or unleaded.

CABP Cubic average boiling point.

Caffeine number A value related to the amount of carcinogenic compounds (high molecular-weight aromatics referred to as tars) in an oil.

CARB California Air Resources Board.

Carbon residue Carbon residue is a measure of the coke-forming tendencies of oil. It is determined by destructive distillation in the absence of air of the sample to a coke residue. The coke residue is expressed as the weight percentage of the original sample. There are two standard ASTM tests, Conradson carbon residue (CCR) and Ramsbottom carbon residue (RCR).

Catalyst A substance that assists a chemical reaction to take place but which is not itself chemically changed as a result.

Catalyst/oil ratio (C/O) The weight of circulating catalyst fed to the reactor of a fluid-bed catalytic cracking unit divided by the weight of hydrocarbons charged during the same interval.

Catalytic cycle stock That portion of a catalytic cracker reactor effluent that is not converted to naphtha and lighter products. This material, generally 340^{+}°F (170^{+}°C), either may be completely or partially recycled. In the latter case, the remainder will be blended to products or processed further.

Cetane index A number calculated from the average boiling point and gravity of a petroleum fraction in the diesel fuel boiling range, which estimates the cetane number of the fraction. An indication of carbon-to-hydrogen ratio.

Cetane number The percentage of pure cetane in a blend of cetane and alpha-methyl-naphthalene that matches the ignition quality of a diesel fuel sample. This quality, specified for middle distillate fuels, is the opposite of the octane number of gasolines. It is an indication of ease of self-ignition.

CFR Combined feed ratio. The ratio of total feed (including recycle) to fresh feed.

CGO Coker gas oil.

Characterization factor An index of feed quality, also useful for correlating data on physical properties. The Watson (UOP) characterization factor is defined as the cube root of the mean average boiling point in degrees Rankine divided by the specific gravity. An indication of carbon-to-hydrogen ratio.

Clarified oil Decanted oil from the FCC unit.

Clay treating An elevated temperature and pressure process usually applied to thermally cracked naphthas to improve stability and color. The stability is increased by the adsorption and polymerization of reactive diolefins in the cracked naphtha. Clay treating is used for treating jet fuel to remove surface-active agents that adversely affect the water separator index specifications.

Clear Without lead. Federal regulations require that fuels containing lead must be dyed.

Cloud point The temperature at which solidifiable compounds (wax) present in the sample begin to crystallize or separate from the solution under a method of prescribed chilling. Cloud point is a typical specification of middle distillate fuels; ASTM D-2500.

Conradson carbon A test used to determine the amount of carbon residue left after the evaporation and pyrolysis of an oil under specified conditions. Expressed as weight percentage; ASTM D-189.

Correlation index (CI) The U.S. Bureau of Mines factor for evaluating individual fractions from crude oil. The CI scale is based upon straight-chain hydrocarbons having a CI value of 0 and benzene having a value of 100. An indication of the hydrogen-to-carbon ratio and the aromaticity of the sample.

Cracking The breaking down of higher molecular-weight hydrocarbons to lighter components by the application of heat. Cracking in the presence of a suitable catalyst produces an improvement in yield and quality over simple thermal cracking.

Crude assay distillation See Fifteen-five (15/5) distillation.

CSO Clarified slurry oil (decanted oil).

Cut That portion of crude oil boiling within certain temperature limits. Usually, the limits are on a crude assay true boiling point (TBP) basis.

Cut point A temperature limit of a cut, usually on a true boiling point basis, although ASTM distillation cut points are not uncommon.

Cycloparaffin A paraffin molecule with a ring structure.

DAO Deasphalted oil. The raffinate product from the propane deasphalting unit.

Decanted oil The bottoms stream from the FCC unit distillation tower after catalyst has been separated from it.

Density The density of crude oil and petroleum fractions is usually specified in °API, specific gravity, or kilograms per cubic meter (kg/m^3). The numerical values of specific gravity and kg/m^3 are equal; that is, a fraction with a specific gravity of 0.873 has a density of 0.873 kg/m^3. The API scale runs opposite to that of specific gravity, with larger values for less dense materials and smaller values for more dense fractions (water = 10°API). By definition, °API is always 15.6°C (60°F) for a liquid.

Dewaxing The removal of wax from lubricating oils, either by chilling and filtering, solvent extraction, or selective hydrocracking.

Diesel index (DI) A measure of the ignition quality of a diesel fuel. Diesel index is defined as

$$\mathrm{DI} = \frac{(^\circ\mathrm{API})(\text{aniline point})}{100}$$

The higher the diesel index, the more satisfactory the ignition quality of the fuel. By means of correlations unique to each crude and manufacturing process, this quality can be used to predict cetane number (if no standardized test for the latter is available).

DIPE Di-isopropyl ether. An oxygenate used in motor fuels.

Distillate Any stream that has been vaporized and then condensed to a liquid.

Doctor test A method for determining the presence of mercaptan sulfur petroleum products. This test is used for products in which a "sweet" odor is desirable for commercial reasons, especially naphtha; ASTM D-484.

DOE U.S. Department of Energy.

Dry gas All C_1 to C_3 material whether associated with a crude or produced as a by-product of refinery processing. Convention often includes hydrogen in dry gas yields.

E85 Fuel containing a blend of 70 to 85% ethanol.

End point The upper temperature limit of a distillation.

Endothermic reaction A reaction in which heat must be added to maintain reactants and products at a constant temperature.

ETBE Ethyl tertiary butyl ether. An oxygenate gasoline blending compound to improve octane and reduce carbon monoxide emissions. Produced by reacting ethanol with isobutylene.

Exothermic reaction A reaction in which heat is evolved. Alkylation, polymerization, and hydrogenation reactions are exothermic.

FBP The final boiling point of a cut, usually on an ASTM distillation basis.

Fifteen-five (15/5) distillation A laboratory batch distillation performed in a 15-theoretical-plate fractionating column with a 5:1 reflux ratio. A good fractionation results in accurate boiling temperatures. For this reason, this distillation is referred to as the true boiling point distillation. This distillation corresponds very closely to the type of fractionation obtained in a refinery.

Fixed carbon The organic portion of the residual coke obtained on the evaporation to dryness of hydrocarbon products in the absence of air.

Flash point The temperature to which a product must be heated under prescribed conditions to release sufficient vapor to form a mixture with air that can be readily ignited. Flash point is used generally as an indication of the fire and explosion potential of a product; ASTM D-56, D-92, D-93, D-134, D-1310.

Flux The addition of a small percentage of a material to a product in order to meet some specification of the final blend.

FOE Barrels of fuel oil equivalent [6.05×10^6 Btu (LHV)].

Foots oil Oil and low-melting materials removed from slack wax to produce a finished wax.

Free carbon The organic materials in tars that are insoluble in carbon disulfide.

Fuel oil equivalent (FOE) The heating value of a standard barrel of fuel oil, equal to 6.05×10^6 Btu (LHV). On a yield chart, dry gas and refinery fuel gas are usually expressed in FOE barrels.

Fungible A hydrocarbon stream that can be pipelined.

FVT The final vapor temperature of a cut. Boiling ranges expressed in this manner are usually on a crude assay, true boiling point basis.

Gas oil Any distillate stream having molecular weights and boiling points higher than heavy naphtha (> 400°F, or 205°C). Frequently, any distillate stream heavier than kerosine. Originally was added to manufactured or city gas to make it burn with a luminous flame. Hence, the name "gas oil."

GHV Gross heating value of fuels. The heat produced by complete oxidation of material at 60°F (25°C) to carbon dioxide and liquid water at 60°F (25°C).

HCGO Heavy coker gas oil.

HCO Heavy FCC cycle gas oil.

Heart cut recycle That unconverted portion of the catalytically cracked material which is recycled to the catalytic cracker. This recycle is usually in the boiling range of the feed and, by definition, contains no bottoms. Recycle allows less severe operation and suppresses the further cracking of desirable products.

Hempel distillation U.S. Bureau of Mines (now Department of Energy) Routine Method of Distillation. Results are frequently used interchangeably with TBP distillation.

HSR Heavy straight-run. Usually a naphtha sidestream from the atmospheric distillation tower.

HVGO Heavy vacuum gas oil. A sidestream from the vacuum distillation tower.

Hydrocarbon Organic molecules containing only hydrogen and carbon.

Hydroskimming refinery A topping refinery with a catalytic reformer.

IBP Initial boiling point of a cut, usually on an ASTM basis.

IPTBE Isopropyl tertiary butyl ether. An oxygenate used in motor fuels.

Isomerate The product of an isomerization process.

Isomerization The rearrangement of straight-chain hydrocarbon molecules to form branched-chain products. Pentanes and hexanes, which are difficult to reform, are isomerized using precious metal catalysts to form gasoline blending components of fairly high octane value. Normal butane may be isomerized to provide a portion of the isobutene feed needed for alkylation processes.

IVT Initial vaporization temperature of a cut, usually based on a crude assay distillation.

Kerosine A middle distillate product composed of material of 300 to 550°F boiling range. The exact cut is determined by various specifications of the finished kerosine. The petroleum industry spells *kerosine* with an "i" to make it consistent with the spelling of *gasoline*.

Lamp sulfur The total amount of sulfur present per unit of liquid product. The analysis is made by burning a sample so that the sulfur content will be converted to sulfur dioxide, which can be quantitatively measured. Lamp sulfur is a critical specification of all motor, tractor, and burner fuels; ASTM D-1266.

LCGO Light coker gas oil.

LCO Light FCC cycle gas oil.

Lead susceptibility The variation of the octane number of a gasoline as a function of the tetraethyl lead content.

LHSV Liquid hour space velocity; volume of feed per hour per volume of catalyst.

LHV Lower heating value of fuels (net heat of combustion). The heat produced by complete oxidation of materials at 60°F (25°C) to carbon dioxide and water vapor at 60°F (25°C).

Light ends Hydrocarbon fractions in the butane and lighter boiling range.

Liquefied petroleum gas (LPG) Liquefied light ends gases used for home heating and cooking. This gas is usually 95% propane, the remainder being split between ethane and butane. It can be any one of several specified mixtures of propane and butane that are sold as liquids under pressure at ambient temperatures.

LNG Liquefied natural gas or methane.

Long resid The bottoms stream from the atmospheric distillation tower.

LPG Liquefied petroleum gas.

LSR Light straight-run. The low-boiling naphtha stream from the atmospheric distillation, usually composed of pentanes and hexanes.

LVGO Light vacuum gas oil; a sidestream from the vacuum distillation tower.

MABP Molal average boiling point:

$$\text{MABP} \sum_{i=1}^{n} 1\ X_i T_{bi}$$

MeABP Mean average boiling point:

$$\text{MeABP} = \frac{(\text{MABP} + \text{CABP})}{2}$$

MHC Mild hydrocracking.

Midboiling point That temperature, usually based on a crude assay distillation, at which one half of the material of a cut has been vaporized.

Middle distillates Atmospheric pipe still cuts boiling in the range of 300 to 700°F vaporization temperature. The exact cut is determined by the specifications of the product.

Midpercent point The vapor temperature at which one half of the material of a cut has been vaporized. Midpercent point is used to characterize a cut in place of temperature limits.

MONC Motor octane number clear (unleaded).

Motor octane number (MON, ASTM ON, F2) A measure of resistance to self-ignition (knocking) of a gasoline under laboratory conditions that correlates with road performance during highway driving conditions. The percentage by volume of isooctane in a mixture of isooctane and n-heptane that knocks with the same intensity as the fuel being tested. A standardized test engine operating under standardized conditions (900 rpm) is used. This test approximates cruising conditions of an automobile; ASTM D-2723.

MPHC Medium pressure hydrocracking or partial conversion hydrocracking.
MTBE Methyl tertiary butyl ether. A high-octane oxygenate gasoline blending stock produced by reacting methanol with isobutylene.
Naphtha A pipe still cut in the range of C_5 to 420°F (216°C). Naphthas are subdivided, according to the actual pipe still cuts, into light, intermediate, heavy, and very heavy virgin naphthas. A typical pipe still operation would be

C_5-160°F (C_5-71°C): light virgin naphtha

160–280°F (71–138°C): intermediate virgin naphtha

280–380°F (138–193°C): heavy virgin naphtha

Naphthas, the major constituents of gasoline, generally need processing to make a suitable quality gasoline.

Napthalene A double-ring aromatic compound in the jet fuel boiling range.
Naphthene A cycloparaffin compound that is a paraffin with a ring structure.
Neutralization number The quantity of acid or base that is required to neutralize all basic or acidic components present in a specified quantity of sample. This is a measure of the amount of oxidation of a product in storage or in service; ASTM D-664, D-974.
Neutral oils Pale or red oils of low viscosity produced from dewaxed oils by distillation.
NGL Natural gas liquids.
Olefin An unsaturated hydrocarbon molecule that has a double bond between two of the carbon atoms in the molecule.
Olefin space velocity The volume of olefin charged per hour to an alkylation reactor.
OPEC Organization of Petroleum Exporting Countries.
Oxygenate Any organic compound containing oxygen. Specifically for the petroleum industry, this term refers to oxygen-containing organic compounds, such as ethers and alcohols, added to fuels to reduce carbon monoxide in the engine exhausts.
Paraffin A saturated hydrocarbon compound in which all carbon atoms in the molecule are connected by single bonds.
Performance rating A method of expressing the quality of a high-octane gasoline relative to isooctane. This rating is used for fuels that are of better quality than isooctane.
Petrolatum Microcrystalline wax.
Pipe still A heater or furnace containing tubes through which oil is pumped while being heated or vaporized. Pipe stills are fired with waste gas, natural gas, or heavy oils and, by providing for rapid heating under conditions of high pressure and temperature, are useful for thermal cracking as well as distillation operations.
Polymerization The combination of two or more unsaturated molecules to form a molecule of higher molecular weight. Propylenes and butylenes are the primary feed material for refinery polymerization processes that use solid or liquid phosphoric acid catalysts.
PON Posted octane number.

Posted octane number Arithmetic average of research octane number and motor octane number. In the United States, this is the octane number that by federal law must be posted on gasoline dispensing pumps.

Pour blending index (PBI) An empirical quantity related to pour point, which allows volumetric blending of pour points of various blend components. This method of blending is most accurate for blending of similar fractions of the same crude.

Pour point The lowest temperature at which a petroleum oil will flow or pour when it is chilled without disturbance at a controlled rate. Pour point is a critical specification of middle distillate products used in cold climates; ASTM D-99.

P-P Propane-propylene fraction

Raffinate The residue recovered from an extraction process. An example is the furfural extraction of aromatics from high molecular-weight distillates. The raffinate is relatively free of aromatics that have poor viscosity–temperature characteristics (low VI; a high rate of change of viscosity with temperature).

Ramsbottom carbon residue A measure of the carbon-forming potential (amount of coke formed) of petroleum fraction. This is determined by a standard laboratory test procedure that subjects the sample to severe thermal cracking conditions. Recommended to replace Conradson carbon; ASTM D-524. Carbon residue expressed in wt% of the original sample.

RBOB Reformulated gasoline blend stocks for oxygenate blending. Needed for blending ethanol into gasoline at the distribution point.

Reconstituted crude A crude to which has been added a specific crude fraction for the purpose of meeting some product volume unattainable with the original crude.

Reduced crude A crude whose API gravity has been reduced by distillation of the lighter lower-boiling constituents.

Reformate A reformed naphtha that is upgraded in octane by means of catalytic or thermal reforming.

Reforming The conversion of naphtha fractions to products of higher octane value. Thermal reforming is essentially a light cracking process applied to heavy naphthas to produce increased yields of hydrocarbons in the gasoline boiling range. Catalytic reforming is applied to various straight-run and cracked naphtha fractions and consists primarily of dehydrogenation of cycloparaffins to aromatics. A number of catalysts, including platinum and platinum–rhenium supported on alumina, are used. A high partial pressure of hydrogen is maintained to prevent the formation of excessive coke and deactivation of the catalyst.

Reformulated gasoline Special blends of gasoline required by EPA for areas that have problems meeting environmental requirements.

Reid vapor pressure (RVP) The vapor pressure at 100°F of a product determined in a volume of air four times the liquid volume. Reid vapor pressure is an indication of the ease of starting and vapor-lock tendency of a motor gasoline as well as explosion and evaporation hazards; ASTM D-323. It is usually expressed in kPa or psig.

Research octane number (RON, CFRR, F-1) The percentage by volume of isooctane in a blend of isooctane and n-heptane that knocks with the same intensity as the fuel being tested. A standardized test engine operating under standardized conditions (600 rpm) is used. Results are comparable to those obtained in an automobile engine operated at low speed or under city driving conditions; ASTM D-2722.

Resid, residuum An undistilled portion of a crude oil. Usually the atmospheric or vacuum tower bottoms.

RFG Reformulated gasoline.

Road octane number The percentage by volume of isooctane that would be required in a blend of isooctane and n-heptane to give incipient knock in an automobile engine operating under the same conditions of engine load, speed, and degree of spark advance as that of the fuel being tested.

RONC Research octane number clear (unleaded).

Salt content Crude oil usually contains salts in solution in water that is emulsified with the crude. The salt content is expressed as the sodium chloride equivalent in pounds per thousand barrels (PTB) of crude oil. Typical values range from 1 to 20 PTB. Although there is no simple conversion from PTB to parts per million by weight (ppm), 1 PTB is roughly equivalent to 3 ppm.

Saponification number An indication of the fraction of fat or fatty oils in a given product. The number of milligrams of potassium hydroxide required to saponify 1 g of the sample.

scf The volume of gas expressed as standard cubic feet. Standard conditions in petroleum and natural gas usage refer to a pressure base of 14.696 psia (101.5 kPa) and a temperature base of 60°F (15°C).

Selectivity The difference between the research octane number and the motor octane number of a given gasoline. Alkylate is an excellent low-sensitivity and reformate a high-sensitivity gasoline component. It is an indication of the sensitivity of the fuel to driving conditions (city vs. highway).

Service factor A quantity that relates the actual on-stream time of a process unit to the total time available for use. Service factors include both planned and unexpected unit shutdowns.

Severity The degree of intensity of the operating conditions of a process unit. Severity may be indicated by clear research octane number of the product (reformer), percentage disappearance of the feed (catalytic cracking), or operating conditions alone (usually the temperature; the higher the temperature, the greater the severity).

Short resid Vacuum tower bottoms.

Slack wax Wax produced in the dewaxing of lube oil base stocks. This wax still contains some oil and must be oiled to produce a finished wax product.

Slurry oil The oil, from the bottoms of the FCC unit fractionating tower, containing FCC catalyst particles carried over by the vapor from the reactor cyclones. The remainder of the FCC bottoms is the decanted oil.

Smoke point A test measuring the burning quality of jet fuels, kerosine, and illuminating oils. It is defined as the height of the flame in millimeters beyond which smoking takes place; ASTM D-1322.

Sour or sweet crude A rather general method for classifying crudes according to sulfur content. Various definitions are available:

Sour crude A crude that contains sulfur in amounts greater than 0.5 to 1.0 wt%, or that contains 0.05 ft^3 or more of hydrogen sulfide (H_2S) per 100 gal, except West Texas crude, which is always considered sour regardless of its hydrogen sulfide content. Arabian crudes are high-sulfur crudes that are not always considered sour because they do not contain highly active sulfur compounds. Original definition was for any crude oil that smelled like rotten eggs.

Sweet crude As evident from the above definition, a sweet crude contains little or no dissolved hydrogen sulfide and relatively small amounts of mercaptans and other sulfur compounds. Original definition was for any crude oil that did not have a bad odor.

Space velocity The volume (or weight) of gas or liquid passing through a given catalyst or reactor space per unit time, divided by the volume (or weight) of catalyst through which the fluid passes. High space velocities correspond to short reaction times. See LHSV and WHSV.

Straight-run gasoline An uncracked gasoline fraction distilled from crude. Straight-run gasolines contain primarily paraffinic hydrocarbons and have lower octane values than cracked gasolines from the same crude feedstock.

Sweetening The removal or conversion to innocuous substances of sulfur compounds in a petroleum product by any of a number of processes (doctor treating, caustic and water washing, and so on).

Synthetic crude Wide boiling-range product of catalytic cracking, coking, hydrocracking, or some other chemical structure change operation.

TA Total alkylate or true alkylate (C_4^+ alkylate).

TAEE Tertiary amyl ethyl ether. A high-octane oxygenate blending stock produced by reacting isoamylene (isopentylene) with ethanol.

Tail gas Light gases (e.g., C_1 to C_3 and H_2) produced as by-products of refinery processing.

TAME Tertiary amyl methyl ether. A high-octane oxygenate gasoline blending compound, produced by reacting isoamylene (isopentylene) with methanol.

TAN Total Acid Number.

TBP distillation See fifteen-five distillation.

Tetraethyl lead An antiknock additive for gasoline that is no longer used in the United States.

Theoretical plate A theoretical contacting unit useful in distillation, absorption, and extraction calculations. Vapors and liquid leaving any such unit are required to be in equilibrium under the conditions of temperature and pressure that apply. An actual fractionator tray or plate is generally less effective than a theoretical plate. The ratio of a number of theoretical plates required to perform a given distillation separation to the number of actual plates used given the overall tray efficiency of the fractionator.

Topping Removal by distillation of the light products and transportation fuel products from crude oil, leaving in the still bottoms all of the components with boiling ranges greater than diesel fuel.

Treat gas Light gases, usually high in hydrogen content, which are required for refinery hydrotreating processes such as hydrodesulfurization. The treat gas for hydrodesulfurization is usually the tail gas from catalytic reforming or the product from a hydrogen unit.

Tube still See pipe still.

ULSD Ultra low-sulfur diesel. Diesel fuel with < 15 ppm sulfur.

U.S. Bureau of Mines Routine Method of Distillation See Hempel distillation.

Vapor-lock index A measure of the tendency of a gasoline to generate excessive vapors in the fuel line, thus causing displacement of a liquid fuel and subsequent interruption of normal engine operation. The vapor-lock index generally is related to RVP and percentage distilled at 158°F (70°C).

VGO Vacuum gas oil. A sidestream from the vacuum distillation tower.

Virgin stocks Petroleum oils that have not been cracked or otherwise subjected to any treatment that would produce appreciable chemical change in their components.

Viscosity Internal resistance to the flow of liquids is expressed as viscosity. The property of liquids under flow conditions that causes them to resist instantaneous change of shape or instantaneous rearrangement of their parts due to internal friction. Viscosity is generally measured as the number of seconds, at a definite temperature, required for a standard quantity of oil to flow through a standard apparatus. Common viscosity scales in use are Saybolt Universal, Saybolt Furol, poises, and kinematic [stokes or centiStokes (cSt)].

VOC Volatile organic compound.

Volatility factor An empirical quantity that indicates good gasoline performance with respect to volatility. It involves actual automobile operating conditions and climatic factors. The volatility factor is generally defined as a function of RVP, percentage distilled at 158°F (70°C), and percentage distilled at 212°F (100°C). This factor is an attempt to predict the vapor-lock tendency of a gasoline.

vppm Parts per million by volume.

VRC Vacuum reduced crude; vacuum tower bottoms.

WABP Weight average boiling point:

$$\text{WABP} = \sum_{i=1}^{n} X_{wi} T_{bi}$$

where

X_{wi} = weight fraction of component i

T_{bi} = average boiling point of component i

WHSV Weight hour space velocity; weight of feed per hour per weight of catalyst.

Wick char A test used as an indication of the burning quality of a kerosine or illuminating oil. It is defined as the weight of deposits remaining on the wick after a specified amount of sample is burned.

wppm Parts per million by weight.

WTI West Texas Intermediate crude oil. One of the crude oils used to indicate prices.

Appendix B

Physical Properties

TABLE B.1
Density Conversion Table

Specific gravity	Density in vacuo			lb/hr* from	Specific gravity	Density in vacuo			lb/hr* from
60/60°F	°API	lb/bbl	lb/gal	bbl/day	60/60°F	°API	lb/bbl	lb/gal	bbl/day
1.165	-10.0	407.8	9.71	16.99	1.092	-2.0	382.6	9.11	15.94
1.163	-9.8	407.1	9.69	16.95	1.090	-1.8	382.0	9.09	15.92
1.161	-9.6	406.5	9.68	16.94	1.089	-1.6	381.4	9.08	15.89
1.159	-9.4	405.8	9.66	16.91	1.087	-1.4	380.8	9.07	15.87
1.157	-9.2	405.1	9.65	16.88	1.085	-1.2	380.3	9.05	15.85
1.155	-9.0	404.5	9.63	16.85	1.084	-1.0	379.7	9.04	15.82
1.153	-8.8	403.8	9.61	16.82	1.082	-0.8	379.1	9.03	15.80
1.151	-8.6	403.2	9.60	16.80	1.080	-0.6	378.5	9.01	15.77
1.149	-8.4	402.5	9.58	16.77	1.079	-0.4	377.9	9.00	15.75
1.147	-8.2	401.9	9.57	16.74	1.077	-0.2	377.4	8.98	15.72
1.145	-8.0	401.2	9.55	16.72	1.076	0.0	376.8	8.97	15.70
1.143	-7.8	400.6	9.54	16.69	1.074	.2	376.2	8.96	15.67
1.142	-7.6	399.9	9.52	16.66	1.073	.4	375.6	8.94	15.65
1.140	-7.4	399.3	9.51	16.64	1.071	.6	375.1	8.93	15.63
1.138	-7.2	398.6	9.49	16.61	1.070	.8	374.5	8.92	15.60
1.136	-7.0	398.0	9.48	16.58	1.068	1.0	373.9	8.90	15.53
1.134	-6.8	397.3	9.46	16.55	1.066	.2	373.4	8.89	15.56
1.132	-6.6	396.7	9.45	16.53	1.065	.4	372.8	8.88	15.53
1.131	-6.4	396.1	9.43	16.50	1.063	.6	372.3	8.86	15.51
1.129	-6.2	395.4	9.42	16.47	1.062	.8	371.7	8.85	15.49
1.127	-6.0	394.8	9.40	16.45	1.060	2.0	371.1	8.84	15.46
1.125	-5.8	394.2	9.39	16.42	1.053	.2	370.6	8.82	15.44
1.123	-5.6	393.6	9.37	16.40	1.057	.4	370.0	8.81	15.42
1.122	-5.4	392.9	9.36	16.37	1.055	.6	369.5	8.80	15.40
1.120	-5.2	392.3	9.34	16.35	1.054	.8	368.9	8.78	15.37
1.118	-5.0	391.7	9.33	16.32	1.052	3.0	368.4	8.77	15.35
1.116	-4.8	391.1	9.31	16.30	1.051	.2	367.8	8.76	15.32
1.115	-4.6	390.5	9.30	16.27	1.049	.4	367.3	8.75	15.30
1.113	-4.4	389.8	9.23	16.24	1.047	.6	366.8	8.73	15.28
1.111	-4.2	389.2	9.27	16.22	1.046	.8	366.2	8.72	15.26
1.109	-4.0	388.6	9.25	16.19	1.044	4.0	365.7	8.71	15.24
1.108	-3.8	388.0	9.24	16.17	1.043	.2	365.1	8.69	15.21
1.106	-3.6	387.4	9.22	16.14	1.041	.4	364.6	8.68	15.19
1.104	-3.4	386.8	9.21	16.12	1.040	.6	364.0	8.67	15.17
1.102	-3.2	386.2	9.19	16.09	1.038	.8	363.5	8.66	15.15
1.101	-3.0	385.6	9.18	16.07	1.037	5.0	363.0	8.64	15.12
1.099	-2.8	385.0	9.16	16.04	1.035	.2	362.4	8.63	15.10
1.097	-2.6	384.4	9.15	16.02	1.034	.4	361.9	8.62	15.08
1.096	-2.4	383.8	9.14	15.99	1.032	.6	361.4	8.60	15.06
1.094	-2.2	383.2	9.12	15.97	1.031	.8	360.9	8.59	15.04

TABLE B.1 (continued)
Density Conversion Table

Specific gravity 60/60°F	Density in vacuo °API	lb/bbl	lb/gal	lb/hr* from bbl/day	Specific gravity 60/60°F	Density in vacuo °API	lb/bbl	lb/gal	lb/hr* from bbl/day
1.029	6.0	360.3	8.58	15.01	0.973	14.0	340.5	8.11	14.19
1.028	.2	359.8	8.57	14.99	0.971	.2	340.1	8.10	14.17
1.026	.4	359.3	8.55	14.97	0.970	.4	339.6	8.09	14.15
1.025	.6	358.8	8.54	14.95	0.969	.6	339.1	8.08	14.13
1.023	.8	358.3	8.53	14.93	0.967	.8	338.7	8.06	14.11
1.022	7.0	357.7	8.52	14.90	0.966	15.0	338.2	8.05	14.09
1.020	.2	357.2	8.51	14.88	0.965	.2	337.8	8.04	14.07
1.019	.4	356.7	8.49	14.86	0.963	.4	337.3	8.03	14.05
1.017	.6	356.2	8.48	14.84	0.962	.6	336.8	8.02	14.03
1.016	.8	355.7	8.47	14.82	0.961	.8	336.4	8.01	14.02
1.014	8.0	355.2	8.46	14.80	0.959	16.0	335.9	8.00	14.00
1.013	.2	354.7	8.44	14.78	0.958	.2	335.5	7.99	13.98
1.011	.4	354.2	8.43	14.76	0.957	.4	335.0	7.98	13.96
1.010	.6	353.7	8.42	14.74	0.955	.6	334.6	7.96	13.94
1.009	.8	353.2	8.41	14.72	0.954	.8	334.1	7.95	13.92
1.007	9.0	352.7	8.40	14.70	0.953	17.0	333.7	7.94	13.90
1.006	.2	352.2	8.38	14.67	0.952	.2	333.2	7.93	13.88
1.004	.4	351.7	8.37	14.65	0.950	.4	332.8	7.92	13.87
1.003	.6	351.2	8.36	14.63	0.949	.6	332.3	7.91	13.85
1.001	.8	350.7	8.35	14.61	0.948	.8	331.9	7.90	13.83
1.000	10.0	350.2	8.34	14.59	0.947	18.0	331.4	7.89	13.81
0.999	10.2	349.7	8.33	14.57	0.945	.2	331.0	7.88	13.79
0.997	10.4	349.2	8.31	14.55	0.944	.4	330.5	7.87	13.77
0.996	10.6	348.7	8.30	14.53	0.943	.6	330.1	7.86	13.75
0.994	10.8	348.2	8.29	14.51	0.942	.8	329.7	7.85	13.74
0.993	11.0	347.7	8.28	14.49	0.940	19.0	329.2	7.84	13.72
0.992	.2	347.2	8.27	14.47	0.939	.2	328.0	7.83	13.70
0.990	.4	346.7	8.26	14.45	0.938	.4	328.4	7.82	13.68
0.989	.6	346.2	8.24	14.43	0.937	.6	327.9	7.81	13.66
0.987	.8	345.8	8.23	14.41	0.935	.8	327.5	7.80	13.65
0.986	12.0	345.3	8.22	14.39	0.934	20.0	327.1	7.79	13.63
0.985	.2	344.8	8.21	14.37	0.933	.2	326.6	7.78	13.61
0.983	.4	344.3	8.20	14.35	0.932	.4	326.2	7.77	13.59
0.982	.6	343.8	8.19	14.33	0.930	.6	325.8	7.76	13.57
0.981	.8	343.4	8.18	14.31	0.929	.8	325.3	7.75	13.55
0.979	13.0	342.9	8.16	14.29	0.928	21.0	324.9	7.74	13.54
0.978	.2	342.4	8.15	14.27	0.927	.2	324.5	7.73	13.52
0.977	.4	341.9	8.14	14.25	0.925	.4	324.0	7.72	13.50
0.975	.6	341.5	8.13	14.23	0.924	.6	323.6	7.71	13.48
0.974	.8	341.0	8.12	14.21	0.923	.8	323.2	7.70	13.47

continued

TABLE B.1 (continued)
Density Conversion Table

Specific gravity 60/60°F	Density in vacuo			lb/hr* from bbl/day	Specific gravity 60/60°F	Density in vacuo			lb/hr* from bbl/day
	°API	lb/bbl	lb/gal			°API	lb/bbl	lb/gal	
0.922	22.0	322.8	7.69	13.45	0.876	30.0	306.8	7.30	12.78
0.921	.2	322.4	7.68	13.43	0.875	.2	306.4	7.30	12.77
0.919	.4	321.9	7.67	13.41	0.874	.4	306.0	7.29	12.75
0.918	.6	321.5	7.66	13.40	0.873	.6	305.7	7.28	12.74
0.917	.8	321.1	7.65	13.38	0.872	.8	305.3	7.27	12.72
0.916	23.0	320.7	7.64	13.36	0.871	31.0	304.9	7.26	12.70
0.915	.2	320.3	7.63	13.35	0.870	.2	304.5	7.25	12.69
0.914	.4	319.9	7.62	13.33	0.869	.4	304.2	7.24	12.67
0.912	.6	319.5	7.61	13.31	0.868	.6	303.8	7.23	12.66
0.911	.8	319.0	7.60	13.29	0.867	.8	303.4	7.22	12.64
0.910	24.0	318.6	7.59	13.27	0.865	32.0	303.0	7.21	12.62
0.909	.2	318.2	7.58	13.26	0.864	.2	302.7	7.20	12.61
0.908	.4	317.8	7.57	13.24	0.863	.4	302.3	7.19	12.60
0.907	.6	317.4	7.56	13.22	0.862	.6	301.9	7.19	12.58
0.905	.8	317.0	7.55	13.21	0.861	.8	301.6	7.18	12.57
0.904	25.0	316.6	7.54	13.19	0.860	33.0	301.2	7.17	12.55
0.903	.2	316.2	7.53	13.17	0.859	.2	300.8	7.16	12.53
0.902	.4	315.8	7.52	13.16	0.858	.4	300.5	7.15	12.52
0.901	.6	315.4	7.51	13.14	0.857	.6	300.1	7.14	12.50
0.900	.8	315.0	7.50	13.12	0.856	.8	299.7	7.14	12.49
0.898	26.0	314.6	7.49	13.11	0.855	34.0	299.4	7.13	12.47
0.897	.2	314.2	7.48	13.09	0.854	.2	299.0	7.12	12.46
0.896	.4	313.8	7.47	13.07	0.853	.4	298.7	7.11	12.45
0.895	.6	313.4	7.46	13.06	0.852	.6	298.3	7.10	12.43
0.894	.8	313.0	7.45	13.04	0.851	.8	297.9	7.09	12.41
0.893	27.0	312.6	7.44	13.02	0.850	35.0	297.6	7.09	12.40
0.892	.2	312.2	7.43	13.01	0.849	.2	297.2	7.08	12.38
0.891	.4	311.8	7.42	12.99	0.848	.4	296.9	7.07	12.37
0.889	.6	311.4	7.41	12.97	0.847	.6	296.5	7.06	12.35
0.888	.8	311.0	7.40	12.96	0.846	.8	296.2	7.05	12.34
0.887	28.0	310.6	7.40	12.95	0.845	36.0	295.8	7.04	12.32
0.886	.2	310.3	7.39	12.93	0.844	.2	295.4	7.04	12.31
0.885	.4	309.9	7.38	12.91	0.843	.4	295.1	7.03	12.30
0.884	.6	309.5	7.37	12.90	0.842	.6	294.8	7.02	12.28
0.883	.8	309.1	7.36	12.88	0.841	.8	294.4	7.01	12.27
0.882	29.0	308.7	7.35	12.86	0.840	37.0	294.0	7.00	12.25
0.881	.2	308.3	7.34	12.85	0.839	.2	293.7	6.99	12.24
0.879	.4	307.9	7.33	12.83	0.838	.4	293.4	6.99	12.21
0.878	.6	307.6	7.32	12.82	0.837	.6	293.0	6.98	12.21
0.877	.8	307.2	7.31	12.80	0.836	.8	292.7	6.97	12.20

TABLE B.1 (continued)
Density Conversion Table

Specific gravity 60/60°F	Density in vacuo °API	lb/bbl	lb/gal	lb/hr* from bbl/day	Specific gravity 60/60°F	Density in vacuo °API	lb/bbl	lb/gal	lb/hr* from bbl/day
0.835	38.0	292.3	6.96	12.18	0.797	46.0	279.1	6.64	11.63
0.834	.2	292.0	6.95	12.17	0.796	.2	278.3	6.64	11.62
0.833	.4	291.6	6.94	12.15	0.795	.4	278.5	6.63	11.60
0.832	.6	291.3	6.94	12.14	0.795	.6	278.2	6.63	11.59
0.831	.8	291.0	6.93	12.12	0.794	.8	277.9	6.62	11.58
0.830	39.0	290.6	6.92	12.11	0.793	47.0	277.6	6.61	11.57
0.829	.2	290.3	6.91	12.10	0.792	.2	277.3	6.60	11.55
0.828	.4	290.0	6.90	12.08	0.791	.4	277.0	6.59	11.54
0.827	.6	289.6	6.89	12.07	0.790	.6	276.7	6.59	11.53
0.826	.8	289.2	6.89	12.05	0.789	.8	276.3	6.58	11.51
0.825	40.0	288.9	6.88	12.04	0.788	48.0	276.0	6.57	11.50
0.824	.2	288.6	6.87	12.02	0.787	.2	275.7	6.56	11.49
0.823	.4	288.2	6.86	12.01	0.787	.4	275.4	6.56	11.47
0.822	.6	287.9	6.85	12.00	0.786	.6	275.1	6.55	11.46
0.821	.8	287.6	6.84	11.93	0.785	.8	274.1	6.54	11.45
0.820	41.0	287.2	6.84	11.97	0.784	49.0	274.5	6.54	11.44
0.819	.2	286.9	6.83	11.95	0.783	.2	274.2	6.53	11.42
0.818	.4	286.6	6.82	11.94	0.782	.4	273.9	6.52	11.41
0.817	.6	286.2	6.81	11.92	0.781	.6	273.6	6.51	11.40
0.817	.8	285.9	6.81	11.91	0.781	.8	273.3	6.51	11.39
0.816	42.0	285.6	6.80	11.90	0.780	50.0	273.0	6.50	11.37
0.815	.2	285.3	6.79	11.89	0.779	.2	272.7	6.49	11.36
0.814	.4	284.9	6.79	11.87	0.778	.4	272.4	6.49	11.35
0.813	.6	284.6	6.78	11.86	0.777	.6	272.1	6.48	11.34
0.812	.8	284.3	6.77	11.85	0.776	.8	271.8	6.47	11.32
0.811	43.0	283.9	6.76	11.83	0.775	51.0	271.5	6.46	11.31
0.810	.2	283.6	6.75	11.82	0.775	.2	271.2	6.46	11.30
0.809	.4	283.3	6.74	11.80	0.774	.4	270.9	6.45	11.29
0.808	.6	283.0	6.74	11.79	0.773	.6	270.6	6.44	11.27
0.807	.8	282.6	6.73	11.77	0.772	.8	270.3	6.44	11.26
0.806	44.0	282.3	6.72	11.76	0.771	52.0	270.0	6.43	11.25
0.805	.2	282.0	6.71	11.75	0.770	.2	269.7	6.42	11.24
0.804	.4	281.7	6.70	11.74	0.769	.4	269.4	6.41	11.22
0.804	.6	281.4	6.70	11.72	0.769	.6	269.1	6.41	11.21
0.803	.8	281.0	6.69	11.71	0.768	.8	268.8	6.40	11.20
0.802	45.0	280.7	6.69	11.70	0.767	53.0	268.5	6.39	11.19
0.801	.2	280.4	6.68	11.68	0.766	.2	268.3	6.39	11.18
0.800	.4	280.1	6.67	11.67	0.765	.4	268.0	6.38	11.17
0.799	.6	279.8	6.66	11.66	0.764	.6	267.7	6.37	11.15
0.798	.8	279.5	6.65	11.65	0.764	.8	267.4	6.37	11.14

continued

TABLE B.1 (continued)
Density Conversion Table

Specific gravity 60/60°F	Density in vacuo			lb/hr* from bbl/day	Specific gravity 60/60°F	Density in vacuo			lb/hr* from bbl/day
	°API	lb/bbl	lb/gal			°API	lb/bbl	lb/gal	
0.763	54.0	267.1	6.36	11.13	0.731	62.0	256.1	6.10	10.67
0.762	.2	266.8	6.35	11.12	0.730	.2	255.8	6.09	10.66
0.761	.4	266.5	6.34	11.10	0.730	.4	255.5	6.08	10.65
0.760	.6	266.2	6.34	11.09	0.729	.6	255.3	6.08	10.64
0.760	.8	265.9	6.33	11.08	0.728	.8	255.0	6.07	10.62
0.759	55.0	265.7	6.33	11.07	0.728	63.0	254.7	6.07	10.61
0.758	.2	265.4	6.32	11.06	0.727	.2	254.5	6.06	10.60
0.757	.4	265.1	6.31	11.05	0.726	.4	254.2	6.05	10.59
0.756	.6	264.8	6.30	11.03	0.725	.6	254.0	6.05	10.58
0.756	.8	264.5	6.30	11.02	0.724	.8	253.7	6.04	10.57
0.755	56.0	264.3	6.29	11.01	0.724	64.0	253.4	6.03	10.56
0.754	.2	264.0	6.29	11.00	0.723	.2	253.2	6.03	10.55
0.753	.4	263.7	6.28	10.99	0.722	.4	252.9	6.02	10.54
0.752	.6	263.4	6.27	10.97	0.722	.6	252.7	6.02	10.53
0.752	.8	263.1	6.27	10.96	0.721	.8	252.4	6.01	10.52
0.751	57.0	262.9	6.26	10.95	0.720	65.0	252.2	6.00	10.51
0.750	.2	262.6	6.26	10.94	0.719	.2	251.9	6.00	10.50
0.749	.4	262.3	6.24	10.93	0.719	.4	251.6	5.99	10.48
0.748	.6	262.0	6.24	10.92	0.718	.6	251.4	5.98	10.47
0.748	.8	261.7	6.23	10.90	0.717	.8	251.1	5.98	10.46
0.747	58.0	261.5	6.23	10.89	0.716	66.0	250.9	5.97	10.45
0.746	.2	261.2	6.22	10.88	0.716	.2	250.6	5.97	10.44
0.745	.4	260.9	6.21	10.87	0.715	.4	250.4	5.96	10.43
0.744	.6	260.6	6.20	10.86	0.714	.6	250.1	5.95	10.42
0.744	.8	260.4	6.20	10.85	0.714	.8	249.9	5.95	10.41
0.743	59.0	260.1	6.19	10.84	0.713	67.0	249.6	5.94	10.40
0.742	.2	259.8	6.19	10.82	0.712	.2	249.4	5.94	10.39
0.741	.4	259.6	6.18	10.81	0.711	.4	249.1	5.93	10.38
0.740	.6	259.3	6.17	10.80	0.711	.6	248.9	5.93	10.37
0.740	.8	259.0	6.17	10.79	0.710	.8	248.6	5.92	10.36
0.739	60.0	258.7	6.16	10.78	0.709	68.0	248.4	5.91	10.35
0.738	.2	258.5	6.15	10.77	0.709	.2	248.1	5.91	10.34
0.737	.4	258.2	6.15	10.76	0.708	.4	247.9	5.90	10.33
0.737	.6	257.9	6.14	10.75	0.707	.6	247.6	5.90	10.32
0.736	.8	257.7	6.14	10.74	0.706	.8	247.4	5.89	10.31
0.735	61.0	257.4	6.13	10.72	0.706	69.0	247.1	5.88	10.30
0.734	.2	257.1	6.12	10.71	0.705	.2	246.9	5.88	10.29
0.734	.4	256.9	6.12	10.70	0.704	.4	246.6	5.87	10.28
0.733	.6	256.6	6.11	10.69	0.704	.6	246.4	5.87	10.27
0.732	.8	256.3	6.10	10.68	0.703	.8	246.1	5.86	10.26

TABLE B.1 (continued)
Density Conversion Table

Specific gravity 60/60°F	Density in vacuo			lb/hr* from bbl/day	Specific gravity 60/60°F	Density in vacuo			lb/hr* from bbl/day
	°API	lb/bbl	lb/gal			°API	lb/bbl	lb/gal	
0.702	70.0	245.9	5.85	10.25	0.646	.5	226.2	5.39	9.42
0.701	.5	245.3	5.84	10.22	0.645	88.0	225.7	5.38	9.40
0.699	71.0	244.7	5.83	10.20	0.643	.5	225.2	5.36	9.38
0.697	.5	244.1	5.81	10.17	0.642	89.0	224.7	5.35	9.36
0.695	72.0	243.5	5.80	10.15	0.640	.5	224.2	5.34	9.34
0.694	.5	242.9	5.78	10.12	0.639	90.0	223.7	5.33	9.32
0.692	73.0	242.3	5.77	10.10	0.637	.5	223.2	5.31	9.30
0.609	.5	241.7	5.75	10.07	0.636	91.0	222.7	5.30	9.28
0.689	74.0	241.1	5.74	10.05	0.635	.5	222.2	5.29	9.26
0.687	.5	240.5	5.73	10.02	0.633	92.0	221.7	5.28	9.24
0.685	75.0	239.9	5.71	10.00	0.632	.5	221.2	5.27	9.22
0.684	.5	239.4	5.70	9.97	0.630	93.0	220.7	5.26	9.20
0.682	76.0	238.8	5.69	9.95	0.629	.5	220.2	5.24	9.18
0.680	.5	238.2	5.67	9.92	0.628	94.0	219.7	5.23	9.16
0.679	77.0	237.6	5.66	9.90	0.626	.5	219.2	5.22	9.14
0.677	.5	237.1	5.64	9.88	0.625	95.0	218/8	5.21	9.12
0.675	78.0	236.5	5.63	9.85	0.623	.5	218.3	5.20	9.10
0.674	.5	235.9	5.62	9.83	0.622	96.0	217.8	5.19	9.03
0.672	79.0	235.4	5.60	9.81	0.621	.5	217.3	5.17	9.06
0.671	.5	234.8	5.59	9.78	0.619	97.0	216.8	5.16	9.04
0.669	80.0	234.3	5.58	9.76	0.618	.5	216.4	5.15	9.02
0.668	.5	233.7	5.56	9.74	0.617	98.0	215.9	5.14	9.00
0.666	81.0	233.2	5.55	9.72	0.615	.5	215.4	5.13	8.98
0.664	.5	232.6	5.54	9.69	0.614	99.0	215.0	5.12	8.96
0.663	82.0	232.1	5.53	9.67	0.613	.5	214.5	5.11	8.94
0.661	.5	231.5	5.51	9.65	0.611	100.0	214.0	5.10	8.92
0.660	83.0	231.0	5.50	9.62	0.610	.5	213.6	5.09	8.90
0.658	.5	230.4	5.49	9.60	0.609	101.0	213.1	5.07	8.83
0.657	84.0	229.9	5.48	9.58	0.607	.5	212.7	5.06	8.86
0.655	.5	229.4	5.46	9.56	0.606	102.0	212.2	5.05	8.84
0.654	85.0	228.9	5.45	9.54	0.605	.5	211.7	5.04	8.82
0.652	.5	228.3	5.44	9.51	0.603	103.0	211.3	5.03	8.80
0.651	86.0	227.8	5.43	9.49	0.602	.5	210.8	5.02	8.78
0.649	.5	227.3	5.41	9.47	0.601	104.0	210.4	5.01	8.77
0.648	87.0	226.8	5.40	9.45	0.600	.5	209.9	5.00	8.75
					0.598	105.0	209.5	4.99	8.73

Source: From ASTM D-1250. Reproduced courtesy of the American Petroleum Institute.

* Multiply barrels/day by the factor in this column corresponding to the API gravity to obtain pounds/hour.

TABLE B.2
Physical Constants of Paraffin Hydrocarbons and Other Components of Natural Gas [NGPA Publication 2145–74(1)

Natural Gas [NGPA Publication 2145–74(1)]

Component	Notes	Methane	Ethane	Propane	Iso-Butane	N-Butane	Iso-Pentane	N-Pentane
Molecular Weight		16.043	30.070	44.097	58.124	58.124	72.151	72.151
Boiling Point @ 14.696 psia, °F		−258.69	−127.48	− 43.67	10.90	31.10	82.12	96.92
Freezing Point @ 14.696 psia, °F		−296.46[d]	−297.89[d]	−305.84[d]	−255.29	−217.05	−255.83	−201.51
Vapor Pressure @ 100°F, psia		(5000)	(800)	190	72.2	51.6	20.44	15.570
Density of Liquid @ 60°F & 14.696 psia								
Specific Gravity @ 60°F/60°F	a,b	0.3[i]	0.3564[h]	0.5077[h]	0.5631[h]	0.5844[h]	0.6247	0.6310
°API	• a,b	340[i]	265.5[h]	147.2[h]	119.8[h]	110.6[h]	95.0	92.7
Lb/gal @ 60°F, wt in vacuum	•	2.5[i]	2.971[h]	4.233[h]	4.695[h]	4.872[h]	5.208	5.261
Lb/gal @ 60°F, wt in air	• c	2.5[i]	2.962[h]	4.223[h]	4.686[h]	4.865[h]	5.199	5.251
Density of Gas @ 60°F & 14.696 psia								
Specific Gravity, Air = 1.00, ideal gas	•	0.5539	1.0382	1.5225	2.0068	2.0068	2.4911	2.4911
Lb/M cu ft, ideal gas	•	42.28	79.24	116.20	153.16	153.16	190.13	190.13
Volume Ratio @ 60°F and 14.696 psia								
Gal/lb mol	•	6.4[i]	10.12[h]	10.42[h]	12.38[h]	11.93[h]	13.85	13.71
Cu ft gas/gal liquid, ideal gas	•	59[i]	37.5[h]	36.43[h]	30.65[h]	31.81[h]	27.39	27.67
Gas vol/liquid vol, ideal gas	•	443[i]	280.5[h]	272.51[h]	229.30[h]	237.98[h]	204.93	207.00
Critical Conditions								
Temperature, °F		−116.63	90.09	206.01	274.98	305.65	369.10	385.7
Pressure, psia		667.8	707.8	616.3	529.1	550.7	490.4	488.6
Gross Heat of Combustion @ 60°F								
Btu/lb liquid	•	—	22,214[d]	21,513[d]	21,091[d]	21,139[d]	20,889	20,928
Btu/lb gas	•	23,885	22,323	21,665	21,237	21,298	21,040	21,089
Btu/cu ft, ideal gas	•p	1009.7	1768.8	2517.5	3252.7	3262.1	4000.3	4009.6
Btu/gal liquid	•	—	65,998[d]	91,065[d]	99,022[d]	102,989[d]	108,790	110,102
Cu ft air to burn 1 cu ft gas — ideal gas	•	9.54	16.70	23.86	31.02	31.02	38.18	38.18
Flammability Limits @ 100°F & 14.696 psia								
Lower, vol % in air		5.0	2.9	2.1	1.8	1.8	1.4	1.4
Upper, vol % in air		15.0	13.0	9.5	8.4	8.4	(8.3)	8.3
Heat of Vaporization @ 14.696 psia								
Btu/lb @ boiling point		219.22	210.41	183.05	157.53	165.65	147.13	153.59
Specific Heat @ 60°F & 14.696 psia								
C_P gas — Btu/lb, °F, ideal gas		0.5266	0.4097	0.3881	0.3872	0.3867	0.3827	0.3883
C_V gas — Btu/lb, °F, ideal gas	•	0.4027	0.3436	0.3430	0.3530	0.3525	0.3552	0.3608
$N = C_P/C_V$	•	1.308	1.192	1.131	1.097	1.097	1.078	1.076
C_P liquid — Btu/lb, °F	•	—	0.9256	0.5920	0.5695	0.5636	0.5353	0.5441
Octane Number								
Motor clear		—	+ .05[f]	97.1	97.6	89.6[j]	90.3	62.6[j]
Research clear		—	+1.6[j,f]	+1.8[j,f]	+0.10[j,f]	93.8[j]	92.3	61.7[j]
Refractive Index n_D @ 68°F		—	—	—	—	1.3326[h]	1.35373	1.35748

NOTES

a. Air saturated hydrocarbons.
b. Absolute values from weights in vacuum.
c. The apparent values for weight in air are shown for users' convenience. All other mass data in this table are on the absolute mass (weight in vacuum) basis.
d. At saturation pressure (triple point).
f. The + sign and number following signify the octane number corresponding to that of 2,2,4 trimethylpentane with the indicated number of ml of TEL added.
h. Saturation pressure at 60°F.
i. Apparent value for methane at 60°F.
j. Average value from octane numbers of more than one sample.
m. Density of liquid, gm/ml at normal boiling point.
n. Heat of sublimation.
p. Gross heat values are reported on a dry basis at 60°F and 14.696 psia based on ideal gas calculation. To convert to water saturation basis, multiply by 0.9826.
s. Extrapolated to room temperature from higher temperature.
• Calculated values. 1969 atomic weights used. See "Constants for Use in Calculations."
( ) Estimated values.

TABLE B.2 (continued)
Physical Constants of Paraffin Hydrocarbons and Other Components of Natural Gas [NGPA Publication 2145–74(1)

N-Hexane	N-Heptane	N-Octane	N-Nonane	N-Decane	Carbon Dioxide	Hydrogen Sulfide	Nitrogen	Oxygen	Air	Water
86.178	100.205	114.232	128.259	142.286	44.010	34.076	28.013	31.999	28.964	18.015
155.72	209.17	258.22	303.47	345.48	−109.3[2]	−76.6[24]	−320.4[2]	−297.4[2]	−317.6[2]	212.0
−139.58	−131.05	−70.18	−64.28	−21.36	–	−117.2[7]	−346.0[24]	−361.8[24]	–	32.0
4.956	1.620	0.537	0.179	0.0597	–	394.0[6]	–	–	–	0.9492[12]
0.6640	0.6882	0.7068	0.7217	0.7342	0.827[h(6)]	0.79[h(6)]	0.808[m(2)]	1.14[m(2)]	0.856[m(8)]	1.000
81.6	74.1	68.7	64.6	61.2	39.6[h]	47.6[h]	43.6[m]	−7.4[m]	33.8[m]	10.0
5.536	5.738	5.893	6.017	6.121	6.89[h]	6.59[h]	6.74[m]	9.50[m]	7.14[m]	8.337
5.526	5.728	5.883	6.008	6.112	6.89[h]	6.58[h]	6.73[m]	9.50[m]	7.13[m]	8.328
2.9753	3.4596	3.9439	4.4282	4.9125	1.5195	1.1765	0.9672	1.1048	1.0000	0.6220
227.09	264.05	301.01	337.98	374.94	115.97	89.79	73.82	84.32	76.32	47.47
15.57	17.46	19.39	21.32	23.24	6.38[h]	5.17[h]	4.16[m]	3.37[m]	4.06[m]	2.16
24.38	21.73	19.58	17.80	16.33	59.5[h]	73.3[h]	91.3[m]	112.7[m]	93.5[m]	175.6
182.37	162.56	146.45	133.18	122.13	444.8[h]	548.7[h]	682.7[m]	843.2[m]	699.5[m]	1313.8
453.7	512.8	564.22	610.68	652.1	87.9[23]	212.7[17]	−232.4[illegible]	−181.1[17]	−221.3[2]	705.6[17]
436.9	396.8	360.6	332	304	1071[17]	1306[17]	493.0[24]	736.9[24]	547[2]	3208[17]
20,784	20,681	20,604	20,544	20,494	–	–	–	–	–	–
20,944	20,840	20,762	20,701	20,649	–	–	–	–	–	–
4756.2	5502.8	6249.7	6996.5	7742.1	–	637[10]	–	–	-	–
115,060	118,668	121,419	123,613	125,444	–	–	–	–	–	–
45.34	52.50	59.65	66.81	73.97	–	7.16	–	–	–	–
1.2	1.0	0.96	0.87[e]	0.78[e]	–	4.30[5]	–	–	–	–
7.7	7.0	–	2.9	2.6	–	45.50	–	–	–	–
143.95	136.01	129.53	123.76	118.68	238.2[n(14)]	235.6[7]	87.8[14]	91.6[14]	92[3]	970.3[12]
0.3864	0.3875	(0.3876)	0.3840	0.3835	0.1991[13]	0.238[4]	0.2482[13]	0.2188[13]	0.2400[9]	0.4446[13]
0.3633	0.3677	0.3702	0.3685	0.3695	0.1539	0.1797	0.1773	0.1567	0.1714	0.3343
1.063	1.054	1.047	1.042	1.038	1.293	1.325	1.400	1.396	1.400	1.330
0.5332	0.5283	0.5239	0.5228	0.5208	–	–	–	–	–	1.0009[7]
26.0	0.0	–	–	–	–	–	–	–	–	–
24.8	0.0	–	–	–	–	–	–	–	–	–
1.37486	1.38764	1.39743	1.40542	1.41189	–	–	–	–	–	1.3330[6]

REFERENCES

1. Values for hydrocarbons were selected or calculated from API Project 44 and are identical to or consistent with ASTM DS 4A, "Physical Constants of Hydrocarbons C1 - C10," 1971. American Society for Testing Materials, Philadelphia.
2. International critical tables.
3. Hodgman, *Handbook of Chemistry & Physics*, 31st edition, 1949.
4. West, J. R., *Chem. Eng. Prog. 44,* 287, 1948.
5. Jones, *Chem. Rev. 22*,1, 1938.
6. Sage and Lacey, API Research Project 37, Monograph, 1955.
7. Perry, *Chemical Engineers' Handbook,* 4th edition, 1963.
8. Matteson and Hanna, *Oil Gas J. 41*,2, 33, 1942.
9. Keenan and Keyes, Thermodynamic Properties of Air, 1947.
12. Keenan and Keyes, Thermodynamic Properties of Steam, 29th ed., 1956.
13. American Petroleum Institute, Project 44.
14. Dreisback, *Physical Properties of Chemical Compounds,* American Chemical Society, 1961.
16. Maxwell, J.B., *Data Book on Hydrocarbons,* Van Nostrand Co., 1950.
17. Kobe, K.A. and Lynn, R.E., Jr., *Chem. Rev. 52*, pp. 117–236, 1953.
23. Din, *Thermodynamic Functions of Gases,* Butterworths, 1956.
24. Thermodynamic Research Center Data Project, Texas A&M University (formerly MCA Research Project).

CONSTANTS FOR USE IN CALCULATIONS

Atomic weights (based on 1969 values, *Pure Applied Chemistry 20*(4), 1969):
C:12.011 H: 1.0080 N: 14.0067 O: 15.9994 S: 32.06
Ideal gas:
1 mol = 379.49 ft^3 at 14.696 psia and 60°F
1 mol = 22.414 liters at 14.696 psia and 32°F
Conversion factors:
1 ft^3 = 28.317 liters 1 gal = 3785.41 milliliters 1 ft^3 = 7.4805 gal
760 mmHg = 14.696 lb/in^2 = 1 atm
1 lb = 453.59 gms 0°F = 459.67° Rankine
Density of water at 60°F = 8.3372 lb/gal = 0.999015 g/cc (weight in vacuum)
Specific gravity at 60°F/60°F × 0.999015 = density at 60°F, g/cc

$$°F = \frac{9}{5}°C + 32 \qquad °API = \frac{141.5}{\text{sp gr at } 60°F/60°F} - 131.5$$

CALCULATED VALUES

Density of liquid at 60°F and 14.696 psia:
lb/gal at 60°F (weight-in-vacuum) = sp gr at 60°F/60°F (weight-in-vacuum) × 8.3372 lb/gal (wt-in-vacuum)
lb/gal at 60°F (weight-in-air) (see ASTM DS 4A, p. 61):

$$\text{gal/lb mol at } 60°F = \frac{\text{mol wt}}{\text{lb/gal at } 60°F \text{ (weight-in-vacuum)}}$$

Density of gas at 60°F and 14.696 psia (ideal gas):

$$\text{sp gr at } 60°F = \frac{\text{mol wt}}{28.964}$$

$$\text{lb/M cu ft} = \frac{(\text{mol wt} \times 1000)}{379.49}$$

$$\text{ft}^3\text{vap/gal liq} = \frac{\text{lb/gal at } 60°F \text{ (wt-in-vac)} \times 379.49}{\text{mol wt}}$$

Ratio gas vol/liq vol:

$$\text{Gas vol/liq vol} = \frac{\text{lb/gal at } 60°F \text{ (wt-in-vac)} \times 379.49 \times 7.4805}{\text{mol wt}}$$

Heat of combustion at 60°F: The heat of combustion in Btu/gal was calculated as follows: The gross heat of combustion in Btu/lb gas less the heat of vaporization at 60°F is the calculated heat of combustion for the liquid state in Btu/lb. The heat of vaporization at 60°F was calculated from the normal boiling point, critical temperature, and heat of vaporization at the normal boiling point using the method of Fishtine (see Reid and Sherwood, *The Properties of Gases and Liquids*, 2nd ed., p. 148). The heat of combustion in Btu/lb liquid multiplied by the weight in vacuum, lb/gal, yields the heat of combustion in Btu/gal at the saturation pressure and 60°F.

Cubic feet of air to burn 1 ft^3 gas (ideal gas)—C_aH_b (see ASTM DS A, p. 63):

$$\frac{\text{ft}^3 \text{ air}}{\text{ft}^3 \text{ gas}} = \frac{(a + b/4)}{0.2095}$$

$$(H_2S + 1.5O_2 \rightarrow H_2O + SO_2)$$

Specific heat at 60°F and 14.696 psia (ideal gas):

$$C_v\,(\text{gas}) = C_p\,(\text{gas}) - \left(\frac{1.98719}{\text{mol wt}}\right) \text{ (for hydrocarbons)}$$

C_v values for nonhydrocarbon components are calculated from C_p and N values.

TABLE B.3
Heats of Combustion of Residual Fuel Oils

Gravity		Density, lb per gal	Total heat of combustion at constant volume, Qv			Net heat of combustion at constant pressure, Qp		
°API at 60°F	Specific at 60°/60°F		Cal per g	Btu per lb	Btu per gal	Cal per g	Btu per lb	Btu per gal
0	1.0760	8.962	9,970	17,950	160,900	9,470	17,050	152,800
1.0	1.0679	8.895	10,010	18,010	160,200	9,500	17,100	152,100
2.0	1.0599	8.828	10,040	18,070	159,500	9,530	17,150	151,400
3.0	1.0520	8.762	10,080	18,140	158,900	9,560	17,210	150,800
4.0	1.0443	8.698	10,110	18,200	158,300	9,590	17,260	150,100
5.0	1.0366	8.634	10,140	18,250	157,600	9,620	17,320	149,500
6.0	1.0291	8.571	10,180	18,320	157,000	9,650	17,370	148,900
7.0	1.0217	8.509	10,210	18,380	156,400	9,670	17,410	148,200
8.0	1.0143	8.448	10,240	18,430	155,700	9,690	17,450	147,400
9.0	1.0071	8.388	10,270	18,490	155,100	9,720	17,500	146,800
10.0	1.0000	8.328	10,300	18,540	154,400	9,740	17,540	146,100
11.0	0.9930	8.270	10,330	18,590	153,700	9,770	17,580	145,500
12.0	0.9861	8.212	10,360	18,640	153,000	9,790	17,620	144,800
13.0	0.9792	8.155	10,390	18,690	152,400	9,810	17,670	144,100
14.0	0.9725	8.099	10,410	18,740	151,800	9,840	17,710	143,500
15.0	0.9659	8.044	10,440	18,790	151,100	9,860	17,750	142,800
16.0	0.9593	7.989	10,470	18,840	150,500	9,880	17,790	142,200
17.0	0.9529	7.935	10,490	18,890	149,900	9,900	17,820	141,500
18.0	0.9465	7;882	10,520	18,930	149,200	9,920	17,860	140,800
19.0	0.9402	7.830	10.540	18,980	148,600	9,940	17,900	140,200
20.0	0.9340	7.778	10,570	19,020	147,900	9,960	17,930	139,500
21.0	0.9279	7.727	10,590	19,060	147,300	9,980	17,960	138,900
22.0	0.9218	7.676	10,620	19,110	146,600	10,000	18,000	138,200
23.0	0.9159	7.627	10,640	19,150	146,000	10,020	18,030	137,600
24.0	0.9100	7.578	10.660	19,190	145,400	10,040	18,070	137,000
25.0	0.9024	7.529	10.680	19,230	144,800	10,050	18,100	136,300
26.0	0.8984	7.481	10,710	19,270	144,100	10,070	18,130	135,700
27.0	0.8927	7.434	10,730	19,310	143,500	10,090	18,160	135,100
28.0	0.8871	7.387	10,750	19,350	142,900	10,110	18,190	134,500
29.0	0.8816	7.341	10,770	19,380	142,300	10,120	18,220	133,800
30.0	0.8762	7.305	10.790	19,420	141,800	10,140	18,250	133,300
31.0	0.8708	7.260	10,810	19,450	141,200	10,150	18,280	132,700
32.0	0.8654	7.215	10,830	19,490	140,600	10,170	18,310	132,100
33.0	0.8602	7.171	10,850	19,520	140,000	10,180	18,330	131,500
34.0	0.8550	7.128	10,860	19,560	139,400	10,200	18,360	130,900

TABLE B.3 (continued)
Heats of Combustion of Residual Fuel Oils

Gravity		Density, lb per gal	Total heat of combustion at constant volume, Qv			Net heat of combustion at constant pressure, Qp		
°API at 60°F	Specific at 60°/60°F		Cal per g	Btu per lb	Btu per gal	Cal per g	Btu per lb	Btu per gal
35.0	0.8498	7.085	10,880	19,590	138,800	10,210	18,390	130,300
36.0	0.8448	7.043	10,900	19,620	138,200	10,230	18,410	129,700
37.0	0.8398	7.001	10,920	19,650	137,600	10,240	18,430	129,100
38.0	0.8348	6.960	10,940	19,680	137,000	10,260	18,460	128,500
39.0	0.8299	6.920	10,950	19,720	136,400	10,270	18,840	127,900
40.0	0.8251	6.879	10,970	19,750	135,800	10,280	18,510	127,300
41.0	0.8203	6.839	10,990	19,780	135,200	10,300	18,530	126,700
42.0	0.8155	6.799	11,000	19,810	134,700	10,310	18,560	126,200
43.0	0.8109	6.760	11,020	19,830	134,100	10,320	18,580	125,600
44.0	0.8063	6.722	11,030	19,860	133,500	10,330	18,600	125,000
45.0	0.8017	6.684	11,050	19,890	132,900	10,340	18,620	124,400
46.0	0.7972	6.646	11,070	19,920	132,400	10,360	18,640	123,900
47.0	0.7927	6.609	11,080	19,940	131,900	10,370	18,660	123,300
48.0	0.7883	6.572	11,100	19,970	131,200	10,380	18,680	122,800
49.0	0.7839	6.536	11,110	20,000	130,700	10,390	18,700	122,200

Source: Guthrie, K.M. (Ed.), *Petroleum Products Handbook*, Copyright 1960, McGraw-Hill Book Co. Used with permission of McGraw-Hill Book Co.

Appendix C

U.S. Bureau of Mines Routine Analyses of Selected Crude Oils

CRUDE PETROLEUM ANALYSIS

Bureau of Mines Bartlesville Laboratory
Sample 55151

IDENTIFICATION

Ten Section field
Stevens, Upper Miocene
7,800-8,400 feet

California
Kern County

GENERAL CHARACTERISTICS

Gravity, specific, 0.854 Gravity, ° API, 34.2 Pour point, ° F., below 5
Sulfur, percent, 0.45 Color, brownish black
Viscosity, Saybolt Universal at 100°F., 43 sec. Nitrogen, percent, 0.289

DISTILLATION, BUREAU OF MINES ROUTINE METHOD

STAGE 1—Distillation at atmospheric pressure, 756 mm. Hg
First drop, 82 ° F.

Fraction No.	Cut temp. ° F.	Percent	Sum, percent	Sp. gr., 60/60° F.	° API, 60° F.	C. I.	Refractive index, n_D at 20° C.	Specific dispersion	S. U. visc., 100° F.	Cloud test, ° F.
1	122	2.6	2.6	0.644	88.2					
2	167	2.3	4.9	.683	75.7	14	1.38469	122.3		
3	212	5.0	9.9	.725	63.7	24	1.40300	124.0		
4	257	7.9	17.8	.751	56.9	27	1.41569	128.6		
5	302	6.2	24.0	.772	51.8	29	1.42785	133.6		
6	347	4.9	28.9	.791	47.4	32	1.43863	135.5		
7	302	4.6	33.5	.808	43.6	33	1.44778	140.5		
8	437	5.2	38.7	.825	40.0	36	1.45638	144.7		
9	482	4.9	43.6	.837	37.6	36	1.46441	152.5		
10	527	6.2	49.8	.852	34.6	39	1.47262	151.8		
STAGE 2—Distillation continued at 40 mm. Hg										
11	392	4.3	54.1	0.867	31.7	42	1.47941	156.0	41	10
12	437	5.2	59.3	.872	30.8	40	1.48461	163.8	49	30
13	482	5.3	64.6	.890	27.5	46	1.49418	170.9	66	55
14	527	3.2	67.8	.897	26.3	46			105	70
15	572	5.4	73.2	.915	23.1	51			200	80
Residuum		25.0	98.2	.984	12.3					

Carbon residue, Conradson: Residuum, 10.5 percent; crude, 3.0 percent.

APPROXIMATE SUMMARY

	Percent	Sp. gr.	° API	Viscosity
Light gasoline	9.9	0.694	72.4	
Total gasoline and naphtha	33.5	0.752	56.7	
Kerosine distillate	5.2	.825	40.0	
Gas oil	18.5	.855	34.0	
Nonviscous lubricating distillate	8.5	.873-.896	30.6-26.4	50-100
Medium lubricating distillate	4.8	.896-.915	26.4-23.1	100-200
Viscous lubricating distillate	2.7	.915-.926	23.1-21.3	Above 200
Residuum	25.0	.984	12.3	
Distillation loss				

U. S. GOVERNMENT PRINTING OFFICE 16—37833-3

CRUDE PETROLEUM ANALYSIS

Bureau of Mines Laramie Laboratory
Sample PC-65-28

IDENTIFICATION

Elk Hills field (Naval Reserve No. 1)
Sub-Scales No. 1 sandstone - Pilocene
3,333-3,370 feet

California
Kern County
SE1/4SW1/4, sec. 8,
T 31 S, R 24 E

GENERAL CHARACTERISTICS

Gravity, specific, 0.896 Gravity, ° API, 26.4 Pour point, ° F., below 5
Sulfur, percent, .51 Color, greenish black
Viscosity, Saybolt Universal at 100°F, 61 sec; at 77°F, 83 sec Nitrogen, percent, 0.398

DISTILLATION, BUREAU OF MINES ROUTINE METHOD

STAGE 1—Distillation at atmospheric pressure, 760 mm. Hg
First drop, 154 ° F.

Fraction No.	Cut temp. ° F.	Percent	Sum, percent	Sp. gr., 60/60° F.	° API, 60° F.	C. I.	Refractive index, n_D at 20° C.	Specific dispersion	S. U. visc., 100° F.	Cloud test, ° F.
1	122	0.6	0.6	0.697	71.5		1.39020	118.3		
2	167	3.3	3.9	.739	60.0	30	1.40622	121.1		
3	212	5.8	9.7	.762	54.2	32	1.41838	122.5		
4	257	5.2	14.9	.782	49.5	34	1.42933	125.8		
5	302	5.1	20.0	.805	44.3	38	1.44218	137.8		
6	347	4.4	24.4	.825	40.0	41	1.45419	141.2		
7	392	5.7	30.1	.840	37.0	43	1.46360	145.8		
8	437	7.4	37.5	.855	34.0	45	1.47338	155.8		
9	482	5.5	43.0	.871	31.0	48	1.48395	170.5		
10	527									
STAGE 2—Distillation continued at 40 mm. Hg										
11	392	1.5	44.5	.890	27.5	53	1.49180	166.4	41	below 5
12	437	5.8	50.3	.892	27.1	50	1.49533	174.4	47	below 5
13	482	5.3	55.6	.906	24.7	53	1.49684	171.6	63	below 5
14	527	4.8	60.4	.931	20.5	62			110	below 5
15	572	5.3	65.7	.938	19.4	62			280	below 5
Residuum		33.8	99.5	.984	12.3					

Carbon residue, Conradson: Residuum, 11.6 percent; crude, 4.3 percent.

APPROXIMATE SUMMARY

	Percent	Sp. gr.	° API	Viscosity
Light gasoline	3.9	0.733	61.5	
Total gasoline and naphtha	24.4	.782	49.5	
Kerosine distillate				
Gas oil	23.9	.863	32.5	
Nonviscous lubricating distillate	8.6	.894-.926	26.8-21.3	50-100
Medium lubricating distillate	3.8	.926-.935	21.3-19.8	100-200
Viscous lubricating distillate	5.0	.935-.942	19.8-18.7	Above 200
Residuum	33.8	.984	12.3	
Distillation loss	.5			

U. S. GOVERNMENT PRINTING OFFICE 16—57835-3

CRUDE PETROLEUM ANALYSIS

Bureau of Mines Bartlesville Laboratory
Sample 55126

IDENTIFICATION

Torrance Field
Del Amo, Miocene
3,100-5,000 feet

California
Los Angeles County

GENERAL CHARACTERISTICS

Gravity, specific, 0.911 Gravity, ° API, 23.8 Pour point, ° F., below 5
Sulfur, percent, 1.84 Color, brownish black
Viscosity, Saybolt Universal at 100°F, 160sec; 130°F, 96 sec Nitrogen, percent, 0.555

DISTILLATION, BUREAU OF MINES ROUTINE METHOD

STAGE 1—Distillation at atmospheric pressure, 746 mm. Hg
First drop, 81 ° F.

Fraction No.	Cut temp. ° F.	Percent	Sum. percent	Sp. gr., 60/60° F.	° API, 60° F.	C. I.	Refractive index, n_D at 20° C.	Specific dispersion	S. U. visc., 100° F.	Cloud test, ° F.
1	122	0.3	0.3	0.675	78.1					
2	167	1.1	1.4	.683	75.7	14	1.39514	126.2		
3	212	2.2	3.6	.725	63.7	24	1.41771	130.2		
4	257	3.7	7.3	.755	55.9	29	1.42942	134.5		
5	302	3.8	11.1	.777	50.6	32	1.43986	138.3		
6	347	3.3	14.4	.796	46.3	34	1.44908	141.7		
7	392	3.5	17.9	.813	42.6	36	1.45776	142.1		
8	437	4.3	22.2	.830	39.0	38	1.46618	148.3		
9	482	4.6	26.8	.843	36.4	39	1.47548	151.0		
10	527	7.4	34.2	.861	32.8	43				
STAGE 2—Distillation continued at 40 mm. Hg										
11	392	1.6	35.8	0.872	30.8	44	1.48248	-	43	10
12	437	5.4	41.2	.884	28.6	46	1.48731	150.6	49	30
13	482	6.0	47.2	.901	25.6	51	1.49769	-	70	45
14	527	5.5	52.7	.910	24.0	52	1.50544	-	125	65
15	572	4.6	57.3	.927	21.1	57			250	80
Residuum		41.9	99.2	1.004	9.4					

Carbon residue, Conradson: Residuum, 13.2 percent; crude, 6.1 percent.

APPROXIMATE SUMMARY

	Percent	Sp. gr.	° API	Viscosity
Light gasoline	3.6	0.708	68.4	
Total gasoline and naphtha	17.9	0.769	52.5	
Kerosine distillate	-	-	-	
Gas oil	21.0	.855	34.0	
Nonviscous lubricating distillate	8.5	.855-.906	28.4-24.7	50-100
Medium lubricating distillate	5.7	.906-.921	24.7-22.1	100-200
Viscous lubricating distillate	4.2	.921-.934	22.1-20.0	Above 200
Residuum	41.9	1.004	9.4	
Distillation loss	.8			

U. S. GOVERNMENT PRINTING OFFICE 16—37635-3

CRUDE PETROLEUM ANALYSIS

Bureau of Mines Bartlesville Laboratory
Sample 58010

IDENTIFICATION

Rangely field
Weber, Pennsylvanian
5,960-6,459 feet

Colorado
Rio Blanco County

GENERAL CHARACTERISTICS

Gravity, specific, 0.851 Gravity, ° API, 34.8 Pour point, ° F., 10
Sulfur, percent, 0.56 Color, greenish black
Viscosity, Saybolt Universal at 100°F., 48 sec. Nitrogen, percent, 0.073

DISTILLATION, BUREAU OF MINES ROUTINE METHOD

STAGE 1—Distillation at atmospheric pressure, 754 mm. Hg
First drop, 88 ° F.

Fraction No.	Cut temp. ° F.	Percent	Sum. percent	Sp. gr., 60/60° F.	° API, 60° F.	C. I.	Refractive index, n_D at 20° C.	Specific dispersion	S. U. visc., 100° F.	Cloud test, ° F.
1	122	1.4	1.4	0.647	87.2					
2	167	2.6	4.0	.670	79.7	7.5	1.37406	125.6		
3	212	2.9	6.9	.709	68.1	16	1.39574	128.0		
4	257	5.2	12.1	.731	62.1	18	1.40817	128.2		
5	302	3.9	16.0	.752	56.7	20	1.41885	131.4		
6	347	5.3	21.3	.772	51.8	23	1.42916	135.2		
7	392	4.8	26.1	.792	47.2	26	1.43936	137.1		
8	437	4.7	30.8	.810	43.2	29	1.44801	138.8		
9	482	5.6	36.4	.824	40.2	30	1.45630	140.5		
10	527	6.6	43.0	.843	36.4	34	1.46581	147.3		

STAGE 2—Distillation continued at 40 mm. Hg

Fraction No.	Cut temp. ° F.	Percent	Sum. percent	Sp. gr., 60/60° F.	° API, 60° F.	C. I.	Refractive index, n_D at 20° C.	Specific dispersion	S. U. visc., 100° F.	Cloud test, ° F.
11	392	2.9	45.9	0.854	34.2	36	1.47399	150.5	40	10
12	437	7.6	53.5	.861	32.8	35	1.47765	162.0	46	24
13	482	6.9	60.4	.880	29.3	41	1.48676	156.8	61	50
14	527	5.8	66.2	.891	27.3	43	1.49293	-	86	60
15	572	5.8	72.0	.902	25.4	45			160	74
Residuum		26.5	98.5	.962	15.6					

Carbon residue, Conradson: Residuum, 7.6 percent; crude, 2.3 percent.

APPROXIMATE SUMMARY

	Percent	Sp. gr.	° API	Viscosity
Light gasoline	6.9	0.682	76.0	
Total gasoline and naphtha	26.1	0.741	59.5	
Kerosine distillate	10.3	.818	41.5	
Gas oil	15.3	.852	34.6	
Nonviscous lubricating distillate	12.7	.865-.893	32.1-27.0	50-100
Medium lubricating distillate	7.6	.893-.907	27.0-24.5	100-200
Viscous lubricating distillate	-	-	-	Above 200
Residuum	26.5	.962	15.6	
Distillation loss	1.5			

U. S. GOVERNMENT PRINTING OFFICE 16 37835-3

CRUDE PETROLEUM ANALYSIS

Bureau of Mines Bartlesville Laboratory
Sample 57069

IDENTIFICATION

Bridgeport field
Bridgeport, Pennsylvanian
906-938 feet

Illinois
Lawrence County

GENERAL CHARACTERISTICS

Gravity, specific, 0.847 Gravity, ° API, 35.6 Pour point, ° F., below 5
Sulfur, percent, 0.21 Color, brownish green
Viscosity, Saybolt Universal at 100°F., 46 sec. Nitrogen, percent, 0.138

DISTILLATION, BUREAU OF MINES ROUTINE METHOD

STAGE 1—Distillation at atmospheric pressure, ... mm. Hg
First drop, ... ° F.

Fraction No.	Cut temp. ° F.	Percent	Sum, percent	Sp. gr., 60/60° F.	° API, 60° F.	C. I.	Refractive index, n_D at 20° C.	Specific dispersion	S. U. visc., 100° F.	Cloud test, ° F.
1	122	2.3	2.3	0.638	90.3					
2	167	2.2	4.5	.671	79.4	8.0	1.37182	121.0		
3	212	4.6	9.1	.712	67.2	18	1.39622	116.9		
4	257	5.7	14.8	.738	60.2	21	1.40852	123.8		
5	302	5.3	20.1	.756	55.7	22	1.41885	128.7		
6	347	5.3	25.4	.776	50.9	24	1.42922	129.4		
7	392	4.4	29.8	.792	47.2	26	1.43930	135.2		
8	437	4.7	34.5	.809	43.4	28	1.44784	133.7		
9	482	4.8	39.3	.823	40.4	30	1.45582	134.3		
10	527	5.9	45.2	.837	37.6	32	1.46433	142.6		
STAGE 2—Distillation continued at 40 mm. Hg										
11	392	3.7	48.9	0.852	34.6	35	1.47330	150.0	40	10
12	437	5.7	54.6	.858	33.4	34	1.47744	--	47	25
13	482	4.5	59.1	.873	30.6	37	1.48469	--	61	50
14	527	4.7	63.8	.885	28.4	40	1.49563	--	89	65
15	572	5.5	69.3	.901	25.6	45			170	75
Residuum		28.3	97.6	.960	15.9					

Carbon residue, Conradson: Residuum, 8.4 percent; crude, 2.7 percent.

APPROXIMATE SUMMARY

	Percent	Sp. gr.	° API	Viscosity
Light gasoline	9.1	0.683	75.7	
Total gasoline and naphtha	29.8	0.739	60.0	
Kerosine distillate	9.5	.816	41.9	
Gas oil	13.7	.847	35.6	
Nonviscous lubricating distillate	9.2	.861-.887	32.8-27.7	50-100
Medium lubricating distillate	6.4	.887-.907	27.7-24.5	100-200
Viscous lubricating distillate	.7	.907-.909	24.5-24.2	Above 200
Residuum	28.3	.960	15.9	
Distillation loss	2.4			

U. S. GOVERNMENT PRINTING OFFICE 10—57835-3

CRUDE PETROLEUM ANALYSIS

Bureau of Mines Bartlesville Laboratory
Sample 61084

IDENTIFICATION

Marcotte field
Arbuckle, Cambro-Ordovician
3,757-3,761 feet

Kansas
Rooks County

GENERAL CHARACTERISTICS

Gravity, specific, 0.897 Gravity, ° API, 26.3 Pour point, ° F., 15
Sulfur, percent, 0.77 Color, brownish black
Viscosity, Saybolt Universal at 100°F., 242 sec.; 130°F., 122 sec. Nitrogen, percent, 0.19

DISTILLATION, BUREAU OF MINES ROUTINE METHOD

STAGE 1—Distillation at atmospheric pressure, 746 mm. Hg
First drop, 185 ° F.

Fraction No.	Cut temp. ° F.	Percent	Sum, percent	Sp. gr., 60/60° F.	° API, 60° F.	C. I.	Refractive index, n_D at 20° C.	Specific dispersion	S. U. visc., 100° F.	Cloud test, ° F.
1	122									
2	167									
3	212									
4	257	2.3	2.3	0.730	62.3	-	1.40764	126.2		
5	302	2.5	4.8	.749	57.4	18	1.41629	124.0		
6	347	3.4	8.2	.770	52.3	22	1.42613	126.1		
7	392	3.6	11.8	.790	47.6	25	1.43574	127.7		
8	437	4.6	16.4	.804	44.5	26	1.44520	129.3		
9	482	4.5	20.9	.821	40.9	29	1.45305	132.8		
10	527	6.1	27.0	.836	37.8	31	1.46130	139.3		
STAGE 2—Distillation continued at 40 mm. Hg										
11	392	2.7	29.7	0.848	35.4	33	1.46983	143.7	39	below 5
12	437	6.4	36.1	.860	33.0	35	1.47567	148.0	46	15
13	482	8.2	44.3	.876	30.0	39	1.48456	155.4	61	30
14	527	4.6	48.9	.893	27.0	44	1.49161	161.2	98	60
15	572	6.7	55.6	.897	26.3	43	1.49562	168.5	155	75
Residuum		43.6	99.2	.974	13.8					

Carbon residue, Conradson: Residuum, 8.5 percent; crude, 4.1 percent.

APPROXIMATE SUMMARY

	Percent	Sp. gr.	° API	Viscosity
Light gasoline	-	-	-	
Total gasoline and naphtha	11.8	0.764	53.7	
Kerosine distillate	9.1	.812	42.8	
Gas oil	14.9	.847	35.6	
Nonviscous lubricating distillate	11.0	.866-.893	31.9-27.0	50-100
Medium lubricating distillate	8.8	.893-.899	27.0-25.9	100-200
Viscous lubricating distillate	-	-	-	Above 200
Residuum	43.6	.974	13.8	
Distillation loss	.8			

U. S. GOVERNMENT PRINTING OFFICE 10—57835-3

CRUDE PETROLEUM ANALYSIS

Bureau of Mines Bartlesville Laboratory
Sample 60052

IDENTIFICATION

Black Bay, West field
9200', Miocene
9,178-9,185 feet

Louisiana
Piaquemines Parish

GENERAL CHARACTERISTICS

Gravity, specific, 0.853 Gravity, ° API, 34.4 Pour point, ° F., below 5
Sulfur, percent, 0.19 Color, brownish green
Viscosity, Saybolt Universal at 100°F, 46 sec. Nitrogen, percent, 0.04

DISTILLATION, BUREAU OF MINES ROUTINE METHOD

STAGE 1—Distillation at atmospheric pressure, 758 mm. Hg
First drop, 113 ° F.

Fraction No.	Cut temp. ° F.	Percent	Sum, percent	Sp. gr., 60/60° F.	° API, 60° F.	C. I.	Refractive index, n_D at 20° C.	Specific dispersion	S. U. visc., 100° F.	Cloud test, ° F.
1	122									
2	167									
3	212	2.6	2.6	0.706	68.9	-	1.39971	129.4		
4	257	3.1	5.7	.739	60.0	21	1.41235	132.0		
5	302	3.7	9.4	.762	54.2	25	1.42308	135.4		
6	347	4.2	13.6	.780	49.9	26	1.43298	137.1		
7	392	5.8	19.4	.796	46.3	28	1.44076	138.4		
8	437	4.9	24.3	.807	43.8	27	1.44701	139.1		
9	482	7.6	31.9	.820	41.1	28	1.45389	140.8		
10	527	9.1	41.0	.834	38.2	30	1.46161	143.0		
STAGE 2—Distillation continued at 40 mm. Hg										
11	392	6.0	47.0	0.846	35.8	32	1.46906	148.8	40	below 5
12	437	8.3	55.3	.854	34.2	32	1.47238	147.4	46	20
13	482	6.8	62.1	.866	31.9	34	1.47868	144.0	58	50
14	527	5.8	67.9	.881	29.1	38	1.48434	-	81	60
15	572	6.1	74.0	.892	27.1	40			135	70
Residuum		24.5	98.5	.940	19.0					

Carbon residue, Conradson: Residuum, 4.6 percent; crude, 1.2 percent.

APPROXIMATE SUMMARY

	Percent	Sp. gr.	° API	Viscosity
Light gasoline	2.6	0.706	68.9	
Total gasoline and naphtha	19.4	0.765	53.5	
Kerosine distillate	12.5	.815	42.1	
Gas oil	21.9	.843	36.4	
Nonviscous lubricating distillate	12.9	.858-.884	33.4-28.6	50-100
Medium lubricating distillate	7.3	.884-.898	28.6-26.1	100-200
Viscous lubricating distillate	-	-	-	Above 200
Residuum	24.5	.940	19.0	
Distillation loss	1.5			

U. S. GOVERNMENT PRINTING OFFICE 16—57535-3

CRUDE PETROLEUM ANALYSIS

Bureau of Mines Bartlesville Laboratory
Sample 54060

IDENTIFICATION

Bayou des Allemands field
Miocene

Louisiana
Lafourche Parish

GENERAL CHARACTERISTICS

Gravity, specific, 0.845 Gravity, ° API, 36.0 Pour point, ° F., 35
Sulfur, percent, 0.20 Color, brownish green
Viscosity, Saybolt Universal at 100°F., 49 sec. Nitrogen, percent, 0.040

DISTILLATION, BUREAU OF MINES ROUTINE METHOD

STAGE 1—Distillation at atmospheric pressure, 743 mm. Hg
First drop, 86 ° F.

Fraction No.	Cut temp. ° F.	Percent	Sum, percent	Sp. gr., 60/60° F.	° API, 60° F.	C. I.	Refractive index, n_D at 20° C.	Specific dispersion	S. U. visc., 100° F.	Cloud test, ° F.
1	122	0.5	0.5	0.670	79.7					
2	167	1.2	1.7	.675	78.1	11				
3	212	1.6	3.3	.722	64.5	23	1.39163	137.0		
4	257	2.7	6.0	.748	57.7	26	1.41725	141.7		
5	302	3.1	9.1	.765	53.5	26	1.42648	142.3		
6	347	3.9	13.0	.778	50.4	25	1.43374	140.7		
7	392	4.7	17.7	.789	47.8	24	1.43962	138.0		
8	437	5.7	23.4	.801	45.2	24	1.44529	137.6		
9	482	8.0	31.4	.814	42.3	25	1.45193	137.4		
10	527	10.7	42.1	.825	40.0	26	1.45884	142.9		
STAGE 2—Distillation continued at 40 mm. Hg										
11	392	5.0	47.1	0.845	36.0	31	1.46614	142.6	40	15
12	437	10.0	57.1	.854	32	32	1.46870	139.9	45	30
13	482	7.8	64.9	.863	32.5	33	1.47403	140.4	56	50
14	527	7.0	71.9	.874	30.4	35			81	65
15	572	6.5	78.4	.889	27.7	39			145	85
Residuum		20.8	99.2	.931	20.5					

Carbon residue, Conradson: Residuum, 3.7 percent; crude, 0.8 percent.

APPROXIMATE SUMMARY

	Percent	Sp. gr.	° API	Viscosity
Light gasoline	3.3	0.697	71.5	
Total gasoline and naphtha	17.7	0.759	54.9	
Kerosine distillate	24.4	.816	41.9	
Gas oil	14.2	.850	35.0	
Nonviscous lubricating distillate	14.1	.858-.878	33.4-29.7	50-100
Medium lubricating distillate	8.0	.878-.895	29.7-26.6	100-200
Viscous lubricating distillate	-	-	-	Above 200
Residuum	20.8	.931	20.5	
Distillation loss	.8			

U. S. GOVERNMENT PRINTING OFFICE 16—57835-3

CRUDE PETROLEUM ANALYSIS

Bureau of Mines Bartlesville Laboratory
Sample 54064

IDENTIFICATION

Sho-Vel-Tum field
Camp area
Springer, Pennsylvanian
6,295-6,385 feet

Oklahoma
Carter County

GENERAL CHARACTERISTICS

Gravity, specific, 0.887 Gravity, ° API, 28.0 Pour point, ° F., 10
Sulfur, percent, 1.41 Color, brownish black
Viscosity, Saybolt Universal at 100°F., 115 sec; 130°F., 81 sec. Nitrogen, percent, 0.318

DISTILLATION, BUREAU OF MINES ROUTINE METHOD

STAGE 1—Distillation at atmospheric pressure, 745 mm. Hg
First drop, 84 ° F.

Fraction No.	Cut temp. ° F.	Percent	Sum, percent	Sp. gr., 60/60° F.	° API, 60° F.	C. I.	Refractive index, n_D at 20° C.	Specific dispersion	S. U. visc., 100° F.	Cloud test, ° F.
1	122	1.3	1.3	0.648	86.9					
2	167	1.5	2.8	.674	78.4	9.4				
3	212	3.3	6.1	.712	67.2	18	1.39123	127.3		
4	257	4.3	10.4	.739	60.0	21	1.40995	127.9		
5	302	4.0	14.4	.758	55.2	23	1.42105	130.1		
6	347	4.1	18.5	.779	50.1	26	1.43181	134.3		
7	392	3.7	22.2	.798	45.8	29	1.44175	136.9		
8	437	4.1	26.3	.814	42.3	31	1.45087	141.2		
9	482	4.8	31.1	.831	38.8	33	1.46025	145.7		
10	527	6.0	37.1	.848	35.4	37	1.46939	155.6		
STAGE 2—Distillation continued at 40 mm. Hg										
11	392	1.1	38.2	0.862	32.7	39	1.47778	153.8	43	10
12	437	4.7	42.9	.873	30.6	41	1.48216	156.6	46	25
13	482	4.6	47.5	.882	28.9	42	1.48952	161.8	58	40
14	527	5.3	52.8	.898	26.1	46			88	55
15	572	5.3	58.1	.911	23.8	49			175	70
Residuum		40.9	99.0	.982	12.6					

Carbon residue, Conradson: Residuum, 11.4 percent; crude, 5.2 percent.

APPROXIMATE SUMMARY

	Percent	Sp. gr.	° API	Viscosity
Light gasoline	6.1	0.689	73.9	
Total gasoline and naphtha	22.2	0.746	58.2	
Kerosine distillate	4.1	.814	42.3	
Gas oil	16.0	.844	36.2	
Nonviscous lubricating distillate	8.6	.854-.871	34.2-31.0	50-100
Medium lubricating distillate	6.1	.871-.891	31.0-27.3	100-200
Viscous lubricating distillate	1.1	.891-.894	27.3-26.8	Above 200
Residuum	40.9	.982	12.6	
Distillation loss	1.0			

U. S. GOVERNMENT PRINTING OFFICE 16—57933-3

CRUDE PETROLEUM ANALYSIS

Bureau of Mines Bartlesville Laboratory
Sample 59172

IDENTIFICATION

Sho-Vel-Tum field
Tatums area
Pennsylvanian

Oklahoma
Garvin County

GENERAL CHARACTERISTICS

Gravity, specific, 0.928 Gravity, ° API, 21.0 Pour point, ° F., below 5
Sulfur, percent, 1.68 Color, brownish black
Viscosity, Saybolt Universal at 100°F., 550 sec; 130°F., 440 sec. Nitrogen, percent, 0.482

DISTILLATION, BUREAU OF MINES ROUTINE METHOD

STAGE 1—Distillation at atmospheric pressure, 743 mm. Hg
First drop, 147 ° F.

Fraction No.	Cut temp. ° F.	Percent	Sum, percent	Sp. gr., 60/60° F.	° API, 60° F.	C. I.	Refractive index, n_D at 20° C.	Specific dispersion	S. U. visc., 100° F.	Cloud test, ° F.
1	122									
2	167									
3	212	2.6	2.6	0.695	72.1	-	1.38886	124.3		
4	257	2.8	5.4	.737	60.5	20	1.40876	124.0		
5	302	2.9	8.3	.758	55.2	23	1.41971	128.8		
6	347	3.5	11.8	.778	50.3	25	1.43164	131.0		
7	392	2.7	14.5	.799	45.6	29	1.44187	135.5		
8	437	3.1	17.6	.817	41.7	32	1.45080	136.7		
9	482	4.4	22.0	.832	38.6	-	1.46062	143.8		
10	527									

STAGE 2—Distillation continued at 40 mm. Hg

Fraction No.	Cut temp. ° F.	Percent	Sum, percent	Sp. gr., 60/60° F.	° API, 60° F.	C. I.	Refractive index, n_D at 20° C.	Specific dispersion	S. U. visc., 100° F.	Cloud test, ° F.
11	392	6.1	28.1	0.859	33.2	-	1.47659	156.3	38	below 5
12	437	4.5	32.6	.878	29.7	43	1.48547	160.2	47	do.
13	482	3.9	36.5	.890	27.5	46	1.49374	157.4	62	do.
14	527	4.8	41.3	.909	24.2	-	1.50253	161.8	105	do.
15	572									
Residuum		55.9	97.2	1.012	8.3					

Carbon residue, Conradson: Residuum, 8.2 percent; crude 5.0 percent.

APPROXIMATE SUMMARY

	Percent	Sp. gr.	° API	Viscosity
Light gasoline	2.6	0.695	72.1	
Total gasoline and naphtha	14.5	0.755	55.9	
Kerosine distillate	3.1	.817	41.7	
Gas oil	13.6	.854	34.2	
Nonviscous lubricating distillate	7.3	.880-.907	29.3-24.5	50-100
Medium lubricating distillate	2.8	.907-.919	24.5-22.5	100-200
Viscous lubricating distillate	-	-	-	Above 200
Residuum	55.9	1.012	8.3	
Distillation loss	2.8			

U. S. GOVERNMENT PRINTING OFFICE 16—57833-3

CRUDE PETROLEUM ANALYSIS

Bureau of Mines Bartlesville Laboratory
Sample 64036

IDENTIFICATION

Sho-Vel-Tum field
Sholem Alechem area
Pennsylvanian
3,468-3,488 feet

Oklahoma
Stephens County

GENERAL CHARACTERISTICS

Gravity, specific, 0.893 Gravity, ° API, 27.0 Pour point, ° F., below 5
Sulfur, percent, 1.34 Color, greenish black
Viscosity, Saybolt Universal at 100°F., 131 sec; 130°F., 84 sec Nitrogen, percent, 0.243

DISTILLATION, BUREAU OF MINES ROUTINE METHOD

STAGE 1—Distillation at atmospheric pressure, 732 mm. Hg
First drop, 77 ° F.

Fraction No.	Cut temp. ° F.	Percent	Sum, percent	Sp. gr., 60/60° F.	° API, 60° F.	C. I.	Refractive index, n_D at 20° C.	Specific dispersion	S. U. visc., 100° F.	Cloud test, ° F.
1	122									
2	167									
3	212	3.2	3.2	0.704	69.5	-	1.39260	128.2		
4	257	3.5	6.7	.737	60.5	20	1.40952	131.7		
5	302	3.7	10.4	.759	54.9	23	1.42062	133.4		
6	347	3.8	14.2	.778	50.4	25	1.43156	135.0		
7	392	3.5	17.7	.798	45.8	29	1.44148	137.7		
8	437	4.0	21.7	.814	42.3	31	1.45105	140.0		
9	482	4.7	26.4	.832	38.6	34	1.46032	148.1		
10	527	5.1	31.5	.847	35.6	36	1.46909	154.9		
STAGE 2—Distillation continued at 40 mm. Hg										
11	392	3.7	35.2	0.870	31.1	43	1.47931	158.1	42	5
12	437	4.7	39.9	.877	29.8	43	1.48496	164.2	50	25
13	482	5.6	45.5	.895	26.6	48	1.49389	-	73	50
14	527	3.8	49.3	.905	24.9	49			115	60
15	572									
Residuum		48.9	98.2	.972	14.1					

Carbon residue, Conradson: Residuum, 9.2 percent; crude, 4.9 percent.

APPROXIMATE SUMMARY

	Percent	Sp. gr.	° API	Viscosity
Light gasoline	3.2	0.704	69.5	
Total gasoline and naphtha	17.7	0.757	55.5	
Kerosine distillate	4.0	.814	42.3	
Gas oil	15.8	.852	34.6	
Nonviscous lubricating distillate	8.2	.877-.901	29.8-25.5	50-100
Medium lubricating distillate	3.6	.901-.909	25.5-24.2	100-200
Viscous lubricating distillate	-	-	-	Above 200
Residuum	48.9	.972	14.1	
Distillation loss	1.8			

U. S. GOVERNMENT PRINTING OFFICE 16—57835-3

CRUDE PETROLEUM ANALYSIS

Bureau of Mines ...Bartlesville... Laboratory
Sample ...62066...

IDENTIFICATION

Hastings, East field
Frio, Oligocene
6,020-6,050 feet

Texas
Brazoria County

GENERAL CHARACTERISTICS

Gravity, specific, 0.871 Gravity, ° API, 31.0 Pour point, ° F., below 5
Sulfur, percent, 0.15 Color, greenish black
Viscosity, Saybolt Universal at 77°F., 62 sec.; 100°F., 55 sec. Nitrogen, percent, 0.02

DISTILLATION, BUREAU OF MINES ROUTINE METHOD

STAGE 1—Distillation at atmospheric pressure, 743 mm. Hg
First drop, 145 ° F.

Fraction No.	Cut temp. ° F.	Percent	Sum, percent	Sp. gr., 60/60° F.	° API, 60° F.	C. I.	Refractive index, n_D at 20° C.	Specific dispersion	S. U. visc., 100° F.	Cloud test, ° F.
1	122									
2	167	1.1	1.1	0.748	57.7	-	1.40061	126.1		
3	212	1.8	2.9	.753	56.4	37	1.40946	129.6		
4	257	1.7	4.6	.757	55.4	30	1.41686	131.3		
5	302	2.7	7.3	.770	52.3	28	1.42613	139.7		
6	347	3.4	10.7	.789	47.8	31	1.43860	142.8		
7	392	5.1	15.8	.813	42.6	36	1.45011	147.6		
8	437	5.9	21.7	.829	39.2	38	1.45805	149.8		
9	482	9.8	31.5	.846	35.8	41	1.46806	153.5		
10	527	10.7	42.2	.860	33.0	42	1.47690	158.6		
STAGE 2—Distillation continued at 40 mm. Hg										
11	392	4.4	46.6	0.871	31.0	44	1.48289	158.7	42	Below 5
12	437	8.7	55.3	.880	29.3	44	1.48436	156.3	49	do.
13	482	6.7	62.0	.891	27.3	46	1.48938	155.1	68	do.
14	527	5.9	67.9	.904	25.0	49	1.49414	153.0	110	do.
15	572	6.6	74.5	.910	24.0	49			225	10
Residuum		23.0	97.5	.942	18.7					

Carbon residue, Conradson: Residuum, 4.3 percent; crude, 1.1 percent.

APPROXIMATE SUMMARY

	Percent	Sp. gr.	° API	Viscosity
Light gasoline	2.9	0.751	56.9	
Total gasoline and naphtha	15.8	0.783	49.2	
Kerosine distillate	-	-	-	
Gas oil	35.6	.855	34.0	
Nonviscous lubricating distillate	12.1	.880-.901	29.3-25.6	50-100
Medium lubricating distillate	6.4	.901-.908	25.6-24.3	100-200
Viscous lubricating distillate	4.6	.908-.913	24.3-23.5	Above 200
Residuum	23.0	.942	18.7	
Distillation loss	2.5			

U. S. GOVERNMENT PRINTING OFFICE 16—87835-2

CRUDE PETROLEUM ANALYSIS

Bureau of Mines Bartlesville Laboratory
Sample 51051

IDENTIFICATION

Cedar Lake field
San Andres, Permian
4,580-4,765 feet

Texas
Gaines County

GENERAL CHARACTERISTICS

Gravity, specific, 0.863 Gravity, ° API, 32.5 Pour point, ° F., below 5
Sulfur, percent, 2.12 Color, greenish black
Viscosity, Saybolt Universal at 100°F, 45 sec. Nitrogen, percent, 0.09

DISTILLATION, BUREAU OF MINES ROUTINE METHOD

STAGE 1—Distillation at atmospheric pressure, 746 mm. Hg
First drop, 88 ° F.

Fraction No.	Cut temp. ° F.	Percent	Sum. percent	Sp. gr., 60/60° F.	° API, 60° F.	C. I.	Refractive index, n_D at 20° C.	Specific dispersion	S. U. visc., 100° F.	Cloud test, ° F.
1	122	1.8	1.8	0.656	84.2		1.39023	139.5		
2	167	6.4	8.2	.697	71.5	20	-	-		
3	212	3.1	11.3	.740	69.5	31	1.42215	146.8		
4	257	4.7	16.0	.761	54.4	32	1.43240	150.7		
5	302	6.4	22.4	.777	50.6	32	1.43892	147.8		
6	347	4.6	27.0	.790	47.6	31	1.44401	145.3		
7	392	4.8	31.8	.801	45.2	30	1.45084	145.9		
8	437	4.3	36.1	.815	42.1	31	1.46039	151.1		
9	482	5.4	41.5	.831	38.8	33	1.47221	162.4		
10	527	6.8	48.3	.849	35.2	37				
STAGE 2—Distillation continued at 40 mm. Hg										
11	392	1.4	49.7	0.862	32.7	39			40	15
12	437	5.7	55.4	.873	30.6	41			44	30
13	482	5.6	61.0	.889	27.7	45			56	50
14	527	5.3	66.3	.899	25.9	47			82	65
15	572	5.5	71.8	.916	23.0	55			150	85
Residuum		27.0	98.8	.987	11.9					

Carbon residue, Conradson: Residuum, 10.7 percent; crude, 3.3 percent.

APPROXIMATE SUMMARY

	Percent	Sp. gr.	° API	Viscosity
Light gasoline	11.3	0.702	70.1	
Total gasoline and naphtha	31.8	0.754	56.2	
Kerosine distillate	4.3	.815	42.1	
Gas oil	19.5	.852	34.6	
Nonviscous lubricating distillate	9.5	.881-.903	29.1-25.2	50-100
Medium lubricating distillate	6.7	.903-.925	25.2-21.5	100-200
Viscous lubricating distillate	-	-	-	Above 200
Residuum	27.0	.987	11.9	
Distillation loss	1.2			

U. S. GOVERNMENT PRINTING OFFICE 16--37835-3

CRUDE PETROLEUM ANALYSIS

Bureau of Mines Bartlesville Laboratory
Sample 56112

IDENTIFICATION

Corsicana field
Wolf City, Upper Cretaceous
1,088-1,116 feet

Texas
Navarro County

GENERAL CHARACTERISTICS

Gravity, specific, 0.834 Gravity, ° API, 38.2 Pour point, ° F., below 5
Sulfur, percent, 0.24 Color, brownish green
Viscosity, Saybolt Universal at 100°F., 43 sec. Nitrogen, percent, 0.000

DISTILLATION, BUREAU OF MINES ROUTINE METHOD

STAGE 1—Distillation at atmospheric pressure, 751 mm. Hg
First drop, 106 ° F.

Fraction No.	Cut temp. ° F.	Percent	Sum. percent	Sp. gr., 60/60° F.	° API, 60° F.	C. I.	Refractive index, n_D at 20° C.	Specific dispersion	S. U. visc., 100° F.	Cloud test, ° F.
1	122									
2	167	1.7	1.7	0.666	81.0					
3	212	3.2	4.9	.701	70.4	12	1.38738	127.2		
4	257	6.2	11.1	.724	63.9	14	1.40466	128.3		
5	302	5.9	17.0	.742	59.2	15	1.41441	122.9		
6	347	7.6	24.6	.762	54.2	18	1.42379	132.2		
7	392	6.8	31.4	.778	50.4	19	1.43298	128.6		
8	437	5.7	37.1	.795	46.5	22	1.44101	125.8		
9	482	7.1	44.2	.809	43.4	23	1.44869	130.9		
10	527	7.4	51.6	.826	39.8	26	1.45727	134.7		
STAGE 2—Distillation continued at 40 mm. Hg										
11	392	4.5	56.1	0.844	36.2	31	1.46666	136.0	41	10
12	437	6.2	62.3	.854	34.2	32	1.47114	139.4	47	25
13	482	5.1	67.4	.866	31.9	34	1.47746	141.3	59	45
14	527	4.8	72.2	.884	28.6	40	1.48359	147.2	84	60
15	572	5.5	77.7	.894	26.8	41	1.49112	142.5	155	75
Residuum		21.6	99.3	.951	17.2					

Carbon residue, Conradson: Residuum, 8.3 percent; crude, 2.0 percent.

APPROXIMATE SUMMARY

	Percent	Sp. gr.	° API	Viscosity
Light gasoline	4.9	0.689	73.9	
Total gasoline and naphtha	31.4	0.743	58.9	
Kerosine distillate	12.8	.803	44.7	
Gas oil	16.5	.839	37.2	
Nonviscous lubricating distillate	10.3	.857-.886	33.6-28.2	50-100
Medium lubricating distillate	6.7	.886-.889	28.2-25.9	100-200
Viscous lubricating distillate	-	-	-	Above 200
Residuum	21.6	.951	17.3	
Distillation loss	.7			

U. S. GOVERNMENT PRINTING OFFICE 16—57835-2

Appendix D

Economic Evaluation Example Problem

UNDERSTANDING "TRUE RATE OF RETURN"

The continuous inflation of our economy and escalation of prices make imperative the consideration of the "time-value" of money when evaluating new investments. The easily understood simple payout and simple percentage return, which were the yardsticks of engineers and management 30 to 35 years ago, have been replaced with new evaluation techniques that account for the time-value of money.

These widely used procedures have many sophisticated names and definitions. All too frequently, the managerial and technical people who use these tools do not clearly understand the actual physical significance of the terms.

A brief review of these definitions, explained with a numerical example, is the best method of providing a clear understanding of the true rate of return (TRR). The terms listed below are all synonymous with true rate of return.

Average annual rate of return (AARR)
Internal rate of return (IRR)
Interest rate of return (IRR)
Discounted cash flow rate of return (DCFRR)
Investor's method rate of return (IMRR)
Profitability index (PI)

A common definition of the terms listed above is "The true rate of return is the [highest, constant] percentage of the outstanding investment which can be realized as a profit [or saving] each year [for the life of the facility] and at the same time provide funds [from the operating income] that will exactly amortize [recover] the investment over the useful life of the facility." This definition is easier to understand if the bracketed words are omitted.

The rate can be calculated for the case of a single lump investment and constant annual cash flow by the following equation:

$$i = \frac{S}{I} - \frac{i}{(1+i)^{T} - 1} \tag{D.1}$$

where

i = annual rate of return (as a fraction)
S = annual cash flow (net after-tax income plus depreciation)
I = investment
T = life of facility (years)

An example will provide tangible significance of the above definition.

Example

Initial (and only) investment (I): $1,000,000
Useful life of facility (T): 3 years
Annual cash flow (S): $500,000 (after-tax income plus depreciation allowed for tax calculation)

Calculate i by solution of the equation, given S/I = 0.5 and T = 3:

$$i = 0.5 - \frac{i}{(1+i)^3 - 1} = 0.234 \tag{D.2}$$

Year	Amt. outstanding on investment A	Cash flow B	Profit realized (23.4% × A) C	Cash flow used to amortize investment (B – C) D	Remaining investment (A – D) E
1	$1,000,000	$500,000	$234,000	$266,000	$734,000
2	734,000	500,000	172,000	328,000	406,000
3	406,000	500,000	94,000	406,000	—
Total		$1,500,000	$500,000	$1,000,000	

From the above example, it can be seen that the total cash flow minus the total profit, taken at a constant percentage (23.4) of outstanding investment, is equal to a total cash allowance, which will exactly recover the investment over the project life.

$$\Sigma B - \Sigma C = \Sigma D = \text{investment} \tag{D.3}$$

The above statement simply says that over the life of the project, the total of the annual net income minus the investment is equal to the total profit. The value of this obvious statement is that it makes the bankers' and accountants' definition of TRR, AARR, IRR, DCFRR, IMRR, and PI understandable. This definition is "The TRR is the particular interest rate that equates that total present value cash investments to the total present value cash flow incomes."

This definition leads to the more common way of determining the TRR. Our previous example can be solved by determining the discount rate at which total cash flows for 3 years will equal the investment.

Year	Cash flow	Discounted factor (23.4%)	Discounted cash flow
1	\$500,000	1/1.234	\$406,000
2	500,000	$1/(1.234)^2$	328,000
3	500,000	$1/(1.234)^3$	266,000
Total			\$1,000,000

A third approach defines the TRR as "The interest rate required for an annuity fund started with the same sum of capital to make the same payments at the same time as the proposed investment." Illustrating this definition with our example, we have the following values:

Year	Principal + interest	Less payments	Balance investment
1	$1{,}000{,}000 \times 1.234 = 1{,}234{,}000$	500,000	734,000
2	$734{,}000 \times 1.234 = 906{,}000$	500,000	406,000
3	$406{,}000 \times 1.234 + 500{,}000$	500,000	—

The above procedure is suitable for evaluation of projects that have varying cash flows and distributed investments.

The examples are based on periodic interest rates and the assumption that the annual cash flow all occurs at the end of the year. For most evaluations, these conditions are sufficiently accurate. A review of alternative procedures to allow for continuous interest and continuous cash has been given by Souders [1].

For those cases where cash flow is constant and the investment is all made at the start of the project, a simple graphical solution for the TRR is possible [2].

A recent article by Reul [3] reviews the relative merits of the several types of investment evaluation techniques. As very clearly illustrated, there is really no acceptable alternative to the true rate of return for comparing the economic performance of various investments.

NOTES

1. Souders, M., *Chem. Eng. Progr.* 62(3), p. 79, 1966.
2. Salmon, R., *Chem. Eng.* 1 Apr., p. 79, 1993.
3. Reul, R., *Chem. Eng.* 22 Apr., p. 212, 1968.

Appendix E

Photographs

E.1 Overall refinery view.

E.2 Crude oil distillation unit. Left: atmospheric distillation tower; right: vacuum distillation tower.

E.3 Delayed coking unit.

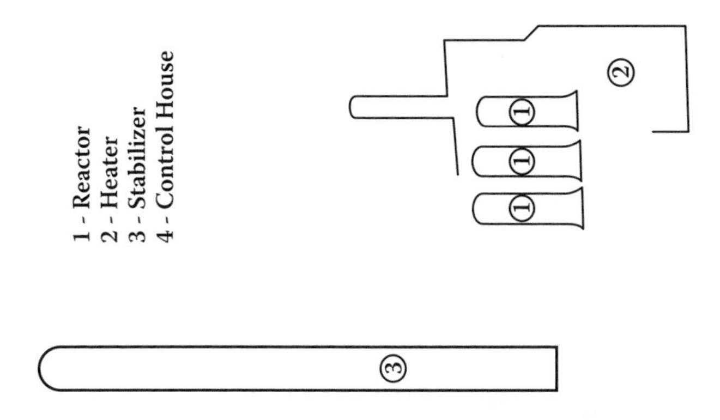

E.4 Semiregenerative Platformer.™ (Photo courtesy of UOP LLC.)

E.5 Powerformer semiregenerative catalytic reforming unit.

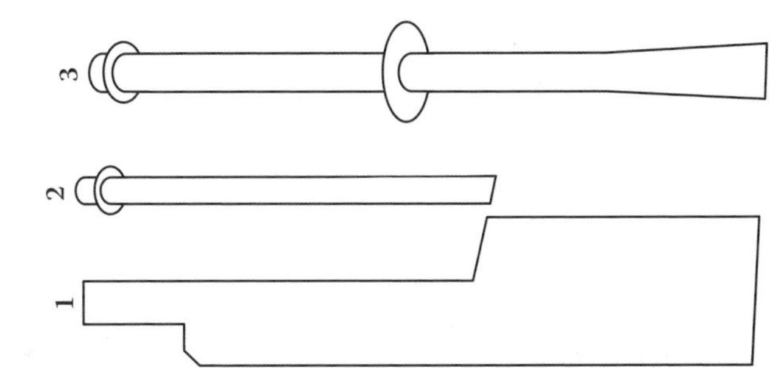

E.6 UOP LLC CCR™ continuous catalyst regeneration catalytic reforming unit. (Photo courtesy of UOP LLC.)

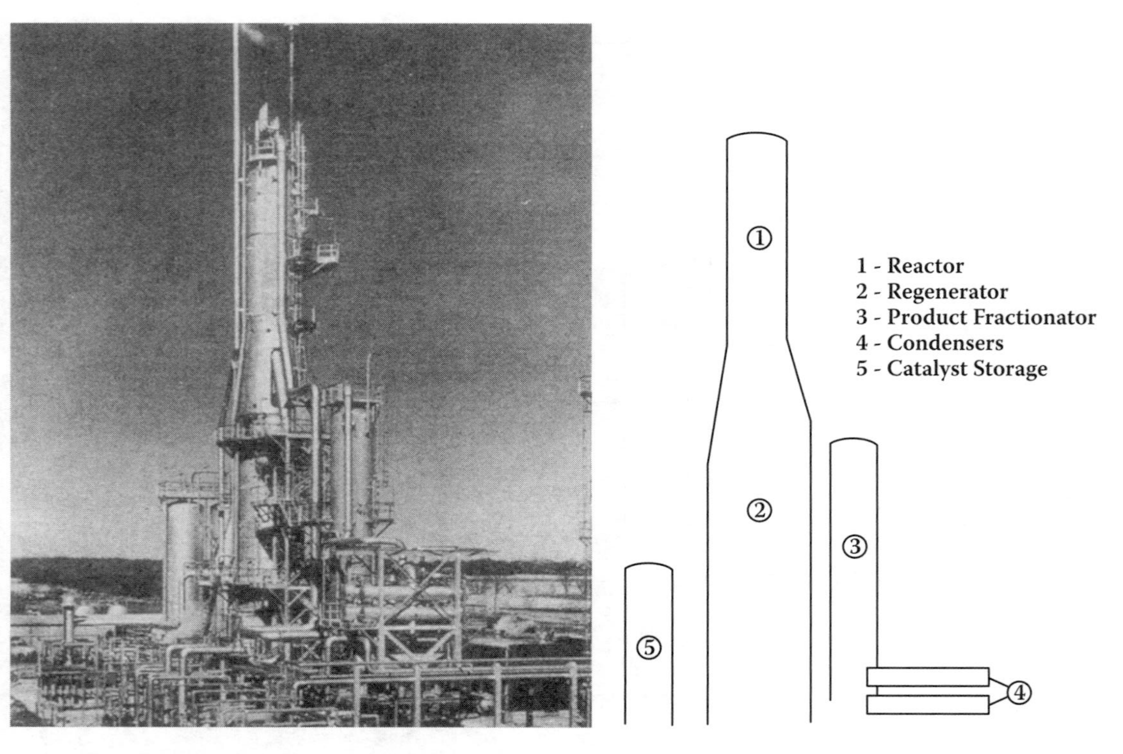

E.7 20,000 BPD fluid catalytic cracking unit. (Photo courtesy of UOP LLC.)

E.8 Exxon Flexicracking™ IIIR fluid catalytic cracking unit. (Photo courtesy of Exxon Research and Engineering Co.)

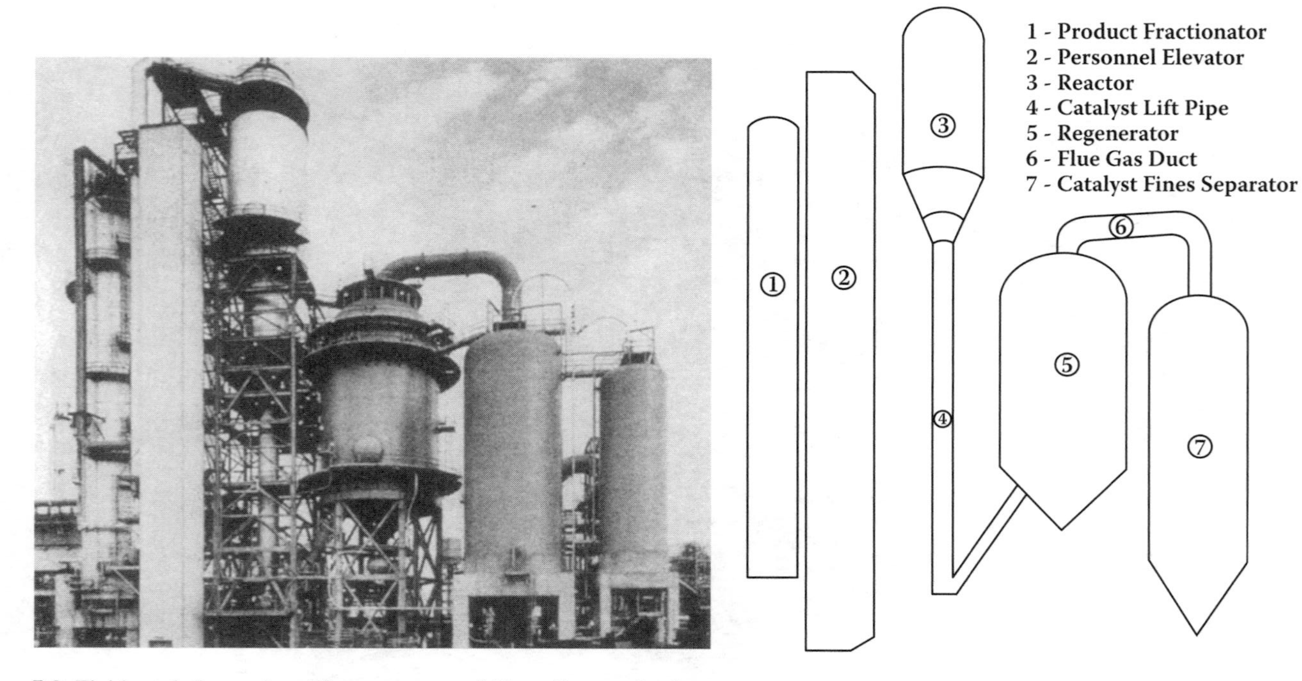

E.9 Fluid catalytic cracker. (Photo courtesy of Fluor Corporation.)

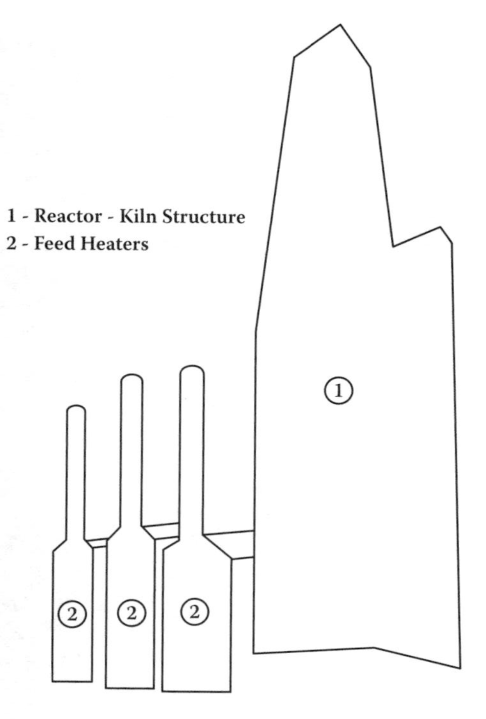

E.10 12,000 BPD TCC™ unit. (Photo by Ray Manley, courtesy of the Stearns Roger Corp.)

E.11 24,000 BPD H-Oil™ unit. (Photo courtesy of Fluor Corporation.)

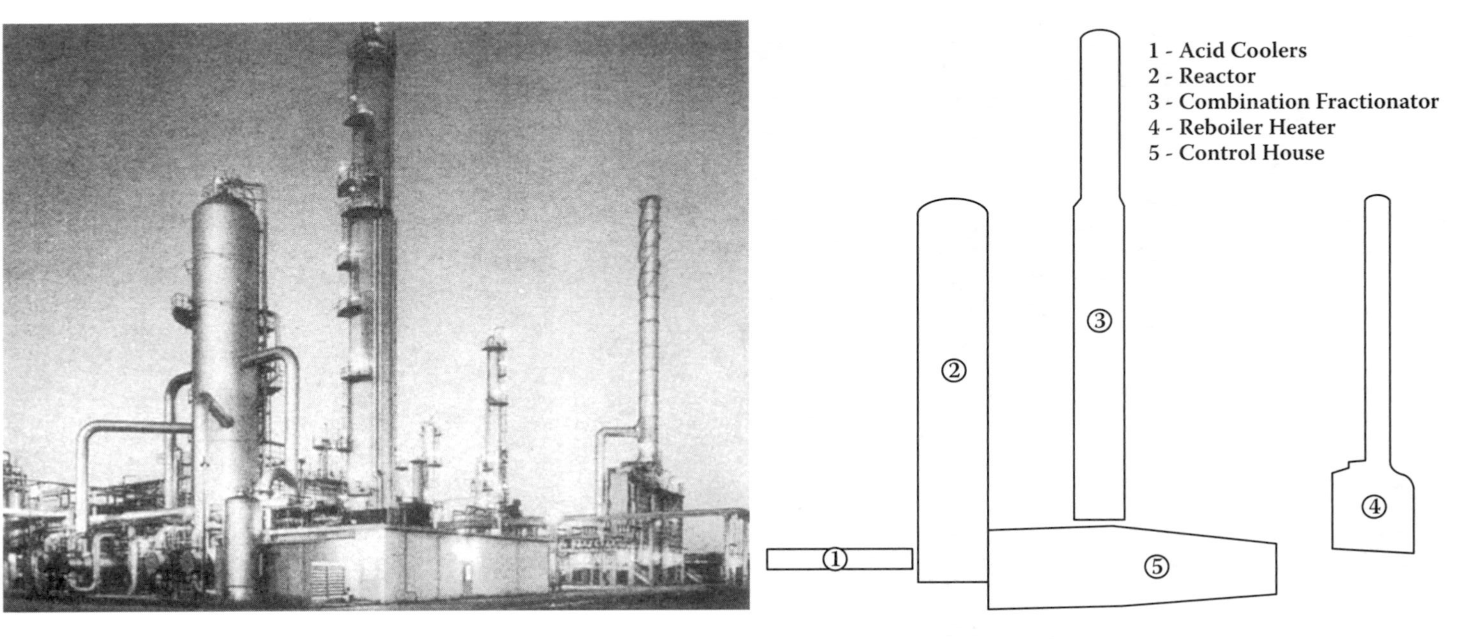

E.12 10,000 BPD hydrofluoric acid alkylation unit. (Photo by Ray Manley, courtesy of the Stearns Roger Corp.)

E.13 Sulfuric acid alkylation unit.

E.14 Stratco™ sulfuric acid alkylation unit. (Photo courtesy of Stratco, Inc.)

E.15 Light ends distillation units.

Index

C

D

E

F

G

H

N

O

P

Q

R

S

T

U

V

W

X

Y

Z

Other Related Titles of Interest

Chemical Process Performance Evaluation
Ali Cinar, Ahmet Palazoglu, and Ferhan Kayihan
ISBN: 0849338069

Engineering Economics and Economic Design for Process Engineers
Thane Brown
ISBN: 0849382122

Industrial Gas Handbook: Gas Separation and Purification
Frank G. Kerry
ISBN: 0849390052

Materials for the Hydrogen Economy
Russell H. Jones and George J. Thomas
ISBN: 0849350247

Petroleum and Gas Field Processing
H.K. Abdel-Aal, Mohamed Aggour, and M.A. Fahim
ISBN: 0824709624

Phase Behavior of Petroleum Reservoir Fluids
Karen Schou Petersen and Peter L. Christensen
ISBN: 0824706943